Response Surface Methodology

Response Surface Methodology:

Process and Product Optimization Using Designed Experiments

RAYMOND II. MYERS
Virginia Polytechnic Institute and State University

DOUGLAS C. MONTGOMERY
Arizona State University

A Wiley-Interscience Publication
JOHN WILEY & SONS, INC.
New York • Chichester • Brisbane • Toronto • Singapore

Library of Congress Cataloging in Publication Data:
Montgomery, Douglas C.
    Response surface methodology: process and product optimization
using designed experiments/Raymond H. Myers and Douglas C.
Montgomery
        p. cm.—(Wiley series in probability and
statistics. Applied probability section)
    "A Wiley-Interscience publication."
    Includes bibliographical references and index.
    ISBN 0-471-58100-3 (acid-free paper)
    1. Experimental design. 2. Response surfaces (Statistics)
I. Myers, Raymond H. II. Title. III. Series: Wiley series in
probability and statistics. Applied probability and statistics.
QA279.M67 1995
519.5—dc20   94-45548

Contents

Preface

This book deals with the exploration and optimization of response surfaces. This is a problem faced by experimenters in many technical fields, where, in general, the response variable of interest is y and there is a set of predictor variables $x_1, x_2, \ldots, x_k$. For example, y might be the viscosity of a polymer and x_1, x_2, and x_3 might be the reaction time, the reactor temperature, and the catalyst feed rate in the process.

In some systems the nature of the relationship between y and the x's might be known "exactly," based on the underlying engineering, chemical, or physical principles. Then we could write a model of the form $y = g(x_1, x_2, \ldots, x_k) + \varepsilon$, where the term ε in this model represents the "error" in the system. This type of relationship is often called a *mechanistic model*.

We consider the more common situation where the underlying mechanism is not fully understood, and the experimenter must approximate the unknown function g with an appropriate empirical model $y = f(x_1, x_2, \ldots, x_k) + \varepsilon$. Usually the function f is a first-order or second order polynomial. This empirical model is called a *response surface model*.

Identifying and fitting from experimental data an appropriate response surface model requires some knowledge of statistical experimental design fundamentals, regression modeling techniques, and elementary optimization methods. This book integrates all three of these topics into what has been popularly called **response surface methodology** (RSM). We assume that the reader has some previous exposure to statistical methods and matrix algebra. Formal coursework in basic principles of experimental design and regression analysis would be helpful, but are not essential, because the important elements of these topics are presented early in the text.

We have used this book in a graduate-level course on RSM for statisticians, engineers, and chemical/ physical scientists. We have also used it in industrial short courses and seminars for individuals with a wide variety of technical backgrounds.

Chapter 1 is an introduction to the general field of RSM, and it describes typical applications such as (a) finding the levels of the predictor variables that result in optimization of the response or (b) discovering what levels for the x's will result in a product satisfying several requirements or specifications such as yield, viscosity, molecular weight, or purity. Chapter 2 is a summary of regression methods useful in response surface work. Chapters 3 and 4 describe two-level factorial and fractional factorial designs. These designs are essential for factor screening (identifying the correct set of process variables to use in the RSM study). They are also basic building blocks for many of the response surface designs discussed later in the text. Chapter 5 presents the method of steepest ascent, a simple but powerful optimization procedure used at the early stages of RSM to move the process from a region of relatively poor performance to one of greater potential. Chapter 6 introduces the analysis and optimization of a second-order response surface, including techniques for the simultaneous optimization of several responses. Chapters 7, 8, and 9 present detailed information on the choice of designs for fitting both first-order and second-order response surfaces. All of the standard designs are discussed, and practical guidance is given on the proper selection of a design. We also discuss and illustrate computer-aided design of experiments for fitting response surfaces. Chapter 10 shows how the robust design methodology of G. Taguchi can be more effectively and efficiently implemented in the context of RSM. Chapters 11 and 12 describe techniques for designing and analyzing experiments with mixtures. A mixture experiment is a special type of response surface problem in which the factors are the components or ingredients of a mixture, and the response depends on the proportionate amounts of the ingredients present. Chapter 13 is an introduction to evolutionary operation, a methodology for carrying experimental design methods on-line into the process as a routine method of plant operation. An extensive set of exercises are provided, along with a reference section.

The text is written to emphasize methods that are useful in industry and that we have found to be helpful in our consulting experience; however, some of the underlying theory is developed so that reader will gain an understanding of the assumptions and conditions necessary to apply the methodology. Some portions of Chapter 9 and 10 require more statistical sophistication than the rest of the book. Instructors in university graduate-level courses may wish to discuss some of the theoretical aspects of the subject more fully, and we have provided a set of technical appendices on a variety of RSM topics that should facilitate that task.

We are grateful to the several classes of students that have used previous versions of the notes and the manuscript that led to this book. Their comments and suggestions for improvement have been invaluable. We are also grateful to John Wiley & Sons, Inc. for permission to use and adapt copyrighted material.

RAYMOND H. MYERS
DOUGLAS C. MONTGOMERY

Blacksburg, Virginia
Tempe, Arizona
July, 1995

Response Surface Methodology

CHAPTER 1

Introduction

1.1 RESPONSE SURFACE METHODOLOGY

Response surface methodology (RSM) is a collection of statistical and mathematical techniques useful for developing, improving, and optimizing processes. It also has important applications in the design, development, and formulation of new products, as well as in the improvement of existing product designs.

The most extensive applications of RSM are in the industrial world, particularly in situations where several input variables potentially influence some performance measure or quality characteristic of the product or process. This performance measure or quality characteristic is called the **response**. It is typically measured on a continuous scale, although attribute responses, ranks, and sensory responses are not unusual. Most real-world applications of RSM will involve more than one response. The input variables are sometimes called **independent variables**, and they are subject to the control of the engineer or scientist, at least for purposes of a test or an experiment.

Figure 1.1 shows graphically the relationship between the response variable yield (y) in a chemical process and the two process variables (or independent variables) reaction time (ξ_1) and reaction temperature (ξ_2). Note that for each value of ξ_1 and ξ_2 there is a corresponding value of yield y, and that we may view these values of the response yield as a surface lying above the time–temperature plane, as in Figure 1.1(a). It is this graphical perspective of the problem environment that has led to the term **response surface methodology**. It is also convenient to view the response surface in the two-dimensional time–temperature plane, as in Figure 1.1(b). In this presentation we are looking down at the time–temperature plane and connecting all points that have the same yield to produce contour lines of constant responses. This type of display is called a **contour plot**.

Clearly, if we could easily construct the graphical displays in Figure 1.1, optimization of this process would be very straightforward. By inspection of the plot, we note that yield is maximized in the vicinity of time $\xi_1 = 2.0$ hr

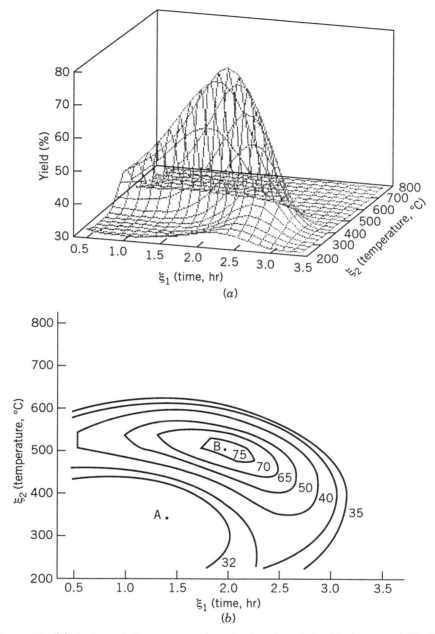

Figure 1.1 (*a*) A theoretical response surface showing the relationship between yield of a chemical process and the process variables reaction time (ξ_1) and reaction temperature (ξ_2). (*b*) A contour plot of the theoretical response surface.

and temperature $\xi_2 = 500°C$. Unfortunately, in most practical situations, the true response function in Figure 1.1 is unknown. The field of response surface methodology consists of the experimental strategy for exploring the space of the process or independent variables (here the variables ξ_1 and ξ_2), empirical statistical modeling to develop an appropriate approximating relationship between the yield and the process variables, and optimization methods for finding the levels or values of the process variables ξ_1 and ξ_2 that produce desirable values of the responses (in this case that maximize yield).

1.1.1   Approximating Response Functions

In general, suppose that the scientist or engineer (whom we will refer to as the *experimenter*) is concerned with a product, process, or system involving a response y that depends on the controllable input variables $\xi_1, \xi_2, \ldots, \xi_k$. The relationship is

$$y = f(\xi_1, \xi_2, \ldots, \xi_k) + \varepsilon \qquad (1.1)$$

where the form of the true response function f is unknown and perhaps very complicated, and ε is a term that represents other sources of variability not accounted for in f. Thus ε includes effects such as measurement error on the response, other sources of variation that are inherent in the process or system (background noise, or common cause variation in the language of statistical process control), the effect of other variables, and so on. We will treat ε as a statistical error, often assuming it to have a normal distribution with mean zero and variance σ^2. If the mean of ε is zero, then

$$E(y) = \eta = E[f(\xi_1, \xi_2, \ldots, \xi_k)] + E(\varepsilon)$$
$$\eta = f(\xi_1, \xi_2, \ldots, \xi_k) \qquad (1.2)$$

The variables $\xi_1, \xi_2, \ldots, \xi_k$ in Equation (1.2) are usually called the **natural variables**, because they are expressed in the natural units of measurement, such as degrees Celsius (°C), pounds per square inch (psi), or g/liter of concentration. In much RSM work it is convenient to transform the natural variables to **coded variables** $x_1, x_2, \ldots, x_k$, where these coded variables are usually defined to be dimensionless with mean zero and the same spread or standard deviation. In terms of the coded variables, the true response function (1.2) is now written as

$$\eta = f(x_1, x_2, \ldots, x_k) \qquad (1.3)$$

Because the form of the true response function f is unknown, we must approximate it. In fact, successful use of RSM is critically dependent upon the experimenter's ability to develop a suitable approximation for f. Usually, a low-order polynomial in some relatively small region of the independent variable space is appropriate. In many cases, either a **first-order** or a **second-order** model is used. For the case of two independent variables, the first-order model in terms of the coded variables is

$$\eta = \beta_0 + \beta_1 x_1 + \beta_2 x_2 \tag{1.4}$$

Figure 1.2 shows the three-dimensional response surface and the two-dimensional contour plot for a particular case of the first-order model, namely,

$$\eta = 50 + 8x_1 + 3x_2$$

In three dimensions, the response surface is a plane lying above the x_1, x_2 space. The contour plot shows that the first-order model can be represented as parallel straight lines of constant response in the x_1, x_2 plane.

The first-order model is likely to be appropriate when the experimenter is interested in approximating the true response surface over a relatively small region of the independent variable space in a location where there is little curvature in f. For example, consider a small region around the point A in Figure 1.1(*b*); the first-order model would likely be appropriate here.

The form of the first-order model in Equation (1.4) is sometimes called a **main effects model**, because it includes only the main effects of the two variables x_1 and x_2. If there is an **interaction** between these variables, it can be added to the model easily as follows:

$$\eta = \beta_0 + \beta_1 x_1 + \beta_2 x_2 + \beta_{12} x_1 x_2 \tag{1.5}$$

This is the first-order model with interaction. Figure 1.3 shows the three-dimensional response surface and the contour plot for the special case

$$\eta = 50 + 8x_1 + 3x_2 - 4x_1 x_2$$

Notice that adding the interaction term $-4x_1 x_2$ introduces curvature in the response function.

Often the curvature in the true response surface is strong enough that the first-order model (even with the interaction term included) is inadequate. A second-order model will likely be required in these situations. For the case of two variables, the second-order model is

$$\eta = \beta_0 + \beta_1 x_1 + \beta_2 x_2 + \beta_{11} x_{11}^2 + \beta_{22} x_{22}^2 + \beta_{12} x_1 x_2 \tag{1.6}$$

This model would likely be useful as an approximation to the true response surface in a relatively small region around the point B in Figure 1.1(*b*), where there is substantial curvature in the true response function f.

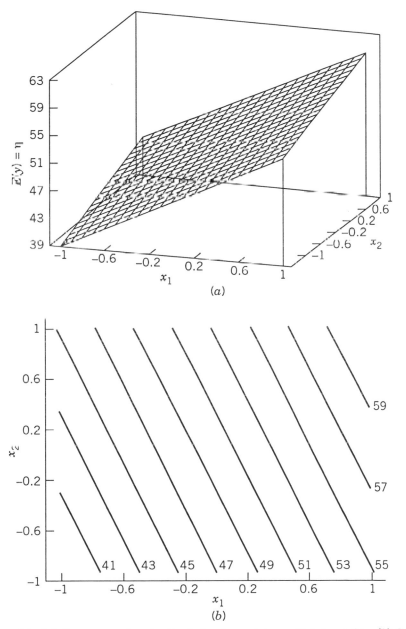

Figure 1.2 (a) Response surface for the first-order model $\eta = 50 + 8x_1 + 3x_2$. (b) Contour plot for the first-order model.

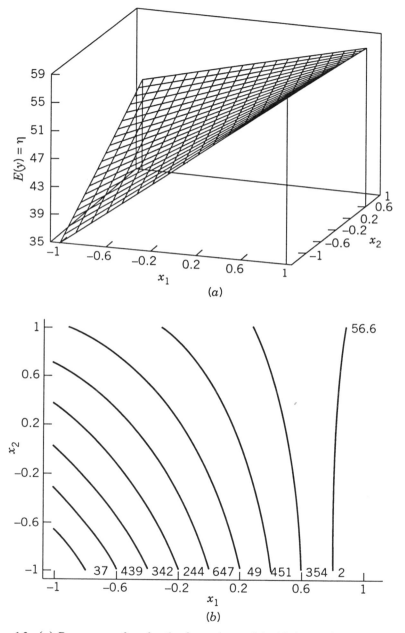

Figure 1.3 (*a*) Response surface for the first-order model with interaction $\eta = 50 + 8x_1 + 3x_2 - 4x_1x_2$. (*b*) Contour plot for the first-order model with interaction.

Figure 1.4 presents the response surface and contour plot for the special case of the second-order model

$$\eta = 50 + 8x_1 + 3x_2 - 7x_{11}^2 - 3x_{22}^2 - 4x_1 x_2$$

Notice the mound-shaped response surface and elliptical contours generated by this model. Such a response surface could arise in approximating a response such as yield, where we would expect to be operating near a maximum point on the surface.

The second-order model is widely used in response surface methodology for several reasons. Among these are the following:

1. The second-order model is very flexible. It can take on a wide variety of functional forms, so it will often work well as an approximation to the true response surface. Figure 1.5 shows several different response surfaces and contour plots that can be generated by a second-order model.

2. It is easy to estimate the parameters (the β's) in the second-order model. The method of least squares, which is presented in Chapter 2, can be used for this purpose.

3. There is considerable practical experience indicating that second-order models work well in solving real response surface problems.

In general, the first-order model is

$$\eta = \beta_0 + \beta_1 x_1 + \beta_2 x_2 + \cdots + \beta_k x_k \tag{1.7}$$

and the second-order model is

$$\eta = \beta_0 + \sum_{j=1}^{k} \beta_j x_j + \sum_{j=1}^{k} \beta_{jj} x_j^2 + \sum\sum_{i<j} \beta_{ij} x_i x_j \tag{1.8}$$

In some situations, approximating polynomials of order greater than two are used. The general motivation for a polynomial approximation for the true response function f is based on the Taylor series expansion around the point $x_{10}, x_{20}, \ldots, x_{k0}$. For example, the first-order model is developed from the first-order Taylor series expansion

$$f \cong f(x_{10}, x_{20}, \ldots, x_{k0}) + \frac{\partial f}{\partial x_1}\bigg|_{\mathbf{x}=\mathbf{x}_0}$$

$$+ \frac{\partial f}{\partial x_2}\bigg|_{\mathbf{x}=\mathbf{x}_0} + \cdots + \frac{\partial f}{\partial x_k}\bigg|_{\mathbf{x}=\mathbf{x}_0} \tag{1.9}$$

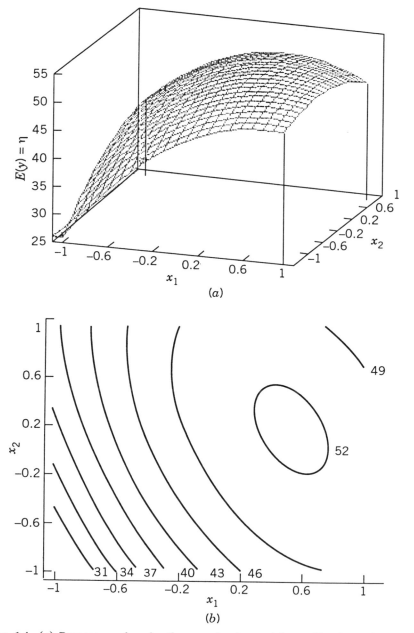

Figure 1.4 (*a*) Response surface for the second-order model $\eta = 50 + 8x_1 + 3x_2 - 7x_1^2 -$
$3x_2^2 - 4x_1x_2$. (*b*) Contour plot for the second order model.

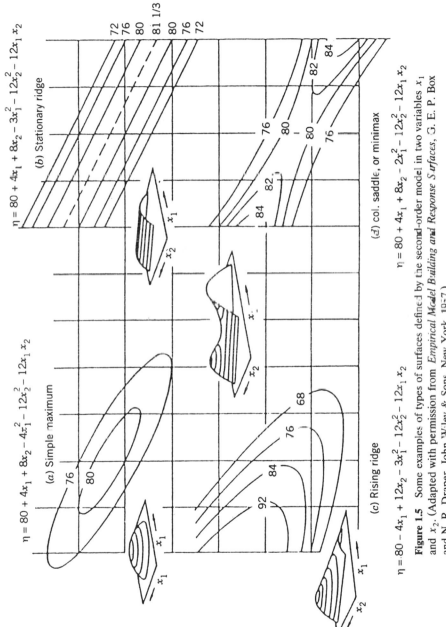

Figure 1.5 Some examples of types of surfaces defined by the second-order model in two variables x_1 and x_2. (Adapted with permission from *Empirical Model Building and Response Surfaces*, G. E. P. Box and N. R. Draper, John Wiley & Sons, New York, 1987.)

$\eta = 80 + 4x_1 + 8x_2 - 4x_1^2 - 12x_2^2 - 12x_1 x_2$

(a) Simple maximum

$\eta = 80 + 4x_1 + 8x_2 - 3x_1^2 - 12x_2^2 - 12x_1 x_2$

(b) Stationary ridge

$\eta = 80 - 4x_1 + 12x_2 - 3x_1^2 - 12x_2^2 - 12x_1 x_2$

(c) Rising ridge

$\eta = 80 + 4x_1 + 8x_2 - 2x_1^2 - 12x_2^2 - 12x_1 x_2$

(d) col. saddle, or minimax

where **x** refers to the vector of independent variables and $\mathbf{x}_0$ is that vector of variables at the specific point $x_{10}, x_{20}, \ldots, x_{k0}$. In Equation (1.9) we have only included the first-order terms in the expansion, thus implying the first-order approximating model in Equation (1.7). If we were to include second-order terms in Equation 1.9, this would lead to the second-order approximating model in Equation (1.8).

Finally, note that there is a close connection between RSM and **linear regression analysis**. For example, consider the model

$$y = \beta_0 + \beta_1 x_1 + \beta_2 x_2 + \cdots + \beta_k x_k + \varepsilon$$

The β's are a set of unknown parameters. To estimate the values of these parameters, we must collect data on the system we are studying. Regression analysis is a branch of statistical model building that uses these data to estimate the β's. Because, in general, polynomial models are linear functions of the unknown β's, we refer to the technique as *linear regression analysis*. We will also see that it is very important to plan the data collection phase of a response surface study carefully. In fact, special types of experimental designs, called **response surface designs**, are valuable in this regard. A substantial part of this book is devoted to response surface designs.

1.1.2 The Sequential Nature of RSM

Most applications of response surface methodology are **sequential** in nature. That is, at first some ideas are generated concerning which factors or variables are likely to be important in the response surface study. This usually leads to an experiment designed to investigate these factors with a view toward eliminating the unimportant ones. This type of experiment is usually called a **screening experiment**. Often at the outset of a response surface study there is a rather long list of variables that could be important in explaining the response. The objective of factor screening is to reduce this list of candidate variables to a relatively few so that subsequent experiments will be more efficient and require fewer runs or tests. We refer to a screening experiment as *phase zero* of a response surface study. You should never undertake a response surface analysis until a screening experiment has been performed to identify the important factors.

Once the important independent variables are identified, *phase one* of the response surface study begins. In this phase, the experimenter's objective is to determine if the current levels or settings of the independent variables result in a value of the response that is near the optimum [such as the point B in Figure 1.1(b)], or if the process is operating in some other region that is (possibly) remote from the optimum [such as the point A in Figure 1.1(b)]. If the current settings or levels of the independent variables are not consistent with optimum performance, then the experimenter must determine a set of adjustments to the process variables that will move the process toward the optimum. This phase of response surface methodology makes considerable use of the first-order model and an optimization technique called the **method of steepest ascent**. These techniques will be discussed and illustrated in Chapter 5.

Phase two of a response surface study begins when the process is near the optimum. At this point the experimenter usually wants a model that will accurately approximate the true response function within a relatively small region around the optimum. Because the true response surface usually exhibits curvature near the optimum (refer to Figure 1.1), a second-order model (or perhaps some higher-order polynomial) will be used. Once an appropriate approximating model has been obtained, this model may be analyzed to determine the optimum conditions for the process. Chapter 6 will present techniques for the analysis of the second-order model and the determination of optimum conditions.

This sequential experimental process is usually performed within some region of the independent variable space called the **operability region**. For the chemical process illustrated in Figure 1.1, the operability region is 0.5 hr $\leq \xi_1 \leq 3.5$ hr and 200°C $\leq \xi_2 \leq 800^\circ$C. Suppose we are currently operating at the levels $\xi_1 = 1.5$ hr and $\xi_2 = 300^\circ$C, shown as point A in Figure 1.6. Now it is unlikely that we would want to explore the entire region of operability with a single experiment. Instead, we usually define a smaller **region of interest** or **region of experimentation** around the point A within the larger region of operability. Typically, this region of experimentation is either a cubodial region, as shown around the point A in Figure 1.6, or a spherical region, as shown around point B.

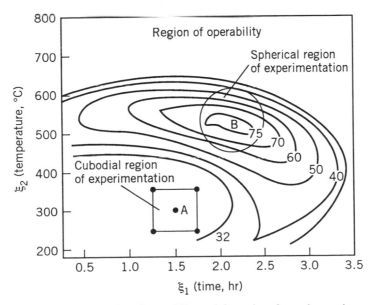

Figure 1.6 The region of operability and the region of experimentation.

The sequential nature of response surface methodology allows the experimenter to learn about the process or system under study as the investigation proceeds. This ensures that over the course of the RSM application the experimenter will learn the answers to questions such as (1) how much replication is necessary, (2) the location of the region of the optimum, (3) the type of approximating function required, (4) the proper choice of experimental designs, and (5) whether or not transformations on the responses or any of the process variables are required. A substantial portion of this book—Chapters 3, 4, 7, and 8—is devoted to experimental designs useful in RSM.

1.1.3 Objectives and Typical Applications of RSM

Response surface methodology is useful in the solution of many types of industrial problems. Generally, these problems fall into three categories:

1. *Mapping a Response Surface Over a Particular Region of Interest.* Consider the chemical process in Figure 1.1(*b*). Normally, this process would operate at a particular setting of reaction time and reaction temperature. However, some changes to these normal operating levels might occasionally be necessary. If the true unknown response function has been approximated over a region around the current operating conditions with a suitable fitted response surface (say a second-order surface), then the process engineer can predict in advance the changes in yield that will result from any readjustments to time and temperature.

2. *Optimization of the Response.* In the industrial world, a very important problem is determining the conditions that optimize the process. In the chemical process of Figure 1.1(*b*), this implies determining the levels of time and temperature that result in maximum yield. An RSM study that begins near point A in Figure 1.1(*b*) would eventually lead the experimenter to the region near point B. A second-order model could then be used to approximate the yield response in a narrow region around point B, and from examination of this approximating response surface the optimum levels or condition for time and temperature could be chosen.

3. *Selection of Operating Conditions to Achieve Specifications or Customer Requirements.* In most response surface problems there are several responses that must in some sense be simultaneously considered. For example, in the chemical process of Figure 1.1, suppose that in addition to yield, there are two other responses, cost and concentration. We would like to maintain yield above 70%, while simultaneously keeping the cost below \$34/ pound; however, the customer has imposed specifications for concentration such that this important physical property must be 65 g/ liter ± 3 g/ liter.

One way that we could solve this problem is to obtain response surfaces for all three responses—yield, cost, and concentration—and then superim-

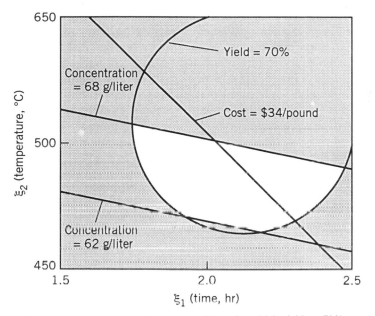

Figure 1.7   The unshaded region showing the conditions for which yield ≥ 70%, cost ≤ $34/ pound, and 62 g/liter ≤ concentration ≤ 68 g/liter.

pose the contours for these responses in the time–temperature plane, as illustrated in Figure 1.7. In this figure we have shown the contours for yield = 70%, cost = $34/ pound, concentration = 62 g/ liter, and concentration = 68 g/ liter. The unshaded region in this figure represents the region containing operating conditions that simultaneously satisfy all requirements on the process.

In practice, complex process optimization problems such as this can often be solved by superimposing appropriate response surface contours. However, it is not unusual to encounter problems with more than two process variables and more complex response requirements to satisfy. In such problems, other optimization methods that are more effective than overlaying contour plots will be necessary. We will discuss methodology for solving these types of problems in Chapter 6.

1.1.4   RSM and the Philosophy of Quality Improvement

During the last 15 years, industrial organizations in the United States and Europe have become keenly interested in quality improvement. Statistical methods, including statistical process control (SPC) and design of experiments, play a key role in this activity. Quality improvement is most effective when it occurs early in the product and process development cycle. Industries such as semiconductors and electronics, aerospace, automotive, biotechnology and pharmaceuticals, medical devices, chemical, and process industries are all examples where experimental design methodology has resulted in products that are easier to manufacture, have higher reliability, have enhanced field performance, and meet or exceed customer requirements.

RSM is an important branch of experimental design in this regard. RSM is a critical technology in developing new processes, optimizing their performance, and improving the design and/or formulation of new products. It is often an important **concurrent engineering tool**, in that product design, process development, quality, manufacturing engineering, and operations personnel often work together in a team environment to apply RSM. The objectives of quality improvement, including reduction of variability and improved product and process performance, can often be accomplished directly using RSM.

1.2 PRODUCT DESIGN AND FORMULATION (MIXTURE PROBLEMS)

Many product design and development activities involve formulation problems, in which two or more ingredients are mixed together. For example, suppose we are developing a new household cleaning product. This product is formulated by mixing several chemical surfactants together. The product engineer or scientist would like to find an appropriate blend of the ingredients so that the grease-cutting capability of the cleaner is good, and so that it generates an appropriate level of foam when in use. In this situation the response variables—namely, grease cutting ability and amount of foam—depend on the percentages or proportions of the individual chemical surfactants (the ingredients) that are present in the product formulation.

There are many industrial problems where the response variables of interest in the product are a function of the proportions of the different ingredients used in its formulation. This is a special type of response surface problem called a **mixture problem**.

While we traditionally think of mixture problems in the product design or formulation environment, they occur in many other settings. Consider plasma etching of silicon wafers, a common manufacturing process in the semiconductor industry. Etching is usually accomplished by introducing a blend of gases inside a chamber containing the wafers. The measured responses include etch rate, the uniformity of the etch, and the selectivity (a measure of the relative etch rates of the different materials on the wafer). All of these responses are a function of the proportions of the different ingredients blended together in the etching chamber.

There are special experimental design techniques and model-building methods for mixture problems. These techniques are discussed in Chapters 11 and 12.

1.3 ROBUST PRODUCT AND PROCESS DESIGN

It is well known that variation in key performance characteristics can result in poor product and process quality. During the 1980s, considerable attention was given to this problem, and methodology was developed for using experimental design, specifically for the following:

1. For designing products or processes so that they are robust to environment conditions.
2. For designing or developing products so that they are robust to component variation.
3. For minimizing variability around a target value.

By *robust*, we mean that the product or process performs consistently on target and is relatively insensitive to factors that are difficult to control.

Professor Genichi Taguchi used the term **robust parameter design** to describe his approach to this important class of industrial problems. Essentially, robust parameter design methodology prefers to reduce product or process variation by choosing levels of controllable factors (or parameters) that make the system insensitive (or robust) to changes in a set of uncontrollable factors that represent most of the sources of variability. Taguchi referred to these uncontrollable factors as *noise factors*. These are the environmental factors such as humidity levels, changes in raw material properties, product aging, and component variability referred to in 1 and 2 above. We usually assume that these noise factors are uncontrollable in the field, but can be controlled during product or process development for purposes of a designed experiment.

Considerable attention has been focused on the methodology advocated by Taguchi, and a number of flaws in his approach have been discovered. However, there are many useful concepts in his philosophy, and it is relatively easy to incorporate these within the framework of response surface methodology. In Chapter 10 we will review and discuss Taguchi's methodology and present several attractive alternatives to robust design that are based on principles and philosophy that are compatible with his, yet which avoid the flaws and controversy that surround his techniques.

CHAPTER 2

Building Empirical Models

2.1 LINEAR REGRESSION MODELS

In the practical application of response surface methodology it is necessary to develop an approximating model for the true response surface. The underlying true response surface is typically driven by some unknown **physical mechanism**. The approximating model is based on observed data from the process or system and is an **empirical model**. Multiple regression is a collection of statistical techniques useful for building the types of empirical models required in response surface methodology (RSM).

As an example, suppose that we wish to develop an empirical model relating the effective life of a cutting tool to the cutting speed and the tool angle. A first-order response surface model that might describe this relationship is

$$y = \beta_0 + \beta_1 x_1 + \beta_2 x_2 + \varepsilon \tag{2.1}$$

where y represents the tool life, x_1 represents the cutting speed, and x_2 represents the tool angle. This is a **multiple linear regression model** with two independent variables. We often call the independent variables **predictor variables** or **regressors**. The term "linear" is used because Equation (2.1) is a linear function of the unknown parameters β_0, β_1, and β_2. The model describes a plane in the two-dimensional x_1, x_2 space. The parameter β_0 defines the intercept of the plane. We sometimes call β_1 and β_2 *partial regression coefficients*, because β_1 measures the expected change in y per unit change in x_1 when x_2 is held constant, and β_2 measures the expected change in y per unit change in x_2 when x_1 is held constant.

In general, the response variable y may be related to k regressor variables. The model

$$y = \beta_0 + \beta_1 x_1 + \beta_2 x_2 + \cdots + \beta_k x_k + \varepsilon \tag{2.2}$$

is called a *multiple linear regression model* with k regressor variables. The

16

parameters β_j, $j = 0, 1, \ldots, k$, are called the *regression coefficients*. This model describes a hyperplane in the k-dimensional space of the regressor variables $\{x_j\}$. The parameter β_j represents the expected change in response y per unit change in x_j when all the remaining independent variables x_i $(i \neq j)$ are held constant.

Models that are more complex in appearance than Equation (2.2) may often still be analyzed by multiple linear regression techniques. For example, considering adding an interaction term to the first-order model in two variables, say

$$y = \beta_0 + \beta_1 x_1 + \beta_2 x_2 + \beta_{12} x_1 x_2 + \varepsilon \qquad (2.3)$$

If we let $x_3 = x_1 x_2$ and $\beta_3 = \beta_{12}$, then Equation (2.3) can be written as

$$y = \beta_0 + \beta_1 x_1 + \beta_2 x_2 + \beta_3 x_3 + \varepsilon \qquad (2.4)$$

which is a standard multiple linear regression model with three regressors As another example, consider the second-order response surface model in two variables:

$$y = \beta_0 + \beta_1 x_1 + \beta_2 x_2 + \beta_{11} x_1^2 + \beta_{22} x_2^2 + \beta_{12} x_1 x_2 + \varepsilon \qquad (2.5)$$

If we let $x_3 = x_1^2$, $x_4 = x_2^2$, $x_5 = x_1 x_2$, $\beta_3 = \beta_{11}$, $\beta_4 = \beta_{22}$, and $\beta_5 = \beta_{12}$, then this becomes

$$y = \beta_0 + \beta_1 x_1 + \beta_2 x_2 + \beta_3 x_3 + \beta_4 x_4 + \beta_5 x_5 + \varepsilon \qquad (2.6)$$

which is a linear regression model. In general, any regression model that is linear in the parameters (the β values) is a linear regression model, regardless of the shape of the response surface that it generates.

In this chapter we will present and illustrate methods for estimating the parameters in multiple linear regression models. This is often called **model fitting**. We will also discuss methods for testing hypotheses and constructing confidence intervals for these models, as well as for checking the adequacy of the model fit. Our focus is primarily on those aspects of regression analysis useful in RSM. For more complete presentations of regression, refer to Montgomery and Peck (1992) and Myers (1990).

2.2 ESTIMATION OF THE PARAMETERS IN LINEAR REGRESSION MODELS

The method of least squares is typically used to estimate the regression coefficients in a multiple linear regression model. Suppose that $n > k$ observations on the response variable are available, say $y_1, y_2, \ldots, y_n$. Along with

Table 2.1   Data for Multiple Linear Regression

y	x_1	x_2	$\cdots$	x_k
y_1	x_{11}	x_{12}	$\cdots$	x_{1k}
y_2	x_{21}	x_{22}	$\cdots$	x_{2k}
$\vdots$	$\vdots$	$\vdots$		$\vdots$
y_n	x_{n1}	x_{n2}	$\cdots$	x_{nk}

each observed response y_i, we will have an observation on each regressor variable, and let x_{ij} denote the ith observation or level of variable x_j. The data will appear as in Table 2.1. We assume that the error term ε in the model has $E(\varepsilon) = 0$ and $\text{Var}(\varepsilon) = \sigma^2$ and that the $\{\varepsilon_i\}$ are uncorrelated random variables.

We may write the model equation [Equation (2.2)] in terms of the observations in Table 2.1 as

$$y_i = \beta_0 + \beta_1 x_{i1} + \beta_2 x_{i2} + \cdots + \beta_k x_{ik} + \varepsilon_i$$

$$= \beta_0 + \sum_{j=1}^{k} \beta_j x_{ij} + \varepsilon_i, \qquad i = 1, 2, \ldots, n \qquad (2.7)$$

The method of least squares chooses the β's in Equation (2.7) so that the sum of the squares of the errors, ε_i, are minimized. The least squares function is

$$L = \sum_{i=1}^{n} \varepsilon_i^2$$

$$= \sum_{i=1}^{n} \left(y_i - \beta_0 - \sum_{j=1}^{k} \beta_j x_{ij} \right)^2 \qquad (2.8)$$

The function L is to be minimized with respect to $\beta_0, \beta_1, \ldots, \beta_k$. The least squares estimators, say $b_0, b_1, \ldots, b_k$, must satisfy

$$\left. \frac{\partial L}{\partial \beta_0} \right|_{b_0, b_1, \ldots, b_k} = -2 \sum_{i=1}^{n} \left(y_i - b_0 - \sum_{j=1}^{k} b_j x_{ij} \right) = 0 \qquad (2.9a)$$

and

$$\left. \frac{\partial L}{\partial \beta_j} \right|_{b_0, b_1, \ldots, b_k} = -2 \sum_{i=1}^{n} \left(y_i - b_0 - \sum_{j=1}^{k} b_j x_{ij} \right) x_{ij} = 0, \qquad j = 1, 2, \ldots, k$$

$$(2.9b)$$

Simplifying Equation (2.9), we obtain

$$nb_0 + b_1 \sum_{i=1}^{n} x_{i1} + b_2 \sum_{i=1}^{n} x_{i2} + \cdots + b_k \sum_{i=1}^{n} x_{ik} = \sum_{i=1}^{n} y_i$$

$$b_0 \sum_{i=1}^{n} + b_1 \sum_{i=1}^{n} x_{i1}^2 + b_2 \sum_{i=1}^{n} x_{i1}x_{i2} + \cdots + b_k \sum_{i=1}^{n} x_{i1}x_{ik} = \sum_{i=1}^{n} x_{i1}y_i$$

$$\vdots \qquad \vdots \qquad \vdots \qquad \qquad \vdots \qquad \vdots \qquad (2.10)$$

$$b_0 \sum_{i=1}^{n} x_{ik} + b_1 \sum_{i=1}^{n} x_{ik}x_{i1} + b_2 \sum_{i-1}^{n} x_{ik}x_{i2} + \cdots + b_k \sum_{i-1}^{n} x_{ik}^2 - \sum_{i=1}^{n} x_{ik}y_i$$

These equations are called the **least squares normal equations**. Note that there are $p = k + 1$ normal equations, one for each of the unknown regression coefficients. The solution to the normal equations will be the least squares estimators of the regression coefficients $b_0, b_1, \ldots, b_k$.

It is simpler to solve the normal equations if they are expressed in matrix notation. We now give a matrix development of the normal equations that parallels the development of Equation (2.10). The model in terms of the observations, Equation (2.7), may be written in matrix notation as

$$\mathbf{y} - \mathbf{X}\boldsymbol{\beta} + \boldsymbol{\epsilon}$$

where

$$\mathbf{y} = \begin{bmatrix} y_1 \\ y_2 \\ \vdots \\ y_n \end{bmatrix}, \qquad \mathbf{X} = \begin{bmatrix} 1 & x_{11} & x_{12} & \cdots & x_{1k} \\ 1 & x_{21} & x_{22} & \cdots & x_{2k} \\ \vdots & \vdots & \vdots & & \vdots \\ 1 & x_{n1} & x_{n2} & \cdots & x_{nk} \end{bmatrix},$$

$$\boldsymbol{\beta} = \begin{bmatrix} \beta_0 \\ \beta_1 \\ \vdots \\ \beta_k \end{bmatrix}, \qquad \text{and} \qquad \boldsymbol{\epsilon} = \begin{bmatrix} \varepsilon_1 \\ \varepsilon_2 \\ \vdots \\ \varepsilon_n \end{bmatrix}$$

In general, $\mathbf{y}$ is an $(n \times 1)$ vector of the observations, $\mathbf{X}$ is an $(n \times p)$ matrix of the levels of the independent variables, $\boldsymbol{\beta}$ is a $(p \times 1)$ vector of the regression coefficients, and $\boldsymbol{\epsilon}$ is an $(n \times 1)$ vector of random errors. We wish to find the vector of least squares estimators, $\mathbf{b}$, that minimizes

$$L = \sum_{i=1}^{n} \varepsilon_i^2 = \boldsymbol{\epsilon}'\boldsymbol{\epsilon} = (\mathbf{y} - \mathbf{X}\boldsymbol{\beta})'(\mathbf{y} - \mathbf{X}\boldsymbol{\beta})$$

Note that L may be expressed as

$$L = \mathbf{y}'\mathbf{y} - \boldsymbol{\beta}'\mathbf{X}'\mathbf{y} - \mathbf{y}'\mathbf{X}\boldsymbol{\beta} + \boldsymbol{\beta}'\mathbf{X}'\mathbf{X}\boldsymbol{\beta}$$

$$= \mathbf{y}'\mathbf{y} - 2\boldsymbol{\beta}'\mathbf{X}'\mathbf{y} + \boldsymbol{\beta}'\mathbf{X}'\mathbf{X}\boldsymbol{\beta} \tag{2.11}$$

since $\boldsymbol{\beta}'\mathbf{X}'\mathbf{y}$ is a (1×1) matrix, or a scalar, and its transpose $(\boldsymbol{\beta}'\mathbf{X}'\mathbf{y})' = \mathbf{y}'\mathbf{X}\boldsymbol{\beta}$ is the same scalar. The least squares estimators must satisfy

$$\left.\frac{\partial L}{\partial \boldsymbol{\beta}}\right|_{\mathbf{b}} = -2\mathbf{X}'\mathbf{y} + 2\mathbf{X}'\mathbf{X}\mathbf{b} = 0$$

which simplifies to

$$\mathbf{X}'\mathbf{X}\mathbf{b} = \mathbf{X}'\mathbf{y} \tag{2.12}$$

Equation (2.12) the set of least squares normal equations in matrix form. It is identical to Equation (2.10). To solve the normal equations, multiply both sides of Equation (2.12) by the inverse of $\mathbf{X}'\mathbf{X}$. Thus, the least squares estimator of $\boldsymbol{\beta}$ is

$$\mathbf{b} = (\mathbf{X}'\mathbf{X})^{-1}\mathbf{X}'\mathbf{y} \tag{2.13}$$

It is easy to see that the matrix form of the normal equations is identical to the scalar form. Writing out Equation (2.12) in detail, we obtain

$$
\begin{bmatrix}
n & \sum_{i=1}^{n} x_{i1} & \sum_{i=1}^{n} x_{i2} & \cdots & \sum_{i=1}^{n} x_{ik} \\
\sum_{i=1}^{n} x_{i1} & \sum_{i=1}^{n} x_{i1}^2 & \sum_{i=1}^{n} x_{i1}x_{i2} & \cdots & \sum_{i=1}^{n} x_{i1}x_{ik} \\
\vdots & \vdots & \vdots & & \vdots \\
\sum_{i=1}^{n} x_{ik} & \sum_{i=1}^{n} x_{ik}x_{i1} & \sum_{i=1}^{n} x_{ik}x_{i2} & \cdots & \sum_{i=1}^{n} x_{ik}^2
\end{bmatrix}
\begin{bmatrix}
b_0 \\ b_1 \\ \vdots \\ b_k
\end{bmatrix}
=
\begin{bmatrix}
\sum_{i=1}^{n} y_i \\ \sum_{i=1}^{n} x_{i1}y_i \\ \vdots \\ \sum_{i=1}^{n} x_{ik}y_i
\end{bmatrix}
$$

If the indicated matrix multiplication is performed, the scalar form of the normal equations [i.e., Equation (2.10)] will result. In this form it is easy to see that $\mathbf{X}'\mathbf{X}$ is a $(p \times p)$ symmetric matrix and $\mathbf{X}'\mathbf{y}$ is a $(p \times 1)$ column vector. Note the special structure of the $\mathbf{X}'\mathbf{X}$ matrix. The diagonal elements of $\mathbf{X}'\mathbf{X}$ are the sums of squares of the elements in the columns of $\mathbf{X}$, and the off-diagonal elements are the sums of cross-products of the elements in the columns of $\mathbf{X}$. Furthermore, note that the elements of $\mathbf{X}'\mathbf{y}$ are the sums of cross-products of the columns of $\mathbf{X}$ and the observations $\{y_i\}$.

The fitted regression model is

$$\hat{\mathbf{y}} = \mathbf{X}\mathbf{b} \tag{2.14}$$

In scalar notion, the fitted model is

$$\hat{y}_i = b_0 + \sum_{i=1}^{k} b_j x_{ij}, \qquad i = 1, 2, \ldots, n$$

The difference between the observation y_i and the fitted value $\hat{y}_i$ is a **residual**, say $e_i = y_i - \hat{y}_i$. The $(n \times 1)$ vector of residuals is denoted by

$$\mathbf{e} = \mathbf{y} - \hat{\mathbf{y}} \tag{2.15}$$

Example 2.1 Transistor gain in an integrated circuit device between emitter and collector (hFE) is related to two variables that can be controlled at the deposition process, emitter drive-in time (ξ_1, in minutes), and emitter dose (ξ_2, in ions $\times 10^{14}$). Fourteen samples were observed following deposition, and the resulting data are shown in Table 2.2. We will fit a linear regression model using gain as the response and emitter drive-in time and emitter dose as the regressor variables.

Columns 1 and 2 of Table 2.2 show the actual or natural unit values of ξ_1 and ξ_2, while columns 3 and 4 contain values of the corresponding coded

Table 2.2 Data on Transistor Gain (y) for Example 2.1

Observation	ξ_1 (drive-in time, minutes)	ξ_2 (dose, ions $\times 10^{14}$)	x_1	x_2	y (gain or hFE)
1	195	4.00	-1	-1	1004
2	255	4.00	1	-1	1636
3	195	4.60	-1	0.6667	852
4	255	4.60	1	0.6667	1506
5	225	4.20	0	-0.4444	1272
6	225	4.10	0	-0.7222	1270
7	225	4.60	0	0.6667	1269
8	195	4.30	-1	-0.1667	903
9	255	4.30	1	-0.1667	1555
10	225	4.00	0	-1	1260
11	225	4.70	0	0.9444	1146
12	225	4.30	0	-0.1667	1276
13	225	4.72	0	1	1225
14	230	4.30	0.1667	-0.1667	1321

variables x_1 and x_2, where

$$x_{i1} = \frac{\xi_{i1} - [\max(\xi_{i1}) + \min(\xi_{i1})]/2}{[\max(\xi_{i1}) - \min(\xi_{i1})]/2}$$

$$= \frac{\xi_{i1} - (255 + 195)/2}{(255 - 195)/2}$$

$$= \frac{\xi_{i1} - 225}{30}$$

$$x_{i2} = \frac{\xi_{i2} - [\max(\xi_{i2}) + \min(\xi_{j2})]/2}{[\max(\xi_{i2}) - \min(\xi_{i2})]/2}$$

$$= \frac{\xi_{i2} - (4.72 + 4.00)/2}{(4.72 - 4.00)/2}$$

$$= \frac{\xi_{i2} - 4.36}{0.36}$$

This coding scheme is widely used in fitting linear regression models, and it results in all the values of x_1 and x_2 falling between -1 and $+1$, as shown in Table 2.2.

We will fit the model

$$y = \beta_0 + \beta_1 x_1 + \beta_2 x_2 + \varepsilon$$

using the coded variables. The **X** matrix and **y** vector are

$$\mathbf{X} = \begin{bmatrix} 1 & -1 & -1 \\ 1 & 1 & -1 \\ 1 & -1 & 0.6667 \\ 1 & 1 & 0.6667 \\ 1 & 0 & -0.4444 \\ 1 & 0 & -0.7222 \\ 1 & 0 & 0.6667 \\ 1 & -1 & -0.1667 \\ 1 & 1 & -0.1667 \\ 1 & 0 & -1 \\ 1 & 0 & 0.9444 \\ 1 & 0 & -0.1667 \\ 1 & 0 & 1 \\ 1 & 1.1667 & -0.1667 \end{bmatrix}, \quad \mathbf{y} = \begin{bmatrix} 1004 \\ 1636 \\ 852 \\ 1506 \\ 1272 \\ 1270 \\ 1269 \\ 903 \\ 1555 \\ 1260 \\ 1146 \\ 1276 \\ 1225 \\ 1321 \end{bmatrix}$$

The $\mathbf{X'X}$ matrix is

$$\mathbf{X'X} = \begin{bmatrix} 1 & 1 & \cdots & 1 \\ -1 & 1 & \cdots & 0.1667 \\ -1 & -1 & \cdots & -0.1667 \end{bmatrix} \begin{bmatrix} 1 & -1 & -1 \\ 1 & 1 & -1 \\ \vdots & \vdots & \vdots \\ 1 & 0.1667 & -0.1667 \end{bmatrix}$$

$$= \begin{bmatrix} 14 & 0.1667 & -0.8889 \\ 0.1667 & 6.027789 & -0.02779 \\ -0.8889 & -0.02779 & 7.055578 \end{bmatrix}$$

and the $\mathbf{X'y}$ vector is

$$\mathbf{X'y} = \begin{bmatrix} 1 & 1 & \cdots & 1 \\ -1 & 1 & \cdots & 0.1667 \\ -1 & -1 & \cdots & -0.1667 \end{bmatrix} \begin{bmatrix} 1004 \\ 1636 \\ \vdots \\ 1321 \end{bmatrix}$$

$$= \begin{bmatrix} 17,495 \\ 2158.211 \\ -1499.74 \end{bmatrix}$$

The least squares estimate of $\boldsymbol{\beta}$ is

$$\mathbf{b} - (\mathbf{X'X})^{-1}\mathbf{X'y}$$

or

$$\mathbf{b} = \begin{bmatrix} 0.072027 & -0.00195 & 0.009067 \\ -0.00195 & 0.165954 & 0.000408 \\ 0.009067 & 0.000408 & 0.142876 \end{bmatrix} \begin{bmatrix} 17,495 \\ 2158.211 \\ -1499.74 \end{bmatrix}$$

$$= \begin{bmatrix} 1242.3057 \\ 323.4366 \\ -54.7691 \end{bmatrix}$$

The least squares fit with the regression coefficients reported to one decimal place is

$$\hat{y} = 1242.3 + 323.4x_1 - 54.8x_2$$

This can be converted into an equation using the natural variables ξ_1 and ξ_2 by substituting the relationships between x_1 and ξ_1 and x_2 and ξ_2 as follows:

$$\hat{y} = 1242.3 + 323.4\left(\frac{\xi_1 - 225}{30}\right) - 54.8\left(\frac{\xi_2 - 4.36}{0.36}\right)$$

Table 2.3   Observations, Fitted Values, Residuals, and Other Summary Information for Example 2.1

Observation	y_i	$\hat{y}_i$	e_i	h_{ij}	r_i	t_i	D_i
1	1004.0	973.7	30.3	0.367	1.092	1.103	0.231
2	1636.0	1620.5	15.5	0.358	0.553	0.535	0.057
3	852.0	882.4	-30.4	0.317	-1.052	-1.057	0.171
4	1506.0	1529.2	-23.2	0.310	-0.801	-0.787	0.096
5	1272.0	1266.7	5.3	0.092	0.160	0.153	0.001
6	1270.0	1281.9	-11.9	0.133	-0.365	-0.350	0.007
7	1269.0	1205.8	63.2	0.148	1.960	2.316	0.222
8	903.0	928.0	-25.0	0.243	-0.823	-0.810	0.072
9	1555.0	1574.9	-19.9	0.235	-0.651	-0.633	0.043
10	1260.0	1297.1	-37.1	0.197	-1.185	-1.209	0.115
11	1146.0	1190.6	-44.6	0.217	-1.442	-1.527	0.192
12	1276.0	1251.4	24.6	0.073	0.730	0.714	0.014
13	1225.0	1187.5	37.5	0.233	1.225	1.256	0.152
14	1321.0	1305.3	15.7	0.077	0.466	0.449	0.006

or

$$\hat{y} = -520.1 + 10.781\xi_1 - 152.15\xi_2$$

Table 2.3 shows the observed values of y_i, the corresponding fitted values $\hat{y}_i$, and the residuals from this model. There are several other quantities given in this table that will be defined and discussed later. Figure 2.1 shows the fitted response surface and the contour plot for this model. The response surface for gain is a plane lying above the time–dose space.

2.3   PROPERTIES OF THE LEAST SQUARES ESTIMATORS AND ESTIMATION OF σ^2

The method of least squares produces an unbiased estimator of the parameter β in the multiple linear regression model. This property may be easily demonstrated by finding the expected value of **b** as follows:

$$\begin{aligned}
E(\mathbf{b}) &= E\left[(\mathbf{X}'\mathbf{X})^{-1}\mathbf{X}'\mathbf{y}\right] \\
&= E\left[(\mathbf{X}'\mathbf{X})^{-1}\mathbf{X}'(\mathbf{X}\boldsymbol{\beta} + \boldsymbol{\varepsilon})\right] \\
&= E\left[(\mathbf{X}'\mathbf{X})^{-1}\mathbf{X}'\mathbf{X}\boldsymbol{\beta} + (\mathbf{X}'\mathbf{X})^{-1}\mathbf{X}'\boldsymbol{\varepsilon}\right] \\
&= \boldsymbol{\beta}
\end{aligned}$$

because $E(\boldsymbol{\varepsilon}) = 0$ and $(\mathbf{X}'\mathbf{X})^{-1}\mathbf{X}'\mathbf{X} = \mathbf{I}$. Thus **b** is an unbiased estimator of

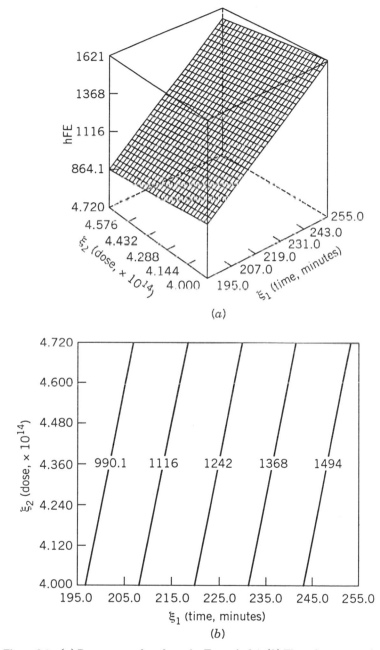

Figure 2.1 (a) Response surface for gain, Example 2.1. (b) The gain contour plot.

β. The variance property of **b** is expressed by the covariance matrix

$$\text{Cov}(\mathbf{b}) = E\{[\mathbf{b} - E(\mathbf{b})][\mathbf{b} - E(\mathbf{b})]'\}$$

The covariance matrix of **b** is a ($p \times p$) symmetric matrix whose jjth element is the variance of ($\mathbf{b}_j$) and whose (i, j)th element is the covariance between b_i and b_j. The *covariance matrix* of **b** is

$$\text{Cov}(\mathbf{b}) = \sigma^2 (\mathbf{X'X})^{-1} \tag{2.16}$$

It is also usually necessary to estimate σ^2. To develop an estimator of this parameter consider the sum of squares of the residuals, say

$$SS_E = \sum_{i=1}^{n} (y_i - \hat{y}_i)^2$$

$$= \sum_{i=1}^{n} e_i^2$$

$$= \mathbf{e'e}$$

Substituting $\mathbf{e} = \mathbf{y} - \hat{\mathbf{y}} = \mathbf{y} - \mathbf{Xb}$, we have

$$SS_E = (\mathbf{y} - \mathbf{Xb})'(\mathbf{y} - \mathbf{Xb})$$

$$= \mathbf{y'y} - \mathbf{b'X'y} - \mathbf{y'Xb} + \mathbf{b'X'Xb}$$

$$= \mathbf{y'y} - 2\mathbf{b'X'y} + \mathbf{b'X'Xb}$$

Because $\mathbf{X'Xb} = \mathbf{X'y}$, this last equation becomes

$$SS_E = \mathbf{y'y} - \mathbf{b'X'y} \tag{2.17}$$

Equation (2.17) is called the **error** or **residual sum of squares**, and it has $n - p$ degrees of freedom associated with it. It can be shown that

$$E(SS_E) = \sigma^2(n - p)$$

so an unbiased estimator of σ^2 is given by

$$\hat{\sigma}^2 = \frac{SS_E}{n - p} \tag{2.18}$$

Example 2.2   We will estimate σ^2 for the regression model from Example 2.1. Because

$$\mathbf{y'y} = \sum_{i=1}^{14} y_i^2 = 22,527,889.0$$

and

$$\mathbf{b'X'y} = \begin{bmatrix} 1242.3 & 323.4 & -54.8 \end{bmatrix} \begin{bmatrix} 17,495 \\ 2158.211 \\ -1499.74 \end{bmatrix} = 22,514,467.9$$

the residual sum of squares is

$$SS_E = \mathbf{y'y} - \mathbf{b'X'y}$$
$$= 22,527,889.0 - 22,514,467.9$$
$$= 13,421.1$$

Therefore, the estimate of σ^2 is computed from Equation (2.18) as follows:

$$\hat{\sigma}^2 = \frac{SS_E}{n-p} = \frac{13,421.1}{14-3} = 1220.1$$

The estimate of σ^2 produced by Equation (2.18) is **model-dependent**. That is, it depends on the form of the model that is fit to the data. To illustrate this point, suppose that we fit a quadratic model to the gain data, say

$$y = \beta_0 + \beta_1 x_1 + \beta_2 x_2 + \beta_{11} x_1^2 + \beta_{22} x_2^2 + \beta_{12} x_1 x_2 + \varepsilon$$

In this model it can be shown that $SS_E = 12,479.8$. Because the number of model parameters, p, equals 6, the estimate of σ^2 based on this model is

$$\hat{\sigma}^2 = \frac{12,479.8}{14-6} = 1559.975$$

This estimate of σ^2 is actually larger than the estimate obtained from the first-order model, suggesting that the first-order model is superior to the quadratic in that there is less unexplained variability resulting from the first-order fit. If replicate runs are available (that is, more than one observation on y at the same x-levels), then a model-independent estimate of σ^2 can be obtained. We will show how to do this in Section 2.7.4.

2.4   HYPOTHESIS TESTING IN MULTIPLE REGRESSION

In multiple linear regression problems, certain tests of hypotheses about the model parameters are helpful in measuring the usefulness of the model. In this section, we describe several important hypothesis-testing procedures. These procedures require that the errors ε_i in the model be normally and independently distributed with mean zero and variance σ^2, abbreviated $\varepsilon \sim \text{NID}\,(0, \sigma^2)$. As a result of this assumption, the observations y_i are normally and independently distributed with mean $\beta_0 + \sum_{j=1}^{k} \beta_j x_{ij}$ and variance σ^2.

2.4.1 Test for Significance of Regression

The test for significance of regression is a test to determine if there is a linear relationship between the response variable y and a subset of the regressor variables $x_1, x_2, \ldots, x_k$. The appropriate hypotheses are

$$H_0: \quad \beta_1 = \beta_2 = \cdots = \beta_k = 0$$
$$H_1: \quad \beta_j \neq 0 \quad \text{for at least one } j \tag{2.19}$$

Rejection of H_0: in (2.19) implies that at least one of the regressor variables $x_1, x_2, \ldots, x_k$ contributes significantly to the model. The test procedure involves partitioning the total sum of squares S_{yy} into a sum of squares due to the model (or to regression) and a sum of squares due to residual (or error), say

$$S_{yy} = SS_R + SS_E \tag{2.20}$$

Now if the null hypothesis $H_0: \beta_1 = \beta_2 = \cdots = \beta_k = 0$ is true, then SS_R/σ^2 is distributed as χ_k^2, where the number of degrees of freedom for χ^2 are equal to the number of regressor variables in the model. Also, we can show that SS_E/σ^2 is distributed as χ_{n-k-1}^2 and that SS_E and SS_R are independent. The test procedure for $H_0: \beta_1 = \beta_2 = \cdots = \beta_k = 0$ is to compute

$$F_0 = \frac{SS_R/k}{SS_E/(n-k-1)} = \frac{MS_R}{MS_E} \tag{2.21}$$

and to reject H_0 if F_0 exceeds $F_{\alpha, k, n-k-1}$. Alternatively, one could use the P-value approach to hypothesis testing and, thus, reject H_0 if the P-value for the statistic F_0 is less than α. The test is usually summarized in a table such as Table 2.4. This test procedure is called an **analysis of variance** because it is based on a decomposition of the total variability in the response variable y.

A computational formula for SS_R may be found easily. We have derived a computational formula for SS_E in equation (2.17)—that is,

$$SS_E = \mathbf{y}'\mathbf{y} - \mathbf{b}'\mathbf{X}'\mathbf{y}$$

Table 2.4   Analysis of Variance for Significance of Regression in Multiple Regression

Source of Variation	Sum of Squares	Degrees of Freedom	Mean Square	F_0
Regression	SS_R	k	MS_R	MS_R/MS_E
Error or residual	SS_E	$n-k-1$	MS_E	
Total	S_{yy}	$n-1$		

Now because $S_{yy} = \sum_{i=1}^{n} y_i^2 - (\sum_{i=1}^{n} y_i)^2/n = \mathbf{y'y} - (\sum_{i=1}^{n} y_i)^2/n$, we may rewrite the foregoing equation as

$$SS_E = \mathbf{y'y} - \frac{\left(\sum_{i=1}^{n} y_i\right)^2}{n} - \left[\mathbf{b'X'y} - \frac{\left(\sum_{i=1}^{n} y_i\right)^2}{n}\right]$$

or

$$SS_E = S_{yy} - SS_R$$

Therefore, the regression sum of squares is

$$SS_R = \mathbf{b'X'y} - \frac{\left(\sum_{i=1}^{n} y_i\right)^2}{n} \tag{2.22}$$

and the error sum of squares is

$$SS_E = \mathbf{y'y} - \mathbf{b'X'y} \tag{2.23}$$

and the total sum of squares is

$$S_{yy} = \mathbf{y'y} - \frac{\left(\sum_{i=1}^{n} y_i\right)^2}{n} \tag{2.24}$$

Example 2.3 We will test for significance of regression using the model fit to the transistor gain data for Example 2.1. Note that

$$S_{yy} = \mathbf{y'y} - \frac{\left(\sum_{i=1}^{14} y_i\right)^2}{14}$$

$$= 22,527,889.0 - \frac{(17,495)^2}{14}$$

$$= 665,387.2$$

$$SS_R = \mathbf{b'X'y} - \frac{\left(\sum_{i=1}^{14} y_i\right)^2}{14}$$

$$= 22,514,467.9 - 21,862,501.8$$

$$= 651,966.1$$

Table 2.5   Test for Significance of Regression, Example 2.3

Source of Variation	Sum of Squares	Degrees of Freedom	Mean Square	F_0	P-Value
Regression	651,996.1	2	325,983.0	267.2	4.74×10^{-10}
Error	13,421.1	11	1220.1		
Total	665,387.2	13			

and

$$SS_E = S_{yy} - SS_R$$

$$= 665,387.2 - 651,966.1$$

$$= 13,421.1$$

The analysis of variance is shown in Table 2.5 If we select $\alpha = 0.05$, then we would reject H_0: $\beta_1 = \beta_2 = 0$ because $F_0 = 267.2 > F_{0.05, 2, 11} = 3.98$. Also, note that the P-value for F_0 (shown in Table 2.5) is considerably smaller than $\alpha = 0.05$.

The coefficient of multiple determination R^2 is defined as

$$R^2 = \frac{SS_R}{S_{yy}} = 1 - \frac{SS_E}{S_{yy}} \tag{2.25}$$

R^2 is a measure of the amount of reduction in the variability of y obtained by using the regressor variables $x_1, x_2, \ldots, x_k$ in the model. From inspection of the analysis of variance identity equation [Equation (2.20)] we see that $0 \leq R^2 \leq 1$. However, a large value of R^2 does not necessarily imply that the regression model is good one. Adding a variable to the model will always increase R^2, regardless of whether the additional variable is statistically significant or not. Thus it is possible for models that have large values of R^2 to yield poor predictions of new observations or estimates of the mean response.

To illustrate, consider the first-order model for the transistor gain data. The value of R^2 for this model is

$$R^2 = \frac{SS_R}{S_{yy}} = \frac{651,966.1}{665,387.2} = 0.9798$$

That is, the first-order model explains about 97.98% of the variability observed in gain. Now if we add quadratic terms to this model, we can show that the value of R^2 increases to 0.9812. This increase in R^2 is relatively small, suggesting that the quadratic terms do not really improve the model.

Because R^2 always increases as we add terms to the model, some regression model builders prefer to use an adjusted R^2 statistic defined as

$$R^2_{adj} = 1 - \frac{SS_E/(n-p)}{S_{yy}/(n-1)} = 1 - \left(\frac{n-1}{n-p}\right)(1 - R^2) \qquad (2.26)$$

In general, the adjusted R^2 statistic will not always increase as variables are added to the model. In fact, if unnecessary terms are added, the value of R^2_{adj} will often decrease.

For example, consider the transistor gain data. The adjusted R^2 for the first-order model is

$$R^2_{adj} = 1 - \left(\frac{n-1}{n-p}\right)(1 - R^2)$$

$$= 1 - \left(\frac{13}{11}\right)(1 - 0.9798)$$

$$= 0.9762$$

which is very close to the ordinary R^2 for the first-order model. When R^2 and R^2_{adj} differ dramatically there is a good chance that nonsignificant terms have been included in the model. Now when the quadratic terms are added to the first-order model, we can show that $R^2_{adj} = 0.9695$; that is, the adjusted R^2 actually decreases when the quadratic terms are included in the model. This is a strong indication that the quadratic terms are unnecessary.

2.4.2 Tests on Individual Regression Coefficients and Groups of Coefficients

We are frequently interested in testing hypotheses on the individual regression coefficients. Such tests would be useful in determining the value of each of the regressor variables in the regression model. For example, the model might be more effective with the inclusion of additional variables, or perhaps with the deletion of one or more of the variables already in the model.

Adding a variable to the regression model always causes the sum of squares for regression to increase and the error sum of squares to decrease. We must decide whether the increase in the regression sum of squares is sufficient to warrant using the additional variable in the model. Furthermore, adding an unimportant variable to the model can actually increase the mean square error, thereby decreasing the usefulness of the model.

The hypotheses for testing the significance of any individual regression coefficient, say β_j, are

$$H_0: \quad \beta_j = 0$$

$$H_1: \quad \beta_j \neq 0$$

If H_0: $\beta_j = 0$ is not rejected, then this indicates that x_j can be deleted from the model. The test statistic for this hypothesis is

$$t_0 = \frac{b_j}{\sqrt{\hat{\sigma}^2 C_{jj}}} \tag{2.27}$$

where C_{jj} is the diagonal element of $(\mathbf{X'X})^{-1}$ corresponding to b_j. The null hypothesis H_0: $\beta_j = 0$ is rejected if $|t_0| > t_{\alpha/2, n-k-1}$. Note that this is really a partial or marginal test, because the regression coefficient b_j depends on all the other regressor variables $x_i (i \neq j)$ that are in the model.

The denominator of Equation (2.27), $\sqrt{\hat{\sigma}^2 C_{jj}}$, is often called the **standard error** of the regression coefficient b_j. That is,

$$se(b_j) = \sqrt{\hat{\sigma}^2 C_{jj}} \tag{2.28}$$

Therefore, an equivalent way to write the test statistic in Equation (2.27) is

$$t_0 = \frac{b_j}{se(b_j)} \tag{2.29}$$

Example 2.4   To illustrate the use of the t-test, consider the regression model for the transistor gain data. We will construct the t-statistic for the hypotheses H_0: $\beta_1 = 0$ and H_0: $\beta_2 = 0$. The main diagonal elements of $(\mathbf{X'X})^{-1}$ corresponding to β_1 and β_2 are $C_{11} = 0.165954$ and $C_{22} = 0.142876$, respectively, so the two t-statistics are computed as follows:

For H_0: $\beta_1 = 0$: $t_0 = \dfrac{b_1}{\sqrt{\hat{\sigma}^2 C_{11}}}$

$$= \frac{323.4}{\sqrt{(1220.1)(0.165954)}} = \frac{323.4}{14.2} = 22.73$$

For H_0: $\beta_2 = 0$: $t_0 = \dfrac{b_2}{\sqrt{\hat{\sigma}^2 C_{22}}} = \dfrac{-54.8}{\sqrt{(1220.1)(0.142876)}}$

$$= \frac{-54.8}{13.2} = -4.15$$

The absolute values of these t-statistics would be compared to $t_{0.025, 11} = 2.201$ (assuming that we select $\alpha = 0.05$). Both t-statistics are larger than this criterion. Consequently we would conclude that $\beta_1 \neq 0$, which implies that x_1 contributes significantly to the model given that x_2 is included, and that $\beta_2 \neq 0$, which implies that x_2 contributes significantly to the model given that x_1 is included.

We may also directly examine the contribution to the regression sum of squares for a particular variable, say x_j, given that other variables $x_i(i \neq j)$ are included in the model. The procedure used to do this is called the **extra sum of squares method**. This procedure can also be used to investigate the contribution of a *subset* of the regressor variables to the model. Consider the regression model with k regressor variables:

$$\mathbf{y} = \mathbf{X}\boldsymbol{\beta} + \boldsymbol{\varepsilon}$$

where $\mathbf{y}$ is $(n \times 1)$, $\mathbf{X}$ is $(n \times p)$, $\boldsymbol{\beta}$ is $(p \times 1)$, $\boldsymbol{\varepsilon}$ is $(n \times 1)$, and $p = k + 1$. We would like to determine if the subset of regressor variables $x_1, x_2, \ldots, x_r$ $(r < k)$ contribute significantly to the regression model. Let the vector of regression coefficients be partitioned as follows:

$$\boldsymbol{\beta} = \begin{bmatrix} \boldsymbol{\beta}_1 \\ \boldsymbol{\beta}_2 \end{bmatrix}$$

where $\boldsymbol{\beta}_1$ is $(r \times 1)$, $\boldsymbol{\beta}_2$ is $[(p - r) \times 1]$. We wish to test the hypotheses

$$
\begin{aligned}
H_0: & \quad \boldsymbol{\beta}_1 = 0 \\
H_1: & \quad \boldsymbol{\beta}_1 \neq 0
\end{aligned}
\tag{2.30}
$$

The model may be written as

$$\mathbf{y} = \mathbf{X}\boldsymbol{\beta} + \boldsymbol{\varepsilon} = \mathbf{X}_1\boldsymbol{\beta}_1 + \mathbf{X}_2\boldsymbol{\beta}_2 + \boldsymbol{\varepsilon} \tag{2.31}$$

where $\mathbf{X}_1$ represents the columns of $\mathbf{X}$ associated with $\boldsymbol{\beta}_1$, and $\mathbf{X}_2$ represents the columns of $\mathbf{X}$ associated with $\boldsymbol{\beta}_2$.

For the **full model** (including both $\boldsymbol{\beta}_1$ and $\boldsymbol{\beta}_2$), we know that $\mathbf{b} = (\mathbf{X}'\mathbf{X})^{-1}\mathbf{X}'\mathbf{y}$. Also, the regression sum of squares for all variables including the intercept is

$$SS_R(\boldsymbol{\beta}) = \mathbf{b}'\mathbf{X}'\mathbf{y} \qquad (p \text{ degrees of freedom})$$

and

$$MS_E = \frac{\mathbf{y}'\mathbf{y} - \mathbf{b}'\mathbf{X}'\mathbf{y}}{n - p}$$

$SS_R(\boldsymbol{\beta})$ is called the regression sum of squares due to $\boldsymbol{\beta}$. To find the contribution of the terms in $\boldsymbol{\beta}_1$ to the regression, fit the model assuming the null hypothesis $H_0: \boldsymbol{\beta}_1 = \mathbf{0}$ to be true. The **reduced model** is found from Equation (2.31) with $\boldsymbol{\beta}_1 = \mathbf{0}$:

$$\mathbf{y} = \mathbf{X}_2\boldsymbol{\beta}_2 + \boldsymbol{\varepsilon} \tag{2.32}$$

The least squares estimator of $\boldsymbol{\beta}_2$ is $\mathbf{b}_2 = (\mathbf{X}_2'\mathbf{X}_2)^{-1}\mathbf{X}_2'\mathbf{y}$, and

$$SS_R(\boldsymbol{\beta}_2) = \mathbf{b}_2'\mathbf{X}_2'\mathbf{y} \qquad (p - r \text{ degrees of freedom}) \qquad (2.33)$$

The regression sum of squares due to $\boldsymbol{\beta}_1$ given that $\boldsymbol{\beta}_2$ is already in the model is

$$SS_R(\boldsymbol{\beta}_1|\boldsymbol{\beta}_2) = SS_R(\boldsymbol{\beta}) - SS_R(\boldsymbol{\beta}_2) \qquad (2.34)$$

This sum of squares has r degrees of freedom. It is the "extra sum of squares" due to $\boldsymbol{\beta}_1$. Note that $SS_R(\boldsymbol{\beta}_1|\boldsymbol{\beta}_2)$ is the increase in the regression sum of squares due to including the variables $x_1, x_2, \ldots, x_r$ in the model. Now $SS_R(\boldsymbol{\beta}_1|\boldsymbol{\beta}_2)$ is independent of MS_E, and the null hypothesis $\boldsymbol{\beta}_1 = \mathbf{0}$ may be tested by the statistic

$$F_0 = \frac{SS_R(\boldsymbol{\beta}_1|\boldsymbol{\beta}_2)/r}{MS_E} \qquad (2.35)$$

If $F_0 > F_{\alpha, r, n-p}$, we reject H_0, concluding that at least one of the parameters in $\boldsymbol{\beta}_1$ is not zero and, consequently, at least one of the variables $x_1, x_2, \ldots, x_r$ in $\mathbf{X}_1$ contributes significantly to the regression model. Some authors call the test in Equation (2.35) a **partial** F-test.

The partial F-test is very useful. We can use it to measure the contribution of x_j as if it were the last variable added to the model by computing

$$SS_R(\beta_j|\beta_0, \beta_1, \ldots, \beta_{j-1}, \beta_{j+1}, \ldots, \beta_k)$$

This is the increase in the regression sum of squares due to adding x_j to a model that already includes $x_1, \ldots, x_{j-1}, x_{j+1}, \ldots, x_k$. Note that the partial F-test on a single variable x_j is equivalent to the t-test in Equation (2.27). However, the partial F-test is a more general procedure in that we can measure the effect of sets of variables. This procedure is used often in response surface work. For example, suppose that we are considering fitting the second-order model

$$y = \beta_0 + \beta_1 x_1 + \beta_2 x_2 + \beta_{11} x_1^2 + \beta_{22} x_2^2 + \beta_{12} x_1 x_2 + \varepsilon$$

and we wish to test the contribution of the second-order terms over and above the contribution from the first-order model. Therefore, the hypotheses of interest are

$$H_0: \quad \beta_{11} = \beta_{22} = \beta_{12} = 0$$

$$H_1: \quad \beta_{11} \neq 0 \text{ and/or } \beta_{22} \neq 0 \text{ and/or } \beta_{12} \neq 0$$

In the notation of this section, $\boldsymbol{\beta}'_1 = [\beta_{11}, \beta_{22}, \beta_{12}]$ and $\boldsymbol{\beta}'_2 = [\beta_0, \beta_1, \beta_2]$, and the columns of $\mathbf{X}_1$ and $\mathbf{X}_2$ are the columns of the original $\mathbf{X}$ matrix associated with the quadratic and linear terms in the model, respectively.

Example 2.5   Consider the transistor gain data in Example 2.1. Suppose that we wish to investigate the contribution of the variable x_2 (dose) to the model. That is, the hypotheses we wish to test are

$$H_0: \quad \beta_2 = 0$$
$$H_1: \quad \beta_2 \neq 0$$

This will require the extra sum of squares due to β_2, or

$$SS_R(\beta_2|\beta_1, \beta_0) = SS_R(\beta_0, \beta_1, \beta_2) - SS_R(\beta_0, \beta_1)$$
$$= SS_R(\beta_1, \beta_2|\beta_0) - SS_R(\beta_1|\beta_0)$$

Now from Example 2.3 where we tested for significance of regression, we have (from Table 2.5)

$$SS_R(\beta_1, \beta_2|\beta_0) = 651{,}966.1$$

This sum of squares has 2 degrees of freedom. The reduced model is

$$y = \beta_0 + \beta_1 x_1 + \varepsilon$$

The least squares fit for this model is

$$\hat{y} = 1245.8 + 323.6 x_1$$

and the regression sum of squares for this model (with 1 degree of freedom) is

$$SS_R(\beta_1|\beta_0) = 630{,}967.9$$

Therefore,

$$SS_R(\beta_2|\beta_0, \beta_1) = 651{,}966.1 - 630{,}967.9$$
$$= 20{,}998.2$$

with $2 - 1 = 1$ degrees of freedom. This is the increase in the regression sum of squares that results from adding x_2 to a model already containing x_1. To test $H_0: \beta_2 = 0$, from the test statistic we obtain

$$F_0 = \frac{SS_R(\beta_2|\beta_0, \beta_1)/1}{MS_E} = \frac{20{,}998.2/1}{1220.1} = 17.21$$

Note that MS_E from the full model (Table 2.5) is used in the denominator of F_0. Now because $F_{0.05, 1, 11} = 4.84$, we would reject H_0: $\beta_2 = 0$ and conclude that x_2 (dose) contributes significantly to the model.

Because this partial F-test involves only a single regressor, it is equivalent to the t-test introduced earlier, because the square of a t random variable with v degrees of freedom is an F random variable with 1 and v degrees of freedom. To see this, recall that the t-statistic for H_0: $\beta_2 = 0$ resulted in $t_0 = -4.15$ and that $t_0^2 = (-4.15)^2 = 17.22 \simeq F_0$.

2.5   CONFIDENCE INTERVALS IN MULTIPLE REGRESSION

It is often necessary to construct confidence interval estimates for the regression coefficients $\{\beta_j\}$ and for other quantities of interest from the regression model. The development of a procedure for obtaining these confidence intervals requires that we assume the errors $\{\varepsilon_i\}$ to be normally and independently distributed with mean zero and variance σ^2, the same assumption made in the section on hypothesis testing (Section 2.4).

2.5.1   Confidence Intervals on the Individual Regression Coefficients

Because the least squares estimator $\hat{\boldsymbol{\beta}}$ is a linear combination of the observations, it follows that $\mathbf{b}$ is normally distributed with mean vector $\boldsymbol{\beta}$ and covariance matrix $\sigma^2(\mathbf{X'X})^{-1}$. Then each of the statistics

$$\frac{b_j - \beta_j}{\sqrt{\hat{\sigma}^2 C_{jj}}}, \qquad j = 0, 1, \ldots, k \tag{2.36}$$

is distributed as t with $n - p$ degrees of freedom, where C_{jj} is the jjth element of the $(\mathbf{X'X})^{-1}$ matrix, and $\hat{\sigma}^2$ is the estimate of the error variance, obtained from Equation (2.18). Therefore, a $100(1 - \alpha)\%$ confidence interval for the regression coefficient β_j, $j = 0, 1, \ldots, k$, is

$$b_j - t_{\alpha/2, n-p} \sqrt{\hat{\sigma}^2 C_{jj}} \le \beta_j \le b_j + t_{\alpha/2, n-p} \sqrt{\hat{\sigma}^2 C_{jj}} \tag{2.37}$$

Note that this confidence interval could also be written as

$$b_j - t_{\alpha/2, n-p} se(b_j) \le \beta_j \le b_j + t_{\alpha/2, n-p} se(b_j)$$

because $se(b_j) = \sqrt{\hat{\sigma}^2 C_{jj}}$.

Example 2.6   We will construct a 95% confidence interval for the parameter β_1 in Example 2.1. Now $b_1 = 323.4$, and because $\hat{\sigma}^2 = 1220.1$ and $C_{11} = 0.165954$, we find that

$$b_1 - t_{0.025, 11}\sqrt{\hat{\sigma}^2 C_{11}} \le \beta_1 \le b_1 + t_{0.025, 11}\sqrt{\hat{\sigma}^2 C_{11}}$$

$$323.4 - 2.201\sqrt{(1220.1)(0.165954)}$$

$$\le \beta_1 \le 323.4 + 2.201\sqrt{(1220.1)(0.165954)}$$

$$323.4 - 2.201(14.2) \le \beta_1 \le 323.4 + 2.201(14.2)$$

and the 95% confidence interval on β_1 is

$$292.1 \le \beta_1 \le 354.7$$

2.5.2   A Joint Confidence Region on the Regression Coefficients β

The confidence intervals in the previous section should be though of as one-at-a-time intervals; that is, the confidence coefficient $1 - \alpha$ applies only to one such interval. Some problems require that several confidence intervals be constructed from the same data. In such cases, the analyst is usually interested in specifying a confidence coefficient that applies to the entire set of confidence intervals. Such intervals are called **simultaneous confidence intervals**.

It is relatively easy to specify a **joint confidence region** for the parameters β in a multiple regression model. We may show that

$$\frac{(\mathbf{b} - \boldsymbol{\beta})'\mathbf{X}'\mathbf{X}(\mathbf{b} - \boldsymbol{\beta})}{pMS_E}$$

has an F distribution with p numerator and $n - p$ denominator degrees of freedom, and this implies that

$$P\left\{ \frac{(\mathbf{b} - \boldsymbol{\beta})'\mathbf{X}'\mathbf{X}(\mathbf{b} - \boldsymbol{\beta})}{pMS_E} \le F_{\alpha, p, n-p} \right\} = 1 - \alpha$$

Consequently a $100(1 - \alpha)\%$ point confidence region for all the parameters in β is

$$\frac{(\mathbf{b} - \boldsymbol{\beta})'\mathbf{X}'\mathbf{X}(\mathbf{b} - \boldsymbol{\beta})}{pMS_E} \le F_{\alpha, p, n-p} \tag{2.38}$$

This inequality describes an elliptically shaped region. Montgomery and Peck (1992) and Myers (1990) demonstrate the construction of this region for $p = 2$. When there are only two parameters, finding this region is relatively simple; however, when more than two parameters are involved, the construction problem is considerably harder.

There are other methods for finding joint simultaneous intervals on regression coefficients. Montgomery and Peck (1992) and Myers (1990) discuss and illustrate many of these methods. They also present methods for finding several other types of interval estimates.

2.5.3 Confidence Interval on thc Mean Response

We may also obtain a confidence interval on the mean response at a particular point, say, $x_{01}, x_{02}, \ldots, x_{0k}$. Define the vector

$$\mathbf{x}_0 = \begin{bmatrix} 1 \\ x_{01} \\ x_{02} \\ \vdots \\ x_{0k} \end{bmatrix}$$

The mean response at this point is

$$\mu_{y|\mathbf{x}_0} = \beta_0 + \beta_1 x_{01} + \beta_2 x_{02} + \cdots + \beta_k x_{0k} = \mathbf{x}_0'\boldsymbol{\beta}$$

The estimated mean response at this point is

$$\hat{y}(\mathbf{x}_0) = \mathbf{x}_0'\mathbf{b} \tag{2.39}$$

This estimator is unbiased, because $E[\hat{y}(\mathbf{x}_0)] = E(\mathbf{x}_0'\mathbf{b}) = \mathbf{x}_0'\boldsymbol{\beta} = \mu_{y|\mathbf{x}_0}$, and the variance of $\hat{y}(\mathbf{x}_0)$ is

$$\text{Var}[\hat{y}(\mathbf{x}_0)] = \sigma^2 \mathbf{x}_0'(\mathbf{X}'\mathbf{X})^{-1}\mathbf{x}_0 \tag{2.40}$$

Therefore, a $100(1 - \alpha)\%$ confidence interval on the mean response at the point $x_{01}, x_{02}, \ldots, x_{0k}$ is

$$\hat{y}(\mathbf{x}_0) - t_{\alpha/2, n-p} \sqrt{\hat{\sigma}^2 \mathbf{x}_0'(\mathbf{X}'\mathbf{X})^{-1}\mathbf{x}_0}$$

$$\leq \mu_{y|\mathbf{x}_0} \leq \hat{y}(\mathbf{x}_0) + t_{\alpha/2, n-p} \sqrt{\hat{\sigma}^2 \mathbf{x}_0'(\mathbf{X}'\mathbf{X})^{-1}\mathbf{x}_0} \tag{2.41}$$

Example 2.7 Suppose that we wish to find a 95% confidence interval on the mean response or the transistor gain problem for the point $\xi_1 = 225$ min and $\xi_2 = 4.36 \times 10^{14}$ ions. In terms of the coded variables x_1 and x_2, this point corresponds to

$$\mathbf{x}_0 = \begin{bmatrix} 1 \\ 0 \\ 0 \end{bmatrix}$$

because $x_{01} = 0$ and $x_{02} = 0$. The estimate of the mean response at this point is computed from Equation (2.39) as

$$\hat{y}(\mathbf{x}_0) = \mathbf{x}_0'\mathbf{b} = \lfloor 1,0,0 \rfloor \begin{bmatrix} 1242.3 \\ 323.4 \\ -54.8 \end{bmatrix} = 1242.3$$

Now from Equation (2.40), we find $\text{Var}[\hat{y}(\mathbf{x}_0)]$ as

$$\text{Var}[\hat{y}(\mathbf{x}_0)] = \sigma^2 \mathbf{x}_0'(\mathbf{X}'\mathbf{X})^{-1}\mathbf{x}_0$$

$$= \sigma^2[1,0,0] \begin{bmatrix} 0.072027 & -0.00195 & 0.009067 \\ -0.00195 & 0.165954 & 0.000408 \\ 0.009067 & 0.000408 & 0.142876 \end{bmatrix} \begin{bmatrix} 1 \\ 0 \\ 0 \end{bmatrix}$$

$$= \sigma^2(0.072027)$$

Using $\hat{\sigma}^2 = MS_E = 1220.1$ and Equation (2.41), we find the confidence interval as

$$\hat{y}(\mathbf{x}_0) - t_{\alpha/2, n-p} \sqrt{\hat{\sigma}^2 \mathbf{x}_0'(\mathbf{X}'\mathbf{X})^{-1}\mathbf{x}_0}$$

$$\leq \mu_{y|\mathbf{x}_0} \leq \hat{y}(\mathbf{x}_0) + t_{\alpha/2, n-p} \sqrt{\hat{\sigma}^2 \mathbf{x}_0'(\mathbf{X}'\mathbf{X})^{-1}\mathbf{x}_0}$$

$$1242.3 - 2.201\sqrt{1220.1(0.072027)}$$

$$\leq \mu_{y|\mathbf{x}_0} \leq 1242.3 + 2.201\sqrt{1220.1(0.072027)}$$

$$1242.3 - 20.6 \leq \mu_{y|\mathbf{x}_0} \leq 1242.3 + 20.6$$

or

$$1221.7 \leq \mu_{y|\mathbf{x}_0} \leq 1262.9$$

2.6 PREDICTION OF NEW RESPONSE OBSERVATIONS

A regression model can be used to predict future observations on the response y corresponding to particular values of the regressor variables, say $x_{01}, x_{02}, \ldots, x_{0k}$. If $\mathbf{x}_0' = [1, x_{01}, x_{02}, \ldots, x_{0k}]$, then a point estimate for the future observation y_0 at the point $x_{01}, x_{02}, \ldots, x_{0k}$ is computed from Equation (2.39):

$$\hat{y}(\mathbf{x}_0) = \mathbf{x}_0'\mathbf{b}$$

A $100(1 - \alpha)\%$ **prediction interval** for this future observation is

$$\hat{y}(\mathbf{x}_0) - t_{\alpha/2, n-p} \sqrt{\hat{\sigma}^2 \left(1 + \mathbf{x}_0'(\mathbf{X}'\mathbf{X})^{-1}\mathbf{x}_0\right)}$$

$$\leq y_0 \leq \hat{y}(\mathbf{x}_0) + t_{\alpha/2, n-p} \sqrt{\hat{\sigma}^2 \left(1 + \mathbf{x}_0'(\mathbf{X}'\mathbf{X})^{-1}\mathbf{x}_0\right)} \qquad (2.42)$$

In predicting new observations and in estimating the mean response at a given point $x_{01}, x_{02}, \ldots, x_{0k}$, one must be careful about extrapolating beyond the region containing the original observations. It is very possible that a model that fits well in the region of the original data will no longer fit well outside of that region. In multiple regression it is often easy to inadvertently extrapolate, since the levels of the variables $(x_{i1}, x_{i2}, \ldots, x_{ik})$, $i = 1, 2, \ldots, n$, jointly define the region containing the data. As an example, consider Figure 2.2, which illustrates the region containing the observations for a two-variable regression model. Note that the point (x_{01}, x_{02}) lies within the ranges of both regressor variables x_1 and x_2, but it is outside the region of the original observations. Thus, either predicting the value of a new observation or estimating the mean response at this point is an extrapolation of the original regression model.

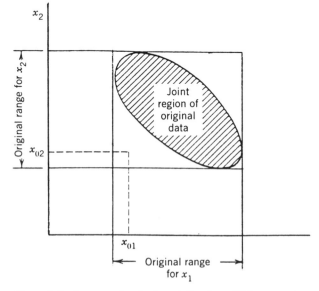

Figure 2.2   An example of extrapolation in multiple regression.

Example 2.8 Suppose that we wish to find a 95% prediction interval on the next observation on transistor gain at the point $\xi_1 = 225$ min and $\xi_2 = 4.36 \times 10^{14}$ ions. In the coded variables, this point is $x_{01} = 0$ and $x_{02} = 0$, so that, as in Example 2.7, $x'_0 = [1, 0, 0]$, and the predicted value of gain at this point is $\hat{y}(x_0) = x'_0$, $b = 1242.3$. From Example 2.7 we know that

$$x'_0(X'X)^{-1}x_0 = 0.072027$$

Therefore, using Equation (2.42) we can find the 95% prediction interval on y_0 as follows:

$$\hat{y}(x_0) - t_{\alpha/2, n-p} \sqrt{\hat{\sigma}^2\left(1 + x'_0(X'X)^{-1}x_0\right)}$$

$$\leq y_0$$

$$\leq \hat{y}(x_0) + t_{\alpha/2, n-p} \sqrt{\hat{\sigma}^2\left(1 + x'_0(X'X)^{-1}x_0\right)}$$

$$1242.3 - 2.201\sqrt{1220.1(1 + 0.072027)}$$

$$\leq y_0$$

$$\leq 1242.3 + 2.201\sqrt{1220.1(1 + 0.072027)}$$

$$1242.3 - 79.6 \leq y_0 \leq 1242.3 + 79.6$$

or

$$1162.7 \leq y_0 \leq 1321.9$$

If we compare the width of the prediction interval at this point with the width of the confidence interval on the mean gain at the same point from Example 2.7, we observe that the prediction interval is much wider. This reflects the fact that it is much more difficult to predict an individual future value of a random variable than it is to estimate the mean of the probability distribution from which that future observation will be drawn.

2.7 MODEL ADEQUACY CHECKING

It is always necessary to (a) examine the fitted model to ensure that it provides an adequate approximation to the true system and (b) verify that none of the least squares regression assumptions are violated. Proceeding with exploration and optimization of a fitted response surface will likely give poor or misleading results unless the model is an adequate fit. In this section we present several technique for checking model adequacy.

2.7.1 Residual Analysis

The residuals from the least squares fit, defined by $e_i = y_i - \hat{y}_i$, $i = 1, 2, \ldots, n$, play an important role in judging model adequacy. The residuals from Example 2.1 are shown in column 3 of Table 2.3.

A check of the normality assumption may be made by constructing a normal probability plot of the residuals, as in Figure 2.3. If the residuals plot approximately along a straight line, then the normality assumption is satisfied. Figure 2.3 reveals no apparent problem with normality. The straight line in this normal probability plot was determined by eye, concentrating on the central portion of the data. When this plot indicates problems with the normality assumption, we often transform the response variable as a remedial measure. For more details, see Montgomery and Peck (1992) and Myers (1990).

Figure 2.4 presents a plot of residuals e_i versus the predicted response $\hat{y}_i$. The general impression is that the residuals scatter randomly on the display, suggesting that the variance of the original observations is constant for all values of y. If the variance of the response depends on the mean level of y, then this plot will often exhibit a funnel-shaped pattern. This is also suggestive of the need for transformation of the response variable y.

It is also useful to plot the residuals in time or run order and versus each of the individual regressors. Nonrandom patterns on these plots would indicate model inadequacy. In some cases, transformations may stabilize the situation. See Montgomery and Peck (1992) and Myers (1990) for more details.

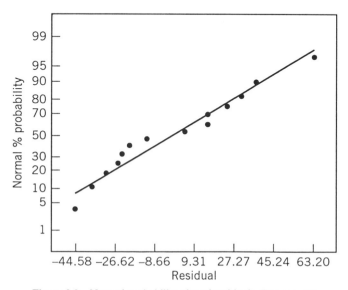

Figure 2.3   Normal probability plot of residuals, Example 2.1.

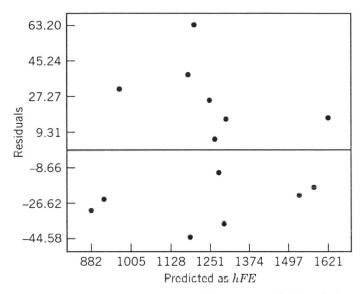

Figure 2.4   Plot of residuals versus predicted response $\hat{y}_i$, Example 2.1.

2.7.2   Scaling Residuals

Standardized and Studentized Residuals
Many response surface analysts prefer to work with **scaled residuals**, in contrast to the ordinary least squares residuals. These scaled residuals often convey more information than do the ordinary residuals.

One type of scaled residual is the **standardized residual**:

$$d_i = \frac{e_i}{\hat{\sigma}}, \qquad i = 1, 2, \ldots, n \tag{2.43}$$

where we generally use $\hat{\sigma} = \sqrt{MS_E}$ in the computation. These standardized residuals have mean zero and approximately unit variance; consequently, they are useful in looking for **outliers**. Most of the standardized residuals should lie in the interval $-3 \le d_i \le 3$, and any observation with a standardized residual outside of this interval is potentially unusual with respect to its observed response. These outliers should be carefully examined, because they may represent something as simple as a data recording error or something of more serious concern, such as a region of the regressor variable space where the fitted model is a poor approximation to the true response surface.

The standardizing process in Equation (2.43) scales the residuals by dividing them by their average standard deviation. In some data sets, residuals may have standard deviations that differ greatly. We now present a scaling that takes this into account.

The vector of fitted values $\hat{y}_i$ corresponding to the observed values y_i is

$$\hat{\mathbf{y}} = \mathbf{Xb}$$

$$= \mathbf{X}(\mathbf{X}'\mathbf{X})^{-1}\mathbf{X}'\mathbf{y}$$

$$= \mathbf{Hy} \qquad (2.44)$$

The $n \times n$ matrix $\mathbf{H} = \mathbf{X}(\mathbf{X}'\mathbf{X})^{-1}\mathbf{X}'$ is usually called the "hat" matrix because it maps the vector of observed values into a vector of fitted values. The hat matrix and its properties play a central role in regression analysis.

The residuals from the fitted model may be conveniently written in matrix notation as

$$\mathbf{e} = \mathbf{y} - \hat{\mathbf{y}} \qquad (2.45)$$

There are several other ways to express the vector of residuals $\mathbf{e}$ that will prove useful, including

$$\mathbf{e} = \mathbf{y} - \mathbf{Xb} \qquad (2.46)$$

$$= \mathbf{y} - \mathbf{Hy}$$

$$= (\mathbf{I} - \mathbf{H})\mathbf{y} \qquad (2.47)$$

The hat matrix has several useful properties. It is symmetric ($\mathbf{H}' = \mathbf{H}$) and idempotent ($\mathbf{HH} = \mathbf{H}$). Similarly the matrix $\mathbf{I} - \mathbf{H}$ is symmetric and idempotent.

The covariance matrix of the residuals is

$$\mathrm{Var}(\mathbf{e}) = \mathrm{Var}[(\mathbf{I} - \mathbf{H})\mathbf{y}]$$

$$= (\mathbf{I} - \mathbf{H})\mathrm{Var}(\mathbf{y})(\mathbf{I} - \mathbf{H})'$$

$$= \sigma^2(\mathbf{I} - \mathbf{H}) \qquad (2.48)$$

because $\mathrm{Var}(\mathbf{y}) = \sigma^2\mathbf{I}$ and $\mathbf{I} - \mathbf{H}$ is symmetric and idempotent. The matrix $\mathbf{I} - \mathbf{H}$ is generally not diagonal, so the residuals have different variances and they are correlated.

The variance of the ith residual is

$$\mathrm{Var}(e_i) = \sigma^2(1 - h_{ii}) \qquad (2.49)$$

where h_{ii} is the ith diagonal element of $\mathbf{H}$. Because $0 \leq h_{ii} \leq 1$, using the residual mean square MS_E to estimate the variance of the residuals actually overestimates $\mathrm{Var}(e_j)$. Furthermore, because h_{ii} is a measure of the location of the ith point in x-space, the variance of e_i depends upon where the point $\mathbf{x}_i$ lies. Generally, residuals near the center of the x-space have larger variance than do residuals at more remote locations. Violations of model assumptions are more likely at remote points, and these violations may be hard to detect from inspection of e_i (or d_i) because their residuals will usually be smaller.

We recommend taking this inequality of variance into account when scaling the residuals. We suggest plotting the **studentized residuals**:

$$r_i = \frac{e_i}{\sqrt{\hat{\sigma}^2(1 - h_{ii})}}, \qquad i = 1, 2, \ldots, n \qquad (2.50)$$

with $\hat{\sigma}^2 = MS_E$ instead of e_i (or d_i). The studentized residuals have constant variance $\text{Var}(r_i) = 1$ regardless of the location of x_i when the form of the model is correct. In many situations the variance of the residuals stabilizes, particularly for large data sets. In these cases there may be little difference between the standardized and studentized residuals. Thus standardized and studentized residuals often convey equivalent information. However, because any point with a large residual and a large h_{ii} is potentially highly influential on the least squares fit, examination of the studentized residuals is generally recommended.

PRESS Residuals

The prediction error sum of squares (PRESS) proposed by Allen (1971, 1974) provides a useful residual scaling. To calculate PRESS, select an observation —for example, i. Fit the regression model to the remaining $n - 1$ observations and use this equation to predict the withheld observation y_i. Denoting this predicted value $\hat{y}_{(i)}$, we may find the prediction error for point i as $e_{(i)} = y_i - \hat{y}_{(i)}$. The prediction error is often called the ith PRESS residual. This procedure is repeated for each observation $i = 1, 2, \ldots, n$, producing a set of n PRESS residuals $e_{(1)}, e_{(2)}, \ldots, e_{(n)}$. Then the PRESS statistic is defined as the sum of squares of the n PRESS residuals as in

$$\text{PRESS} = \sum_{i=1}^{n} e_{(i)}^2 = \sum_{i=1}^{n} \left[y_i - \hat{y}_{(i)} \right]^2 \qquad (2.51)$$

Thus PRESS uses each possible subset of $n - 1$ observations as an estimation data set, and every observation in turn is used to form a prediction data set.

It would initially seem that calculating PRESS requires fitting n different regressions. However, it is possible to calculate PRESS from the results of a single least squares fit to all n observations. It turns out that the ith PRESS residual is

$$e_{(i)} = \frac{e_i}{1 - h_{ii}} \qquad (2.52)$$

Thus because PRESS is just the sum of the squares of the PRESS residuals, a simple computing formula is

$$\text{PRESS} = \sum_{i=1}^{n} \left(\frac{e_i}{1 - h_{ii}} \right)^2 \tag{2.53}$$

From Equation (2.52) it is easy to see that the PRESS residual is just the ordinary residual weighted according to the diagonal elements of the hat matrix h_{ii}. Data points for which h_{ii} are large will have large PRESS residuals. These observations will generally be **high influence** points. Generally, a large difference between the ordinary residual and the PRESS residuals will indicate a point where the model fits the data well, but a model built without that point predicts poorly. In the next section we will discuss some other measures of influence.

The variance of the ith PRESS residual is

$$\text{Var}[e_{(i)}] = \text{Var}\left[\frac{e_i}{1 - h_{ii}} \right]$$

$$= \frac{1}{(1 - h_{ii})^2} \left[\sigma^2 (1 - h_{ii}) \right]$$

$$= \frac{\sigma^2}{1 - h_{ii}}$$

so that standardized PRESS residual is

$$\frac{e_{(i)}}{\sqrt{\text{Var}[e_{(i)}]}} = \frac{e_i / (1 - h_{ii})}{\sqrt{\sigma^2 / (1 - h_{ii})}}$$

$$= \frac{e_i}{\sqrt{\sigma^2 (1 - h_{ii})}}$$

which if we use MS_E to estimate σ^2 is just the studentized residual discussed previously.

Finally, we note that PRESS can be used to compute an approximate R^2 for prediction, say

$$R^2_{\text{prediction}} = 1 - \frac{\text{PRESS}}{S_{yy}} \tag{2.54}$$

This statistic gives some indication of the predictive capability of the regression model. For the transistor gain model we can compute the PRESS residuals using the ordinary residuals and the values of h_{ii} found in Table 2.3. The corresponding value of the PRESS statistic is PRESS = 22,225.0. Then

$$R^2_{\text{prediction}} = 1 - \frac{\text{PRESS}}{S_{yy}}$$

$$= 1 - \frac{22{,}225.0}{665{,}387.2}$$

$$= 0.9666$$

Therefore we could expect this model to "explain" about 96.66% of the variability in predicting new observations, as compared to the approximately 97.98% of the variability in the original data explained by the least squares fit. The overall predictive capability of the model based on this criterion seems very satisfactory.

R Student

The studentized residual r_i discussed above is often considered an outlier diagnostic. It is customary to use MS_E as an estimate of σ^2 in computing r_i. This is referred to as internal scaling of the residual because MS_E is an internally generated estimate of σ^2 obtained from fitting the model to all n observations. Another approach would be to use an estimate of σ^2 based on a data set with the ith observation removed. Denote the estimate of σ^2 so obtained by $S^2_{(i)}$. We can show that

$$S^2_{(i)} = \frac{(n - p)MS_E - e_i^2/(1 - h_{ii})}{n - p - 1} \tag{2.55}$$

The estimate of σ^2 in Equation (2.55) is used instead of MS_E to produce an externally studentized residual, usually called R-student, given by

$$t_i = \frac{e_i}{\sqrt{S^2_{(i)}(1 - h_{ii})}}, \qquad i = 1, 2, \ldots, n \tag{2.56}$$

In many situations, t_i will differ little from the studentized residual r_i. However, if the ith observation is influential, then $S^2_{(i)}$ can differ significantly from MS_E, and thus the R-student will be more sensitive to this point. Furthermore, under the standard assumptions, t_i has a t_{n-p-1}-distribution. Thus R-student offers a more formal procedure for outlier detection via hypothesis testing. One could use a simultaneous inference procedure called the *Bonferroni approach* and compare all n values of $|t_i|$ to $t_{(\alpha/2n), n-p-1}$ to provide guidance regarding outliers. However, it is our view that a formal approach is usually not necessary and that only relatively crude cutoff values need be considered. In general, a diagnostic view as opposed to a strict statistical hypothesis-testing view is best. Furthermore, detection of outliers needs to be considered simultaneously with detection of influential observations.

Example 2.9   Table 2.3 presents the studentized residuals r_i and the R-student values t_i defined in Equations (2.50) and (2.56). None of these values are large enough to cause any concern regarding outliers.

Figure 2.5 is a normal probability plot of the studentized residuals. It conveys exactly the same information as the normal probability plot of the ordinary residuals e_i in Figure 2.3. This is because most of the h_{ii} values are relatively similar and there are no unusually large residuals. In some applications, however, the h_{ii} can differ considerably, and in those cases plotting the studentized residuals is the best approach.

2.7.3   Influence Diagnostics

We occasionally find that a small subset of the data exerts a disproportionate influence on the fitted regression model. That is, parameter estimates or predictions may depend more on the influential subset than on the majority of the data. We would like to locate these influential points and assess their impact on the model. If these influential points are "bad" values, then they should be eliminated. On the other hand, there may be nothing wrong with these points, but if they control key model properties, we would like to know

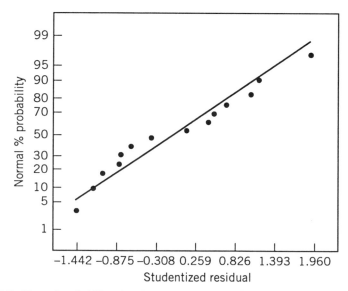

Figure 2.5   Normal probability plot of the studentized residuals for the transistor gain data.

it, because it could affect the use of the model. In this section we describe and illustrate several useful measure of influence.

Leverage Points

The disposition of points in x-space is important in determining model properties. In particular remote observations potentially have disproportionate leverage on the parameter estimates, predicted values, and the usual summary statistics.

The hat matrix $\mathbf{H} = \mathbf{X}(\mathbf{X}'\mathbf{X})^{-1}\mathbf{X}'$ is very useful in identifying influential observations. As noted earlier, $\mathbf{H}$ determines the variances and covariances of $\hat{\mathbf{y}}$ and $\mathbf{e}$, because $\mathrm{Var}(\hat{\mathbf{y}}) = \sigma^2\mathbf{H}$ and $\mathrm{Var}(\mathbf{e}) = \sigma^2(\mathbf{I} - \mathbf{H})$. The elements h_{ij} of $\mathbf{H}$ may be interpreted as the amount of leverage exerted by y_j on $\hat{y}_i$. Thus inspection of the elements of $\mathbf{H}$ can reveal points that are potentially influential by virtue of their location in x-space. Attention is usually focused on the diagonal elements h_{ii}. Because $\sum_{i=1}^{n}h_{ii} = \mathrm{rank}(\mathbf{H}) = \mathrm{rank}(\mathbf{X}) = p$, the average size of the diagonal element of the $\mathbf{H}$ matrix is p/n. As a rough guideline, then, if a diagonal element h_{ii} is greater than $2p/n$, observation i is a high-leverage point. To apply this to the transistor gain data in Example 2.1, note that $2p/n = 2(3)/14 = 0.43$. Table 2.3 gives the hat diagonals h_{ii} for the first-order model; and because none of the h_{ii} exceed 0.43, we would conclude that there are no leverage points in these data. Further properties and uses of the elements of the hat matrix in regression diagnostics are discussed by Belsley et al. (1980).

Influence on Regression Coefficients

The hat diagonals will identify points that are potentially influential due to their location in x-space. It is desirable to consider both the location of the point and the response variable in measuring influence. Cook (1977, 1979) has suggested using a measure of the squared distance between the least squares estimate based on all n points $\mathbf{b}$, and the estimate obtained by deleting the ith point, say $\mathbf{b}_{(i)}$. This distance measure can be expressed in a general form as

$$D_i(\mathbf{M}, c) = \frac{(\mathbf{b}_{(i)} - \mathbf{b})'\mathbf{M}(\mathbf{b}_{(i)} - \mathbf{b})}{c}, \qquad i = 1, 2, \ldots, n \qquad (2.57)$$

The usual choices of $\mathbf{M}$ and c are $\mathbf{M} = \mathbf{X}'\mathbf{X}$ and $c = pMS_E$, so that Equation (2.57) becomes

$$D_i(\mathbf{M}, c) \equiv D_i = \frac{(\mathbf{b}_{(i)} - \mathbf{b})'\mathbf{X}'\mathbf{X}(\mathbf{b}_{(i)} - \mathbf{b})}{pMS_E}, \qquad i = 1, 2, \ldots, n \quad (2.58)$$

Points with large values of D_i have considerable influence on the least squares estimates $\mathbf{b}$. The magnitude of D_i may be assessed by comparing it to

$F_{\alpha, p, n-p}$. If $D_i \simeq F_{0.5, p, n-p}$, then deleting point i would move **b** to the boundary of a 50% confidence region for β based on the complete data set.* This is a large displacement and indicates that the least squares estimate is sensitive to the ith data point. Because $F_{0.5, p, n-p} \simeq 1$, we usually consider points for which $D_i > 1$ to be influential. Practical experience has shown the cutoff value of 1 works well in identifying influential points.

The D_i statistic may be rewritten as

$$D_i = \frac{r_i^2}{p} \frac{\text{Var}[\hat{y}(x_i)]}{\text{Var}(e_i)} = \frac{r_i^2}{p} \frac{h_{ii}}{(1 - h_{ii})}, \qquad i = 1, 2, \ldots, n \qquad (2.59)$$

Thus we see that, apart from the constant p, D_i is the product of the square of the ith studentized residual and $h_{ii}/(1 - h_{ii})$. This ratio can be shown to be the distance from the vector $\mathbf{x}_i$ to the centroid of the remaining data. Thus D_i is made up of a component that reflects how well the model fits the ith observation y_i and a component that measures how far that point is from the rest of the data. Either component (or both) may contribute to a larger value of D_i.

Table 2.3 presents the values of D_i for the first-order model fit to the transistor gain data in Example 2.1. None of these values of D_i exceed 1, so there is no strong evidence of influential observations in these data.

2.7.4 Testing for Lack of Fit

In RSM, usually we are fitting the regression model to data from a designed experiment. It is frequently useful to obtain two or more observations (replicates) on the response at the same settings of the independent or regressor variables. When this has been done we may conduct a formal test for the lack of fit on the regression model. For example, consider the data in Figure 2.6. There is some indication that the straight-line fit is not very satisfactory, and it would be helpful to have a statistical test to determine if there is systematic curvature present.

The lack-of-fit test requires that we have **true replicates** on the response y for at least one set of levels on the regressors $x_1, x_2, \ldots, x_k$. These are not just duplicate readings or measurements of y. For example, suppose that y is product viscosity and there is only one regressor x (temperature). True replication consists of running n_i separate experiments (usually in random order) at $x = x_i$ and observing viscosity, not just running a single experiment at x_i and measuring viscosity n_i times. The readings obtained from the latter procedure provide information only on the variability of the method of measuring viscosity. The error variance σ^2 includes measurement error, variability in the process over time, and the variability associated with

---

*The distance measure D_i is not an F random variable, but is compared to an F-value because of the similarity of D_i to the normal theory confidence ellipsoid [Equation (2.38)].

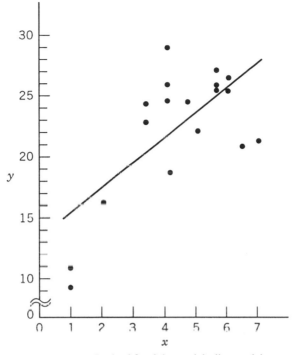

Figure 2.6   Lack of fit of the straight-line model.

reaching and maintaining the same temperature level in different experiments. These replicate points are used to obtain a model-independent estimate of σ^2.

Suppose that we have n_i observations on the response at the ith level of the regressors x_i, $i = 1, 2, \ldots, m$. Let y_{ij} denote that jth observation on the response at x_i, $i = 1, 2, \ldots, m$, and $j = 1, 2, \ldots, n_i$. There are $n = \sum_{i=1}^{m} n_i$ total observations. The test procedure involves partitioning the residual sum of squares into two components, say

$$SS_E = SS_{PE} + SS_{LOF}$$

where SS_{PE} is the sum of squares due to pure error and SS_{LOF} is the sum of squares due to lack of fit.

To develop this partitioning of SS_E, note that the (ij)th residual is

$$y_{ij} - \hat{y}_i = \left(y_{ij} - \bar{y}_i \right) + \left(\bar{y}_i - \hat{y}_i \right) \tag{2.60}$$

where $\bar{y}_i$ is the average of the n_i observations at x_i. Squaring both sides of

Equation (2.60) and summing over i and j yields

$$\sum_{i=1}^{m} \sum_{j=1}^{n_i} \left(y_{ij} - \hat{y}_i \right)^2 = \sum_{i=1}^{m} \sum_{j=1}^{n_i} \left(y_{ij} - \bar{y}_i \right)^2 + \sum_{i=1}^{m} n_i \left(\bar{y}_i - \hat{y}_i \right)^2 \quad (2.61)$$

The left-hand side of Equation (2.61) is the usual residual sum of squares. The two components on the right-hand side measure pure error and lack of fit. We see that the pure error sum of squares

$$SS_{PE} = \sum_{i=1}^{m} \sum_{j=1}^{n_i} \left(y_{ij} - \bar{y}_i \right)^2 \quad (2.62)$$

is obtained by computing the corrected sum of squares of the repeat observations at each level of $\mathbf{x}$ and then pooling over the m levels of $\mathbf{x}$. If the assumption of constant variance is satisfied, this is a **model-independent** measure of pure error, because only the variability of the y's at each $\mathbf{x}_i$ level is used to compute SS_{PE}. Because there are $n_i - 1$ degrees of freedom for pure error at each level $\mathbf{x}_i$, the total number of degrees of freedom associated with the pure error sum of squares is

$$\sum_{i=1}^{m} (n_i - 1) = n - m \quad (2.63)$$

The sum of squares for lack of fit

$$SS_{LOF} = \sum_{i=1}^{m} n_i \left(\bar{y}_i - \hat{y}_i \right)^2 \quad (2.64)$$

is a weighted sum of squared deviations between the mean response $\bar{y}_i$ at each $\mathbf{x}_i$ level and the corresponding fitted value. If the fitted values $\hat{y}_i$ are close to the corresponding average responses $\bar{y}_i$, then there is a strong indication that the regression function is linear. If the $\hat{y}_i$ deviate greatly from the $\bar{y}_i$, then it is likely that the regression function is not linear. There are $m - p$ degrees of freedom associated with SS_{LOF}, because there are m levels of $\mathbf{x}$ and p degrees of freedom are lost because p parameters must be estimated for the model. Computationally we usually obtain SS_{LOF} by subtracting SS_{PE} from SS_E.

The test statistic for lack of fit is

$$F_0 = \frac{SS_{LOF}/(m - p)}{SS_{PE}/(n - m)} = \frac{MS_{LOF}}{MS_{PE}} \quad (2.65)$$

The expected value of MS_{PE} is σ^2, and the expected value of MS_{LOF} is

$$E(MS_{LOF}) = \sigma^2 + \frac{\sum_{i=1}^{m} n_i \left[E(y_i) - \beta_0 - \sum_{j=1}^{k} \beta_j x_{ij} \right]^2}{m-2} \qquad (2.66)$$

If the true regression function is linear, then $E(y_i) = \beta_0 + \sum_{j=1}^{k} \beta_j x_{ij}$, and the second term of Equation (2.66) is zero, resulting in $E(MS_{LOF}) = \sigma^2$. However, if the true regression function is not linear, then $E(y_i) \neq \beta_0 + \sum_{j=1}^{k} \beta_j x_{ij}$, and $E(MS_{LOF}) > \sigma^2$. Furthermore, if the true regression function is linear, then the statistic F_0 follows the $F_{m-p,\,n-m}$ distribution. Therefore, to test for lack of fit, we would compute the test statistic F_0 and conclude that the regression function is not linear if $F_0 > F_{\alpha,\,m-p,\,n-m}$.

This test procedure may be easily introduced into the analysis of variance conducted for significance of regression. If we conclude that the regression function is not linear, then the tentative model must be abandoned and attempts made to find a more appropriate equation. Alternatively, if F_0 does not exceed $F_{\alpha,\,m-p,\,n-m}$, there is no strong evidence of lack of fit, and MS_{PE} and MS_{LOF} are often combined to estimate σ^2.

Ideally, we find that the F-ratio for lack of fit is not significant and the hypothesis of significance of regression is rejected. Unfortunately, this does not guarantee that the model will be satisfactory as a prediction equation. Unless the variation of the predicted values is large relative to the random error, the model is not estimated with sufficient precision to yield satisfactory predictions. That is, the model may have been fitted to the errors only. Some analytical work has been done on developing criteria for judging the adequacy of the regression model from a prediction point of view. See Box and Wetz (1973), Ellerton (1978), Gunst and Mason (1979), Hill, et al. (1978), and Suich and Derringer (1977). The Box and Wetz work suggests that the observed F-ratio must be at least four or five times the critical value from the F-table if the regression model is to be useful as a predictor—that is, if the spread of predicted values is to be large relative to the noise.

A relatively simple measure of potential prediction performances is found by comparing the range of the fitted values $\hat{y}$ (i.e., $\hat{y}_{max} - \hat{y}_{min}$) to their average standard error. It can be shown that, regardless of the form of the model, the average variance of the fitted values is

$$\overline{\text{Var}(\hat{y})} = \frac{1}{n} \sum_{i=1}^{n} \text{Var}[\hat{y}(\mathbf{x}_i)] = \frac{p\sigma^2}{n} \qquad (2.67)$$

where p is the number of parameters in the model. In general, the model is not likely to be a satisfactory predictor unless the range of the fitted values $\hat{y}_i$ is large relative to their average estimated standard error $\sqrt{(p\hat{\sigma}^2)/n}$, where $\hat{\sigma}^2$ is a model-independent estimate of the error variance.

Example 2.10 The data in Figure 2.6 are shown below:

x	1.0	1.0	2.0	3.3	3.3	4.0	4.0	4.0	4.7	5.0
y	10.84	9.30	16.35	22.88	24.35	24.56	25.86	29.46	24.59	22.25

x	5.6	5.6	5.6	6.0	6.0	6.5	6.9
y	25.90	27.20	25.61	25.45	26.56	21.03	21.46

The straight-line fit is $\hat{y} = 13.301 + 2.108x$, with $S_{yy} = 487.6126$, $SS_R = 234.7087$, and $SS_E = 252.9039$. Note that there are 10 distinct levels of x, with repeat points at $x = 1.0$, $x = 3.3$, $x = 4.0$, $x = 5.6$, and $x = 6.0$. The pure sum of squares is computed using the repeat points as follows:

Level of x	$\sum_j (y_{ij} - \bar{y}_i)^2$	Degrees of Freedom
1.0	1.1858	1
3.3	1.0805	1
4.0	11.2467	2
5.6	1.4341	2
6.0	0.6161	1
Total	15.5632	7

The lack-of-fit sum of squares is found by subtraction as

$$SS_{LOF} = SS_E - SS_{PE}$$

$$= 252.9039 - 15.5632 = 237.3407$$

with $m - p = 10 - 2 = 8$ degrees of freedom. The analysis of variance incorporating the lack-of-fit test is shown in Table 2.6. The lack-of-fit test statistic is $F_0 = 13.34$, and because the P-value for this test statistic is very small, we reject the hypothesis that the tentative model adequately describes the data.

Table 2.6 Analysis of Variance for Example 2.10

Source of Variation	Sum of Squares	Degrees of Freedom	Mean Square	F_0	P-value
Regression	234.7087	1	234.7087		
Residual	252.9039	15	16.8603		
Lack of fit	237.3407	8	29.6676	13.34	0.0013
Pure error	15.5632	7	2.2233		
Total	487.6126	16			

2.8 FITTING A SECOND-ORDER MODEL

Many applications of response surface methodology involve fitting and checking the adequacy of a second-order model. In this section we present a complete example of this process.

Table 2.7 presents the data resulting from an investigation into the effect of two variables, reaction temperature (ξ_1), and reactant concentration (ξ_2), on the percent conversion of a chemical process (y). The process engineers had used an approach to improving this process based on designed experiments. The first experiment was a screening experiment involving several factors that isolated temperature and concentration as the two most important variables. Because the experimenters thought that the process was operating in the vicinity of the optimum, they elected to fit a quadratic model relating yield to temperature and concentration.

Panel A of Table 2.7 shows the levels used for ξ_1 and ξ_2 in the natural units of measurements. Panel B shows the levels in terms of coded variables x_1 and x_2. Figure 2.7 shows the experimental design in Table 2.7 graphically. This design is called a **central composite design**, and it is widely used for fitting a second-order response surface. Notice that the design consists of four runs at the corners of a square, plus four runs at the center of this square, plus four axial runs. In terms of the coded variables the corners of the square are $(x_1, x_2) = (-1, -1), (1, -1), (1, 1), (1, 1)$; the center points are at $(x_1, x_2) = (0, 0)$; and the axial runs are at $(x_1, x_2) = (-1.414, 0), (1.414, 0), (0, -1.414), (0, 1.414)$.

We will fit the second-order model

$$y = \beta_0 + \beta_1 x_1 + \beta_2 x_2 + \beta_{11} x_1^2 + \beta_{22} x_2^2 + \beta_{12} x_1 x_2 + \varepsilon$$

Table 2.7 Central Composite Design for Chemical Process Example

		A		B		
		Temperature (°C)	Conc. (%)			
Observation	Run	ξ_1	ξ_2	x_1	x_2	y
1	4	200	15	−1	−1	43
2	12	250	15	1	−1	78
3	11	200	25	−1	1	69
4	5	250	25	1	1	73
5	6	189.65	20	−1.414	0	48
6	7	260.35	20	1.414	0	78
7	1	225	12.93	0	−1.414	65
8	3	225	27.07	0	1.414	74
9	8	225	20	0	0	76
10	10	225	20	0	0	79
11	9	225	20	0	0	83
12	2	225	20	0	0	81

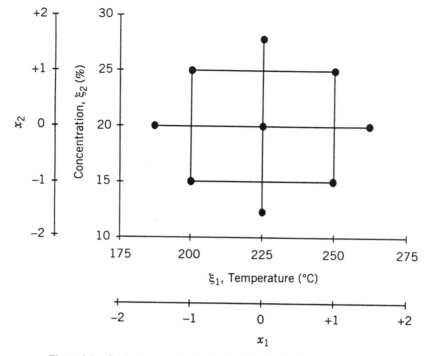

Figure 2.7 Central composite design for the chemical process example.

using the coded variables. The **X** matrix and **y** vector for this model are

$$
\mathbf{X} = \begin{array}{c}
\begin{array}{ccccc} x_1 & x_2 & x_1^2 & x_2^2 & x_1x_2 \end{array} \\
\begin{bmatrix}
1 & -1 & -1 & 1 & 1 & 1 \\
1 & 1 & -1 & 1 & 1 & -1 \\
1 & -1 & 1 & 1 & 1 & -1 \\
1 & 1 & 1 & 1 & 1 & 1 \\
1 & -1.414 & 0 & 2 & 0 & 0 \\
1 & 1.414 & 0 & 2 & 0 & 0 \\
1 & 0 & -1.414 & 0 & 2 & 0 \\
1 & 0 & 1.414 & 0 & 2 & 0 \\
1 & 0 & 0 & 0 & 0 & 0 \\
1 & 0 & 0 & 0 & 0 & 0 \\
1 & 0 & 0 & 0 & 0 & 0 \\
1 & 0 & 0 & 0 & 0 & 0
\end{bmatrix}, \quad
\mathbf{y} = \begin{bmatrix}
43 \\ 78 \\ 69 \\ 73 \\ 48 \\ 76 \\ 65 \\ 74 \\ 76 \\ 79 \\ 83 \\ 81
\end{bmatrix}
\end{array}
$$

Notice that we have shown the variables associated with each column above that column in the X-matrix. The entries in the columns associated with x_1^2 and x_2^2 are found by squaring the entries in columns x_1 and x_2, respectively, and the entries in the x_1x_2 column are found by multiplying each entry from

x_1 by the corresponding entry from x_2. The $\mathbf{X'X}$ matrix and $\mathbf{X'y}$ vector are

$$\mathbf{X'X} = \begin{bmatrix} 12 & 0 & 0 & 8 & 8 & 0 \\ 0 & 8 & 0 & 0 & 0 & 0 \\ 0 & 0 & 8 & 0 & 0 & 0 \\ 8 & 0 & 0 & 12 & 4 & 0 \\ 8 & 0 & 0 & 4 & 12 & 0 \\ 0 & 0 & 0 & 0 & 0 & 4 \end{bmatrix}, \quad \mathbf{X'y} = \begin{bmatrix} 845.000 \\ 78.592 \\ 33.726 \\ 511.000 \\ 541.000 \\ -31.000 \end{bmatrix}$$

and from $\mathbf{b} = (\mathbf{X'X})^{-1}\mathbf{X'y}$ we obtain

$$\mathbf{b} = \begin{bmatrix} 79.75 \\ 9.83 \\ 4.22 \\ -8.88 \\ -5.13 \\ -7.75 \end{bmatrix}$$

Therefore the fitted model for percent conversion is

$$\hat{y} = 79.75 + 9.83x_1 + 4.22x_2 - 8.88x_1^2 - 5.13x_2^2 - 7.75x_1x_2$$

In terms of the natural variables, the model is

$$\hat{y} = -1105.56 + 8.0242\xi_1 + 22.994\xi_2 + 0.0142\xi_1^2 + 0.20502\xi_2^2 + 0.062\xi_1\xi_2$$

Table 2.8 shows the analysis of variance for this model. Because the experimental design has four replicate runs, the residual sum of squares can be partitioned into pure error and lack of fit components, as discussed in Section 2.7.4. The lack-of-fit test in Table 2.8 is testing the lack of fit for the quadratic model. The P-value for this test is large ($P = 0.8120$), implying that

Table 2.8 Analysis of Variance for the Chemical Process Example

Source of Variation	Sum of Squares	Degrees of Freedom	Mean Square	F_0	P-Value
Regression	1733.6	5	346.71	58.86	< 0.0001
$SS_R(\beta_1, \beta_2 \mid \beta_0)$	(914.4)	(2)	(475.20)		
$SS_R(\beta_{11}, \beta_{22}, \beta_{12} \mid \beta_0, \beta_1, \beta_2)$	(819.2)	(3)	(273.10)		
Residual	35.3	6	5.89		
Lack of Fit	(8.5)	(3)	(2.83)	0.3176	0.8120
Pure Error	(26.8)	(3)	(8.92)		
Total	1768.9	11			

$R^2 = 0.9800$	$R^2_{adj} = 0.9634$	PRESS = 108.7

the quadratic model is adequate. Therefore the residual mean square with 6 degrees of freedom is used for the remaining analysis. The F-test for significance of regression is $F_0 = 58.86$; and because the P-value is very small, we would reject the hypothesis H_0: $\beta_1 = \beta_2 = \beta_{11} = \beta_{22} = \beta_{12} = 0$, concluding that at least some of these parameters are nonzero. This table also shows the sum of squares for testing the contribution of only the linear terms to the model $[SS_R(\beta_1, \beta_2|\beta_0) = 918.4$ with 2 degrees of freedom], and the sum of squares for testing the contribution of the quadratic terms given that the model already contains the linear terms $[SS_R(\beta_{11}, \beta_{22}, \beta_{12}|\beta_0, \beta_1, \beta_2) = 819.2$ with 3 degrees of freedom]. Comparing both of the corresponding mean squares to the residual mean square gives the following F-statistic:

$$F_0 = \frac{SS_R(\beta_1, \beta_2|\beta_0)/2}{MS_E} = \frac{914.2/2}{5.89} = \frac{475.2}{5.89} = 80.68$$

for which the P-value is $P = 4.6 \times 10^{-5}$, and

$$F_0 = \frac{SS_R(\beta_{11}, \beta_{22}, \beta_{12}|\beta_0, \beta_1, \beta_2)/3}{MS_R} = \frac{819.2/3}{5.89} = \frac{273.1}{5.89} = 46.37$$

for which the P-value is $P = 0.0002$. Therefore, both the linear and quadratic terms contribute significantly to the model. Table 2.9 shows t-tests on each individual variable. All t-values are large enough for us to conclude that there are no nonsignificant terms in the model. If some of these t-statistics had been small, some analysts would drop the nonsignificant variables for the model, resulting on a reduced quadratic model for the process. It is also possible to employ variable selection methods (see Appendix 1) to determine the subset of variables to include in the model. Generally, we prefer to fit the full quadratic model whenever possible, unless there are large differences between the full and reduced model in terms of PRESS and adjusted R^2. Table 2.8 indicates that the R^2 and adjusted R^2 values for this model are

Table 2.9   Tests on the Individual Variables, Chemical Process Quadratic Model

Variable	Coefficient Estimate	Standard Error	t for H_0 Coefficient $= 0$	P-Value
Intercept	79.75	1.21	65.72	
x_1	9.83	0.86	11.45	0.0001
x_2	4.22	0.86	4.913	0.0027
x_1^2	-8.88	0.96	-9.250	0.0001
x_2^2	-5.13	0.96	-5.341	0.0018
$x_1 x_2$	-7.75	1.21	-6.386	0.0007

satisfactory. The $R^2_{\text{prediction}}$ based on PRESS is

$$R^2_{\text{prediction}} = 1 - \frac{\text{PRESS}}{S_{yy}}$$

$$= 1 - \frac{108.7}{1768.9}$$

$$= 0.9385$$

indicating that the model will probably explain a high percentage (about 94%) of the variability in new data.

Table 2.10 contains the observed and predicted values of percent conversion, the residuals, and other diagnostic statistics for this model. None of the studentized residuals or the values of R-student are large enough to indicate any potential problem with outliers. Notice that the hat diagonals h_{ii} take on only two values, either 0.625 or 0.250. The values of $h_{ii} = 0.625$ are associated with the four runs at the corners of the square in the design and the four axial runs. All eight of these points are equidistant from the center of the design; this is why all of the h_{ii} values are identical. The four center points all have $h_{ii} = 0.250$. Figures 2.8, 2.9, and 2.10 show a normal probability plot of the studentized residuals, a plot of the studentized residuals versus the predicted values $\hat{y}_i$, and a plot of the studentized residuals versus run order. None of these plots reveal any model inadequacy.

Plots of the conversion response surface and the contour plot, respectively, for the fitted model are shown in Figure 2.11(a) and 2.11(b). The response

Table 2.10 Observed Values, Predicted Values, Residuals, and Other Diagnostics for the Chemical Process Example

Obs Order	Actual Value	Predicted Value	Residual	h_{ij}	Student Residual	Cook's D	R-Student
1	43.00	43.96	−0.96	0.625	−0.643	0.115	−0.609
2	78.00	79.11	−1.11	0.625	−0.745	0.154	−0.714
3	69.00	67.89	1.11	0.625	0.748	0.155	0.717
4	73.00	72.04	0.96	0.625	0.646	0.116	0.612
5	48.00	48.11	−0.11	0.625	−0.073	0.001	−0.067
6	76.00	75.90	0.10	0.625	−0.073	0.001	−0.067
7	65.00	63.54	1.46	0.625	0.982	0.268	0.979
8	74.00	75.46	−1.46	0.625	−0.985	0.269	−0.982
9	76.00	79.75	−3.75	0.250	−1.784	0.177	−2.377
10	79.00	79.75	−0.75	0.250	−0.357	0.007	−0.329
11	83.00	79.75	3.25	0.250	1.546	0.133	1.820
12	81.00	79.75	1.25	0.250	0.595	0.020	0.560

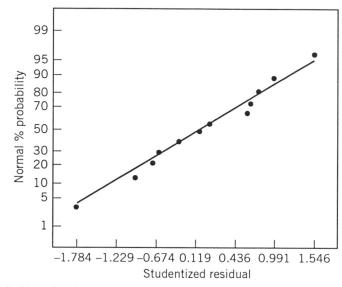

Figure 2.8   Normal probability plot of the studentized residuals, chemical process example.

surface plots indicate that the maximum percent conversion is at about 245°C and 20% concentration.

   In many response surface problems the experimenter is interested in predicting the response y or estimating the mean response at a particular point in the process variable space. The response surface plots in Figure 2.11 give a graphical display of these quantities. Typically, the variance of the

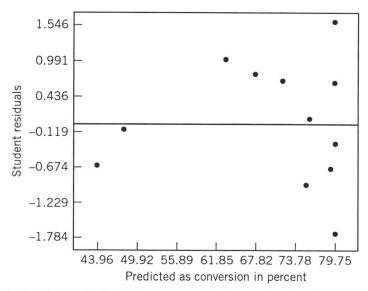

Figure 2.9   Plot of studentized residuals versus predicted conversion, chemical process example.

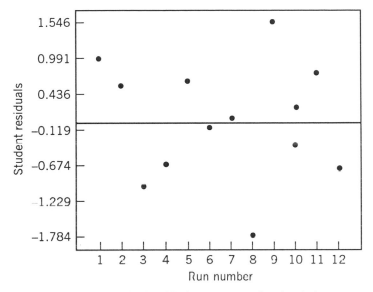

Figure 2.10 Plot of the studentized residuals versus run order, chemical process example.

prediction is also of interest, because this is a direct measure of the likely error associated with the point estimate produced by the model. Recall from Equation (2.40) that the variance of the estimate of the mean response at the point x_0 is given by $\text{Var}[\hat{y}(x_0)] = \sigma^2 x_0'(X'X)^{-1}x_0$. Plots of $\sqrt{\text{Var}[\hat{y}(x_0)]}$, with σ^2 estimated by the mean square error $MS_E = 5.89$ for this model for all values of x_0 in the region of experimentation, are presented in Figure 2.12(*a*) and 2.12(*b*). Both the response surface in Figure 2.12(*a*) and the contour plot of constant $\sqrt{\text{Var}[\hat{y}(x_0)]}$ in Figure 2.12(*b*) show that the $\sqrt{\text{Var}[\hat{y}(x_0)]}$ is the same for all points x_0 that are the same distance from the center of the design. This is a result of the spacing of the axial runs in the central composite design at 1.414 units from the origin (in the coded variables), and is a design property called **rotatability**. This is a very important property for a second-order response surface design, and it will be discussed in more detail in Chapter 7.

2.9 QUALITATIVE REGRESSOR VARIABLES

The regression models employed in RSM usually involved **quantitative** variables—that is, variables that are measured on a numerical scale. For example, variables such as temperature, pressure, distance, and age are quantitative variables. Occasionally, we need to incorporate **qualitative** variables in a regression model. For example, suppose that one of the variables in a regression model is the machine from which each observation y_i is taken. Assume that only two machines are involved. We may wish to assign different

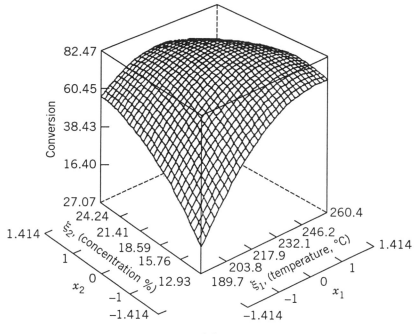

(a)

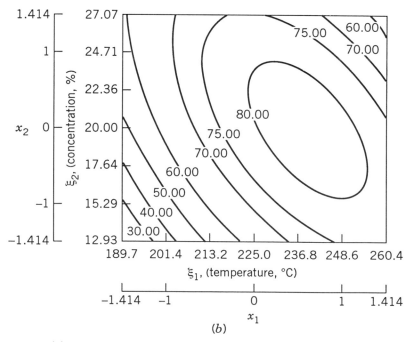

(b)

Figure 2.11 (a) Response surface of predicted conversion. (b) Contour plot of predicted conversion.

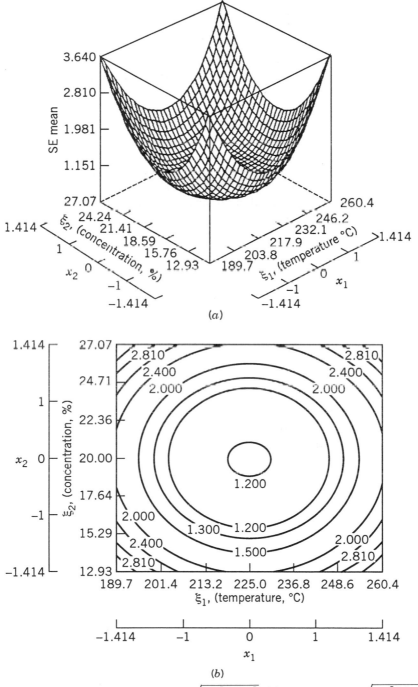

Figure 2.12 (a) Response Surface plot of $\sqrt{\text{Var}[\hat{y}(\mathbf{x}_0)]}$. (b) Contour plot of $\sqrt{\text{Var}[\hat{y}(\mathbf{x}_0)]}$.

levels to the two machines to account for the possibility that each machine may have a different effect on the response.

The usual method of accounting for the different levels of a qualitative variable is by using **indicator variables**. For example, to introduce the effect of two different machines into a regression model, we could define an indicator variable as follows:

$$x = 0 \quad \text{if the observation is from machine 1}$$
$$x = 1 \quad \text{if the observation is from machine 2}$$

In general, a qualitative variable with t levels is represented by $t - 1$ indicator variables, which are assigned the values either 0 or 1. Thus, if there were three machines, the different levels would be accounted for by two indicator variables defined as follows:

x_1	x_2	
0	0	if the observation if from machine 1
1	0	if the observation is from machine 2
0	1	if the observation is from machine 3

Indicator variables are also referred to as dummy variables. The following example illustrates some of the uses of indicator variables. For other applications, see Montgomery and Peck (1992) and Myers (1990).

Example 2.11 [Adapted from Montgomery and Peck (1992)] A mechanical engineer is investigating the surface finish of metal parts produced on a lathe and its relationship to the speed [in revolutions per minute (RPM)] of the lathe. The data are shown in Table 2.11. Note that the data have been collected using two different types of cutting tools. Because it is likely that the type of cutting tool affects the surface finish, we will fit the model

$$y = \beta_0 + \beta_1 x_1 + \beta_2 x_2 + \varepsilon$$

where y is the surface finish, x_1 is the lathe speed in RPM, and x_2 is an indicator variable denoting the type of cutting tool used; that is,

$$x_2 = \begin{cases} 0 & \text{for tool type 301} \\ 1 & \text{for tool type 416} \end{cases}$$

The parameters in this model may be easily interpreted. If $x_2 = 0$, then the model becomes

$$y = \beta_0 + \beta_1 x_1 + \varepsilon$$

which is a straight-line model with slope β_1 and intercept β_0. However, if

$x_2 = 1$, the model becomes

$$y = \beta_0 + \beta_1 x_2 + \beta_2(1) + \varepsilon = (\beta_0 + \beta_2) + \beta_1 x_1 + \varepsilon$$

which is a straight-line model with slope β_1 and intercept $\beta_0 + \beta_2$. Thus, the model $y = \beta_0 + \beta_1 x + \beta_2 x_2 + \varepsilon$ implies that surface finish is linearly related to lathe speed and that the slope β_1 does not depend on the type of cutting tool used. However, the type of cutting tool does affect the intercept, and β_2 indicates the change in the intercept associated with a change in tool type from 302 to 416.

The **X** matrix and **y** vector for this problem are as follows:

$$\mathbf{X} = \begin{bmatrix} 1 & 225 & 0 \\ 1 & 200 & 0 \\ 1 & 250 & 0 \\ 1 & 245 & 0 \\ 1 & 235 & 0 \\ 1 & 237 & 0 \\ 1 & 265 & 0 \\ 1 & 259 & 0 \\ 1 & 221 & 0 \\ 1 & 218 & 0 \\ 1 & 224 & 1 \\ 1 & 212 & 1 \\ 1 & 248 & 1 \\ 1 & 260 & 1 \\ 1 & 243 & 1 \\ 1 & 238 & 1 \\ 1 & 224 & 1 \\ 1 & 251 & 1 \\ 1 & 232 & 1 \\ 1 & 216 & 1 \end{bmatrix}, \quad \mathbf{y} = \begin{bmatrix} 45.44 \\ 42.03 \\ 50.10 \\ 48.75 \\ 47.92 \\ 47.79 \\ 52.26 \\ 50.52 \\ 45.58 \\ 44.78 \\ 33.50 \\ 31.23 \\ 37.52 \\ 37.13 \\ 34.70 \\ 33.92 \\ 32.13 \\ 35.47 \\ 33.49 \\ 32.29 \end{bmatrix}$$

The fitted model is

$$\hat{y} = 14.27620 + 0.14115x_1 - 13.28020x_2$$

The analysis of variance for this model is shown in Table 2.12. Note that the hypothesis $H_0: \beta_1 = \beta_2 = 0$ (significance of regression) is rejected. This table also contains the sums of squares

$$SS_R = SS_R(\beta_1, \beta_2 | \beta_0)$$

$$= SS_R(\beta_1 | \beta_0) + SS_R(\beta_2 | \beta_1, \beta_0)$$

Table 2.11 Surface Finish Data for Example 2.11

Observation Number, i	Surface Finish, y_i	RPM	Type of Cutting Tool
1	45.44	225	302
2	42.03	200	302
3	50.10	250	302
4	48.75	245	302
5	47.92	235	302
6	47.79	237	302
7	52.26	265	302
8	50.52	259	302
9	45.58	221	302
10	44.78	218	302
11	33.50	224	416
12	31.23	212	416
13	37.52	248	416
14	37.13	260	416
15	34.70	243	416
16	33.92	238	416
17	32.13	224	416
18	35.47	251	416
19	33.49	232	416
20	32.29	216	416

so that a test of the hypothesis H_0: $\beta_2 = 0$ can be made. This hypothesis is also rejected, so we conclude that tool type has an effect on surface finish.

It is also possible to use indicator variables to investigate whether tool type affects both the slope and intercept. Let the model be reformulated as

$$y = \beta_0 + \beta_1 x_1 + \beta_2 x_2 + \beta_3 x_1 x_2 + \varepsilon$$

where x_2 is the indicator variable. Now if tool type 302 is used, then $x_2 = 0$,

Table 2.12 Analysis of Variance of Example 2.11

Source of Variation	Sum of Squares	Degrees of Freedom	Mean Square	F_0	P-Value
Regression	1012.0595	1	506.0297	1103.69	1.0175×10^{-18}
$SS_R(\beta_1\|\beta_0)$	(130.6091)	(1)	130.6091	284.87	4.6980×10^{-12}
$SS_R(\beta_2\|\beta_1, \beta_0)$	(881.4504)	(1)	881.4504	1922.52	6.2439×10^{-19}
Error	7.7943	17	0.4508		
Total	1019.8538	19			

and the model is

$$y = \beta_0 + \beta_1 x_1 + \varepsilon$$

If tool type 416 is used, then $x_2 = 1$, and the model becomes

$$y = \beta_0 + \beta_1 x_1 + \beta_2 + \beta_3 x_1 + \varepsilon$$
$$= (\beta_0 + \beta_2) + (\beta_1 + \beta_3) x_1 + \varepsilon$$

Note that β_2 is the change in the intercept, and β_3 is the change in slope produced by a change in tool type.

Another method of analyzing this data set is to fit separate regression models to the data for each tool type. However, the indicator variable approach has several advantages. First, only one regression model must be estimated. Second, by pooling the data on both tool types, more degrees of freedom for error are obtained. Third, tests of both hypotheses on the parameters β_2 and β_3 are just special cases of the extra sum-of-squares method.

EXERCISES

2.1 A study was performed on wear of a bearing y and its relationship to ξ_1 − oil viscosity and ξ_2 − load. The following data were obtained:

y	ξ_1	ξ_2
193	1.6	851
230	15.5	816
172	22.0	1058
91	43.0	1201
113	33.0	1357
125	40.0	1115

(a) Fit a multiple linear regression model to these data, using coded variables x_1 and x_2, where the coded variables are defined so that $-1 \le x_i \le 1$, $i = 1, 2$.

(b) Convert the model in part (a) to a model in the natural variables ξ_1 and ξ_2.

(c) Use the model to predict wear when $\xi_1 = 25$ and $\xi_2 = 1000$.

(d) Fit a multiple linear regression model with an interaction term to these data. Use the coded variables defined in part (a).

(e) Use the model in (d) to predict wear when $\xi_1 = 25$ and $\xi_2 = 1000$. Compare this prediction with the predicted value from part (b) above.

2.2 Consider the regression models developed in Exercise 2.1.

(a) Test for significance of regression from the first-order model in part (a) of Exercise 2.1 using the analysis of variance with $\alpha = 0.05$. What are your conclusions?

(b) Use the extra sum of squares method to investigate adding the interaction term to the model [part (d) of Exercise 2.1]. With $\alpha = 0.05$, what are your conclusions about this term?

2.3 The pull strength of a wire bond is an important characteristic. The table below gives information on pull strength (y), die height (x_1), post height (x_2), loop height (x_3), wire length (x_4), bond width on the die (x_5), and bond width on the post (x_6).

y	x_1	x_2	x_3	x_4	x_5	x_6
8.0	5.2	19.6	29.6	94.9	2.1	2.3
8.3	5.2	19.8	32.4	89.7	2.1	1.8
8.5	5.8	19.6	31.0	96.2	2.0	2.0
8.8	6.4	19.4	32.4	95.6	2.2	2.1
9.0	5.8	18.6	28.6	86.5	2.0	1.8
9.3	5.2	18.8	30.6	84.5	2.1	2.1
9.3	5.6	20.4	32.4	88.8	2.2	1.9
9.5	6.0	19.0	32.6	85.7	2.1	1.9
9.8	5.2	20.8	32.2	93.6	2.3	2.1
10.0	5.8	19.9	31.8	86.0	2.1	1.8
10.3	6.4	18.0	32.6	87.1	2.0	1.6
10.5	6.0	20.6	33.4	93.1	2.1	2.1
10.8	6.2	20.2	31.8	83.4	2.2	2.1
11.0	6.2	20.2	32.4	94.5	2.1	1.9
11.3	6.2	19.2	31.4	83.4	1.9	1.8
11.5	5.6	17.0	33.2	85.2	2.1	2.1
11.8	6.0	19.8	35.4	84.1	2.0	1.8
12.3	5.8	18.8	34.0	86.9	2.1	1.8
12.5	5.6	18.6	34.2	83.0	1.9	2.0

(a) Fit a multiple linear regression model using x_2, x_3, x_4, and x_5 as the regressors.

(b) Test for significance of regression using the analysis of variance with $\alpha = 0.05$. What are your conclusions?

(c) Use the model from part (a) to predict pull strength when $x_2 = 20$, $x_3 = 30$, $x_4 = 90$, and $x_5 = 2.0$.

2.4 An engineer at a semiconductor company wants to model the relationship between the device gain or $hFE(y)$ and three parameters: emitter-RS (x_1), base-RS (x_2), and emitter-to-base-RS (x_3). The data are shown below:

x_1 Emitter-RS	x_2 Base-RS	x_3 B-E-RS	y hFE-1M-5V
14.620	226.00	7.000	128.40
15.630	220.00	3.375	52.62
14.620	217.40	6.375	113.90
15.000	220.00	6.000	98.01
14.500	226.50	7.625	139.90
15.250	224.10	6.000	102.60
16.120	220.50	3.375	48.14
15.130	223.50	6.125	109.60
15.500	217.60	5.000	82.68
15.130	228.50	6.625	112.60
15.500	230.20	5.750	97.52
16.120	226.50	3.750	59.06
15.130	226.60	6.125	111.80
15.630	225.60	5.375	89.09
15.380	234.00	8.875	171.90
15.500	230.00	4.000	66.80
14.250	224.30	8.000	157.10
14.500	240.50	10.870	208.40
14.620	223.70	7.375	133.40

(a) Fit a multiple linear regression model to the data.

(b) Predict hFE when $x_1 = 14.5$, $x_2 = 220$, and $x_3 = 5.0$.

(c) Test for significance of regression using the analysis of variance with $\alpha = 0.05$. What conclusions can you draw?

2.5 The electric power consumed each month by a chemical plant is thought to be related to the average ambient temperature (x_1), the number of days in the month (x_2), the average product purity (x_3), and the tons of product produced (x_4). The past year's historical data are available and are presented in the following table.

y	x_1	x_2	x_3	x_4
240	25	24	91	100
236	31	21	90	95
290	45	24	88	110
274	60	25	87	88
301	65	25	91	94
316	72	26	94	99
300	80	25	87	97
296	84	25	86	96
267	75	24	88	110
276	60	25	91	105
288	50	25	90	100
261	38	23	89	98

(a) Fit a multiple linear regression model to the data.

(b) Predict power consumption for a month in which $x_1 = 75°F$, $x_2 = 24$ days, $x_3 = 90\%$, and $x_4 = 98$ tons.

2.6 Heat treating is often used to carburize metal parts, such as gears. The thickness of the carburized layer is considered an important feature of the gear, and it contributes to the overall reliability of the part. Because of the critical nature of this feature, two different lab tests are performed on each furnace load. One test is run on a sample pin that accompanies each load. The other test is a destructive test, where an actual part is cross-sectioned. This test involved running a carbon analysis on the surface of both the gear pitch (top of the gear tooth) and the gear root (between the gear teeth). The data below are the results of the pitch carbon analysis test catch for 32 parts.

TEMP	SOAKTIME	SOAKPCT	DIFFTIME	DIFFPCT	PITCH
1650	0.58	1.10	0.25	0.90	0.013
1650	0.66	1.10	0.33	0.90	0.016
1650	0.66	1.10	0.33	0.90	0.015
1650	0.66	1.10	0.33	0.95	0.016
1600	0.66	1.15	0.33	1.00	0.015
1600	0.66	1.15	0.33	1.00	0.016
1650	1.00	1.10	0.50	0.80	0.014
1650	1.17	1.10	0.58	0.80	0.021
1650	1.17	1.10	0.58	0.80	0.018
1650	1.17	1.10	0.58	0.80	0.019

1650	1.17	1.10	0.58	0.90	0.021
1650	1.17	1.10	0.58	0.90	0.019
1650	1.17	1.15	0.58	0.90	0.021
1650	1.20	1.15	1.10	0.80	0.025
1650	2.00	1.15	1.00	0.80	0.025
1650	2.00	1.10	1.10	0.80	0.026
1650	2.20	1.10	1.10	0.80	0.024
1650	2.20	1.10	1.10	0.80	0.025
1650	2.20	1.15	1.10	0.80	0.024
1650	2.20	1.10	1.10	0.90	0.025
1650	2.20	1.10	1.10	0.90	0.027
1650	2.20	1.10	1.50	0.90	0.026
1650	3.00	1.15	1.50	0.80	0.029
1650	3.00	1.10	1.50	0.70	0.030
1650	3.00	1.10	1.50	0.75	0.028
1650	3.00	1.15	1.66	0.85	0.032
1650	3.33	1.10	1.50	0.80	0.033
1700	4.00	1.10	1.50	0.70	0.039
1650	4.00	1.10	1.50	0.70	0.040
1650	4.00	1.15	1.50	0.85	0.035
1700	12.50	1.00	1.50	0.70	0.056
1700	18.50	1.00	1.50	0.70	0.068

(a) Fit a linear regression model relating the results of the pitch carbon analysis test (PTCH) to the five regressor variables.

(b) Test for significance of regression. Use $\alpha = 0.05$.

2.7 A regression model $y = \beta_0 + \beta_1 x_1 + \beta_2 x_2 + \beta_3 x_3 + \varepsilon$ has been fit to a sample of $n = 25$ observations. The calculated t-ratios $b_j / se(b_j)$, $j = 1, 2, 3$ are as follows: For β_1, $t_0 = 4.82$; for $\beta_2, t_0 = 8.21$; and for β_3, $t_0 = 0.98$.

(a) Find P-values for each of the t-statistics.

(b) Using $\alpha = 0.05$, what conclusions can you draw about the regressor x_3? Does it seem likely that this regressor contributes significantly to the model?

2.8 Consider the electric power consumption data in Exercise 2.5.

(a) Estimate σ^2 for the model fit in Exercise 2.5.

(b) Use the t-test to assess the contribution of each regressor to the model. Using $\alpha = 0.01$, what conclusions can you draw?

2.9 Consider the bearing wear data in Exercise 2.1

(a) Estimate σ^2 for the no-interaction model.

(b) Compute the t-statistic for each regression coefficient. Using $\alpha = 0.05$, what conclusions can you draw?

(c) Use the extra sum of squares method to investigate the usefulness of adding x_2 = load to the model that already contains x_1 = oil viscosity. Use $\alpha = 0.05$.

2.10 Consider the wire bond pull strength data in Exercise 2.3.

(a) Estimate σ^2 for this model.

(b) Find the standard errors for each of the regression coefficients.

(c) Calculate the t-test statistic for each regression coefficient. Using $\alpha = 0.05$, what conclusions can you draw? Do all variables contribute to the model?

2.11 Reconsider the semiconductor data in Exercise 2.4.

(a) Estimate σ^2 for the model you have fit to the data.

(b) Find the standard errors of the regression coefficients.

(c) Calculate the t-test statistic for each regression coefficient. Using $\alpha = 0.05$, what conclusions can you draw?

2.12 Exercise 2.6 presents data on heat treating gears.

(a) Estimate σ^2 for the model.

(b) Find the standard errors of the regression coefficients.

(c) Evaluate the contribution of each regressor to the model using the t-test with $\alpha = 0.05$.

(d) Fit a new model to the response PITCH using new regressors x_1 = SOAKTIME × SOAKPCT and x_2 = DIFFTIME × DIFFPCT.

(e) Test the model in part (d) for significance of regression using $\alpha = 0.05$. Also calculate the t-test for each regressor and draw conclusions.

(f) Estimate σ^2 for the model from part (d) and compare this to the estimate of σ^2 obtained in part (b) above. Which estimate is smaller? Does this offer any insight regarding which model might be preferable?

2.13 Consider the wire bond pull strength data in Exercise 2.3.

(a) Find 95% confidence intervals on the regression coefficients.

(b) Find a 95% confidence interval on mean pull strength when $x_2 = 20$, $x_3 = 30$, $x_4 = 90$, and $x_5 = 2.0$.

2.14 Consider the semiconductor data in Exercise 2.4.

    (a) Find 99% confidence intervals on the regression coefficients.

    (b) Find a 99% prediction interval on hFE when $x_1 = 14.5$, $x_2 = 220$, and $x_3 = 5.0$.

    (c) Find a 99% confidence interval on mean hFE when $x_1 = 14.5$, $x_2 = 220$, and $x_3 = 5.0$.

2.15 Consider the heat-treating data from Exercise 2.6.

    (a) Find 95% confidence intervals on the regression coefficients.

    (b) Find a 95% interval on mean PITCH on TEMP = 1650, SOAK-TIME = 1.00, SOAKPCT = 1.10, DIFFTIME = 1.00, and DIFF-PCT = 0.80.

2.16 Reconsider the heat treating in Exercise 2.6 and 2.12, where we fit a model to PITCH using regressors x_1 = SOAKTIME × SOAKPCT and x_2 = DIFFTIME × DIFFPCT.

    (a) Using the model with regressors x_1 and x_2, find a 95% confidence interval on mean PITCH when SOAKTIME = 1.00, SOAKPCT – 1.10, DIFFTIME = 1.00, and DIFFPCT = 0.80.

    (b) Compare the length of this confidence interval with the length of the confidence interval on mean PITCH at the same point from Exercise 2.15 part (b), where an additive model in SOAKTIME, SOAKPCT, DIFFTIME, and DIFFPCT was used. Which confidence interval is shorter? Does this tell you anything about which model is preferable?

2.17 For the regression model for the wire bond pull strength data in Exercise 2.3.

    (a) Plot the residuals versus $\hat{y}$ and versus the regressors used in the model. What information is provided by these plots?

    (b) Construct a normal probability plot of the residuals. Are there reasons to doubt the normality assumption for this model?

    (c) Are there any indications of influential observations in the data?

2.18 Consider the semiconductor hFE data in Exercise 2.4.

    (a) Plot the residuals from this model versus $\hat{y}$. Comment on the information in this plot.

    (b) What is the value of R^2 for this model?

(c) Refit the model using ln *hFE* as the response variable.

(d) Plot the residuals versus predicted ln *hFE* for the model in part (c) above. Does this give any information about which model is preferable?

(e) Plot the residuals from the model in part (d), versus the regressor x_3. Comment on this plot.

(f) Refit the model to ln *hFE* using x_1, x_2, and $1/x_3$ as the regressors. Comment on the effect of this change in the model.

2.19 Consider the regression model for the heat-treating data in Exercise 2.6.

(a) Calculate the percent of variability explained by this model.

(b) Construct a normal probability plot for the residuals. Comment on the normality assumption.

(c) Plot the residuals versus $\hat{y}$ and interpret the display.

(d) Calculate Cook's distance for each observation and provide an interpretation of this statistic.

2.20 In Exercise 2.12 we fit a model to the response PITCH in the heat treating data of Exercise 2.6 using new regressors x_1 = SOAKTIME × SOAKPCT and x_2 = DIFFTIME × DIFFPCT.

(a) Calculate the R^2 for this model and compare it to the value of R^2 from the original model in Exercise 2.6. Does this provide some information about which model is preferable?

(b) Plot the residuals from this model versus $\hat{y}$ and on a normal probability scale. Comment on model adequacy.

(c) Find the values of Cook's distance measure. Are any observations unusually influential?

2.21 An article entitled "A Method for Improving the Accuracy of Polynomial Regression Analysis" in the *Journal of Quality Technology* (1971, pp. 149–155) reported the following data on y = ultimate shear strength of a rubber compound (psi) and x = cure temperature (°F).

y	770	800	840	810	735	640	590	560
x	280	284	292	295	298	305	308	315

(a) Fit a second-order polynomial to this data.

(b) Test for significance of regression, using $\alpha = 0.05$.

(c) Test the hypothesis that $\beta_{11} = 0$, using $\alpha = 0.05$.

(d) Compute the residuals and use them to evaluate model adequacy.

(e) Suppose that the following additional observations are available: $(x = 284, y = 815)$, $(x = 295, y = 830)$, $(x = 305, y = 660)$, and $(x = 315, y = 545)$. Test the second-order model for lack of fit.

2.22 Consider the following data, which result from an experiment to determine the effect of x = test time in hours at a particular temperature to y = change in oil viscosity.

y	4.42h	1.39h	-1.55h	-1.89h	-2.43h	-3.15	-4.05	-5.15h	-6.43h	-7.89
x	.25	.50	.75	1.00	1.25	1.50	1.75	2.00	2.25	2.50

(a) Fit a second-order polynomial to the data.

(b) Test for significance of regression, using $\alpha = 0.05$.

(c) Test the hypothesis that $\beta_{11} = 0$, using $\alpha = 0.05$.

(d) Compute the residuals and use them to evaluate model adequacy.

2.23 When fitting polynomial regression models we often subtract $\bar{x}$ from each x value to produce a "centered" regressor $x' = x - \bar{x}$. This reduces the effects of dependencies among the model terms and often leads to more accurate estimates of the regression coefficients. Using the data from Exercise 2.21, fit the model $y = \beta_0^* + \beta_1^* x' + \beta_{11}^*(x')^2 + \varepsilon$. Use the results to estimate the coefficients in the uncentered model $y = \beta_0 + \beta_1 x + \beta_{11} x^2 + \varepsilon$.

2.24 Suppose that we use a standardized variable $x' = (x - \bar{x})/s_x$, where s_x is the standard deviation of x, in constructing a polynomial regression model. Using the data in Exercise 2.21 and the standardized variable approach, fit the model $y = \beta_0^* + \beta_1^* x' + \beta_{11}^*(x')^2 + \varepsilon$.

(a) What value of y do you predict when $x = 285°F$?

(b) Estimate the regression coefficients in the unstandardized model $y = \beta_0 + \beta_1 x + \beta_{11} x^2 + \varepsilon$.

(c) What can you say about the relationship between SS_E and R^2 for the standardized and unstandardized models?

(d) Suppose that $y' = (y - \bar{y})/s_y$ is used in the model along with x'. Fit the model and comment on the relationship between SS_E and R^2 in the standardized model and the unstandardized model.

2.25 An article in the *Journal of Pharmaceuticals Sciences* (vol. 80, 1991, pp. 971–977) presents data on the observed mole fraction solubility of a solute at a constant temperature to the dispersion, dipolar, and hydrogen bonding Hansen partial solubility parameters. The data are as follows:

Obs	y	x_1	x_2	x_3
1	0.22200	7.3	0.0	0.0
2	0.39500	8.7	0.0	0.3
3	0.42200	8.8	0.7	1.0
4	0.43700	8.1	4.0	0.2
5	0.42800	9.0	0.5	1.0
6	0.46700	8.7	1.5	2.8
7	0.44400	9.3	2.1	1.0
8	0.37800	7.6	5.1	3.4
9	0.49400	10.0	0.0	0.3
10	0.45600	8.4	3.7	4.1
11	0.45200	9.3	3.6	2.0
12	0.11200	7.7	2.8	7.1
13	0.43200	9.8	4.2	2.0
14	0.10100	7.3	2.5	6.8
15	0.23200	8.5	2.0	6.6
16	0.30600	9.5	2.5	5.0
17	0.09230	7.4	2.8	7.8
18	0.11600	7.8	2.8	7.7
19	0.07640	7.7	3.0	8.0
20	0.43900	10.3	1.7	4.2
21	0.09440	7.8	3.3	8.5
22	0.11700	7.1	3.9	6.6
23	0.07260	7.7	4.3	9.5
24	0.04120	7.4	6.0	10.9
25	0.25100	7.3	2.0	5.2
26	0.00002	7.6	7.8	20.7

where y is the negative logarithm of the mole fraction solubility, x_1 is the dispersion Hansen partial solubility, x_2 is the dipolar partial solubility, and x_3 is the hydrogen bonding partial solubility.

(a) Fit the model $y = \beta_0 + \beta_1 x_1 + \beta_2 x_2 + \beta_3 x_3 + \beta_{12} x_1 x_2 + \beta_{13} x_1 x_3 + \beta_{23} x_2 x_3 + \beta_{11} x_1^2 + \beta_{22} x_2^2 + \beta_{33} x_3^2 + \varepsilon$.

(b) Test for significance of regression, using $\alpha = 0.05$.

(c) Plot the residuals and comment on model adequacy.

(d) Use the extra sum of squares method to test the contribution of the second-order terms, using $\alpha = 0.05$.

2.26 Below are data on y = green liquor (g/liter) and x = paper machine speed (ft/min) from a kraft paper machine (the data were read from a graph in an article in the *Tappi Journal*, March, 1986).

y	16.0h	15.8h	15.6h	15.5h	14.8h	14.0h	13.5h	13.0h	12.0h	11.0
x	1700	1720	1730	1740	1750	1760	1770	1780	1790	1795

(a) Fit the model $y = \beta_0 + \beta_1 x + \beta_2 x^2 + \varepsilon$ using least squares.

(b) Test for significance of regression using $\alpha = 0.05$. What are your conclusions?

(c) Test the contribution of the quadratic term to the model, over the contribution of the linear term, using an F-statistic. If $\alpha = 0.05$, what conclusion can you draw?

(d) Plot the residuals from this model versus $\hat{y}$. Does the plot reveal any inadequacies?

(e) Construct a normal probability plot of the residuals. Comment on the normality assumption.

2.27 Consider a multiple regression model with k regressors. Show that the test statistic for significance of regression can be written as

$$F_k = \frac{R^2/k}{(1 - R^2)/(n - k - 1)}$$

Suppose that $n = 20$, $k = 4$, and $R^2 = 0.90$. If $\alpha = 0.05$, what conclusion would you draw about the relationship between y and the four regressors?

2.28 A regression model is used to relate a response y to $k = 4$ regressors. What is the smallest value of R^2 that will result in a significant regression if $\alpha = 0.05$? Use the results of the previous exercise. Are you surprised by how small the value of R^2 is?

2.29 Show that we can express the residuals from a multiple regression model as

$$\mathbf{e} = (\mathbf{I} - \mathbf{H})\mathbf{y}, \text{ where } \mathbf{H} = \mathbf{X}(\mathbf{X}'\mathbf{X})^{-1}\mathbf{X}'.$$

2.30 Show that the variance of the ith residual e_i in a multiple regression model is $\sigma^2(1 - h_{ii})$ and that the covariance between e_i and e_j is $-\sigma^2 h_{ij}$, where the h's are the elements of $\mathbf{H} = \mathbf{X}(\mathbf{X}'\mathbf{X})^{-1}\mathbf{X}'$.

2.31 Consider the multiple linear regression model $\mathbf{y} = \mathbf{X}\boldsymbol{\beta} + \boldsymbol{\varepsilon}$. If $\mathbf{b}$ denotes the least squares estimator of $\boldsymbol{\beta}$, show that $\mathbf{b} = \boldsymbol{\beta} + \mathbf{R}\boldsymbol{\varepsilon}$, where $\mathbf{R} = (\mathbf{X}'\mathbf{X})^{-1}\mathbf{X}'$.

2.32 **Constrained Least Squares**. Suppose we wish to find the least squares estimator of $\boldsymbol{\beta}$ in the model $\mathbf{y} = \mathbf{X}\boldsymbol{\beta} + \boldsymbol{\varepsilon}$ subject to a set of equality constraints, say $\mathbf{T}\boldsymbol{\beta} = \mathbf{c}$.

Show that the estimator is

$$\mathbf{b}_c = \mathbf{b}(\mathbf{X}'\mathbf{X})^{-1}\mathbf{T}'\left[\mathbf{T}(\mathbf{X}'\mathbf{X})^{-1}\mathbf{T}'\right]^{-1}(\mathbf{c} - \mathbf{Tb})$$

where $\mathbf{b} = (\mathbf{X}'\mathbf{X})^{-1}\mathbf{X}'\mathbf{y}$.

CHAPTER 3

Two-Level Factorial Designs

3.1 INTRODUCTION

Factorial designs are widely used in experiments involving several factors where it is necessary to investigate the joint effects of the factors on a response variable. By joint factor effects, we typically mean main effects and interactions. A very important special case of the factorial design is that where each of the k factors of interest has only two levels. Because each replicate of such a design has exactly 2^k experimental trials or runs, these designs are usually called 2^k **factorial designs**. These designs are the subject of this chapter.

The class of 2^k factorial designs are very important in response surface work. Specifically, they find applications in three areas:

1. A 2^k design (and a fractional version discussed in the next chapter) is useful at the start of a response surface study where screening experiments should be performed to identify the important process or system variables.

2. A 2^k design is often used to fit a first-order response surface model and to generate the factor effect estimates required to perform the method of steepest ascent.

3. The 2^k design is a basic building block used to create other response surface designs. For example, if you augment a 2^2 design with axial runs as in Figure 2.7 of Chapter 2, then a central composite design results. As we will subsequently see, the central composite design is one of the most important designs for fitting second-order response surface models.

3.2 THE 2^2 DESIGN

The simplest design in the 2^k-series is one with only two factors, say A and B, each run at two levels. This design is called a 2^2 factorial design. The levels of the factors may be arbitrarily called "low" and "high." These two levels may be quantitative, such as two values of temperature or pressure; or they may be qualitative, such as two machines, or two operators. In most response surface studies the factors and their levels are quantitative.

As an example, consider an investigation into the effect of the concentration of the reactant and the feed rate on the viscosity of product from a chemical process. Let the reactant concentration be factor A, and let the two levels of interest be 15% and 25%. The feed rate is factor B, with the low level being 20 lb/ hr and the high level being 30 lb/ hr. The experiment is replicated four times, and the data are as follows:

Factor Level	Replicate				
Combination	I	II	III	IV	Total
A low, B low	145	148	147	140	580
A high, B low	158	152	155	152	617
A low, B high	135	138	141	139	553
A high, B high	150	152	146	149	597

The four factor-level or treatment combinations in this design are shown graphically in Figure 3.1. By convention, we denote the effect of factor by a capital Latin letter. That is, A refers to the effect of factor A, B refers to the effect of factor B, and AB refers to the AB interaction. In the 2^2 design the low and high levels of A and B are denoted by $-$ and $+$, respectively, on the A- and B-axes. Thus, $-$ on the A-axis represents the low level of concentration (15%), and $+$ represents the high level (25%). Similarly, $-$ on

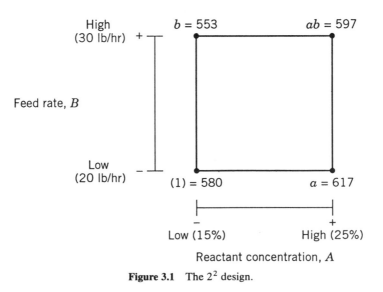

Figure 3.1   The 2^2 design.

the B-axis represents the low level of feed rate, and $+$ denotes the high level.

The four factor-level combinations in the design are usually represented by lowercase letters, as shown in Figure 3.1. We see from the figure that the high level of any factor at a point in the design is denoted by the corresponding lowercase letter and that the low level of a factor at a point in the design is denoted by the absence of the corresponding letter. Thus, a represents the combination of factor levels with A at the high level and B at the low level, b represents A at the low level and B at the high level, and ab represents both factors at the high level. By convention, (1) is used to denote the run where both factors are at the low level. This notation is used throughout the 2^k series.

Define the **average effect of a factor** as the change in response produced by a change in the level on that factor averaged over the levels of the other factor. Also the symbols (1), a, b, and ab now represent the total of all n replicates taken at the point in the design, as illustrated in Figure 3.1. The formulas for the effects of A, B and AB may be easily derived. The effect of A can be found as the difference in the average response of the two points on the right-hand side of the square in Figure 3.1 (call this average $\bar{y}_{A+}$, because it is the average response at the points where A is at the high level) and the two points on the left hand side (or $\bar{y}_{A-}$). That is,

$$A = \bar{y}_{A+} - \bar{y}_{A-}$$

$$= \frac{ab + a}{2n} - \frac{b + (1)}{2n}$$

$$= \frac{1}{2n}[ab + a - b - (1)] \tag{3.1}$$

The effect of factor B is found as the difference between the average of the response at the two points on the top of the square $(\bar{y}_{B+})$ and the average of the response at the two points on the bottom $(\bar{y}_{B-})$, or

$$B = \bar{y}_{B+} - \bar{y}_{B-}$$

$$= \frac{ab + b}{2n} - \frac{a + (1)}{2n}$$

$$= \frac{1}{2n}[ab + b - a(1)] \tag{3.2}$$

Finally, the interaction effect AB is the average of the responses on the right-to-left diagonal points in the square [ab and (1)] minus the average of the responses on the left-to-right diagonal points (a and b), or

$$AB = \frac{ab + (1)}{2n} - \frac{a + b}{2n}$$

$$= \frac{1}{2n}[ab + (1) - a - b] \qquad (3.3)$$

Using the example data in Figure 3.1, we may estimate the effects as

$$A = \frac{1}{2(4)}(597 + 617 - 553 - 580) = 10.125$$

$$B = \frac{1}{2(4)}(597 + 553 - 617 - 580) = -5.875$$

$$AB = \frac{1}{2(4)}(597 + 580 - 617 - 553) = 0.875$$

The effect of A (reactant concentration) is positive; this suggests that increasing A from the low level (15%) to the high level (25%) will increase the viscosity. The effect of B (feed rate) is negative; this suggests that increasing the feed rate will decrease the viscosity. The interaction effect appears to be small relative to the two main effects.

In many experiments involving 2^k designs, we will examine the magnitude and direction of the factor effects to determine which variables are likely to be important. The analysis of variance can generally be used to confirm this interpretation. In the 2^k design, there are some special, efficient methods for performing the calculations in the analysis of variance.

Consider the sums of squares for A, B, and AB. Note from Equation (3.1) that a **contrast** is used in estimating A, namely,

$$\text{Contrast}_A = ab + a - b - (1) \qquad (3.4)$$

We usually call this contrast the total effect of A. From Equations (3.2) and (3.3), we see that contrasts are also used to estimate B and AB. Furthermore, these three contrasts are orthogonal. The sum of squares for any contrast is equal to the contrast squared divided by the number of observations in each total in the contrast times the sum of the squares of the contrast coefficients. Consequently, we have

$$SS_A = \frac{[ab + a - b - (1)]^2}{n \cdot 4} \qquad (3.5)$$

$$SS_B = \frac{[ab + b - a - (1)]^2}{n \cdot 4} \qquad (3.6)$$

and

$$SS_{AB} = \frac{[ab + (1) - a - b]^2}{n \cdot 4} \tag{3.7}$$

as the sums of squares for A, B, and AB.

Using the data in Figure 3.1, we may find the sums of the squares from Equations (3.5), (3.6), and (3.7) as

$$SS_A - \frac{(81)^2}{4(4)} - 410.0625$$

$$SS_B = \frac{(47)^2}{4(4)} = 138.0625$$

and

$$SS_{AB} = \frac{(7)^2}{4(4)} = 3.0625$$

The total sum of squares is found in the usual way; that is,

$$SS_T = \sum_{i=1}^{2} \sum_{j=1}^{2} \sum_{k=1}^{n} y_{ijk}^2 - \frac{y_{...}^2}{4n} \tag{3.8}$$

In general, SS_T has $4n - 1$ degrees of freedom. The error sum of squares, with $4(n - 1)$ degrees of freedom, is usually computed by subtraction as

$$SS_E = SS_T - SS_A - SS_B - SS_{AB} \tag{3.9}$$

For the data in Figure 3.1, we obtain

$$SS_T = \sum_{i=1}^{2} \sum_{j=1}^{2} \sum_{k=1}^{4} y_{ijk}^2 - \frac{y_{...}^2}{4(4)}$$

$$= 334{,}527.0000 - 344{,}275.5625 = 651.4375$$

and

$$SS_E = SS_T - SS_A - SS_B - SS_{AB}$$

$$= 651.4375 - 410.0625 - 138.0625 - 3.0625$$

$$= 100.2500$$

using SS_A, SS_B, and SS_{AB} computed previously. The complete analysis of

Table 3.1 Analysis for Data in Figure 3.1

Source of Variation	Sum of Squares	Degrees of Freedom	Mean Square	F_0	P-Value
A	410.0625	1	410.0625	49.08	1.42×10^{-5}
B	138.0625	1	138.0625	16.53	0.0016
AB	3.0625	1	3.0625	0.37	0.5562
Error	100.2500	12	8.3542		
Total	651.4375	15			

variance is summarized in Table 3.1. Both main effects are statistically significant and the interaction is not significant. This confirms our initial interpretation of the data based on the magnitudes of the factor effects.

It is often convenient to write down the factor level or treatment combinations in the order (1), a, b, ab, as in the column labeled "treatment combination" in Table 3.2. This is referred to as **standard order**. The columns labeled A and B in this table contain the $+$ and $-$ signs corresponding to the four runs at the corners of the square in Figure 3.1. Note that the contrast coefficients for estimating the interaction effect are just the product of the corresponding coefficients for the two main effects. The column labeled I in Table 3.2 represents the total or average of all of the observations in the entire experiment. Notice that the column corresponding to I has only plus signs. The row designators are the treatment combinations. To find the contrast for estimating any effect, simply multiply the signs in the appropriate column of the table by the corresponding treatment combination and add. For example, to estimate A, the contrast is $-(1) + a - b + ab$, which agrees with Equation (3.1).

The Regression Model
It is easy to convert the effect estimates in a 2^k factorial design into a regression model that can be used to predict the response at any point in the space spanned by the factors in the design. For the chemical process experiment, the first-order regression model is

$$y = \beta_0 + \beta_1 x_1 + \beta_2 x_2 + \varepsilon \qquad (3.10)$$

Table 3.2 Signs for Calculating Effects in the 2^2 Design

Treatment Combination	Factorial Effect			
	I	A	B	AB
(1)	$+$	$-$	$-$	$+$
a	$+$	$+$	$-$	$-$
b	$+$	$-$	$+$	$-$
ab	$+$	$+$	$+$	$+$

where x_1 is the coded variable that represents the reactant concentration, x_2 is the coded variable that represents the feed rate, and the β's are the regression coefficients. The relationship between the natural variables, the reactant concentration and the feed rate, and the coded variables is

$$x_1 = \frac{\text{Conc} - (\text{Conc}_{\text{low}} + \text{Conc}_{\text{high}})/2}{(\text{Conc}_{\text{high}} - \text{Conc}_{\text{low}})/2}$$

and

$$x_2 = \frac{\text{Feed} - (\text{Feed}_{\text{low}} + \text{Feed}_{\text{high}})/2}{(\text{Feed}_{\text{high}} - \text{Feed}_{\text{low}})/2}$$

When the natural variables have only two levels, this coding will produce the ± 1 notation for the levels of the coded variables. To illustrate this for our example, note that

$$x_1 = \frac{\text{Conc} - (15 + 25)/2}{(25 - 15)/2}$$

$$= \frac{\text{Conc} - 20}{5}$$

Thus, if the concentration is at the high level (Conc $= 25\%$), then $x_1 = +1$, whereas if the concentration is at the low level (Conc $= 15\%$), then $x_1 = -1$. Furthermore,

$$x_2 = \frac{\text{Feed} - (20 + 30)/2}{(30 - 20)/2}$$

$$= \frac{\text{Feed} - 25}{5}$$

Thus, if the feed rate is at the high level (Feed $= 30$ lb/hr), then $x_2 = +1$, whereas if the feed rate is at the low level (Feed $= 20$ lb/hr), then $x_2 = -1$.
   The fitted regression model is

$$\hat{y} = 146.6875 + \left(\frac{10.125}{2}\right)x_1 + \left(\frac{-5.875}{2}\right)x_2$$

$$= 146.6875 + 5.0625x_1 - 2.9375x_2$$

where the intercept is the grand average of all 16 observations, and the regression coefficients b_1 and b_2 are one-half the corresponding factor effect estimates. The reason that the regression coefficient is one-half the effect estimate is that a regression coefficient measures the effect of a unit change in x on the mean of y, and the effect estimate is based on a two-unit change (from -1 to $+1$). This model explains about 84% of the variability in viscosity, because

$$R^2 = SS_{\text{model}} / SS_{\text{total}} = (SS_A + SS_B) / SS_T$$

$$= (410.0625 + 138.0625) / 651.4375$$

$$= 548.1250 / 651.4375 = 0.8414$$

This regression model is really a first-order response surface model for the process. If the interaction term in the analysis of variance (Table 3.1) were significant, the regression model would have been

$$y = \beta_0 + \beta_1 x_2 + \beta_2 x_1 + \beta_{12} x_1 x_2 + \varepsilon$$

and the estimate of the regression coefficient β_{12} would be one-half the AB interaction effect, or $b_{12} = 0.875 / 2$. Clearly this coefficient is small relative to b_1 and b_2, and so this is further evidence that interaction contributes very little to explaining viscosity in this process.

The regression coefficient estimates obtained above are least squares estimates. To verify this, write out the model in Equation (3.10) in matrix form, that is,

$$\mathbf{y} = \mathbf{X}\boldsymbol{\beta} + \boldsymbol{\varepsilon}$$

where

$$
\mathbf{y} = \begin{bmatrix} 145 \\ 148 \\ 147 \\ 140 \\ 158 \\ 152 \\ 155 \\ 152 \\ 135 \\ 138 \\ 141 \\ 139 \\ 150 \\ 152 \\ 146 \\ 149 \end{bmatrix},
\quad
\mathbf{X} = \begin{bmatrix} 1 & -1 & -1 \\ 1 & -1 & -1 \\ 1 & -1 & -1 \\ 1 & -1 & -1 \\ 1 & 1 & -1 \\ 1 & 1 & -1 \\ 1 & 1 & -1 \\ 1 & 1 & -1 \\ 1 & -1 & 1 \\ 1 & -1 & 1 \\ 1 & -1 & 1 \\ 1 & -1 & 1 \\ 1 & 1 & 1 \\ 1 & 1 & 1 \\ 1 & 1 & 1 \\ 1 & 1 & 1 \end{bmatrix},
\quad
\boldsymbol{\beta} = \begin{bmatrix} \beta_0 \\ \beta_1 \\ \beta_2 \end{bmatrix},
\quad
\boldsymbol{\varepsilon} = \begin{bmatrix} \varepsilon_1 \\ \varepsilon_2 \\ \varepsilon_3 \\ \varepsilon_4 \\ \varepsilon_5 \\ \varepsilon_6 \\ \varepsilon_7 \\ \varepsilon_8 \\ \varepsilon_9 \\ \varepsilon_{10} \\ \varepsilon_{11} \\ \varepsilon_{12} \\ \varepsilon_{13} \\ \varepsilon_{14} \\ \varepsilon_{15} \\ \varepsilon_{16} \end{bmatrix}
$$

The least squares estimates of β_0, β_1, and β_2 are found from

$$\mathbf{b} = (\mathbf{X'X})^{-1}\mathbf{X'y}$$

or

$$\begin{bmatrix} b_0 \\ b_1 \\ b_2 \end{bmatrix} = \begin{bmatrix} 16 & 0 & 0 \\ 0 & 16 & 0 \\ 0 & 0 & 16 \end{bmatrix}^{-1} \begin{bmatrix} 2347 \\ 81 \\ -47 \end{bmatrix}$$

$$= \frac{1}{16}\mathbf{I}_3 \begin{bmatrix} 2347 \\ 81 \\ -47 \end{bmatrix}$$

where $\mathbf{I}_3$ is a 3×3 identity matrix. This last expression reduces to

$$\begin{bmatrix} b_0 \\ b_1 \\ b_2 \end{bmatrix} = \begin{bmatrix} 2347/16 \\ 81/16 \\ -47/16 \end{bmatrix}$$

or $b_0 = 2347/16 = 146.6875$, $b_1 = 5.0625$, and $b_2 = -2.9375$. Therefore, the least squares estimates of β_1 and β_2 are exactly one-half of the factor effect estimates. This will always be the case in any 2^k factorial design.

Residual Analysis
It is easy to compute the residuals from a 2^k design via the regression model. For the chemical process experiment, the regression model is

$$\hat{y} = 146.6875 + \left(\frac{10.125}{2}\right)x_1 - \left(\frac{5.875}{2}\right)x_2$$

This model can be used to generate the predicted values of y at the four points in the design. For example, when the reactant concentration is at the low level ($x_1 = -1$) and the feed rate is at the low level ($x_2 = -1$) the predicted viscosity is

$$\hat{y} = 146.6875 + \left(\frac{10.125}{2}\right)(-1) - \left(\frac{5.875}{2}\right)(-1)$$

$$= 144.563$$

There are four observations at this treatment combination, and the residuals are

$$e_1 = 145 - 144.563 = 0.438$$
$$e_2 = 148 - 144.563 = 3.438$$
$$e_3 = 152 - 144.563 = 2.438$$
$$e_4 = 140 - 144.563 = -4.563$$

The remaining predicted values and residuals are calculated similarly. For the high level of the reactant concentration and the low level of the feed rate:

$$\hat{y} = 146.6875 + \left(\frac{10.125}{2}\right)(+1) + \left(\frac{-5.875}{2}\right)(-1)$$
$$= 154.688$$

and

$$e_5 = 158 - 154.688 = -3.688$$
$$e_6 = 152 - 154.688 = -2.688$$
$$e_7 = 155 - 154.688 = -0.313$$
$$e_8 = 152 - 154.688 = -2.688$$

For the low level of the reactant concentration and the high level of the feed rate:

$$\hat{y} = 146.6875 + \left(\frac{10.125}{2}\right)(-1) + \left(\frac{-5.875}{2}\right)(+1)$$
$$= 138.688$$

and

$$e_9 = 135 - 138.688 = -3.688$$
$$e_{10} = 138 - 138.688 = -0.688$$
$$e_{11} = 141 - 138.688 = 2.313$$
$$e_{12} = 139 - 138.688 = 0.313$$

Finally, for the high level of both factors:

$$\hat{y} = 146.6875 + \left(\frac{10.125}{2}\right)(+1) + \left(\frac{-5.875}{2}\right)(+1)$$
$$= 148.812$$

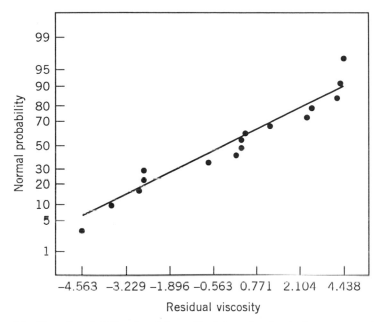

Figure 3.2   Normal probability plot of the residuals from the chemical process experiment.

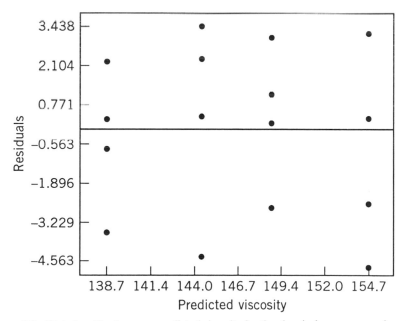

Figure 3.3   Plot of residuals versus predicted viscosity for the chemical process experiment.

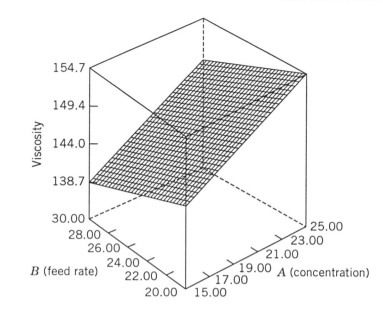

(a)

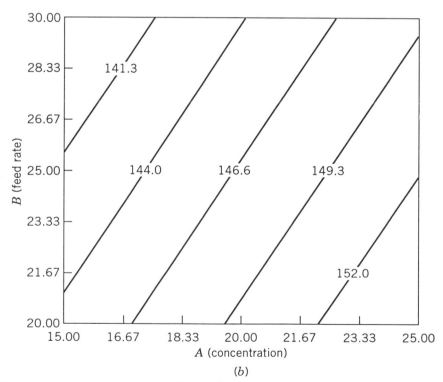

(b)

Figure 3.4 (a) Response surface plot of viscosity from the chemical process experiment. (b) The contour plot.

and

$$e_{13} = 150 - 148.812 = 1.188$$

$$e_{14} = 152 - 148.812 = 3.188$$

$$e_{15} = 146 - 148.812 = -2.813$$

$$e_{16} = 149 - 148.812 = 0.188$$

Figure 3.2 presents a normal probability plot of these residuals, and Figure 3.3 is a plot of the residuals versus the predicted values $\hat{y}$. These plots appear satisfactory, so we have no reason to suspect problems with the validity of our conclusions.

The Response Surface

Figure 3.4(a) presents a three-dimensional response surface plot of viscosity, based on the first-order model using reactant concentration and feed rate as the regressors, and Figure 3.4(b) is the contour plot. Because the model is first-order, the fitted response surface is a plane. Based on examination of this fitted surface, it is clear that viscosity increases as reactant concentration increases and feed rate decreases. Often a fitted surface such as this can be used to determine an appropriate **direction of potential improvement** for a process. A formal method for doing this called **method of steepest ascent** will be introduced in Chapter 5.

3.3 THE 2^3 DESIGN

Suppose that three factors, A, B, and C, each at two levels, are of interest. The design is called a 2^3 *factorial design* and the eight treatment combinations can now be displayed graphically as a cube, as shown in Figure 3.5. Extending the notation discussed in Section 3.2, we write the treatment combinations in standard order as (1), a, b, ab, c, ac, bc, and abc. Remember that these symbols also represent the total of all n observations taken at that particular treatment combinations.

There are actually three different notations that are widely used for the runs in the 2^k designs. The first is the "$+$" and "$-$" notation, often called the **geometric notation**. The second is the use of lowercase letters to identify the treatment combinations. The final notation uses 1 and 0 to denote high and low factor levels, respectively, instead of $+$ and $-$. These different

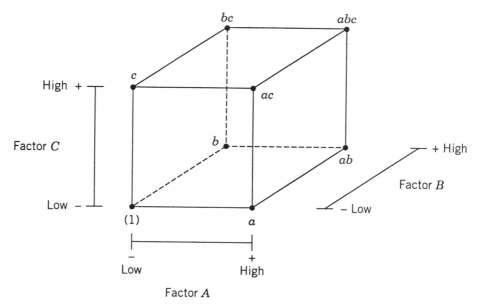

Figure 3.5   The 2^3 factorial design.

notations are illustrated below for the 2^3 design:

Run	A	B	C	Treatment Combinations	A	B	C
1	−	−	−	(1)	0	0	0
2	+	−	−	a	1	0	0
3	−	+	−	b	0	1	0
4	+	+	−	ab	1	1	0
5	−	−	+	c	0	0	1
6	+	−	+	ac	1	0	1
7	−	+	+	bc	0	1	1
8	+	+	+	abc	1	1	1

While some authors have advocated the 0, 1 notation, we prefer the geometric $(-, +)$ notation because it facilitates the translation of the analysis of variance results into a regression model. This geometric notation is widely used in response surface methodology.

There are seven degrees of freedom between the eight treatment combinations in the 2^3 design. Three degrees of freedom are associated with the main effects of A, B, and C. Four degrees of freedom are associated with interactions; one each with AB, AC, and BC and one with ABC.

Consider estimating the main effects. First, consider estimating the main effect of A. This effect estimate can be expressed as a contrast between the

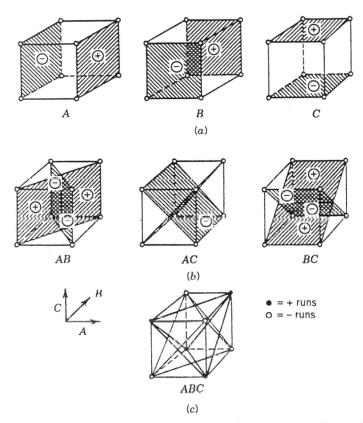

Figure 3.6 Geometric presentation of contrasts corresponding to the main effects and interaction in the 2^3 design. (*a*) Main effects. (*b*) Two-factor interactions. (*c*) Three-factor interaction.

four treatment combinations in the right face of the cube in Figure 3.6(*a*) (where A is at the high level) and the four in the left face (where A is at the low level). That is, the A effect is just the average of the four runs where A is at the high level ($\bar{y}_{A+}$) minus the average of the four runs where A is at the low level ($\bar{y}_{A-}$), or

$$A = \bar{y}_{A+} - \bar{y}_{A-}$$

$$= \frac{a + ab + ac + abc}{4n} - \frac{(1) + b + c + bc}{4n}$$

This equation can be rearranged as

$$A = \frac{1}{4n}[a + ab + ac + abc - (1) - b - c - bc] \qquad (3.11)$$

In a similar manner, the effect of B is the difference in averages between the four treatment combinations in the front face of the cube and the four in the back. This yields

$$B = \bar{y}_{B+} - \bar{y}_{B-}$$

$$= \frac{1}{4n}[b + ab + bc + abc - (1) - a - c - ac] \qquad (3.12)$$

The effect of C is the difference in averages between the four treatments combinations in the top face of the cube and the four in the bottom; that is,

$$C = \bar{y}_{C+} - \bar{y}_{C-}$$

$$= \frac{1}{4n}[c + ac + bc + abc - (1) - a - b - ab] \qquad (3.13)$$

The two-factor interaction effects may be computed easily. A measure of the AB interaction is the difference between the average A effects at the two levels of B. By convention, one-half of this difference is called the AB interaction. This is expressed symbolically as follows:

B	Average A Effect
High ($+$)	$\dfrac{[(abc - bc) + (ab - b)]}{2n}$
Low ($-$)	$\dfrac{\{(ac - c) + [a - (1)]\}}{2n}$
Difference	$\dfrac{[abc - bc + ab - b - ac + c - a + (1)]}{2n}$

Because the AB interaction is one-half of this difference, we have

$$AB = \frac{[abc - bc + ab - b - ac + c - a + (1)]}{4n} \qquad (3.14)$$

We could write Equation (3.14) as follows:

$$AB = \frac{abc + ab + c + (1)}{4n} - \frac{bc + b + ac + a}{4n}$$

In this form, the AB interaction is easily seen to be the difference in averages between runs on two diagonal planes in the cube in Figure 3.6(b).

Using similar logic and referring to Figure 3.6(b), the AC and BC interactions are

$$AC = \frac{1}{4n}[(1) - a + b - ab - c + ac - bc + abc] \qquad (3.15)$$

and

$$BC = \frac{1}{4n}[(1) + a - b - ab - c - ac + bc + abc] \qquad (3.16)$$

The ABC interaction is defined as the average difference between the AB interaction for the two different levels of C. Thus,

$$ABC = \frac{1}{4n}\{[abc - bc] - [ac - c] - [ab - b] + [a - (1)]\}$$

$$= \frac{1}{4n}[abc - bc - ac + c - ab + b + a - (1)] \qquad (3.17)$$

As before, we can think of the ABC interaction as the difference in two averages. If the runs in the two averages are isolated, they define the vertices of the two tetrahedra that comprise the cube in Figure 3.6(c).

In Equations (3.11) through (3.17), the quantities in brackets are contrasts in the treatment combinations. A table of plus and minus signs can be developed for the contrasts and is shown in Table 3.3. Signs for the main effects are determined by associating a plus with the high level and a minus with the low level. Once the signs for the main effects have been established, the signs for the interaction columns can be obtained by multiplying the appropriate preceding columns, row by row. For example, the signs in the AB column are the product of the A and B column signs in each row. The contrast for any effect can be obtained from this table.

Table 3.3 Algebraic Signs for Calculating Effects in the 2^3 Design

Treatment Combination	Factorial Effect							
	I	A	B	AB	C	AC	BC	ABC
(1)	+	−	−	+	−	+	+	−
a	+	+	−	−	−	−	+	+
b	+	−	+	−	−	+	−	+
ab	+	+	+	+	−	−	−	−
c	+	−	−	+	+	−	−	+
ac	+	+	−	−	+	+	−	−
bc	+	−	+	−	+	−	+	−
abc	+	+	+	+	+	+	+	+

Table 3.3 has several interesting properties: (1) Except for column I, every column has an equal number of plus and minus signs. (2) The sum of the products of the signs in any two columns is zero. (3) Column I multiplied by any column leaves that column unchanged. That is, I is an identity element. (4) The product of any two columns yields a column in the table. For example, $A \times B = AB$, and

$$AB \times B = AB^2 = A$$

We see that the exponents in the products are formed by using modulus 2 arithmetic. (That is, the exponent can only be zero or one; if it is greater than one, it is reduced by multiples of two until it is either zero or one.) All of these properties are implied by the orthogonality of the contrasts used to estimate the effects.

Sums of squares for the effects are easily computed, because each effect has a corresponding single-degree-of-freedom contrast. In the 2^3 design with n replicates, the sum of squares for any effect is

$$SS = \frac{(\text{Contrast})^2}{8n} \tag{3.18}$$

Example 3.1 An electrical engineer is investigating a plasma etching process used in semiconductor manufacturing. He is studying the effects of three factors—anode—cathode gap (A), C_2F_6 gas flow rate (B), and the power applied to the cathode (C)—on the etch rate. Each factor is run at two levels. Two replicates of a 2^3 factorial design are shown in Table 3.4, along with the resulting response variable values. The design is shown graphically in Figure 3.7.

Table 3.4 The 2^3 Factorial Design for Example 3.1

A (Gap)	B (C_2F_6 Flow)	C (Power)	Etch Rate (Å/ min)		Totals
			Replicate I	Replicate II	
-1	-1	-1	247	400	$(1) = 647$
1	-1	-1	470	446	$a = 916$
-1	1	-1	429	405	$b = 834$
1	1	-1	435	445	$ab = 880$
-1	-1	1	837	850	$c = 1687$
1	-1	1	551	670	$ac = 1221$
-1	1	1	775	865	$bc = 1640$
1	1	1	660	530	$abc = 1190$

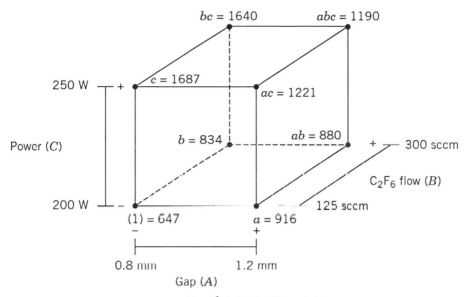

Figure 3.7 The 2^3 design for Example 3.1.

Using the totals shown in Table 3.4 for each of the eight runs in the design, we may calculate the effect estimates as follows:

$$A = \frac{1}{4n}\left[a - (1) + ab - b + ac - c + abc - bc\right]$$

$$= \frac{1}{8}[916 - 647 + 880 - 834 + 1221 - 1687 + 1190 - 1640]$$

$$= \frac{1}{8}[-601] = -75.125$$

$$B = \frac{1}{4n}\left[b + ab + bc + abc - (1) - a - c - ac\right]$$

$$= \frac{1}{8}[834 + 880 + 1640 + 1190 - 647 - 916 - 1687 - 1221]$$

$$= \frac{1}{8}[73] = 9.125$$

$$C = \frac{1}{4n}\left[c + ac + bc + abc - (1) - a - b - ab\right]$$

$$= \frac{1}{8}[1687 + 1221 + 1640 + 1190 - 647 - 916 - 834 - 886]$$

$$= \frac{1}{8}[2461] = 307.625$$

$$AB = \frac{1}{4n}[ab - a - b + (1) + abc - bc - ac + c]$$

$$= \frac{1}{8}[880 - 916 - 834 + 647 + 1190 - 1640 - 1221 + 1687]$$

$$= \frac{1}{8}[-207] = -25.875$$

$$AC = \frac{1}{4n}[(1) - a + b - ab - c + ac - bc + abc]$$

$$= \frac{1}{8}[647 - 916 + 834 - 880 - 1687 + 1221 - 1640 + 1190]$$

$$= \frac{1}{8}[-1231] = -153.875$$

$$BC = \frac{1}{4n}[(1) + a - b - ab - c - ac + bc + abc]$$

$$= \frac{1}{8}[647 + 916 - 834 - 880 - 1687 - 1221 + 1640 + 1190]$$

$$= \frac{1}{8}[-229] = -28.625$$

and

$$ABC = \frac{1}{4n}[abc - bc - ac + c - ab + b + a - (1)]$$

$$= \frac{1}{8}[1190 - 1640 - 1221 + 1687 - 880 + 834 + 916 - 647]$$

$$= \frac{1}{8}[239] = 29.875$$

The largest effects are for gap ($A = -75.125$), power ($C = 307.625$), and the gap–power interaction ($AC = -153.875$).

The analysis of variance may be used to confirm the magnitude of these effects. From Equation (3.18), the sums of squares are

$$SS_A = \frac{(-601)^2}{16} = 22,575.1$$

$$SS_B = \frac{(73)^2}{16} = 333.1$$

$$SS_C = \frac{(2461)^2}{16} = 378,532.6$$

$$SS_{AB} = \frac{(-207)^2}{16} = 2678.1$$

$$SS_{AC} = \frac{(-1231)^2}{16} = 94,710.1$$

$$SS_{BC} = \frac{(-229)^2}{16} = 3277.6$$

and

$$SS_{ABC} = \frac{(239)^2}{16} = 3570.1$$

The analysis of variance is summarized in Table 3.5. Based on the P-values shown in this table, we concluded that gap (A) and power (C) are statistically significant effects, as is the gap–power (AC) interaction.

Table 3.5 Analysis of Variance for Example 3.5

Source of Variation	Sum of Squares	Degrees of Freedom	Mean Square	F_0	P-Value
A	22,575.1	1	22,575.1	5.64	0.0448
B	333.1	1	333.1	0.08	0.7802
C	378,532.6	1	378,532.6	94.65	1.04×10^{-5}
AB	2678.1	1	2678.1	0.67	0.4369
AC	94,710.1	1	94,710.1	23.68	0.0012
BC	3277.6	1	3277.6	0.82	0.3918
ABC	3570.1	1	3570.1	0.89	0.3724
Error	31,995.5	8	3999.4		
Total	537,671.9	15			

The regression model for this process, based on the above analysis, is

$$\hat{y} = b_0 + b_1 x_1 + b_3 x_3 + b_{13} x_1 x_3$$

$$= 563.438 + \left(\frac{-75.125}{2}\right) x_1 + \left(\frac{307.625}{2}\right) x_3 + \left(\frac{-153.875}{2}\right) x_1 x_3$$

$$= 563.438 - 37.563 x_1 + 153.813 x_3 - 76.938 x_1 x_3$$

where the variables x_1 and x_3 represent the design factors A and C, respectively, on the coded $(-1, +1)$ scale. This regression model can be used to generate predicted values at each corner of the design, from which residuals can be obtained. The reader should obtain these residuals and graphically analyze them.

The factor effect estimates (and the regression model) indicate that etch rate increases as the gap (A) decreases and as the power (C) increases, and that these factors interact. The two-factor interaction graph, shown in Figure 3.8, is helpful in the practical interpretation of the results. This graph was constructed by plotting average etch rate version gap (A) for each power setting and connecting the points for the low- and high-power settings to give the two curves shown in the figure. Inspection of the interaction graph indicates that changes in the anode–cathode gap produces a much larger change in etch rate at the high-power setting than at the low-power setting.

Figure 3.9 is a response surface plot and contour plot of etch rate as a function of gap and power setting. These plots were obtained from the fitted

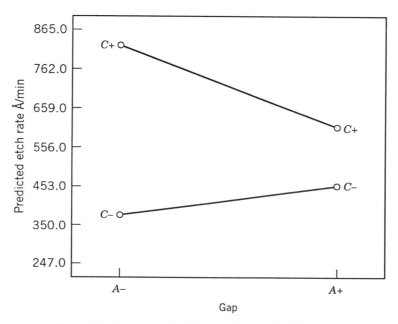

Figure 3.8 Gap–power (AC) interaction graph, Example 3.1.

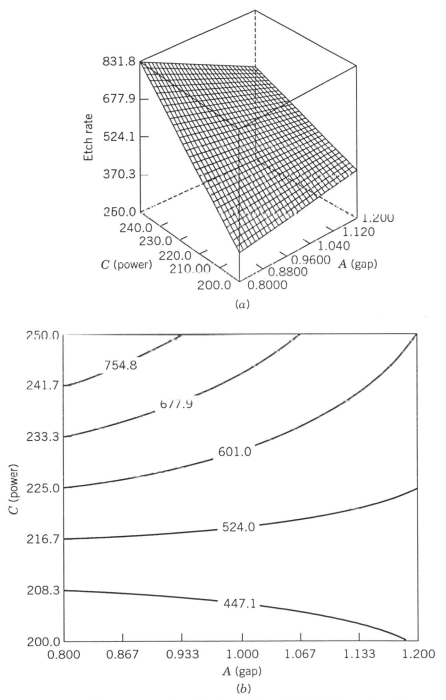

Figure 3.9 (*a*) Response surface of etch rate, Example 3.1. (*b*) The contour plot.

model. The impact of the strong interaction on this process is very clear; notice that the response surface is not a plane (or that the lines in the contour plot are not parallel straight lines). In processes such as this one, the engineer is usually trying to select the appropriate factor settings (in this case gap and power) to obtain a desired etch rate. The response surface and contour plot would be most helpful for this analysis.

Other Methods for Judging the Significance of Effects

The analysis of variance is a formal way to determine which factor effects are nonzero. There are two other methods that are useful. In the first method, we calculate the **standard error** of the effects and compare the magnitude of the effects to their standard errors. The second method, which we will illustrate in Section 3.5, uses **normal probability plots** to assess the importance of the effects.

The standard error of an effect is easy to find. If we assume that there are n replicates at each of the 2^k runs or treatment combinations in the design, and if $y_{i1}, y_{i2}, \ldots, y_{in}$ are the observations at the ith run, then

$$S_i^2 = \frac{1}{n-1} \sum_{j=1}^{n} \left(y_{ij} - \bar{y}_i \right)^2, \qquad i = 1, 2, \ldots, 2^k$$

is an estimate of the variance at the ith run. The 2^k variance estimates can be combined to give an overall variance estimate:

$$S^2 = \frac{1}{2^k(n-1)} \sum_{i=1}^{2^k} \sum_{j=1}^{n} \left(y_{ij} - \bar{y}_i \right)^2$$

This is also the variance estimate given by the error mean square in the analysis of variance. The variance of each effect estimate is

$$\text{Var(Effect)} = \text{Var}\left(\frac{\text{Contrast}}{n2^{k-1}} \right)$$

$$= \frac{1}{\left(n2^{k-1} \right)^2} \text{Var(contrast)}$$

Each contrast is a linear combination of 2^k treatment totals, and each total consists of n observations. Therefore,

$$\text{Var(Contrast)} = n2^k \sigma^2$$

and the variance of an effect is

$$\text{Var(Effect)} = \frac{1}{\left(n2^{k-1} \right)^2} n2^k \sigma^2$$

$$= \frac{1}{n2^{k-2}} \sigma^2 \qquad\qquad (3.19)$$

The estimated standard error would be found by replacing σ^2 by its estimate S^2 and taking the square root of Equation (3.19).

To illustrate this method, consider the etch rate experiment in Example 3.1. The mean square error is $MS_E = 3999.4$. Therefore, the standard error of each effect is (using $S^2 = MS_E$)

$$se(\text{Effect}) = \sqrt{\frac{1}{n2^{k-2}}S^2}$$

$$= \sqrt{\frac{1}{2(2^{3-2})}3999.4}$$

$$= 31.62$$

Two standard error limits on the effect estimates are then

A:	-75.125 ± 63.24
B:	9.125 ± 63.24
C:	307.625 ± 63.24
AB:	-25.875 ± 63.24
AC:	-153.875 ± 63.24
BC:	-28.625 ± 63.24
ABC:	29.875 ± 63.24

These intervals are approximate 95% confidence intervals. This analysis indicates that A, C, and the AC interaction are important effects, because they are the only factor effect estimates for which the intervals do not include zero.

3.4   THE GENERAL 2^k DESIGN

The methods of analysis that we have presented thus far may be generalized to the case of a 2^k factorial design; that is, a design with k factors each at two levels. The statistical model for a 2^k design would include k main effects, $\binom{k}{2}$ two-factor interactions, $\binom{k}{3}$ three-factor interactions, ..., and one k-factor interaction. That is, for a 2^k design the complete model would contain $2^k - 1$ effects. The notation introduced earlier for treatment combinations is also used here. For example, in a 2^5 design abd denotes the treatment combination with factors A, B, and D at the high level and factors C and E at the low level. The treatment combinations may be written in standard order by introducing the factors, one at a time, with each new factor being successively combined with those that precede it. For example, the standard order for a 2^4 design is (1), a, b, ab, c, ac, bc, abc, d, ad, bd, abd, cd, acd, bcd, and $abcd$.

To estimate an effect or to compute the sum of squares for an effect, we must first determine the contrast associated with that effect. This can always be done by using a table of plus and minus signs, such as Table 3.2 or Table 3.3. However, for large values of k, this is awkward. An alternative to this procedure is the tabular algorithm devised by Yates (1937) for calculating the contrasts in a 2^k design. For a description of this algorithm and an example, see Montgomery (1991a). While using this algorithm can be simpler than using the table of plus and minus signs, it is still awkward, and it is not appropriate if the number of observations at each treatment combination are not the same or if some observations are missing. Most modern computer programs generate the effect estimates, the regression coefficients, and the sums of squares for the analysis of variance using the method of least squares.

3.5 A SINGLE REPLICATE OF THE 2^k DESIGN

For even a moderate number of factors, the total number of treatment combinations in a 2^k factorial design is large. For example, a 2^5 design has 32 treatment combinations, a 2^6 design has 64 treatment combinations, and a 2^{10} design has 1024 treatment combinations. Because resources are usually limited, the number of replicates that the experimenter can employ may be restricted. Frequently, available resources only allow a single replicate of the design to be run, unless the experimenter is willing to omit some of the original factors.

A **single replicate** of a 2^k design is sometimes called an **unreplicated factorial.** With only one replicate, there is no estimate of error. One approach to the analysis of an unreplicated factorial is to assume that certain high-order interactions are negligible and combine their mean squares to estimate the error. This is an appeal to the **sparsity of effects principle**; that is, most systems are dominated by some of the main effects and low-order interactions, and most high-order interactions are negligible.

When analyzing data from unreplicate factorial designs, occasionally very-high-order interactions occur. The use of an error mean square obtained by pooling high-order interactions is inappropriate in these cases. A method of analysis attributed to Daniel (1959) provides a simple way to overcome this problem. Daniel suggests plotting the estimates of the effects on normal probability paper. The effects that are negligible are normally distributed, with mean zero and variance σ^2, and will tend to fall along a straight line on this plot, whereas significant effects will have nonzero means and will not lie along the straight line. We will illustrate this method in the following example.

Table 3.6   Pilot Plant Filtration Rate Experiment

Run Number	Factor A	B	C	D	Treatment Combination	Filtration Rate (gal/hr)
1	−	−	−	−	(1)	45
2	+	−	−	−	a	71
3	−	+	−	−	b	48
4	+	+	−	−	ab	65
5	−	−	+	−	c	68
6	+	−	+	−	ac	60
7	−	+	+	−	bc	80
8	+	+	+	−	abc	65
9	−	−	−	+	d	43
10	+	−	−	+	ad	100
11	−	+	−	+	bd	45
12	+	+	−	+	abd	104
13	−	−	+	+	cd	75
14	+	−		+	acd	86
15	−	+	+	+	bcd	70
16	+	+	+	+	abcd	96

Example 3.2   A Single Replicate of the 2^4 Design [from Montgomery (1991a)]   A chemical product is produced in a pressure vessel. A factorial experiment is carried out in the pilot plant to study the factors thought to influence the filtration rate of this product. The four factors are temperature (A), pressure (B), concentration of formaldehyde (C), and stirring rate (D). Each factor is present at two levels, and the data obtained from a single replicate of the 2^4 experiment are shown in Table 3.6 and Figure 3.10. The 16 runs are made in random order. The process engineer is interested in maximizing the filtration rate. Current process conditions give filtration rates

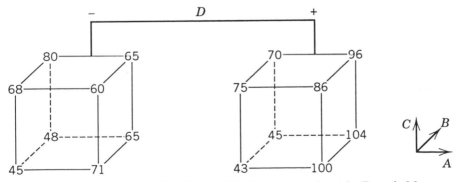

Figure 3.10   Data from the pilot plant filtration rate experiment for Example 3.2.

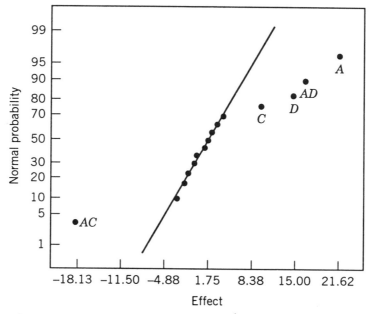

Figure 3.11 Normal probability plot for the 2^4 factorial in Example 3.2.

of around 75 gal/hr. The process also currently uses the concentration of formaldehyde, factor C, at the high level. The engineer would like to reduce the formaldehyde concentration as much as possible but has been unable to do so because it always results in lower filtration rates.

We will begin the analysis of this data by plotting the effect estimates on normal probability paper. The table of plus and minus signs for the contrast constants for the 2^4 design are shown in the Table 3.7. From these contrasts, we may estimate the 15 factorial effect estimates shown at the foot of each column in Table 3.7. The normal probability plot of these effects is shown in Figure 3.11. All of the effects that lie along the line are negligible, whereas the large effects are far from the line. The important effects that emerge from this analysis are the main effects of A, C, and D and the AC and AD interactions. Table 3.8 summarizes the analysis of variance for this design. In this analysis of variance we have pooled the nonsignificant effects to form the error term. The F-tests reveal that the effects of A, C, D and the AC and AD interactions are large.

The main effects of A, C, and D are plotted in Figure 3.12(a). All three effects are positive, and if we considered only these main effects, we would run all three factors at the high level to maximize the filtration rate. However, it is always necessary to examine any interactions that are important. Remember that main effects do not have much meaning when they are involved in significant interactions.

Table 3.7 Contrast Constants and Effect Estimates for the 2⁴ Design, Example 3.2

Observations	A	B	AB	C	AC	BC	ABC	D	AD	BD	ABD	CD	ACD	BCD	ABCD
(1) = 45	−	−	+	−	+	+	−	−	+	+	−	+	−	−	+
a = 71	+	−	−	−	−	+	+	−	−	+	+	+	+	−	−
b = 48	−	+	−	−	+	−	+	−	+	−	+	+	−	+	−
ab = 65	+	+	+	−	−	−	−	−	−	−	−	+	+	+	+
c = 68	−	−	+	+	−	−	+	−	+	+	−	−	+	+	−
ac = 60	+	−	−	+	+	−	−	−	−	+	+	−	−	+	+
bc = 80	−	+	−	+	−	+	−	−	+	−	+	−	+	−	+
abc = 65	+	+	+	+	+	+	+	−	−	−	−	−	−	−	−
d = 43	−	−	+	−	+	+	−	+	−	−	+	−	+	+	−
ad = 100	+	−	−	−	−	+	+	+	+	−	−	−	−	+	+
bd = 45	−	+	−	−	+	−	+	+	−	+	−	−	+	−	+
abd = 104	+	+	+	−	−	−	−	+	+	+	+	−	−	−	−
cd = 75	−	−	+	+	−	−	+	+	−	−	+	+	−	−	+
acd = 86	+	−	−	+	+	−	−	+	+	−	−	+	+	−	−
bcd = 70	−	+	−	+	−	+	−	+	−	+	−	+	−	+	−
abcd = 96	+	+	+	+	+	+	+	+	+	+	+	+	+	+	+
Effect Estimates	21.625	3.125	0.125	9.875	−18.125	2.375	1.875	14.625	16.625	−0.375	4.125	−1.125	−1.625	−2.625	1.375

107

Table 3.8   Analysis of Variance, Example 3.2

Source of Variation	Sum of Squares	Degrees of Freedom	Mean Square	F_0	P-Value
A	1870.563	1	1870.563	95.86	1.93×10^{-6}
C	390.062	1	390.062	19.99	0.0012
D	855.562	1	855.562	43.85	0.0001
AC	1314.062	1	1314.062	67.34	9.42×10^{-6}
AD	1105.563	1	1105.563	56.66	2.00×10^{-5}
Error	195.125	10	19.513		
Total	5730.937	15			

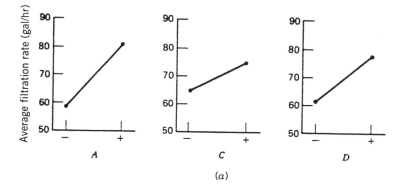

(a)

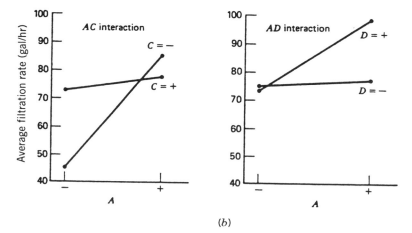

(b)

Figure 3.12   Main effect and interaction plots for Example 3.2. (a) Main effect plots. (b) Interaction plots.

Table 3.9 **Analysis of Variance for the Pilot Plant Filtration Rate Experiment in A, C, and D**

Source of Variation	Sum of Squares	Degrees of Freedom	Mean Square	F_0	P-Value
A	1870.56	1	1870.56	83.36	1.67×10^{-5}
C	390.06	1	390.06	17.38	0.0031
D	855.56	1	855.56	38.13	0.0003
AC	1314.06	1	1314.06	58.56	0.0001
AD	1105.56	1	1105.56	49.27	0.0001
CD	5.06	1	5.06	0.23	0.6476
ACD	10.56	1	10.56	0.47	0.5121
Error	179.52	8	22.44		
Total	5730.94	15			

The AC and AD interactions are plotted in Figure 3.12(b). These interactions are the key to solving the problem. Note from the AC interaction that the temperature effect is very small when the concentration is at the high level and very large when the concentration is at the low level, with the best results obtained with low concentration and high temperature. The AD interaction indicates that stirring rate D has little effect at low temperature but has a large positive effect at high temperature. Therefore, the best filtration rates would appear to be obtained when A and D are at the high level and C is at the low level. This would allow the reduction of the formaldehyde concentration to a lower level, another objective of the experimenter.

Design Projection

Another interpretation of the data in Figure 3.11 is possible. Because B (pressure) is not significant and all interactions involving B are negligible, we may discard B from the experiment so that the design becomes a 2^3 factorial in A, C, and D with two replicates. This is easily seen from examining only columns A, C, and D in Table 3.6 and noting that those columns form two replicates of a 2^3 design. The analysis of variance for the data using this simplifying assumption is summarized in Table 3.9. The conclusions that we would draw from this analysis are essentially unchanged from those of Example 3.2. Note that by projecting the single replicate of the 2^4 into a replicated 2^3, we now have both an estimate of the ACD interaction and an estimate of error based on replication.

The concept of projecting an unreplicated factorial into a replicated factorial in fewer factors is very useful. In general, if we have a single replicate of a 2^k design and if $h(h < k)$ factors are negligible and can be dropped, then the original data correspond to a full two-level factorial in the remaining $k - h$ factors with 2^h replicates.

The Regression Model and Diagnostic Checking

The usual diagnostic checks should be applied to the residuals from a 2^k design. Our analysis indicates that the only significant effect are $A = 21.625$, $C = 9.875$, $D = 14.625$, $AC = -18.125$, and $AD = 16.625$. If this is true, then the estimated filtration rate over the region spanned by the design is given by

$$\hat{y} = 70.06 + \left(\frac{21.625}{2}\right)x_1 + \left(\frac{9.875}{2}\right)x_3 + \left(\frac{14.625}{2}\right)x_4$$
$$- \left(\frac{18.125}{2}\right)x_1 x_3 + \left(\frac{16.625}{2}\right)x_1 x_4$$

where 70.06 is the average response and the coded variables x_1, x_3, x_4 take on values in the interval from -1 to $+1$. The predicted filtration rate at the run where all three factors are at the low level is

$$\hat{y} = 70.06 + \left(\frac{21.625}{2}\right)(-1) + \left(\frac{9.875}{2}\right)(-1) + \left(\frac{14.625}{2}\right)(-1)$$
$$- \left(\frac{18.125}{2}\right)(-1)(-1) + \left(\frac{16.625}{2}\right)(-1)(-1)$$
$$= 46.22$$

Because the observed value is 45, the residual is $e = y - \hat{y} = 45 - 46.22 = -1.22$. The values of y, $\hat{y}$, and $e = y - \hat{y}$ for all 16 observations follow:

	y	$\hat{y}$	$e = y - \hat{y}$
(1)	45	46.22	-1.22
a	71	69.39	1.61
b	48	46.22	1.78
ab	65	69.39	-4.39
c	68	74.23	-6.23
ac	60	61.14	-1.14
bc	80	74.23	5.77
abc	65	61.14	3.86
d	43	44.22	-1.22
ad	100	100.65	3.35
bd	45	44.22	0.78
abd	104	100.65	3.35
cd	75	72.23	2.77
acd	86	92.40	-6.40
bcd	70	72.23	-2.23
$abcd$	96	92.40	3.60

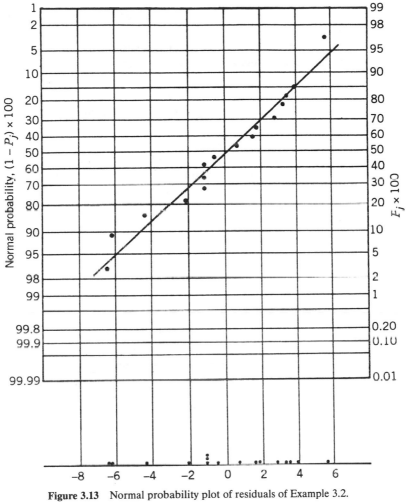

Figure 3.13 Normal probability plot of residuals of Example 3.2.

A normal probability plot of these residuals is shown in Figure 3.13. The points on this plot lie reasonably close to a straight line, lending support to our conclusion that A, C, D, AC, and AD are the only significant effects and that the underlying assumptions of the analysis are satisfied.

3.6 THE ADDITION OF CENTER POINTS TO THE 2^k DESIGN

A potential concern in the use of two-level factorial designs is the assumption of linearity in the factor effects. Of course, perfect linearity is unnecessary, and the 2^k system will work quite well even when the linearity assumption holds only very approximately. There is, however, a method of replicating certain points in a 2^k factorial that will provide **protection against curvature** as well as allow an **independent estimate of error** to be obtained. The method consists of adding **center points** to the 2^k design. These consists of n_C replicates run at the point $x_i = 0$ $(i = 1, 2, \ldots, k)$. One important reason for adding the replicate runs at the center is that center points do not impact the usual effect estimates in a 2^k design. We assume that the k factors are quantitative.

To illustrate the approach, consider a 2^2 design with one observation at each of the factorial points $(-, -)$, $(+, -)$, $(-, +)$, and $(+, +)$ and n_C observations at center points $(0, 0)$. Figure 3.14 illustrates the situation. Let $\bar{y}_F$ be the average of the four runs at the four factorial points and let $\bar{y}_C$ be the average of the n_C runs at the center point. If the difference $\bar{y}_F - \bar{y}_C$ is large, then curvature is present.

A single-degree-of-freedom sum of squares for curvature is given by

$$SS_{\text{Curvature}} = \frac{n_F n_C (\bar{y}_F - \bar{y}_C)^2}{n_f + n_C} \tag{3.20}$$

where, in general, n_F is the number of factorial design points. This quantity may be compared to the error mean square to test for curvature. More specifically, when points are added to the center of the 2^k design, then the model we may entertain is the second-order response surface model

$$y = \beta_0 + \sum_{j=1}^{k} \beta_j x_j + \sum_{i<j} \sum \beta_{ij} x_i x_j + \sum_{j=1}^{k} \beta_{jj} x_j^2 + \varepsilon$$

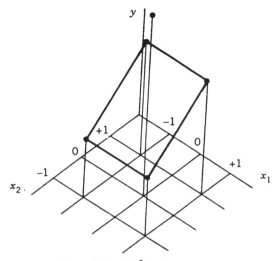

Figure 3.14   A 2^2 design with center.

where β_{ij} are pure quadratic effects. We can show that the test for curvature actually tests the hypotheses

$$H_0: \quad \sum_{j=1}^{k} \beta_{jj} = 0$$

$$H_1: \quad \sum_{j=1}^{k} \beta_{jj} \neq 0$$

Furthermore, if the factorial points in the design are unreplicated, one may use the n_C center points to construct an estimate of error with $n_C - 1$ degrees of freedom.

Example 3.3 A chemical engineer is studying the yield of a process. There are two variables of interest; reaction time and reaction temperature. Because she is uncertain about the assumption of linearity over the region of exploration, the engineer decides to conduct a 2^2 design (with a single replicate of each factorial run) augmented with five center points. The design and the yield data are shown in Figure 3.15.

Table 3.10 summarizes the analysis of variance for this experiment. The mean square error is calculated from the center points as follows:

$$MS_E = \frac{SS_E}{n_C - 1}$$

$$= \frac{\sum_{\text{Center points}} (y_i - \bar{y}_c)^2}{n_c - 1}$$

$$= \frac{\sum_{i=1}^{5} (y_i - 40.46)^2}{4}$$

$$= \frac{0.1720}{4}$$

$$= 0.0430$$

The average of the points in the factorial portion of the design is $\bar{y}_F = 40.425$, and the average of the points at the center is $\bar{y}_C = 40.46$. The difference $\bar{y}_F - \bar{y}_C = 40.425 - 40.46 = -0.035$ appears to be small. The curvature sum of squares in the analysis of variance table is computed from Equation (3.20)

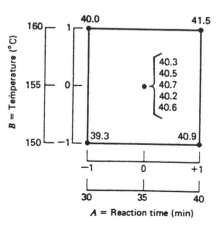

Figure 3.15 The 2^2 design with five center points for Example 3.3.

Table 3.10   Analysis of Variance of Example 3.3

Source of Variability	Sum of Squares	Degrees of Freedom	Mean Square	F_0
A (time)	2.4025	1	2.4025	55.87
B (temperature)	0.4225	1	0.4225	9.83
AB	0.0025	1	0.0025	0.06
Curvature	0.0027	1	0.0027	0.006
Error	0.1720	4	0.0430	
Total	3.0022	8		

as follows:

$$SS_{\text{Curvature}} = \frac{n_F n_C (\bar{y}_F - \bar{y}_C)^2}{n_F + n_C}$$

$$= \frac{(4)(5)(-0.035)^2}{4 + 5}$$

$$= 0.0027$$

The analysis of variance indicates that both factors exhibit significant main effects, that there is no interaction, and that there is no evidence of curvature in the response over the region of exploration. That is, the null hypothesis $H_0: \sum_{j=1}^{2} \beta_{jj} = 0$ cannot be rejected.

3.7   BLOCKING IN THE 2^k FACTORIAL DESIGN

There are many situations in which it is impossible to perform all of the runs in a factorial experiment under homogeneous conditions. As examples, these conditions might be a single time period (such as one day), a single batch of homogeneous raw material, or using only one operator, and so on. In other cases, it might be desirable to deliberately vary the experimental conditions to ensure that the treatments are equally effective (one might say robust) across many situations likely to be encountered in practice. For example, a chemical engineer may run a pilot plant experiment with several batches of raw material because he knows that different raw material batches of different quality grades will be used in the full-scale process.

The design technique used in these situations is called **blocking**. In this section we show how to arrange the 2^k factorial design in blocks.

3.7.1 Blocking the Replicated Design

Suppose that the 2^k factorial design has been replicated n times. A simple way to incorporate nonhomogeneous conditions in this situation is to consider each set of these conditions as a block and to run each replicate of the design in a separate block. The runs in each block would be made in random order.

Example 3.4 Consider the chemical process experiment described in Section 3.2. Suppose that only four experimental trials could be easily made from a single batch of raw material. Therefore, four batches of raw material will be required in order to run all four replicates. Table 3.11 shows the design, assuming that each replicate is run in a single batch of material. Within each material batch, the four runs are made in random order.

The analysis of variance for this design is shown in Table 3.12. All of the sums of squares in this table are calculated as in a standard (unblocked) 2^k factorial, except the sum of squares for blocks. If we let B_i ($i = 1, 2, 3, 4$) represent the four block totals (see Table 3.11), then

$$
SS_{\text{Blocks}} = \sum_{i=1}^{4} \frac{B_i^2}{4} - \frac{y_{...}^2}{16}
$$

$$
= \frac{(588)^2 + (590)^2 + (589)^2 + (580)^2}{4} - \frac{(2347)^2}{16}
$$

$$
= 15.6875
$$

Table 3.11   Chemical Process Experiment in Four Blocks

	Block 1	Block 2	Block 3	Block 4
	(1) = 145	(1) = 148	(1) = 147	(1) = 140
	a = 158	a = 152	a = 155	a = 152
	b = 135	b = 138	b = 141	b = 139
	ab = 150	ab = 152	ab = 146	ab = 149
Block totals	588	590	589	580

**Table 3.12   Analysis of Variance for the Chemical Process Experiment
in Four Blocks, Example 3.4**

Source of Variation	Sum of Squares	Degrees of Freedom	Mean Square	F_0	P-Value
Blocks	15.6875	3	5.2292		
A	410.0625	1	410.0625	43.64	0.0001
B	138.0625	1	138.0625	14.69	0.0040
AB	3.0625	1	3.0625	0.33	0.5820
Error	84.5625	9	9.3959		
Total	651.4375	15			

There are three degrees of freedom among the four blocks. Table 3.12
indicates that the conclusions from this experiment are identical to those in
Section 3.2 and that the block effect is relatively small.

3.7.2   Confounding in the 2^k Design

In many situations, it is impossible to perform a complete replicate of a
factorial design in one block. **Confounding** is a design technique for arrang-
ing a complete factorial experiment in blocks, where the block size is smaller
than the number of treatment combinations in one replicate. The technique
causes information about certain treatment effects (usually high-order interactions) to be indistinguishable from, or **confounded** with, blocks. In this
section we introduce confounding systems for the 2^k factorial design. Note
that even though the designs presented are incomplete block designs because
each block does not contain all the treatments or treatment combinations,
the special structure of the 2^k factorial system allows a simplified method of
design construction and analysis.

We consider the construction and analysis of the 2^k factorial design is 2^p
incomplete blocks, where $p < k$. Consequently, these designs can be run in
two blocks, four blocks, eight blocks, and so on.

Two Blocks

Suppose that we wish to run a single replicate of the 2^2 design. Each of the
$2^2 = 4$ treatment combinations requires some quantity of raw material, for
example, and each batch of raw material is only large enough for two
treatment combinations to be tested. Thus, two batches of raw material are
required. If batches of raw material are considered as blocks, then we must
assign two of the four treatment combinations to each block.

Figure 3.16 shows one possible design for this problem. The geometric
view, Figure 3.16(a), indicates that treatment combinations on opposing
diagonals are assigned to different blocks. Notice from Figure 3.16(b) that
block 1 contains the treatment combinations (1) and ab and that block 2

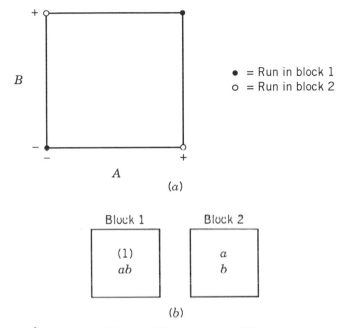

Figure 3.16   A 2^2 design in two blocks. (*a*) Geometric view. (*b*) Assignment of the four runs to two blocks.

contains *a* and *b*. Of course the order in which the treatment combinations are run within a block is randomly determined. We would also randomly decide which block to run first. Suppose we estimate the main effect of *A* and *B* just as if no blocking had occurred. The effect estimates are

$$A = \tfrac{1}{2}[ab + a - b - (1)]$$
$$B = \tfrac{1}{2}[ab + b - a - (1)]$$

Note that both of the effect estimates *A* and *B* are unaffected by blocking because in each estimate there is one plus and one minus treatment combination from each block. That is, any difference between block 1 and block 2 will cancel out.

Now consider the *AB* interaction

$$AB = \tfrac{1}{2}[ab + (1) - a - b]$$

Because the two treatment combinations with the plus sign [*ab* and (1)] are in block 1 and the two with the minus sign [*a* and *b*] are in block 2, the block effect and the *AB* interaction are identical. That is, *AB* is confounded with blocks.

Table 3.13   Table of Plus and Minus Signs for the 2^2 Design

Treatment Combination	Factorial Effect			
	I	A	B	AB
(1)	+	−	−	+
a	+	+	−	−
b	+	−	+	−
ab	+	+	+	+

The reason for this is apparent from the table of plus and minus signs for the 2^2 design. This was originally given as Table 3.2, but for convenience it is reproduced as Table 3.13. From this table, we see that all treatment combinations that have a plus on AB are assigned to block 1, whereas all treatment combinations that have a minus sign on AB are assigned to block 2. This approach can be used to confound any effect (A, B, or AB) with blocks. For example, if (1) and b had been assigned to block 1 and a and ab to block 2, then the main effect A would have been confounded with blocks. The usual practice is to confound the highest-order interaction with blocks.

This scheme can be used to confound any 2^k design in two blocks. As a second example, consider a 2^3 design run in two blocks. Suppose we wish to confound the three-factor interaction ABC with blocks. From the table of plus and minus signs shown in Table 3.3, we assign the treatment combinations that are minus on ABC to block 1 and those that are plus on ABC to block 2. The resulting design is shown in Figure 3.17. Once again, we emphasize that the treatment combinations within a block are run in random order.

Example 3.5   Consider the situation described in Example 3.2. Recall that four factors—temperature (A), pressure (B), concentration of formaldehyde (C), and stirring rate (D)—are studied in a pilot plant to determine their effect on product filtration rate. Suppose now that the $2^4 = 16$ treatment combinations cannot all be run on one day. The experimenter can make 8 runs per day, so a 2^4 design confounded in two blocks seems appropriate. It is logical to confound the highest-order interaction $ABCD$ with blocks. The design is shown in Figure 3.18. Because the block totals are 566 and 555, the sum of squares for blocks is

$$SS_{\text{Blocks}} = \frac{(566)^2 + (555)^2}{8} - \frac{(1121)^2}{16} = 7.5625$$

which is identical to the sum of squares for $ABCD$. A normal probability plot

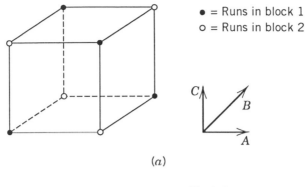

● = Runs in block 1
○ = Runs in block 2

(a)

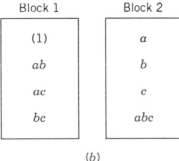

Block 1	Block 2
(1)	a
ab	b
ac	c
bc	abc

(b)

Figure 3.17 A 2^3 design in two blocks. (a) Geometric view. (b) Assignment of the 8 runs to two blocks.

of the remaining effects indicates that only factors A, C, D and the AC and CD interactions are important. Therefore, the error sum of squares is formed by pooling the remaining effects. The resulting analysis of variance is shown in Table 3.14. The practical conclusions are identical to those found in the original Example 3.2.

Four or More Blocks
It is possible to construct 2^k factorial designs confounded in four blocks of $2^k / 4 = 2^{k-2}$ = observations each. These designs are particularly useful in situations where the number of factors is moderately large, say $k \geq 4$, and block sizes are relatively small.

As an example, consider the 2^5 design. If each block will hold only eight runs, then four blocks must be used. The construction of this design is relatively straightforward. Select two effects to be confounded with blocks, say ADE and BCE. Then construct a table of plus and minus signs for the 2^5 design. If you consider only the ADE and BCE columns in such a table, there will be exactly four different sign patterns for the 32 observations in

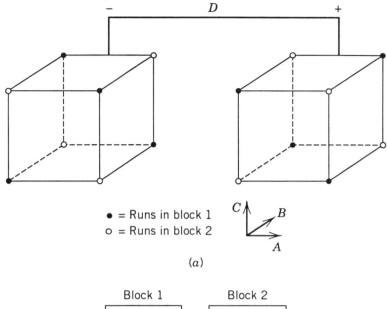

● = Runs in block 1
○ = Runs in block 2

(a)

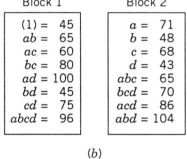

(b)

Figure 3.18 The 2^4 design in two blocks for Example 3.5. (a) Geometric view. (b) Assignment of the 16 runs to two blocks.

Table 3.14 Analysis of Variance, Example 3.5

Source of Variation	Sum of Squares	Degrees of Freedom	Mean Squares	F_0	P-Value
Blocks (*ABCD*)	7.563	1			
A	1870.563	1	1870.563	89.76	5.60×10^{-6}
C	390.062	1	390.062	18.72	0.0019
D	855.562	1	855.562	41.05	0.0001
AC	1314.062	1	1314.062	63.05	2.35×10^{-5}
AD	1105.562	1	1105.563	53.05	4.65×10^{-5}
Error	187.562	9	20.840		
Total	5730.937	15			

	Block 1		Block 2		Block 3		Block 4	
ADE	−		+		−		+	
BCE	−		−		+		+	

Block 1		Block 2		Block 3		Block 4	
(1)	abc	a	bc	b	abce	e	abcde
ad	ace	d	abde	abd	ae	ade	bd
bc	cde	abc	ce	c	bcde	bce	ac
abcd	bde	bcd	acde	acd	de	ab	cd

Figure 3.19   The 2^5 design in four blocks with ADE, BCE, and $ABCD$ confounded.

these two columns. The sign patterns are as follows:

ADE	BCE
−	−
+	−
−	+
+	+

The design would be constructed by placing all the runs that are $(-, -)$ in block 1, those that are $(+, -)$ in block 2, those that are $(-, +)$ in block 3, and those that are $(+, +)$ in block 4. The design is shown in Figure 3.19.

With a little reflection we realize that another effect in addition to ADE and BCE must be confounded with blocks. Because there are four blocks with three degrees of freedom between them, and because ADE and BCE have only one degree of freedom each, clearly an additional effect with one degree of freedom must be confounded. This effect is the *generalized interaction* of ADE and BCE, which is defined as the product of ADE and BCE modulus 2. Thus, in our example the generalized interaction $(ADE)(BCE) = ABCDE^2 = ABCE$ is also confounded with blocks. It is easy to verify this by referring to a table of plus and minus signs for the 2^5 design. Inspection of such a table would reveal that the treatment combinations are assigned to the blocks as follows:

Treatment Combinations in	Sign on ADE	Sign on BCE	Sign on $ABCD$
Block 1	−	−	+
Block 2	+	−	−
Block 3	−	+	−
Block 4	+	+	+

Notice that the product of signs of any two effects for a particular block (e.g., ADE and BCE) yields the sign of the other effect for that block (in this case $ABCD$). Thus ADE, BCE, and $ABCD$ are all confounded with blocks.

The general procedure for constructing a 2^k design confounded in four blocks is to choose two effects to generate the blocks, automatically confounding a third effect that is the generalized interaction of the first two. Then, the design is constructed by assigning the treatment combinations that have the sign pattern $(-, -), (+, +), (+, -), (-, +)$ in the two original columns to four blocks. In selecting effects to be confounded with blocks, care must be exercised to obtain a design that does not confound effects that may be of interest. For example, in a 2^5 design we might choose to confound $ABCDE$ and ABD, which automatically confounds CE, an effect that is probably of interest. A better choice is to confound ADE and BCE, which automatically confounds $ABCD$. It is preferable to sacrifice information on the three-factor interactions ADE and BCE instead of the two-factor interaction CE.

We could extend this procedure to even larger numbers of blocks. It is possible to construct a 2^k factorial design in 2^p blocks (where $p < k$) of 2^{k-p} treatment combinations each. To do this, select p independent effects to confound with blocks, where "independent" means that no chosen effect is the generalized interaction of the others. The plus and minus sign patterns on these p columns are used to assign the treatment combinations to the blocks. In addition, exactly $2^p - p - 1$ other effects will be confounded with blocks, these effects being the generalized interactions of the p effects initially chosen. The choice of these p effects is critical, because the confounding pattern in the design depends on them. Table 3.15, from Montgomery (1991a), presents a selection of useful designs for $k \le 7$ factors. To illustrate the use of this table, suppose we want to run a 2^6 design in $2^3 = 8$ blocks of $2^{6-3} = 8$ runs each. Table 3.15 indicates that we should choose $ABEF$, $ABCD$, and ACE as the $p = 3$ independent effects to construct the blocks. Then $2^p - p - 1 = 2^3 - 3 - 1 = 4$ other effects are also confounded. These are the generalized interaction of these three, or

$$(ABEF)(ABCD) = A^2B^2CDEF \quad = CDEF$$

$$(ABEF)(ACE) = A^2BCE^2F \quad = BCF$$

$$(ABCD)(ACE) = A^2BC^2ED \quad = BDE$$

$$(ABEF)(ABCD)(ACE) = A^3B^2C^2DE^2F = ADF$$

Notice that this design would confound three four-factor interactions and four three-factor interactions with blocks. This is the best possible blocking arrangement for this design.

There are several computer programs for constructing and analyzing two-level factorial designs. Most of these programs have the capability to generate the confounded versions of these designs discussed in this section. The authors of these programs have typically used Table 3.15, or a similar one, as the basis for their designs.

Table 3.15 Suggested Blocking Arrangements for the 2^k Factorial Design

Number of Factors, k	Number of Blocks, 2^p	Block Size, 2^{k-p}	Effects Chosen to Generate the Blocks	Interactions Confounded with Blocks
3	2	4	ABC	ABC
	4	2	AB, AC	AB, AC, BC
4	2	8	ABCD	ABCD
	4	4	ABC, ACD	ABC, ACD, BD
	8	2	AB, BC, CD	AB, BC, CD, AC, BD, AD, ABCD
5	2	16	ABCDE	ABCDE
	4	8	ABC, CDE	ABC, CDE, ABDE
	8	4	ABE, BCE, CDE	ABE, BCE, CDE, AC, ABCD, BD, ADE
	16	2	AB, AC, CD, DE	All 2-factor and 4-factor interactions (15 effects)
6	2	32	ABCDEF	ABCDEF
	4	16	ABCF, CDEF	ABCF, CDEF, ABDE
	8	8	ABEF, ABCD, ACE	ABEF, ABCD, ACE, BCF, BDE, CDEF, ADF
	16	4	ABF, ACF, BDF, DEF	ABE, ACF, BDF, DEF, BC, ABCD, ABDE, AD, ACDE, CE, BDF, BCDEF, ABCEF, AEF, BE
	32	2	AB, BC, CD, DE, EF	All 2-factor, 4-factor, and 6-factor interactions (31 effects)
7	2	64	ABCDEFG	ABCDEFG
	4	32	ABCFG, CDEFG	ABCFG, CDEFG, ABDE
	8	16	ABC, DEF, AFG	ABC, DEF, AFG, ABCDEF, DCFG, ADEG, BCDEFG
	16	8	ABD, EFG, CDE, ADG	ABCD, EFG, CDE, ADG, ABCDEFG, ABE, BCG, CDFG, ADEF, ACEG, ABFG, BCEF, BDEG, ACF, BDF
	32	4	ABG, BCG, CDG, DEG, EFG	ABG, BCG, CDG, DEG, EFG, AC, BD, DE, DF, AE, BE, ABCD, ABDE, ABEF, BCDE, BCEF, CDEF, ABCDEFG, ADG, ACDEG, ACEFG, AEDFG, ABCEG, BEG, BDEFG, CFG, ADEF, ACDF, ABCF, AFG
	64	2	AB, BC, CD, DE, EF, FG	All 2-factor, 4-factor, and 6-factor interactions (63 effects)

123

EXERCISES

3.1 A router is used to cut registration notches on a printed circuit board. The vibration level at the surface of the board as it is cut is considered to be a major source of dimensional variation in the notches. Two factors are thought to influence vibration: bit size (A) and cutting speed (B). Two bit sizes ($\frac{1}{16}$ and $\frac{1}{8}$ inch) and two speeds (40 and 90 rpm) are selected, and four boards are cut at each set of conditions shown below. The response variable is vibration measured as the resultant vector of three accelerometers (x, y, and z) on each test circuit board.

A	B	Treatment Combination	Replicate I	II	III	IV
−	−	(1)	18.2	18.9	12.9	14.4
+	−	a	27.2	24.0	22.4	22.5
−	+	b	15.9	14.5	15.1	14.2
+	+	ab	41.0	43.9	36.3	39.9

(a) Analyze the data from this experiment.

(b) Plot the residuals on normal probability paper and versus the predicted vibration level. Interpret these plots.

(c) Draw the AB interaction plot. Interpret this plot. What levels of bit size and speed would you recommend for routine operation?

(d) Construct a contour plot of vibration as a function of speed and bit size.

3.2 An engineer is interested in the effects of cutting speed (A), tool geometry (B), and cutting angle (C) on the life (in hours) of a machine tool. Two levels of each factor are chosen, and two replicates of a 2^3 factorial design are run. The results follow:

A	B	C	Treatment Combination	Replicate I	II
−	−	−	(1)	22	31
+	−	−	a	32	43
−	+	−	b	35	34
+	+	−	ab	55	47
−	−	+	c	44	45
+	−	+	ac	40	37
−	+	+	bc	60	50
+	+	+	abc	39	41

    (a) Estimate the factor effects. Which effects appear to be large?

    (b) Use the analysis of variance to confirm your conclusions for part (a).

    (c) Analyze the residuals. Are there obvious problems?

    (d) What levels of A, B, and C would you recommend, based on the data from this experiment?

3.3 An industrial engineer employed by a beverage bottler is interested in the effects of two different types of 32-ounce bottles on the time to deliver 12-bottle cases of the product. The two bottle types are glass and plastic. Two workers are used to perform a task consisting of moving 40 cases of the product 50 ft on a standard type of hand truck and stacking the cases in a display. Four replicates of a 2^2 factorial design are performed, and the times observed are listed in the following table. Analyze the data and draw appropriate conclusions. Analyze the residuals and comment on the model's adequacy.

Bottle Type	Worker			
	1		2	
Glass	5.12	4.89	6.65	6.24
	4.98	5.00	5.49	5.55
Plastic	4.95	4.43	5.28	4.91
	4.27	4.25	4.75	4.71

3.4 Find the two standard error limits for the factor effects in Exercise 3.1. Does the results of this analysis agree with the conclusions from the analysis of variance?

3.5 Find the two standard error limits for the factor effects in Exercise 3.2. Does the results of this analysis agree with the conclusions from the analysis of variance?

3.6 An experiment was performed to improve the yield of a chemical process. Four factors were selected, and one replicate of a completely randomized experiment was run. The results are shown in the following table:

Treatment Combination	Yield	Treatment Combination	Yield
(1)	90	d	98
a	64	ad	62
b	81	bd	87
ab	63	abd	75
c	77	cd	99
ac	61	acd	69
bc	88	bcd	87
abc	53	abcd	60

(a) Estimate the factor effects. Construct a normal probability plot of these effects. Which effects appear large?

(b) Prepare an analysis of variance table using the information obtained in part (a).

(c) Plot the residuals versus the predicted yield and on normal probability paper. Does the residual analysis appear satisfactory?

(d) Construct a contour plot of yield as a function of the important process variables.

3.7 Consider the design of Exercise 3.6. Suppose that four additional runs were made at the center of the region of experimentation. The response values at these center points were 94, 90, 99, and 87. Use this additional information to test for curvature in the response function. What are your conclusions?

3.8 An article in the *AT & T Technical Journal* (March/April 1986, Vol. 65, pp. 39–50) describes the application of two-level factorial designs to integrated circuit manufacturing. A basic processing step is to grow an epitaxial layer on polished silicon wafers. The wafers mounted on a susceptor are positioned inside a bell jar, and chemical vapors are introduced. The susceptor is rotated and heat is applied until the epitaxial layer is thick enough. An experiment was run using two factors: arsenic flow rate (A) and deposition time (B). Four replicates were run, and the epitaxial layer thickness was measured (in μm).

The data are shown below.

A	B	I	II	III	IV		Low (−)	High (+)
			Replicate				Factor Levels	
−	−	14.037	16.165	13.972	13.907	A	55%	59%
+	−	13.880	13.860	14.032	13.914			
−	+	14.821	14.757	14.843	14.878	B	Short	Long
+	+	14.88	14.921	14.415	14.932			

(a) Estimate the factor effects.
(b) Conduct the analysis of variance. Which factors are important?
(c) Analyze the residuals. Are there any residuals that should cause concern?
(d) Discuss how you might deal with the potential outlier found in part (c).
(e) Construct a contour plot of the layer thickness response surface as a function of flow rate and deposition time.

3.9 A nickel–titanium alloy is used to make components for jet turbine aircraft engines. Cracking is a potentially serious problem in the final part, because it can lead to nonrecoverable failure. A test is run at the parts producer to determine the effect of four factors on cracks. The four factors are pouring temperature (A), titanium content (B), heat treatment method (C), and amount of grain refiner used (D). Two replicates of a 2^4 design are run, and the length of crack (in mm) induced in a sample coupon subjected to a standard test is measured. The data are shown below:

A	B	C	D	Treatment Combination	I	II
					Replicate	
−	−	−	−	(1)	1.71	1.91
+	−	−	−	a	1.42	1.48
−	+	−	−	b	1.35	1.53
+	+	−	−	ab	1.67	1.55
−	−	+	−	c	1.23	1.38
+	−	+	−	ac	1.25	1.26
−	+	+	−	bc	1.46	1.42
+	+	+	−	abc	1.29	1.27
−	−	−	−	d	2.04	2.19
+	−	−	+	ad	1.86	1.85
−	+	−	+	bd	1.79	1.95
+	+	−	+	abd	1.42	1.59
−	−	+	+	cd	1.81	1.92
+	−	+	+	acd	1.34	1.29
−	+	+	+	bcd	1.46	153
+	+	+	+	abcd	1.38	1.35

(a) Estimate the factor effects. Which factor effects appear to be large?

(b) Conduct an analysis of variance. Do any of the factors affect cracking?

(c) Analyze the residuals from this experiments.

(d) Is there an indication that any of the factors affect the variability in cracking?

(e) What recommendations would you make regarding process operations?

3.10 An article in *Solid State Technology* ("Orthogonal Design for Process Optimization and its Application in Plasma Etching," May 1987, pp. 127–132) describes the application of factorial designs in developing a nitride etch process on a single-wafer plasma etcher. The process uses C_2F_6 as the reactant gas. Four factors are of interest: anode–cathode gap (A), pressure in the reactor chamber (B), C_2F_6 gas flow (C), and power applied to the cathode (D). The response variable of interest is the etch rate for silicon nitride. A single replicate of a 2^4 design is run, and the data are shown below:

Actual Run Number	Etch Run Order	A	B	C	D	Rate (Å/min)	Factor Levels Low (−)	High (+)
1	13	−	−	−	−	500	A (cm) 0.8	1.20
2	8	+	−	−	−	669	B (mTorr) 4.5	550
3	12	−	+	−	−	604	C (SCCM) 125	200
4	9	+	+	−	−	650	D (W) 275	325
5	4	−	−	+	−	633		
6	15	+	−	+	−	642		
7	16	−	+	+	−	601		
8	3	+	+	+	−	635		
9	1	−	−	−	+	1037		
10	14	+	−	−	+	749		
11	5	−	+	−	+	1052		
12	10	+	+	−	+	868		
13	11	−	−	+	+	1075		
14	2	+	−	+	+	860		
15	7	−	+	+	+	1063		
16	6	+	+	+	+	729		

(a) Estimate the factor effects. Construct a normal probability plot of the factor effects. Which effects appear large?

(b) Conduct an analysis of variance to confirm your findings for part (a).

(c) Analyze the residuals from this experiment. Comment on the model's adequacy.

(d) If not all the factors are important, project the 2^4 design into a 2^k design with $k < 4$ and conduct the analysis of variance.

(e) Draw graphs to interpret any significant interactions.

(f) Plot the residuals versus the actual run order. What problems might be revealed by this plot?

3.11 Consider the single replicate of the 2^4 design in Example 3.10. Suppose we had arbitrarily decided to analyze the data assuming that all three- and four-factor interactions were negligible. Conduct this analysis and compare your results with those obtained in the example. Do you think that it is a good idea to arbitrarily assume interactions to be negligible even if they are relatively high-order ones?

3.12 An experiment was run in a semiconductor fabrication plant in an effort to increase the number of good chips produced per wafer. Five factors, each at two levels, were studied. The factors (and levels) were A = aperture setting (small, large), B = exposure time (20% below nominal), C = development time (30 sec 45 sec), D = mask dimension (small, large), and E = etch time (14.5 min, 15.5 min). The unreplicated 2^5 design shown below as run.

(1)	15	d	16	e	14	de	12
a	20	ad	21	ae	23	ade	19
b	70	bd	65	be	71	bde	60
ab	112	abd	100	abe	110	$abde$	106
c	30	cd	35	ce	31	cde	32
ac	40	acd	45	ace	43	$acde$	39
bc	79	bcd	85	bce	90	$bcde$	82
abc	125	$abcd$	120	$abcd$	130	$abcde$	128

(a) Construct a normal probability plot of the effect estimates. Which effects appear to be large?

(b) Conduct an analysis of variance to confirm your findings for part (a).

(c) Plot the residuals on normal probability paper. Is the plot satisfactory?

(d) Plot the residuals versus the predicted yields and versus each of the five factors. Comment on the plots.

(e) Interpret any significant interactions.

(f) What are your recommendations regarding process operating conditions?

(g) Project the 2^5 design in this problem into a 2^k design in the important factors. Sketch the design and shown the average and range of yields at each run. Does this sketch aid in data interpretation?

(h) Construct a contour plot showing how yield changes in terms of the important factors.

3.13 After running a 2^4 factorial design, an engineer obtained the following factor effect estimates:

$$A = -1.3 \qquad AB = -2.9 \qquad ABC = 0.5$$
$$B = 10.1 \qquad AC = 1.9 \qquad ABD = 1.8$$
$$C = 7.5 \qquad AD = -4.0 \qquad ACD = -2.7$$
$$D = -1.4 \qquad BC = -5.1 \qquad BCD = 2.8$$
$$BD = 2.2 \qquad ABCD = -0.6$$
$$CD = -0.2$$

Use a normal probability plot to interpret these effects.

3.14 Consider the data from the experiment in Exercise 3.10. Suppose that the last run made in this experiment (run order number 16) was lost; in fact, the wafer was broken so that no reading on etch rate could be obtained. One logical approach to analyzing these data is to estimate the missing value that makes the highest-order interaction effect estimate zero. Apply this technique to these data and compare your results with those obtained in the analysis of the full data set in Exercise 3.10.

3.15 **Continuation of Exercise 3.14.** Reconsider the situation described in Exercise 3.14.

(a) Analyze the data by fitting a regression model in the main effects to the 15 observations.

(b) Investigate the adequacy of this model. What are your conclusions?

(c) Build a model for the 15 observations with stepwise regression methods, using the main effects and two-factor interactions as the candidate variables.

(d) Comment on the model obtained above and compare them to those from Exercise 3.10 and 3.14.

3.16 Consider the experiment described in Exercise 3.1. Suppose that only one replicate (4 runs) could be obtained in a single 4-hr time period, and the experimenters were concerned about unknown factors that could vary from one time period to another.

(a) Set up a design in four blocks that will minimize the time effects.

(b) Analyze the data assuming that the design had been run as in part (a). Compare the results of your analysis with the original analysis in Exercise 3.1.

(c) Suppose that the experimenter's concerns were realized and that as a result of the time effect, the observations in replicate IV were (1) = 34.4, a = 44.5, b = 34.2, and ab = 59.9 instead of the values shown in Exercise 3.1. Reanalyze these new data, assuming that the blocked design from part (a) had been used. How has the time effect impacted your conclusions?

3.17 Consider the experiment described in Exercise 3.2. Suppose that this experiment had been run with each replicate considered as a block. Analyze the data and draw conclusions. How do the results of your analysis compare with the analysis of the original Exercise 3.2?

3.18 Consider only the data from the first replicate of Exercise 3.2. Set up a design to run these observations in two blocks of four runs each with ABC confounded with blocks. Analyze the data.

3.19 Consider the data from both replicates of Exercise 3.2. Suppose that only four runs could be made under the same conditions, so that it is necessary to run this design in blocks of four runs.

(a) Set up a design that confounds ABC in both replicates. Analyze the data.

(b) Set up a design that confounds ABC in the first replicate and AB in the second replicate. Analyze this design, using the data from replicate I to draw conclusions about the AB effect, and the data from replicate II to draw conclusions about the ABC effect (this is a design technique called *partial confounding*).

3.20 Consider the data from Exercise 3.6. Construct a design with two blocks of eight runs each, with $ABCD$ confounded with blocks. Analyze the data.

3.21 **Continuation of Exercise 3.20**. Reconsider the situation in Exercise 3.20, and assume that the block effect causes all observations in the second block to be reduced by 20. Analyze the data that result. Is the block effect significant? How are the other conclusions affected?

3.22 Consider the situation described in Exercise 3.12. Set up a design in two blocks for 16 observations each of this problem. Analyze the data that result.

3.23   Consider the situation described in Exercise 3.12. Set up a design in four blocks of 8 runs each, with $ACDE$ and BCE confounded with blocks. What is the other confounded effect? Analyze the data and draw conclusions.

3.24   Construct a 2^6 design in 8 blocks of 8 runs each. Show the complete set of effects that are confounded with blocks.

3.25   Suppose that an experimenter has run a 2^3 factorial design (8 trials) in random order. He fits the following model to the data:

$$y = \beta_0 + \beta_1 x_1 + \beta_2 x_2 + \beta_3 x_3 + \varepsilon$$

(a)   Suppose that the true model is

$$E(y) = \beta_0 + \beta_1 x_2 + \beta_2 x_2 + \beta_3 x_3 + \beta_{12} x_1 x_2$$

Does the interaction term result in biased estimates of β_0, β_1, β_2, and β_3 in the fitted first-order model?

(b)   Suppose that the true model is

$$E(y) = \beta + \beta_1 x_1 + \beta_2 x_2 + \beta_3 x_3 + \beta_{12} x_1 x_2 + \beta_{11} x_1^2$$

Determine the bias induced in the least squares estimates of β_0, β_1, β_2, and β_3 by the failure to include the terms $\beta_{12} x_1 x_2$ and $\beta_{11} x_1^2$ in the fitted model.

3.26   An experimenter has run a 2^3 design in standard order (no randomization). Unknown to the experimenter a fourth variables x_4 takes on the values shown below:

Run	x_1	x_2	x_3	x_4
1	-1	-1	-1	-4
2	1	-1	-1	-3
3	-1	1	-1	-2
4	1	1	-1	-1
5	-1	-1	1	1
6	1	-1	1	2
7	-1	1	1	3
8	1	1	1	4

Suppose that the experimenter fits the first-order model

$$y = \beta_0 + \beta_1 x_1 + \beta_2 x_2 + \beta_3 x_3 + \varepsilon$$

Find the bias in the least squares estimates of β_0, β_1, β_2, and β_3 that results from not including the term $\beta_4 x_4$ in the model.

3.27 Consider the runs from a 2^3 factorial design shown below, in random order.

Run Order	A	B	C	Observation
1	+	+	−	$ab + 20$
2	+	−	−	$a + 20$
3	−	−	−	$(1) + 20$
4	+	l	+	$abc + 20$
5	−	+	−	b
6	+	−	+	ac
7	−	−	+	c
8	−	+	+	bc

The first four observations are impacted by a block effect of 20 units.

(a) Find the impact of the block effect on the estimates of the main effects of A, B, and C.

(b) Suppose that the design had been set up with ABC confounded with blocks, and once again, the first four runs with are impacted by a block effect of 20 units. What is the impact of this block effect on the estimates of A, B, and C?

3.28 Consider a 2^3 design with a center point that has been replicated n_C times. Show that the runs at the center do not impact the estimates of any of the factorial effects. Do the center points impact the estimate of the overall mean? Will these results generalize to any 2^k design?

CHAPTER 4

Two-Level Fractional Factorial Designs

4.1 INTRODUCTION

As the number of factors in a 2^k factorial design increases, the number of runs required for a complete replicate of the design rapidly outgrows the resources of most experimenters. For example, a complete replicate of the 2^6 design requires 64 runs. In this design only 6 of the 63 degrees of freedom are used to estimate the main effects, and only 15 degrees of freedom are used to estimate the two-factor interactions. The remaining 42 degrees of freedom are associated with three-factor and higher interactions.

If the experimenter can reasonably assume that certain high-order interactions are negligible, then information on the main effects and low-order interactions may be obtained by running only a fraction of the complete factorial experiment. These **fractional factorial designs** are among the most widely used types of design in industry.

A major use of fractional factorials is in **screening experiments**. These are experiments in which many factors are considered with the purpose of identifying those factors (if any) that have large effects. Remember that screening experiments are usually performed early in a response surface study when it is likely that many of the factors initially considered have little or no effect on the response. The factors that are identified as important are then investigated more thoroughly in subsequent experiments.

The successful use of fractional factorial designs is based on three key ideas:

1. **The Sparsity-of-Effect Principle.** When there are several variables, the system or process is likely to be driven primarily by some of the main effects and low-order interactions.

2. **The Projection Property.** Fractional factorial designs can be projected into stronger (larger) designs in the subset of significant factors.

3. **Sequential Experimentation.** It is possible to combine the runs of two (or more) fractional factorials to assemble sequentially a larger design to estimate the factor effects and interactions of interest.

We will focus on these principles in this chapter and illustrate them with several examples.

4.2 THE ONE-HALF FRACTION OF THE 2^k DESIGN

Consider the situation in which three factors, each at two levels, are of interest, but the experimenters do not wish to run all $2^3 = 8$ treatment combinations. Suppose they consider a design with four runs. This suggests a one-half fraction of a 2^3 design. Because the design contains $2^{3-1} = 4$ treatment combinations, a one-half fraction of the 2^3 design is often called a 2^{3-1} design.

The table of plus and minus signs for the 2^3 design is shown in Table 4.1. Suppose we select the four treatment combinations a, b, c, and abc as our one-half fraction. These runs are shown in the top half of Table 4.1 and in Figure 4.1(a). Notice that the 2^{3-1} design is formed by selecting only those treatment combinations that have a plus in the ABC column. Thus, ABC is called the **generator** of this particular fraction. Sometimes we will refer to a generator such as ABC as a **word**. Furthermore, the identity column I is also always plus, so we call

$$I = ABC$$

the **defining relation** for our design. In general, the defining relation for a fractional factorial will always be the set of all columns that are equal to the identity column I.

Table 4.1 Plus and Minus Signs for the 2^3 Factorial Design

Treatment Combination	Factorial Effect							
	I	A	B	C	AB	AC	BC	ABC
a	+	+	−	−	−	−	+	+
b	+	−	+	−	−	+	−	+
c	+	−	−	+	+	−	−	+
abc	+	+	+	+	+	+	+	+
ab	+	+	+	−	+	−	−	−
ac	+	+	−	+	−	+	−	−
bc	+	−	+	+	−	−	+	−
(1)	+	−	−	−	+	+	+	−

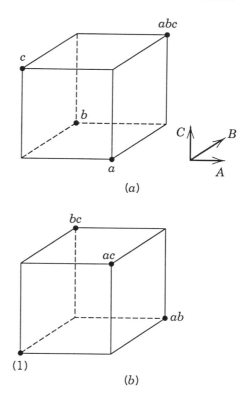

Figure 4.1 The two one-half fractions of the 2^3 design. (*a*) The principal fraction, $I = +ABC$. (*b*) The alternate fraction, $I = -ABC$.

The treatment combinations in the 2^{3-1} design yield three degrees of freedom that we may use to estimate the main effects. Referring to Table 4.1, we note that the linear combinations of the observations used to estimate the main effects of A, B, and C are

$$l_A = \tfrac{1}{2}(a - b - c + abc)$$

$$l_B = \tfrac{1}{2}(-a + b - c + abc)$$

$$l_C = \tfrac{1}{2}(-a - b + c + abc)$$

It is also easy to verify that the linear combinations of the observations used to estimate the two-factor interactions are

$$l_{BC} = \tfrac{1}{2}(a - b - c + abc)$$

$$l_{AC} = \tfrac{1}{2}(-a + b - c + abc)$$

$$l_{AB} = \tfrac{1}{2}(-a - b + c + abc)$$

Thus, $l_A = l_{BC}$, $l_B = l_{AC}$, and $l_C = l_{AB}$; consequently, it is impossible to differentiate between A and BC, B and AC, and C and AB. In fact, when we estimate A, B, and C we are really estimating $A + BC$, $B + AC$, and $C + AB$. Two or more effects that have this property are called **aliases**. In our example, A and BC are aliases, B and AC are aliases, and C and AB are aliases. We indicate this by the notation $l_A \rightarrow A + BC$, $l_B \rightarrow B + AC$, and $l_C \rightarrow C + AB$.

The alias structure for this design may be easily determined by using the defining relation $I = ABC$. Multiplying any column by the defining relation yields the aliases for that effect. In our example, this yields as the alias of A

$$A \cdot I = A \cdot ABC = A^2 BC$$

or, because the square of any column is just the identity I,

$$A = BC$$

Similarly, we find the aliases of B and C as

$$B \cdot I = B \cdot ABC$$
$$B = AB^2C = AC$$

and

$$C \cdot I = C \cdot ABC$$
$$C = ABC^2 = AB$$

This one-half fraction, with $I = +ABC$, is usually called the **principal fraction**.

Now suppose that we had selected the other one-half fraction, that is, the treatment combinations in Table 4.1 associated with a minus sign in the ABC column. This **alternate** or *complementary fraction* is shown in Figure 4.1(b). Notice that it consists of the runs labeled (1), ab, ac, and bc. The defining relation for this design is

$$I = -ABC$$

The linear combination of the observations, say l'_A, l'_B, and l'_C, from the alternate fraction gives us the following alias relationships:

$$l'_A \rightarrow A - BC$$
$$l'_B \rightarrow B - AC$$
$$l'_C \rightarrow C - AB$$

Thus, when we estimate A, B, and C with this particular fraction, we are really estimating $A - BC$, $B - AC$, and $C - AB$.

In practice, it does not matter which fraction is actually used. Both fractions belong to the same **family**; that is, the two one-half fractions form a complete 2^3 design. This is easily seen by referring to Figure 4.1(a) and 4.1(b).

Suppose that after running one of the one-half fractions of the 2^3 design, the other one was also run. Thus, all 8 runs associated with the full 2^3 are now available. We may now obtain de-aliased estimates of all the effects by analyzing the 8 runs as a full 2^3 design in two blocks of 4 runs each. This could also be done by adding and subtracting the linear combination of effects from the two individual fractions. For example, consider $l_A \rightarrow A + BC$ and $l'_A \rightarrow A - BC$. This implies that

$$\tfrac{1}{2}(l_A + l'_A) = \tfrac{1}{2}(A + BC + A - BC) \rightarrow A$$

and

$$\tfrac{1}{2}(l_A - l'_A) = \tfrac{1}{2}(A + BC - A + BC) \rightarrow BC$$

Thus, for all three pairs of linear combinations, we would obtain the following:

i	From $\tfrac{1}{2}(l_i + l'_i)$	From $\tfrac{1}{2}(l_i - l'_i)$
A	A	BC
B	B	AC
C	C	AB

Design Resolution

The preceding 2^{3-1} design is called a **resolution III design**. In such a design, main effects are aliased with two-factor interactions. A design is of resolution R if no p-factor effect is aliased with another effect containing less than $R - p$ factors. We usually employ a Roman numeral subscript to denote design resolution; thus, the one-half fraction of the 2^3 design with the defining relation $I = ABC$ (or $I = -ABC$) is a 2_{III}^{3-1} design.

Designs of resolution III, IV, and V are particularly important. The definitions of these designs and an example of each follow.

1. **Resolution III Designs.** These are designs in which no main effects are aliased with any other main effect, but main effects are aliased with two-factor interactions and two-factor interactions may be aliased with each other. The 2^{3-1} design in Table 4.1 is of resolution III (2_{III}^{3-1}).

2. **Resolution IV Designs.** These are designs in which no main effect is aliased with any other main effect or with any two-factor interaction, but two-factor interactions are aliased with each other. A 2^{4-1} design with $I = ABCD$ is of resolution IV (2_{IV}^{4-1}).

3. **Resolution V Designs.** These are designs in which no main effect or two-factor interaction is aliased with any other main effect or two-factor interaction, but two-factor interactions are aliased with three-factor interactions. A 2^{5-1} design with $I = ABCDE$ is of resolution V (2_V^{5-1}).

In general, the resolution of a two-level fractional factorial design is equal to the smallest number of letters in any word in the defining relation. Consequently, we could call the preceding design types three-letter, four-letter, and five-letter designs, respectively. We usually like to employ fractional designs that have the highest possible resolution consistent with the degree of fractionation required. The higher the resolution, the less restrictive the assumptions that are required regarding which interactions are negligible in order to obtain a unique interpretation of the data.

Constructing One-Half Fractions
A one-half fraction of the 2^k design of the highest resolution may be constructed by writing down a *basic design* consisting of the runs for a full 2^{k-1} factorial and then adding the kth factor by identifying its plus and minus levels with the plus and minus signs of the highest-order interaction $ABC \ldots (K-1)$. Therefore, the 2_{III}^{3-1} fractional factorial is obtained by writing down the full 2^2 factorial as the basic design and then equating factor C to the AB interaction. The alternate fraction would be obtained by equating factor C to the $-AB$ interaction. This approach is illustrated in Table 4.2. Notice that the basic design always has the right number of runs (rows), but it is missing one column. The generator $K = ABC \ldots (K-1)$ defines the product of plus and minus signs to use in each row to produce the levels for the kth factor.

Note that any interaction effect could be used to generate the column for the kth factor. However, using any effect other than $ABC \ldots (K-1)$ as the generator will not produce a design of the highest possible resolution.

Another way to view the construction of a one-half fraction is to partition the runs into two blocks with the highest-order interaction $ABC \ldots K$

Table 4.2 The Two One-Half Fractions of the 2^3 Design

Run	Full 2^2 Factorial (Basic Design)		$2_{III}^{3-1}, I = ABC$			$2_{III}^{3-1}, I = -ABC$		
	A	B	A	B	$C = AB$	A	B	$C = -AB$
1	$-$	$-$	$-$	$-$	$+$	$-$	$-$	$-$
2	$+$	$-$	$+$	$-$	$-$	$+$	$-$	$+$
3	$-$	$+$	$-$	$+$	$-$	$-$	$+$	$+$
4	$+$	$+$	$+$	$+$	$+$	$+$	$+$	$-$

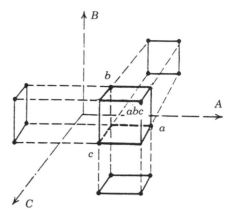

Figure 4.2 Projection of a 2_{III}^{3-1} design into three 2^2 designs.

confounded. Each block is a 2^{k-1} fractional factorial design of the highest resolution.

Projection of Fractions into Factorials

Any fractional factorial design of resolution R contains complete factorial designs (possibly replicated factorials) in any subset of $R - 1$ factors. This is an important and useful concept. For example, if an experimenter has several factors of potential interest but believes that only $R - 1$ of them have important effects, then a fractional factorial design of resolution R is the appropriate choice of design. If the experimenter is correct, then the fractional factorial design of resolution R will project into a full factorial in the $R - 1$ significant factors. This process is illustrated in Figure 4.2 for the 2_{III}^{3-1} design, which projects into a 2^2 design in every subset of two factors.

Because the maximum possible resolution of a one-half fraction of the 2^k design is $R = k$, every 2^{k-1} design will project into a full factorial in any $(k - 1)$ of the original k factors. Furthermore, a 2^{k-1} design may be projected into two replicates of a full factorial in any subset of $k - 2$ factors, four replicates of a full factorial in any subset of $k - 3$ factors, and so on.

Sequences of Fractional Factorials

Using fractional factorial designs often leads to great economy and efficiency in experimentation, particularly if the runs can be made sequentially. For example, suppose that we were investigating $k = 4$ factors ($2^4 = 16$ runs). It is almost always preferable to run a 2_{IV}^{4-1} fractional design (8 runs), analyze the results, and then decide on the best set of runs to perform next. If it is necessary to resolve ambiguities, we can always run the alternate fraction and complete the 2^4 design. When this method is used to complete the design, both one-half fractions represent blocks of the complete design with the highest-order interaction confounded with blocks (here $ABCD$) would be confounded. Thus, sequential experimentation has the result of losing information only on the highest-order interaction. Alternatively, in many cases we learn enough from the one-half fraction to proceed to the next stage of

Table 4.3   The 2_{IV}^{4-1} Design with the Defining Relation $I = ABCD$

Run	Basic Design				Treatment Combination	Filtration Rate
	A	B	C	$D = ABC$		
1	$-$	$-$	$-$	$-$	(1)	45
2	$+$	$-$	$-$	$+$	ad	100
3	$-$	$+$	$-$	$+$	bd	45
4	$+$	$+$	$-$	$-$	ab	65
5	$-$	$-$	$+$	$+$	cd	75
6	$+$	$-$	$+$	$-$	ac	60
7	$-$	$+$	$+$	$-$	bc	80
8	$+$	$+$	$+$	$+$	$abcd$	96

experimentation, which might involve adding or removing factors, changing responses, or varying some of the factors over new ranges.

Example 4.1   Consider the filtration rate experiment in Example 3.2. The original design, shown in Table 3.6, is a single replicate of the 2^4 design. In this example, we found that the main effects A, C, and D and the interactions AC and AD were different from zero. We will now return to this experiment and simulate what would have happened if a half-fraction of the 2^4 design had been run instead of the full factorial.

We will use the 2^{4-1} design with $I = ABCD$, because this choice of generator will result in a design of the highest possible resolution (IV). To construct the design, we first write down the basic design, which is a 2^3 design, as shown in the first three columns of Table 4.3. Because the generator $ABCD$ is positive, this 2_{IV}^{4-1} design is the principal fraction. The design is shown graphically in Figure 4.3.

Using the defining relation, we note that each main effect is aliased with a three-factor interaction; that is, $A = A^2BCD = BCD$, $B = AB^2CD = ACD$,

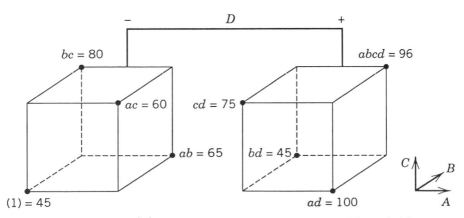

Figure 4.3   The 2_{IV}^{4-1} design for the filtration rate experiment of Example 4.1.

$C = ABC^2D = ABD$, and $D = ABCD^2 = ABC$. Furthermore, every two-factor interaction is aliased with another two-factor interaction. These alias relationships are $AB = CD$, $AC = BD$, and $BC = AD$. The four main effects plus the three two-factor interaction alias pairs account for the seven degrees of freedom for the design.

At this point, we would normally randomize the eight runs and perform the experiment. Because we have already run the full 2^4 design, we will simply select the eight observed filtration rates from Example 3.2 that correspond to the runs in the 2_{IV}^{4-1} design. These observations are shown in the last column of Table 4.3 and are also shown in Figure 4.3.

The estimates of the effects obtained from this 2_{IV}^{4-1} design are shown in Table 4.4. To illustrate the calculations, the linear combination of observations associated with the A effect is

$$l_A = \tfrac{1}{4}(-45 + 100 - 45 + 65 - 75 + 60 - 80 + 96) = 19.00 \rightarrow A + BCD$$

whereas for the AB effect, we would obtain

$$l_{AB} = \tfrac{1}{4}(45 - 100 - 45 + 65 + 75 - 60 - 80 + 96) = -1.00 \rightarrow AB + CD$$

From inspection of the information in Table 4.4, it is not unreasonable to conclude that the main effects A, C, and D are large and that the AC and AD interactions are also significant. This agrees with the conclusions from the analysis of the complete 2^4 design in Example 3.2.

Because factor B is not significant, we may drop it from consideration. Consequently, we may project this 2_{IV}^{4-1} design into a single replicate of the 2^3 design in factors A, C, and D, as shown in Figure 4.4. Visual examination of this cube plot makes us more comfortable with the conclusions reached above. Notice that if the temperature (A) is at the low level, the concentration (C) has a large positive effect, whereas if the temperature is at the high level, the concentration has a very small effect. This is likely due to an AC interaction. Furthermore, if the temperature is at the low level, the effect of the stirring rate (D) is negligible, whereas if the temperature is at the high level, the stirring rate has a large positive effect. This is likely due to the AD interaction tentatively identified previously.

Table 4.4   Estimates of Effects and Aliases from Example 4.1[a]

Estimate	Alias Structure
$l_A = \quad 19.00$	$l_A \rightarrow \mathbf{A} + BCD$
$l_B = \quad\ \ 1.50$	$l_B \rightarrow B + ACD$
$l_C = \quad 14.00$	$l_C \rightarrow \mathbf{C} + ABD$
$l_D = \quad 16.50$	$l_D \rightarrow \mathbf{D} + ABC$
$l_{AB} = \ -1.00$	$l_{AB} \rightarrow AB + CD$
$l_{AC} = -18.50$	$l_{AC} \rightarrow \mathbf{AC} + BD$
$l_{BC} = \quad 19.00$	$l_{BC} \rightarrow BC + \mathbf{AD}$

[a]Significant effects are shown in boldface type.

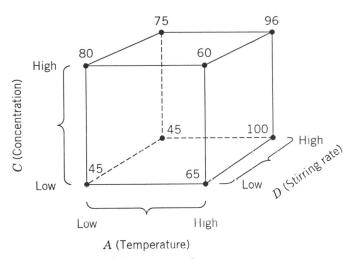

Figure 4.4 Projection of the 2_{IV}^{4-1} design into 2^3 design in A, C, and D for Example 4.1.

Now suppose that the experimenter decided to run the alternate fraction, given by $I = -ABCD$. It is straightforward to show that the design and the responses are as follows:

Run	A	B	C	$D = -ABC$	Treatment Combination	Filtration Rate
1	$-$	$-$	$-$	$+$	d	43
2	$+$	$-$	$-$	$-$	a	71
3	$-$	$+$	$-$	$-$	b	48
4	$+$	$+$	$-$	$+$	abd	104
5	$-$	$+$	$+$	$-$	c	68
6	$+$	$-$	$+$	$+$	acd	86
7	$-$	$+$	$+$	$+$	bcd	70
8	$+$	$+$	$+$	$-$	abc	65

The linear combinations of observations obtained from this alternate fraction are

$$l'_A = 24.25 \rightarrow A - BCD$$
$$l'_B = 4.75 \rightarrow B - ACD$$
$$l'_C = 5.75 \rightarrow C - ABD$$
$$l'_D = 12.75 \rightarrow D - ABC$$
$$l'_{AB} = 1.25 \rightarrow AB - CD$$
$$l'_{AC} = -17.75 \rightarrow AC - BD$$
$$l'_{BC} = -14.25 \rightarrow BC - AD$$

These estimates may be combined with those obtained from the original one-half fraction to yield the following estimates of the effects:

i	From $\frac{1}{2}(l_i + l'_i)$	From $\frac{1}{2}(l_i + l'_i)$
A	$21.63 \rightarrow A$	$-2.63 \rightarrow BCD$
B	$3.13 \rightarrow B$	$-1.63 \rightarrow ACD$
C	$9.88 \rightarrow C$	$4.13 \rightarrow ABD$
D	$14.63 \rightarrow D$	$1.88 \rightarrow ABC$
AB	$0.13 \rightarrow AB$	$-1.13 \rightarrow CD$
AC	$-18.13 \rightarrow AC$	$-0.38 \rightarrow BD$
BC	$2.38 \rightarrow BC$	$16.63 \rightarrow AD$

These estimates agree exactly with those from the original analysis of the data as a single replicate of a 2^4 factorial design, as reported in Example 3.2.

Example 4.2 Five factors in a manufacturing process for an integrated circuit were investigated in a 2^{5-1} design with the objective of learning how these factors affect the resistivity of the wafer. The five factors were $A =$ implant dose, $B =$ temperature, $C =$ time, $D =$ oxide thickness, and $E =$ furnace position. Each factor was run at two levels. The construction of the 2^{5-1} design is shown in Table 4.5. Notice that the design was constructed by

Table 4.5   A 2^{5-1} Design for Example 4.2

Run	A	B	C	D	$E = ABCD$	Treatment Combination	Resistivity
1	$-$	$-$	$-$	$-$	$+$	e	15.1
2	$+$	$-$	$-$	$-$	$-$	a	20.6
3	$-$	$+$	$-$	$-$	$-$	b	68.7
4	$+$	$+$	$-$	$-$	$+$	abe	101.0
5	$-$	$-$	$+$	$-$	$-$	c	32.9
6	$+$	$-$	$+$	$-$	$+$	ace	46.1
7	$-$	$+$	$+$	$-$	$+$	bce	87.5
8	$+$	$+$	$+$	$-$	$-$	abc	119.0
9	$-$	$-$	$-$	$+$	$-$	d	11.3
10	$+$	$-$	$-$	$+$	$+$	ade	19.6
11	$-$	$+$	$-$	$+$	$+$	bde	62.1
12	$+$	$+$	$-$	$+$	$-$	abd	103.2
13	$-$	$-$	$+$	$+$	$+$	cde	27.1
14	$+$	$-$	$+$	$+$	$-$	acd	40.3
15	$-$	$+$	$+$	$+$	$-$	bcd	87.7
16	$+$	$+$	$+$	$+$	$+$	$abcde$	128.3

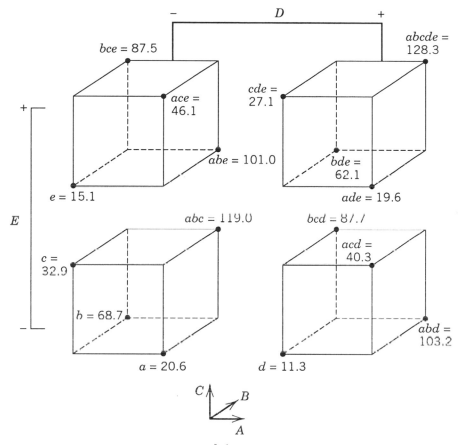

Figure 4.5   The 2_V^{5-1} design for Example 4.2.

writing down the basic design having 16 runs (a 2^4 design in A, B, C, and D), selecting $ABCDE$ as the generator, and then setting the levels of the fifth factor $E = ABCD$. Figure 4.5 gives a pictorial representation of the design.

The defining relation for the design is $I = ABCDE$. Consequently, every main effect is aliased with a four-factor interaction:

$$l_A \rightarrow A + BCDE$$

$$l_B \rightarrow B + ACDE$$

$$l_C \rightarrow C + ABDE$$

$$l_D \rightarrow D + ABCE$$

$$l_E \rightarrow E + ABCD$$

Every two-factor interaction is aliased with a three-factor interaction:

$$l_{AB} \rightarrow AB + CDE \qquad l_{BD} \rightarrow BD + ACE$$
$$l_{AC} \rightarrow AC + BDE \qquad l_{BE} \rightarrow BE + ACD$$
$$l_{AD} \rightarrow AD + BCE \qquad l_{CD} \rightarrow CD + ABE$$
$$l_{AE} \rightarrow AE + BCD \qquad l_{CE} \rightarrow CE + ABD$$
$$l_{BC} \rightarrow BC + ADE \qquad l_{DE} \rightarrow DE + ABC$$

This is a resolution V design. We would expect this 2_V^{5-1} design to provide excellent information concerning the main effects and all two-factor interactions.

The estimates of the effects are

$$l_A \rightarrow A + BCDE = 23.2125$$
$$l_B \rightarrow B + ACDE = 68.0625$$
$$l_C \rightarrow C + ABDE = 20.9125$$
$$l_D \rightarrow D + ABCE = -1.4125$$
$$l_E \rightarrow E + ABCD = 0.3875$$

$$l_{AB} \rightarrow AB + CDE = 13.1625 \qquad l_{BD} \rightarrow BD + ACE = 2.6875$$
$$l_{AC} \rightarrow AC + BDE = 1.4125 \qquad l_{BE} \rightarrow BE + ACD = -0.3125$$
$$l_{AD} \rightarrow AD + BCE = 2.5875 \qquad l_{CD} \rightarrow CD + ABE = 0.8875$$
$$l_{AE} \rightarrow AE + BCD = 2.5875 \qquad l_{CE} \rightarrow CE + ABD = 1.8875$$
$$l_{BC} \rightarrow BC + ADE = 0.9675 \qquad l_{DE} \rightarrow DE + ABC = -1.7375$$

The effects A, B, C, and AB seem large. Figure 4.6 presents a normal probability plot of the effect estimates from this experiment. This plot confirms that the main effects of A, B, and C, and the AB interaction are large. Remember that, because of aliasing, these effects are really $A + BCDE$, $B + ACDE$, $C + ABDE$, and $AB + CDE$. However, because it seems plausible that three-factor and higher interactions are negligible, we feel safe in concluding that only A, B, C, and AB are important effects.

Table 4.6 summarizes the analysis of variance for this experiment. The model sum of squares is $SS_{\text{Model}} = SS_A + SS_B + SS_C + SS_{AB} = 23{,}127.63$, and this accounts for over 99% of the total variability in the response. The regression model that would be used to predict resistivity over the space of the three factors A, B, and C is

$$\hat{y} = 60.65625 + 11.60625x_1 + 34.03125x_2 + 10.45625x_3 + 6.58125x_1x_2$$

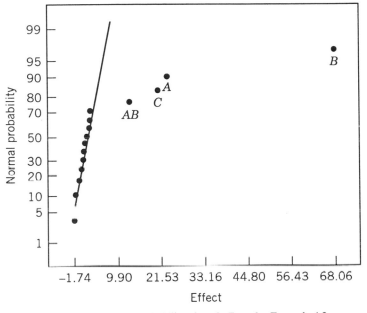

Figure 4.6 Normal probability plot of effects for Example 4.2.

where x_1, x_2, and x_3 are coded variables on the interval $-1, +1$ that represent A, B, and C.

Figure 4.7 presents a normal probability plot of the residuals from this model, and Figure 4.8 is a plot of the residuals versus the predicted values. Both plots are satisfactory.

The three factors A, B, and C have large positive effects. The AB or implant dose–temperature interaction is plotted in Figure 4.9. This plot indicates that the implant dose has a much stronger effect on resistivity when temperature is at the high level.

Table 4.6 Analysis of Variance for Example 4.2

Source of Variation	Sum of Squares	Degrees of Freedom	Mean Square	F_0
A	2155.28	1	2155.28	193.20
B	18,530.02	1	18,530.02	1791.24
C	1749.33	1	1749.33	184.61
AB	693.01	1	693.01	73.78
Error	132.59	11	12.05	
Total	23,260.22	15		

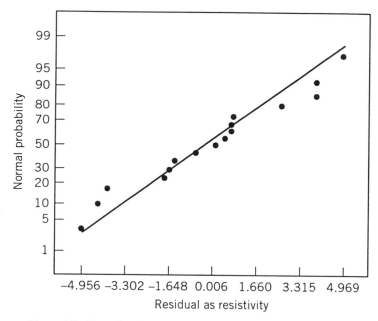

Figure 4.7   Normal probability plot of the residuals for Example 4.2.

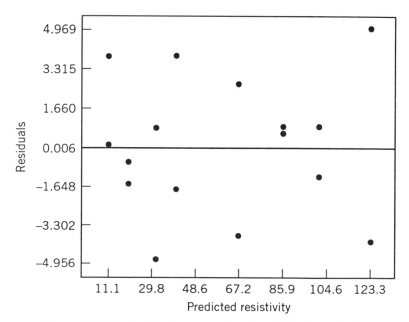

Figure 4.8   Plot of residuals versus predicted resistivity, Example 4.2.

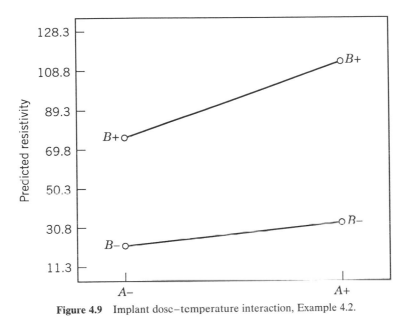

Figure 4.9 Implant dose–temperature interaction, Example 4.2.

The 2^{5-1} design will collapse into two replicates of a 2^3 design in any three of the original five factors. (Looking at Figure 4.5 will help you visualize this.) Figure 4.10 is a cube plot in the factors A, B, and C with the average resistivity superimposed on the eight corners. It is clear from inspection of the cube plot that highest resistivity values are achieved with A, B, and C all at the high level. Factors D and E have little effect on resistivity and may be set to values that optimize other objectives (such as cost).

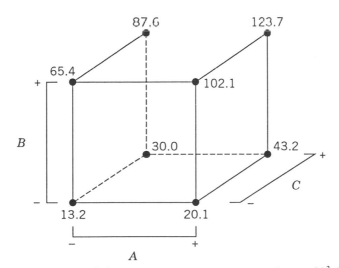

Figure 4.10 Projection of the 2_V^{5-1} design in Example 4.2 into two replicates of 2^3 design in the factors A, B, and C.

4.3   THE ONE-QUARTER FRACTION OF THE 2^k DESIGN

For a moderately large number of factors, smaller fractions of the 2^k design are frequently useful. Consider a one-quarter fraction of the 2^k design. This design contains 2^{k-2} runs and is usually called a 2^{k-2} fractional factorial.

The 2^{k-2} design may be constructed by first writing down a basic design consisting of the runs associated with a full factorial in $k - 2$ factors and then associating the two additional columns with generators that are appropriately chosen interactions involving the first $k - 2$ factors. Thus, a one-quarter fraction of the 2^k design has two generators. If P and Q represent the generators chosen, then $I = P$ and $I = Q$ are called **generating relations** for the design. The signs of P and Q (either $+$ or $-$) determine which one of the one-quarter fractions is produced. All four fractions associated with the choice of generators $\pm P$ and $\pm Q$ are members of the same family. The fraction for which both P and Q are positive is the **principal fraction**.

The **complete defining relation** for the design consists of all the columns that are equal to the identity column I. These will consist of P, Q, and their **generalized interaction** PQ; that is, the defining relation is $I = P = Q = PQ$. We call the elements P, Q, and PQ in the defining relation **words**. The aliases of any effect are produced by the multiplication of the column for that effect by each word in the defining relation. Clearly, each effect has three aliases. The experimenter should be careful in choosing the generators so that potentially important effects are not aliased with each other.

As an example, consider the 2^{6-2} design. Suppose we choose $I = ABCE$ and $I = BCDF$ as the design generators. Now the generalized interaction of the generators $ABCE$ and $BCDF$ is $ADEF$; therefore, the complete defining relation for this design is

$$I = ABCE = BCDF = ADEF$$

Consequently, this design is of resolution IV. To find the aliases of any effect (e.g., A), multiply that effect by each word in the defining relation. For A, this produces

$$A = BCE = ABCDF = DEF$$

It is easy to verify that every main effect is aliased by three-factor and five-factor interactions, whereas two-factor interactions are aliased with each other and with higher-order interactions. Thus, when we estimate A, for example, we are really estimating $A + BCE + DEF + ABCDF$. The complete alias structure of this design is shown in Table 4.7. If three-factor and higher interactions are negligible, this design gives clear estimates of the main effects.

To construct the design, first write down the basic design, which consists of the 16 runs for a full $2^{6-2} = 2^4$ design in A, B, C, and D. Then the two factors E and F are added by associating their plus and minus levels with the

Table 4.7 Alias Structure for the 2_{IV}^{6-2} Design with $I = ABCE = BCDF = ADEF$

$A = BCE = DEF = ABCDF$	$AB = CE = ACDF = BDEF$
$B = ACE = CDF = ABDEF$	$AC = BE = ABDF = CDEF$
$C = ABE = BDF = ACDEF$	$AD = EF = ABDF = CDEF$
$D = BCF = AEF = ABCDE$	$AE = BC = DF = ABCDEF$
$E = ABC = ADF = BCDEF$	$AF = DE = BCEF = ABCD$
$F = BCD = ADE = ABCEF$	$BD = CF = ACDE = ABEF$
	$BF = CD = ACEF = ABDE$
$ABD = CDE = ACF = BEF$	
$ACD = BDE = ABF = CEF$	

plus and minus signs of the interactions ABC and BCD, respectively. This procedure is shown in Table 4.8.

There are, of course, three alternate fractions of this particular 2_{IV}^{6-2} design. They are the fractions with generating relationships $I = ABCE$ and $I = -BCDF$, $I = -ABCE$ and $I = BCDF$; and $I = -ABCE$ and $I = -BCDF$. These fractions may be easily constructed by the method shown in Table 4.8. For example, if we wish to find the fraction for which $I = ABCE$ and $I = -BCDF$, then in the last column of Table 4.8 we set $F = -BCD$, and the column of levels for factor F becomes

$$+ + - - - - + + - - + + + + - -$$

Table 4.8 Construction of the 2_{IV}^{6-2} Design with the Generators $I = ABCE$ and $I = BCDF$

Run	A	B	C	D	$E = ABC$	$F = BCD$
1	−	−	−	−	−	−
2	+	−	−	−	+	−
3	−	+	−	−	+	+
4	+	+	−	−	−	+
5	−	−	+	−	+	+
6	+	−	+	−	−	+
7	−	+	+	−	−	−
8	+	+	+	−	+	−
9	−	−	−	+	−	+
10	+	−	−	+	+	+
11	−	+	−	+	+	−
12	+	+	−	+	−	−
13	−	−	+	+	+	−
14	+	−	+	+	−	−
15	−	+	+	+	−	+
16	+	+	+	+	+	+

The complete defining relation for this alternative fraction is $I = ABCE = -BCDF = -ADEF$. Certain signs in the alias structure in Table 4.7 are now changed; for instance, the aliases of A are $A = BCE = -DEF = -ABCDF$. Thus, the linear combination of the observations l_A actually estimates $A + BCE - DEF - ABCDF$.

Finally, note that the 2_{IV}^{6-2} fractional factorial collapses to a single replicate of a 2^4 design in any subset of four factors that is not a word in the defining relation. It also collapses to a replicated one-half fraction of a 2^4 in any subset of four factors that is a word in the defining relation. Thus, the design in Table 4.9 becomes two replicates of a 2^{4-1} in the factors $ABCE$, $BCDF$, and $ABEF$, because these are the words in the defining relation. There are 12 other combinations of the six factors, such as $ABCD$, $ABDF$, and so on, for which the design projects to a single replicate of the 2^4. This design also collapses to two replicates of a 2^3 in any subset of three of the six factors or four replicates of a 2^2 in any subset of two factors.

In general, any 2^{k-2} fractional factorial design can be collapsed into either a full factorial or a fractional factorial in some subset of $r \leq k - 2$ of the original factors. Those subsets of variables that form full factorials in $r = k - 2$ are not words in the complete defining relation. Any subset of $r = 3$ factors will give a full factorial design.

Table 4.9   A 2_{IV}^{6-2} Design for the Injection Molding Experiment in Example 4.3

Run	A	B	C	D	$E = ABC$	$F = BCD$	Observed Shrinkage ($\times 10$)
1	−	−	−	−	−	−	6
2	+	−	−	−	+	−	10
3	−	+	−	−	+	+	32
4	+	+	−	−	−	+	60
5	−	−	+	−	+	+	4
6	+	−	+	−	−	+	15
7	−	+	+	−	−	−	26
8	+	+	+	−	+	−	60
9	−	−	−	+	−	+	8
10	+	−	−	+	+	+	12
11	−	+	−	+	+	−	34
12	+	+	−	+	−	−	60
13	−	−	+	+	+	−	16
14	+	−	+	+	−	−	5
15	−	+	+	+	−	+	37
16	+	+	+	+	+	+	52
17	0	0	0	0	0	0	29
18	0	0	0	0	0	0	34
19	0	0	0	0	0	0	26
20	0	0	0	0	0	0	30

Example 4.3 Parts manufactured in an injection molding process are showing excessive shrinkage. This is causing problems in assembly operations downstream from the injection molding area. A quality improvement team has decided to use a designed experiment to study the injection molding process so that shrinkage can be reduced. The team decides to investigate six factors—mold temperature (A), screw speed (B), holding time (C), cycle time (D), gate size (E), and holding pressure (F)—each at two levels, with the objective of learning how each factor affects shrinkage and, also, something about how the factors interact.

The team decides to use the 16-run two-level fractional factorial design in Table 4.8 along with four center points, so that the linearity of the response function can be investigated via a curvature test. The design is shown in Table 4.9, along with the observed shrinkage ($\times 10$) for the test part produced at each of the 20 runs in the design.

A normal probability plot of the effect estimates from this experiment is shown in Figure 4.11. The only large effects are A (mold temperature), B (screw speed), and the AB interaction. In light of the alias relationships in Table 4.7, it seems reasonable to adopt these conclusions tentatively. Table 4.10 shows the analysis of variance for this experiment. The pure error sum of squares is computed from the replicate runs at the center, and the lack-of-fit sum of squares is formed by pooling all the effects that appeared small on the normal probability plot. Notice that the lack-of-fit and curvature tests have small F-ratios, so the model is correctly specified.

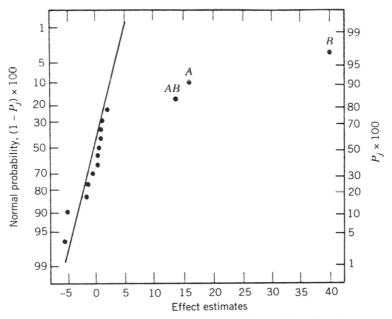

Figure 4.11 Normal probability plot of effects for Example 4.3.

Table 4.10   Analysis of Variance for the Injection Molding Experiment Example 4.3

Source of Variation	Sum of Squares	Degrees of Freedom	Mean Square	F_0	P-Value
A	770.063	1	770.063	41.033	1.18×10^{-5}
B	5076.563	1	5076.563	270.505	5.25×10^{-11}
AB	564.063	1	564.063	30.056	0.0001
Curvature	19.012	1	19.012	1.742	0.2786
Residual	281.500	15	18.767		
Lack-of-fit	(248.750)	(12)	20.729	1.899	0.3277
Pure error	(32.750)	(3)	10.917		
Total	6711.200	19			

The plot of the AB interaction in Figure 4.12 shows that the process is very insensitive to temperature if the screw speed is at the low level but very sensitive to temperature if the screw speed is at the high level. With the screw speed at the low level, the process should produce an average shrinkage of around 10% regardless of the temperature level chosen.

Based on this initial analysis, the team decides to set both the mold temperature and the screw speed at the low level. This set of conditions will reduce the mean shrinkage of parts to around 10%. However, the variability

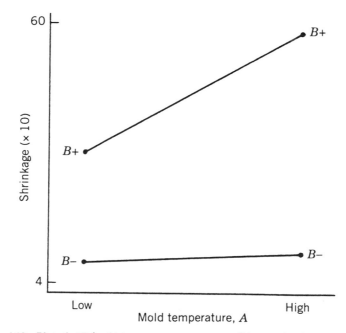

Figure 4.12   Plot of AB (mold temperature–screw speed) interaction for Example 4.3.

in shrinkage from part to part is still a potential problem. In effect, the mean shrinkage can be substantially reduced by the above modifications; however, the part-to-part variability in shrinkage over a production run could still cause problems in assembly. One way to address this issue is to see if any of the process factors affect the variability in parts shrinkage.

Figure 4.13 presents the normal probability plot of the residuals. This plot appears satisfactory. The plots of residuals versus each factor were then constructed. One of these plots, that for residuals versus factor C (holding time), is shown in Figure 4.14. The plot reveals that there is much less scatter in the residuals at the low holding time than at the high holding time. These residuals were obtained in the usual way from the model for predicted shrinkage

$$\hat{y} = b_0 + b_1 x_1 + b_2 x_2 + b_{12} x_1 x_2$$

$$= 27.3125 + 6.9375 x_1 + 17.8125 x_2 + 5.9375 x_1 x_2$$

where x_1, x_2, and $x_1 x_2$ are the coded variables that correspond to the factors A and B and the AB interaction. The residuals are then

$$e = y - \hat{y}$$

The regression model used to produce the residuals essentially removes the

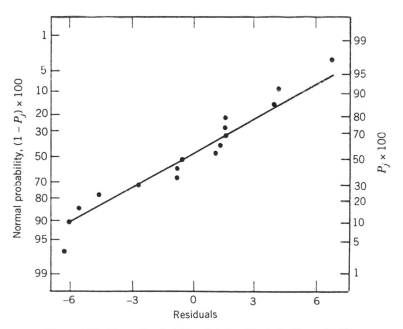

Figure 4.13   Normal probability plot of residuals for Example 4.3.

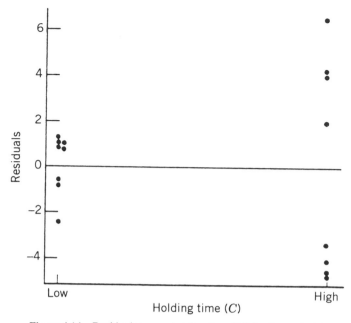

Figure 4.14   Residuals versus holding time (C) for Example 4.4.

location effects of A, B, and AB from the data; the residuals therefore contain information about unexplained variability. Figure 4.14 indicates that there is a pattern in the variability and that the variability in shrinkage of parts may be smaller when the holding time is at the low level.

Figure 4.15 shows the data from this experiment projected onto a cube in the factors A, B, and C. The average observed shrinkage and the range of observed shrinkage are shown at each corner of the cube. From inspection of this figure, we see that running the process with the screw speed (B) at the low level is the key to reducing average parts shrinkage. If B is low, virtually any combination of temperature (A) and holding time (C) will result in low values of average part shrinkage. However, from examining the ranges of the shrinkage values at each corner of the cube, it is immediately clear that setting the holding time (C) at the low level is the only reasonable choice if we wish to keep the part-to-part variability in shrinkage low during a production run.

In many response surface problems it will be useful to actually build a model for both the mean response and a measure of the variability in response (such as the range, the variance, or the standard deviation). Then the optimization problem is a tradeoff between these two criteria, such as minimizing the variability while simultaneously moving the mean response to a desired or target value.

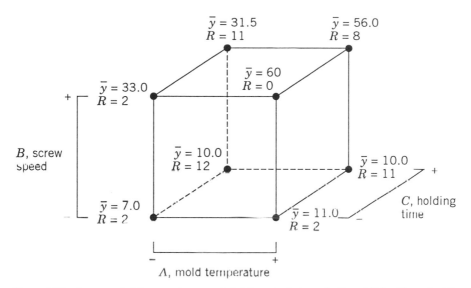

Figure 4.15 Average shrinkage and range of shrinkage in factors A, B, and C for Example 4.3.

4.4 THE GENERAL 2^{k-p} FRACTIONAL FACTORIAL DESIGN

A 2^k fractional factorial design containing 2^{k-p} runs is called a $1/2^p$ fraction of the 2^k design or, more simply, a 2^{k-p} fractional factorial design. These designs require the selection of p independent design generators. The defining relation for the design consists of the p generators initially chosen and their $2^p - p - 1$ generalized interactions. In this section we discuss the construction and analysis of these designs.

The alias structure may be found by multiplying each effect column by the defining relations. Care should be exercised in choosing the generators so that effects of potential interest are not aliased with each other. Each effect has $2^p - 1$ aliases. For moderately large values of k, we usually assume higher-order interactions (say, third- or fourth-order and higher) to be negligible, and this greatly simplifies the alias structure.

It is important to select the p generators for a 2^{k-p} fractional factorial design in such a way that we obtain the best possible alias relationships. A reasonable criterion is to select the generators such that the resulting 2^{k-p} design has the highest possible resolution. Table 4.11 presents a selection of 2^{k-p} fractional factorial designs for $k \leq 11$ factors and up to $n \leq 128$ runs. The suggested generators in this table will result in a design of the highest possible resolution. The alias relationships for all of the designs in Table 4.11 for which $n \leq 64$ are given in Appendix Table XII(a–v) of Montgomery (1991a).

Table 4.11   Selected 2^{k-p} Fractional Factorial Designs

Number of Factors k	Fraction	Number of Runs	Design Generators
3	2_{III}^{3-1}	4	$C = \pm AB$
4	2_{IV}^{4-1}	8	$D = \pm ABC$
5	2_{V}^{5-1}	16	$E = \pm ABCD$
	2_{III}^{5-2}	8	$D = \pm AB$
			$E = \pm AC$
6	2_{VI}^{6-1}	32	$F = \pm ABCDE$
	2_{IV}^{6-2}	16	$E = \pm ABC$
			$F = \pm BCD$
	2_{III}^{6-3}	8	$D = \pm AB$
			$E = \pm AC$
			$F = \pm BC$
7	2_{VII}^{7-1}	64	$G = \pm ABCDEF$
	2_{IV}^{7-2}	32	$F = \pm ABCD$
			$G = \pm ABDE$
	2_{IV}^{7-3}	16	$E = \pm ABC$
			$F = \pm BCD$
			$G = \pm ACD$
	2_{III}^{7-4}	8	$D = \pm AB$
			$E = \pm AC$
			$F = \pm BC$
			$G = \pm ABC$
8	2_{V}^{8-2}	64	$G = \pm ABCD$
			$H = \pm ABEF$
	2_{IV}^{8-3}	32	$F = \pm ABC$
			$G = \pm ABD$
			$H = \pm BCDE$
	2_{IV}^{8-4}	16	$E = \pm BCD$
			$F = \pm ACD$
			$G = \pm ABC$
			$H = \pm ABD$
9	2_{VI}^{9-2}	128	$H = \pm ACDFG$
			$J = \pm BCEFG$
	2_{IV}^{9-3}	64	$G = \pm ABCD$
			$H = \pm ACEF$
			$J = \pm CDEF$
	2_{IV}^{9-4}	32	$F = \pm BCDE$
			$G = \pm ACDE$
			$H = \pm ABDE$
			$J = \pm ABCE$
	2_{III}^{9-5}	16	$E = \pm ABC$
			$F = \pm BCD$
			$G = \pm ACD$
			$H = \pm ABD$
			$J = \pm ABCD$

Table 4.11 (*Continued*)

Number of Factors k	Fraction	Number of Runs	Design Generators
10	2_V^{10-3}	128	$H = \pm ABCG$ $J = \pm ACDE$ $K = \pm ACDF$
	2_{IV}^{10-4}	64	$G = \pm BCDF$ $H = \pm ACDF$ $J = \pm ABDE$ $K = \pm ABCE$
	2_{IV}^{10-5}	32	$F = \pm ABCD$ $G = \pm ABCE$ $H = \pm ABDE$ $J = \pm ACDE$ $K = \pm BCDE$
	2_{III}^{10-6}	16	$E = \pm ABC$ $F = \pm BCD$ $G = \pm ACD$ $H = \pm ABD$ $J = \pm ABCD$ $K = \pm AB$
11	2_{IV}^{11-5}	64	$G = \pm CDE$ $H = \pm ABCD$ $J = \pm ABF$ $K = \pm BDEF$ $L = \pm ADEF$
	2_{IV}^{11-6}	32	$F = \pm ABC$ $G = \pm BCD$ $H = \pm CDE$ $J = \pm ACD$ $K = \pm ADE$ $L = \pm BDE$
	2_{III}^{11-7}	16	$E = \pm ABC$ $F = \pm BCD$ $G = \pm ACD$ $H = \pm ABD$ $J = \pm ABCD$ $K = \pm AB$ $L = \pm AC$

Example 4.4 To illustrate the use of Table 4.11, suppose that we have seven factors and that we are interested in estimating the seven main effects and getting some insight regarding the two-factor interactions. We are willing to assume that three-factor and higher interactions are negligible. This information suggests that a resolution IV design would be appropriate.

Table 4.11 shows that there are two resolutions IV fractions available: the 2_{IV}^{7-2} design with 32 runs and the 2_{IV}^{7-3} with 16 runs. Consider the 16 run design. This is a one-eighth fraction with generators $E = \pm ABC$, $F = \pm BCD$, and $G = \pm ACD$, and if we choose the principal fraction the complete defining relation is

$$I = ABCE = BCDF = ADEF = ACDG = BDEG = ABFG = CEFG$$

Now because the design is of resolution IV, it is clear that the main effects will be aliased with at worst a three-factor interaction. So if we can safely ignore these higher-order interactions, this design will give excellent information on the main effects. The two-factor interaction alias chains are

$$AB = CE = FG$$
$$AC = BE = DG$$
$$AD = EF = CG$$
$$AE = BC = DF$$
$$AF = DE = BG$$
$$AG = CD = BF$$
$$BD = CF = EG$$

when ignoring three-factor and higher interactions. Thus this design would likely be satisfactory to satisfy the experimenter's objectives.

The complete design is shown in Table 4.12. Notice that it was constructed by starting with the 16-run 2^4 design in A, B, C, and D as the basic design and then adding the three columns $E = ABC$, $F = BCD$, and $G = ACD$ as suggested in Table 4.11.

Projection of the 2^{k-p} Fractional Factorial

The 2^{k-p} design collapses into either a full factorial or a fractional factorial in any subset of $r \leq k - p$ of the original factors. Those subsets of factors providing fractional factorials are subsets appearing as words in the complete defining relation. Furthermore, a design of resolution R will project into a full factorial in any subset of $R - 1$ factors. These results are particularly useful in screening experiments when we suspect at the outset of the experiment that most of the original factors will have small effects. The original factors will have small effects. The original 2^{k-p} fractional factorial can then be projected into a full factorial, say, in the most interesting factors. Conclusions drawn from designs of this type should be considered tentative and subject to further analysis. It is usually possible to find alternative explanations of the data involving higher-order interactions.

As an example, consider the 2_{IV}^{7-3} design from Example 4.4. This is a 16-run design involving seven factors. It will project into a full factorial in any four of the original seven factors that is not a word in the complete defining

Table 4.12 A 2_{IV}^{7-3} Fractional Factorial Design

Run	Basic Design				$E = ABC$	$F = BCD$	$G = ACD$
	A	B	C	D			
1	−	−	−	−	−	−	−
2	+	−	−	−	+	−	+
3	−	+	−	−	+	+	−
4	+	+	−	−	−	+	+
5	−	−	+	−	+	+	+
6	+	−	+	−	−	+	−
7	−	+	+	−	−	−	+
8	+	+	+	−	+	−	−
9	−	−	+	+	−	+	+
10	+	−	−	+	+	+	−
11	−	+	−	+	+	−	+
12	+	+	−	+	−	−	−
13	−	−	+	+	+	−	−
14	+	−	+	+	−	−	+
15	−	+	+	+	−	+	−
16	+	+	+	−	+	+	+

relation (see Table 4.12). Thus, there are 28 subsets of four factors that would form 2^4 designs. One combination that is obvious upon inspecting Table 4.12 is A, B, C, and D. It will also form a replicated full factorial in any subset of three of the original seven factors.

To illustrate the usefulness of this projection properly, suppose that we are conducting an experiment to improve the performance of a chemical process and the seven factors are:

1. Concentration of reactant A
2. Temperature
3. Feed rate
4. Time
5. Agitation rate
6. Concentration of reactant B
7. Catalyst type

We are fairly certain that time, temperature, catalyst type, and concentration of reactant A will affect performance and that these factors may interact. The role of the other three factors is less well known, but it is likely that they are negligible. A reasonable strategy would be to assign these four factors to columns A, B, C, and D, respectively, in Table 4.12. The remaining three factors would be assigned to columns E, F, and G, respectively. If we are correct and the "minor variables" E, F, and G are negligible, we will be left with a full 2^k design in the key process variables.

4.5   RESOLUTION III DESIGNS

As indicated earlier, the sequential use of fractional factorial designs is very useful, often leading to great economy and efficiency in experimentation. We now illustrate these ideas using the class of resolution III designs. Box and Hunter (1961a, b) are excellent references for the technical details.

It is possible to construct resolution III designs for investigating up to $k = N - 1$ factors in only N runs, where N is a multiple of 4. These designs are frequently useful in industrial experimentation. Designs in which N is a power of 2 can be constructed by the methods presented earlier in this chapter, and these are presented first. Of particular importance are designs requiring 4 runs for up to 3 factors, 8 runs for up to 7 factors, and 16 runs for up to 15 factors. If $k = N - 1$, the fractional factorial design is said to be saturated.

A design for analyzing up to three factors in four runs is the 2_{III}^{3-1} design, presented in Section 4.2. Another very useful saturated fractional factorial is a design for studying up to seven factors in eight runs; that is, the 2_{III}^{7-4} design. This design is a one-sixteenth fraction of the 2^7. It may be constructed by first writing down as the basic design the plus and minus levels for a full 2^3 design in A, B, and C and then associating the levels of four additional factors with the interactions of the original three as follows: $D = AB$, $E = AC$, $F = BC$, and $G = ABC$. Thus, the generators for this design are $I = ABD$, $I = ACE$, $I = BCF$, and $I = ABCG$. The design is shown in Table 4.13.

The complete defining relation for this design is obtained by multiplying the four generators ABD, ACE, BCF, and $ABCG$ together two at a time,

Table 4.13   The Saturated 2_{III}^{7-4} Design with the Generators $I = ABD$, $I = ACE$, $I = BCF$, and $I = ABCG$

Run	A	B	C	$D = AB$	$E = AC$	$F = BC$	$G = ABC$	
1	−	−	−	+	+	+	−	*def*
2	+	−	−	−	−	+	+	*afg*
3	−	+	−	−	+	−	+	*beg*
4	+	+	−	+	−	−	−	*abd*
5	−	−	+	+	−	−	+	*cdg*
6	+	−	+	−	+	−	−	*ace*
7	−	+	+	−	−	+	−	*cdg*
8	+	+	+	+	+	+	+	*abcdefg*

three at a time, and four at a time, yielding

$$I = ABD = ACE = BCF = ABCG = BCDE = ACDF = CDG$$
$$= ABEF = BEG = AFG = DEF = ADEG = CEFG = BDFG = ABCDEFG$$

To find the aliases of any effect, simply multiply the effect by each word in the defining relation. For example, the aliases of B are

$$B = AD = ABCE = CF = ACG = CDE = ABCDF = BCDG = AEF = EG$$
$$= ABFG = BDEF - ABDEG - BCEFG - DFG = ACDEFG$$

This design is a one-sixteenth fraction; and because the signs chosen for the generators are positive, this is the principal fraction. It is also of resolution III because the smallest number of letters in any word of the defining contrast is three. Any one of the 16 different 2_{III}^{7-4} designs in this family could be constructed by using the generators with one of the 16 possible arrangements of signs in $I = \pm ABC, I = \pm ACE, I = +BCF, I = \pm ABCG$.

The seven degrees of freedom in this design may be used to estimate the seven main effects. Each of these effects has 15 aliases; however, if we assume that three-factor and higher interactions are negligible, then considerable simplification in the alias structure results. Making this assumption, each of the linear combinations associated with the seven main effects in this design actually estimates the main effect and three two-factor interactions:

$$l_A \rightarrow A + BD + CE + FG$$
$$l_B \rightarrow B + AD + CF + EG$$
$$l_C \rightarrow C + AE + BF + DG$$
$$l_D \rightarrow D + AB + CG + EF \quad\quad\quad\quad (4.1)$$
$$l_E \rightarrow E + AC + BG + DF$$
$$l_F \rightarrow F + BC + AG + DE$$
$$l_G \rightarrow G + CD + BE + AF$$

In obtaining these aliases, we have ignored the three-factor and higher-order interactions, assuming that they will be negligible in most practical applications where a design of this type would be considered.

The saturated 2_{III}^{7-4} design in Table 4.13 can be used to obtain resolution III designs for studying fewer than seven factors in eight runs. For example, to generate a design for six factors in eight runs, simply drop any one column in Table 4.13, for example, column G. This produces the design shown in Table 4.14.

Table 4.14   A 2_{III}^{6-3} Design with the Generators $I = ABD$, $I = ACE$, and $I = BCF$

Run	A	B	C	$D = AB$	$E = AC$	$F = BC$	
1	−	−	−	+	+	+	def
2	+	−	−	−	−	+	af
3	−	+	−	−	+	−	be
4	+	+	−	+	−	−	abd
5	−	−	+	+	−	−	cd
6	+	−	+	−	+	−	ace
7	−	+	+	−	−	+	bcf
8	+	+	+	+	+	+	abcdef

It is easy to verify that this design is also of resolution III; in fact, it is a 2_{III}^{6-3} or a one-eighth fraction of the 2^6 design. The defining relation for the 2_{III}^{6-3} design is equal to the defining relation for the original 2_{III}^{7-4} design with any words containing the letter G deleted. Thus, the defining relation for our new design is

$$I = ABD = ACE = BCF = BCDE = ACDF = ABEF = DEF$$

In general, when d factors are dropped to produce a new design, the new defining relation is obtained as those words in the original defining relation that do not contain any dropped letters. When constructing designs by this method, care should be exercised to obtain the best arrangement possible. If we drop columns B, D, F, and G from Table 4.13, we obtain a design for three factors in eight runs, yet the treatment combinations correspond to two replicates of a 2^2 design. The experimenter would probably prefer to run a full 2^3 design in A, C, and E.

It is also possible to obtain a resolution III design for studying up to 15 factors in 16 runs. This saturated 2_{III}^{15-11} design can be generated by first writing down the 16 treatment combinations associated with a 2^4 design in A, B, C, and D and then equating 11 new factors with the 2-, 3-, and 4-factor interactions of the original 4. In this design, each of the 15 main effects is aliased with seven 2-factor interactions. A similar procedure can be used for the 2_{III}^{31-26} design, which allows up to 31 factors to be studied in 32 runs.

Sequential Use of Fractions to Separate Effects
By combining fractional factorial designs in which certain signs are switched, we can systematically isolate effects of potential interest. The alias structure for any fraction with the signs for one or more factors reversed is obtained by making changes of sign on the appropriate factors in the alias structure of the original fraction. This general procedure is called **fold-over**, and it is used in resolution III designs to break the links between main effects and two-factor interactions. In a **full fold-over**, we add to a resolution III fractional a second

fraction in which the signs for all the factors are reversed. We may now use the combined design to estimate all the main effects clear of any two-factor interactions. The following example illustrates the technique.

Example 4.5 A human performance analyst is conducting an experiment to study eye focus time and has built an apparatus in which several factors can be controlled during the test. The factors he initially regards as important are acuity or sharpness of vision (A), distance from target to eye (B), target shape (C), illumination level (D), target size (E), target density (F), and subject (G). Two levels of each factor are considered. He suspects that only a few of these seven factors are of major importance and that high-order interactions between the factors can be neglected. On the basis of this assumption, the analyst decides to run a screening experiment to identify the most important factors and then to concentrate further study on those. To screen these seven factors, he runs the treatment combinations from the 2_{III}^{7-4} design in Table 4.13 in random order, obtaining the focus times in milliseconds, as shown in Table 4.15.

Seven main effects and their aliases may be estimated from these data. From Equation (4.1), we see that the effects and their aliases are

$$l_A = \quad 20.63 \rightarrow A + BD + CE + FG$$

$$l_B = \quad 38.38 \rightarrow B + AD + CF + EG$$

$$l_C = -0.28 \rightarrow C + AE + BF + DG$$

$$l_D = \quad 28.88 \rightarrow D + AB + CG + EF$$

$$l_E = -0.28 \rightarrow E + AC + BG + DF$$

$$l_F = -0.63 \rightarrow F + BC + AG + DE$$

$$l_G = -2.43 \rightarrow G + CD + BE + AF$$

Table 4.15 A 2_{III}^{7-4} Design for the Eye Focus Time Experiment

Run	A	B	C	$D = AB$	$E = AC$	$F = BC$	$G = ABC$		Time
1	−	−	−	+	+	+	−	def	85.5
2	+	−	−	−	−	+	+	afg	75.1
3	−	+	−	−	+	−	+	beg	93.2
4	+	+	−	+	−	−	−	abd	145.4
5	−	−	+	+	−	−	+	cdg	83.7
6	+	−	+	−	+	−	−	ace	77.6
7	−	+	+	−	−	+	−	bcf	95.0
8	+	+	+	+	+	+	+	abcdefg	141.8

For example,

$$l_A = \tfrac{1}{4}(-85.5 + 75.1 - 93.2 + 145.4 - 83.7 + 77.6 - 95.0 + 141.8)$$

$$= 20.63$$

The three largest effects are l_A, l_B, and l_D. The simplest interpretation of the data is that the main effects of A, B, and D are all significant. However, this interpretation is not unique, because one could also logically conclude that A, B, and the AB interaction, or perhaps B, D, and the BD interaction, or perhaps A, D, and the AD interaction are the true effects.

Notice that ABD is a word in the defining relation for this design. Therefore, this 2_{III}^{7-4} design does not project into a full 2^3 factorial in A, B, and D; instead, it projects into two replicates of a 2^{3-1} design, as shown in Figure 4.16. Because the 2^{3-1} design is a resolution III design, A will be aliased with BD, B will be aliased with AD, and D will be aliased with AB, so the interactions cannot be separated from the main effects. The analyst here may have been unlucky. If he had assigned illumination level to column C in Table 4.13 instead of D, the design would have projected into a 2^3 design.

To separate the main effects and the two-factor interactions, a second fraction is run with all the signs reversed. This fold-over design is shown in Table 4.16 along with the observed responses. Notice that when we fold over a resolution III design in this manner, we (in effect) change the signs on the generators that have an odd number of letters. The effects estimated by this

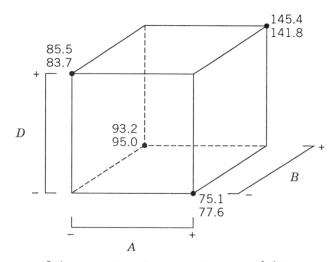

Figure 4.16   The 2_{III}^{7-4} design projected into two replicates of a 2_{III}^{3-1} design in A, B, and D.

Table 4.16 Fold-Over of the 2_{III}^{7-4} Design in Table 4.15

Run	A	B	C	$D = -AB$	$E = -AC$	$F = -BC$	$G = ABC$		Time
1	+	+	+	−	−	−	+	$abcg$	91.3
2	−	+	+	+	+	−	−	$bcde$	136.7
3	+	−	+	+	−	+	−	$acdf$	82.4
4	−	−	+	−	+	+	+	$cefg$	73.4
5	+	+	−	−	+	+	−	$abef$	94.1
6	−	+	−	+	−	+	+	$bdfg$	143.8
7	+	−	−	+	+	−	+	$adeg$	87.3
8	−	−	−	−	−	−	−	(1)	71.9

fraction are

$$l'_A = -17.68 \rightarrow A - BD - CE - FG$$

$$l'_B = 37.73 \rightarrow B - AD - CF - EG$$

$$l'_C = -3.33 \rightarrow C - AE - BF - DG$$

$$l'_D = 29.88 \rightarrow D \quad AB - CG - EF$$

$$l'_E = 0.53 \rightarrow E - AC - BG - DF$$

$$l'_F = 1.63 \rightarrow F - BC - AG - DE$$

$$l'_G = 2.68 \rightarrow G - CD - BE - AF$$

By combining the effect estimates from this second fraction with the effect estimates from the original eight runs, we obtain the following estimates of the effects:

i	From $\frac{1}{2}(l_i + l'_i)$	From $\frac{1}{2}(l_i - l'_i)$
A	$A = 1.48$	$BD + CE + FG = 19.15$
B	$B = 38.05$	$AD + CF + EG = 0.33$
C	$C = -1.80$	$AE + BF + DG = 1.53$
D	$D = 29.38$	$AB + CG + EF = -0.50$
E	$E = 0.13$	$AC + BG + DF = -0.40$
F	$F = 0.50$	$BC + AG + DE = -1.53$
G	$G = 0.13$	$CD + BE + AF = -2.55$

The two largest effects are B and D. Furthermore, the third largest effect is $BD + CE + FG$, so it seems reasonable to attribute this to the BD interaction. The analyst used the two factors distance (B) and illumination level (D) in subsequent experiments with the other factors A, C, E, and F at standard settings and verified the results obtained here. He decided to use subjects as blocks in these new experiments rather than ignore a potential subject effect, because several different subjects had to be used to complete the experiment.

While the full fold-over strategy illustrated in Example 4.5 is used very frequently, there is another variation of this technique that occasionally proves helpful. In a **single-factor fold-over** we add to a fractional factorial design of resolution III a second fraction of the same size with the signs for only one of the factors reversed. In the combined design, we will be able to estimate the main effect of the factor for which the signs were reversed as well as all two-factor interactions involving that factor.

To illustrate, consider the 2_{III}^{7-4} design in Table 4.13. Suppose that along with this principal fraction a second fractional design with the signs reversed in the column for factor D is also run. That is, the column for D in the second fraction is

$$- + + - - + + -$$

The effects that may be estimated from the first fraction are shown in Equation (4.1), and from the second fraction we obtain

$$l'_A = \rightarrow A - BD + CE + FG$$
$$l'_B = \rightarrow B - AD + CF + EG$$
$$l'_C = \rightarrow C + AE + BF - DG$$
$$l'_D = \rightarrow D - AB - CG - EF$$

that is,

$$l'_{-D} = \rightarrow -D + AB + CG + EF$$
$$l'_E = \rightarrow \quad E + AC + BG - DF$$
$$l'_F = \rightarrow \quad F + BC + AG - DE$$
$$l'_G = \rightarrow \quad G - CD + BE + AF$$

assuming that three-factor and higher interactions are insignificant. Now from the two linear combinations of effects $\frac{1}{2}(l_i + l'_i)$ and $\frac{1}{2}(l_i - l'_i)$ we obtain

i	$\frac{1}{2}(l_i + l'_i)$	$\frac{1}{2}(l_i - l'_i)$
A	$A + CE + FG$	BD
B	$B + CF + EG$	AD
C	$C + AE + BF$	DG
D	D	$AB + CG + EF$
E	$E + AC + BG$	DF
F	$F + BC + AG$	DE
G	$G + BE + AF$	CD

Thus, we have isolated the main effect of D and all of its two-factor interactions.

The Defining Relation for a Fold-Over Design

Combining fractional factorial designs via fold over as demonstrated in Example 4.5 is a very useful technique. It is often of interest to know the defining relation for the combined design. Fortunately, this can be easily determined. Each separate fraction will have $L + U$ words used as generators: L words of like sign and U words of unlike sign. The combined design will have $L + U - 1$ words used as generators. These will be the L words of like sign and the $U - 1$ words consisting of independent even products of the words of unlike sign. (Even products are words taken two at a time, four at a time, and so forth.)

To illustrate this procedure, consider the design in Example 4.5. For the first fraction, the generators are

$$I = ABD, \qquad I = ACE, \qquad I = BCF, \quad \text{and} \quad I = ABCG$$

and for the second fraction, they are

$$I = -ABD, \qquad I = -ACE, \qquad I = -BCF, \quad \text{and} \quad I = ABCG$$

Notice that in the second fraction we have switched the signs on the generators with an odd number of letters. Also, notice that $L + U = 1 + 3 = 4$. The combined design will have $I = ABCG$ (the like sign word) as a generator, and two words that are independent even products of the words of unlike sign. For example, take $I = ABD$ and $I = ACE$; then $I = (ABD)(ACE) = BCDE$ is a generator of the combined design. Also, take $I = ABD$ and $I = BCF$; then $I = (ABD)(BCF) = ACDF$ is a generator of the combined design. The complete defining relation for the combined design is

$$I = ABCG = BCDE = ACDF = ADEG = BDFG = ABEF = CEFG$$

Because the defining relation for the combined design contains only four-letter words, the combined design is of resolution IV.

Plackett–Burman Designs

These designs, attributed to Plackett and Burman (1946), are two-level fractional designs for studying up to $k = N - 1$ variables in N runs, where N is a multiple of 4. If N is a power of 2, these designs are identical to those presented earlier in this section. However, for $N = 12, 20, 24, 28$, and 36, the Plackett–Burman designs are sometimes of interest.

The upper half of Table 4.17 presents rows of plus and minus signs that are used to construct the Plackett–Burman designs for $N = 12, 20, 24$, and 36, whereas the lower half of the table presents blocks of plus and minus

Table 4.17   Plus and Minus Signs for the Plackett–Burman Designs

$k = 11, N = 12$ $+ + - + + + - - - + -$
$k = 19, N = 20$ $+ + - - + + + + - + - + - - - - + + -$
$k = 23, N = 24$ $+ + + + + - + - + + - - + + - - + - + - - - -$
$k = 35, N = 36$ $- + - + + + - - - + + + + + - + + + - - + - - - - + - + - + + - - + -$

$k = 27, N = 28$		
$+ - + + + + - - -$	$- + - - - + - - +$	$+ + - + - + + - +$
$+ + - + + + - - -$	$- - + + - - + - -$	$- + + + + - + + -$
$- + + + + + - - -$	$+ - - - + - - + -$	$+ - + - + + - + +$
$- - - + - + + + +$	$- - + - + - - - +$	$+ - + + + - + - +$
$- - - + + - + + +$	$+ - - - - + + - -$	$+ + - - + + + + -$
$- - - - + + + + +$	$- + - + - - - + -$	$- + + + - + - + +$
$+ + + - - - + - +$	$- - + - - + - + -$	$+ - + + - + + + -$
$+ + + - - - + + -$	$+ - - + - - - - +$	$+ + - + + - - + +$
$+ + + - - - - + +$	$- + - - + - + - -$	$- + + - + + + - +$

signs for constructing the design for $N = 28$. The designs for $N = 12, 20, 24$, and 36 are obtained by writing the appropriate row in Table 4.17 as a column (or row). A second column for (or row) is then generated from this first one by moving the elements of the column (or row) down (or to the right) one position and placing the last element in the first position. A third column (or row) is produced from the second similarly, and the process continued until column (or row) k is generated. A row of minus signs is then added, completing the design. For $N = 28$, the three blocks X, Y, and Z are written down in the order

$$
\begin{array}{ccc}
X & Y & Z \\
Z & X & Y \\
Y & Z & X
\end{array}
$$

and a row of minus signs is added to these 27 rows. The design for $N = 12$ runs and $k = 11$ factors is shown in Table 4.18.

The Plackett–Burman design for $N = 12, 20, 24, 28$, and 36 have very messy alias structures. For example, in the 12-run design every main effect is partially aliased with every two-factor interaction not involving itself. For example, the AB interaction is aliased with the nine main effects $C, D, \ldots, K$. Furthermore, each main effect is partially aliased with 45 two-factor interactions. In the larger designs, the situation is even more complex. We advise experimenter to use these designs very carefully.

The projection properties of the Plackett–Burman designs are not terribly attractive. For example, consider the 12-run design in Table 4.18. This design will project into three replicates of a full 2^2 design in any 2 of the original 11 factors. However, in 3 factors, the projected design is a full 2^3 factorial plus a 2_{III}^{3-1} fractional factorial [see Figure 4.17(a)]. The four-dimensional projections are shown in Figure 4.17(b). Notice that these 3- and 4-factor projec-

Table 4.18 Plackett-Burman Design for $N = 12$, $k = 11$

Run	A	B	C	D	E	F	G	H	I	J	K
1	+	−	+	−	−	−	+	+	+	−	+
2	+	+	−	+	−	−	−	+	+	+	−
3	−	+	+	−	+	−	−	−	+	+	+
4	+	−	+	+	−	+	−	−	−	+	+
5	+	+	−	+	+	−	+	−	−	−	+
6	+	+	+	−	+	+	−	+	−	−	−
7	−	+	+	+	−	+	+	−	+	−	−
8	−	−	+	+	+	−	+	+	−	+	−
9	−	−	−	+	+	+	−	+	+	−	+
10	+	−	−	−	+	+	+	−	+	+	−
11	−	+	−	−	−	+	+	+	−	+	+
12	−	−	−	−	−	−	−	−	−	−	−

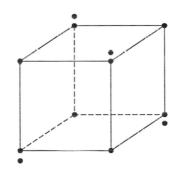

(a)

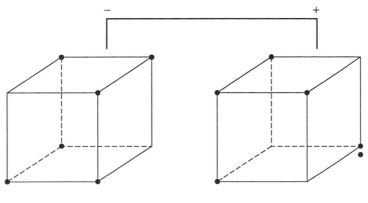

(b)

Figure 4.17 Projection of the 12-run Plackett–Burman design into three- and four-factor designs. (a) Projection into three factors. (b) Projection into four factors.

tions are not balanced designs. This would complicate their interpretation. Plackett–Burman designs with $N = 12$, 20, 24, 28, and 36 are often called **nongeometric** Plackett–Burman designs.

4.6   RESOLUTION IV AND V DESIGNS

A 2^{p-k} fractional factorial design is of resolution IV if the main effects are clear of two-factor interactions, but some two-factor interactions are aliased with each other. Thus, if three-factor and higher interactions are suppressed, the main effects may be estimated directly in a 2_{IV}^{k-p} design. An example is the 2_{IV}^{6-2} design in Table 4.9. Furthermore, the two combined fractions of the 2_{III}^{7-4} design in Example 4.5 yield a 2_{IV}^{7-3} design.

Any 2_{IV}^{k-p} design must contain at least $2k$ runs. Resolution IV designs that contain exactly $2k$ runs are called **minimal designs**. Resolution IV designs may be obtained from resolution III designs by the process of fold-over. Recall that to fold over a design, simply add to the original fraction a second fraction with all the signs reversed. Then the plus signs in the identity column I in the first fraction could be switched in the second fraction, and a $(k + 1)$st factor associated with this column. The result is a 2_{IV}^{k+1-p} fractional factorial design. The process is demonstrated in Table 4.19 for the 2_{III}^{3-1} design. It is easy to verify that the resulting design is a 2_{IV}^{4-1} design with defining relation $I = ABCD$.

It is also possible to fold over resolution IV designs to separate two-factor interactions that are aliased with each other. To fold over a resolution IV design, run a second fraction in which the sign is reversed on every design generator that has an **even number of letters**. To illustrate, consider the 2_{IV}^{6-2} design used for the injection molding experiment in Example 4.3. The generators for the design in Table 4.11 are $I = ABCE$ and $I = BCDF$. The

Table 4.19   A 2_{IV}^{4-1} Design Obtained by Fold-Over

D I	A	B	C
Original 2_{III}^{3-1} with $I = ABC$			
+	−	−	+
+	+	−	−
+	−	+	−
+	+	+	+
Second 2_{III}^{3-1} with Signs Switched			
−	+	+	−
−	−	+	+
−	+	−	+
−	−	−	−

second fraction would use the generators $I = -ABCE$ and $I = -BCDF$, and the single generator for the combined design would be $I = ADEF$. Thus, the combined design is still a resolution IV fractional factorial design. However, the alias relationships will be much simpler than in the original 2_{IV}^{6-2} fractional. In fact, the only two-factor interactions that will be aliased are $AD = EF$, $AE = DF$, and $AF = DE$. All the other two-factor interactions can be estimated from the combined design.

Notice that when you start with a resolution III design, the fold-over procedure guarantees that the combined design will be of resolution IV, thereby ensuring that all the main effects can be separated from their two-factor interaction aliases. When folding over a resolution IV design, we will not necessarily separate all the two-factor interactions. In fact, if the original fraction has an alias structure with more than two two factor interactions in any alias chain, folding over will not completely separate all the two-factor interactions. Notice that in the foregoing example, the 2_{IV}^{6-2} has one such two-factor interaction alias chain.

Resolution V designs are fractional factorials in which the main effects and the two-factor interactions do not have other main effects and two-factor interactions as their aliases. These are very powerful designs, allowing the unique estimation of all the main effects and two factor interactions provided that all the three-factor and higher interactions are negligible. The smallest word in the defining relation of such a design must have five letters. The 2^{5-1} with the generating relation $I = ABCDE$ is of resolution V. Another example is the 2_V^{8-2} design with the generating relations $I = ABCDG$ and $I - ABEFH$. Further examples of these designs are given by Box and Hunter (1961b).

4.7 SUMMARY

This chapter has introduced the 2^{k-p} fractional factorial design. We have emphasized the use of these designs in screening experiments to identify quickly and efficiently the subset of factors that are active and to provide some information on interaction. The projection property of these designs makes it possible in many cases to examine the active factors in more detail. Sequential assembly of these designs via fold-over is a very effective way to gain additional information about interactions that an initial experiment may identify as possibly important.

In practice, 2^{k-p} fractional factorial designs with $N = 4$, 8, 16, and 32 runs are highly useful. Table 4.20 summarizes these designs, identifying how many factors can be used with each design to obtain various types of screening experiments. For example, the 16-run design is a full factorial for 4 factors, a one-half fraction for 5 factors, resolution IV fractional design for 6–8 factors, and a resolution III fraction for 9–15 factors. All of these

Table 4.20 Useful Factorial and Fractional Factorial Designs from the 2^{k-p} System[a]

Design Type	Number of Runs			
	4	8	16	32
Full factorial	2	3	4	5
Half-fraction	3	4	5	6
Resolution IV fraction	—	4	6–8	7–16
Resolution III fraction	3	5–7	9–15	17–31

[a]The numbers in the cells are the number of factors in the experiment.

designs may be constructed using the methods discussed in this chapter. Montgomery (1991a) gives the alias relationships for many of these designs.

EXERCISES

4.1 Suppose that in the chemical process development experiment described in Exercise 3.6, it was only possible to run a one-half fraction of the 2^4 design. Construct the design and perform the statistical analysis.

4.2 Suppose that in Exercise 3.9, only a one-half fraction of the 2^4 design could be run. Construct the design and perform the analysis.

4.3 Consider the plasma etch experiment described in Exercise 3.10. Suppose that only a one-half fraction of the design could be run. Set up the design and analyze the data.

4.4 Example 4.2 describes a process improvement study in the manufacturing process of an integrated circuit. Suppose that only eight runs could be made in this process. Set up an appropriate 2^{5-2} design and find the alias structure. Use the data from Example 4.2 as the observations in this design and estimate the factor effects. What conclusions can you draw?

4.5 **Continuation of Exercise 4.4.** Suppose you have made the eight runs in the 2^{5-2} design in Exercise 4.4. What additional runs would be required to identify the factor effects that are of interest? What are the alias relationships in the combined design?

4.6 R. D. Snee ("Experimenting with a Large Number of Variables," in *Experiments in Industry: Design, Analysis and Interpretation of Results*, by R. D. Snee, L. B. Hare, and J. B. Trout, editors, ASQC, 1985) describes an experiment in which a 2^{5-1} design with $I = ABCDE$ was used to investigate that effects of five factors on the color of a chemical product. The factors are A = solvent/reactant, B = catalyst/reactant, C = temperature, D = reactant purity, and D = reactant pH. The results obtained were as follows:

$$e = -0.63 \qquad d = 6.79$$
$$a = 2.51 \qquad ade = 5.47$$
$$b = -2.68 \qquad bde = 3.45$$
$$abe = 1.66 \qquad abd = 5.68$$
$$c = 2.06 \qquad cde = 5.22$$
$$ace = 1.22 \qquad acd = 4.38$$
$$bce = -2.09 \qquad bcd = 4.30$$
$$abc = 1.93 \qquad abcde = 4.05$$

(a) Prepare a normal probability plot of the effects. Which effects seem active?

(b) Calculate the residuals. Construct a normal probability plot of the residuals and plot the residuals versus the fitted values. Comment on the plots.

(c) If any factors are negligible, collapse the 2^{5-1} design into a full factorial in the active factors. Comment on the resulting design, and interpret the results.

4.7 An article in the *Journal of Quality Technology* (Vol. 17, 1985, pp. 198–206) describes the use of a replicated fractional factorial to investigate the effect of five factors on the free height of leaf springs used in a automotive application. The factors are A = furnace temperature, B = heating time, C = transfer time, D = hold down time, and E = quench oil temperature. The data are shown below:

A	B	C	D	E	Free Height		
−	−	−	−	−	7.78	7.78	7.81
+	−	−	+	−	8.15	8.18	7.88
−	+	−	+	−	7.50	7.56	7.50
+	+	−	−	−	7.59	7.56	7.75
−	−	+	+	−	7.54	8.00	7.88
+	−	+	−	−	7.69	8.09	8.06
−	+	+	−	−	7.56	7.52	7.44
+	+	+	+	−	7.56	7.81	7.69
−	−	−	−	+	7.50	7.25	7.12
+	−	−	+	+	7.88	7.88	7.44
−	+	−	+	+	7.50	7.56	7.50
+	+	−	−	+	7.63	7.75	7.56
−	−	+	+	+	7.32	7.44	7.44
+	−	+	−	+	7.56	7.69	7.62
−	+	+	−	+	7.18	7.18	7.25
+	+	+	+	+	7.81	7.50	7.59

(a) Write out the alias structure for this design. What is the resolution of this design?

(b) Analyze the data. What factors influence the mean free height?

(c) Calculate the range and standard deviation of the free height for each run. Is there any indication that any of these factors affects variability in the free height?

(d) Analyze the residuals from this experiment, and comment on your findings.

(e) Is this the best possible design for five factors in 16 runs? Specifically, can you find a fractional design for five factors in 16 runs with a higher resolution than this one?

4.8 Reconsider the experiment in Exercise 4.6. Suppose that along with the original 16 runs the experimenter had run four center points with the following observed responses: 5.20, 4.98, 5.26, and 5.40.

(a) Use the center points to obtain an estimate of pure error.

(b) Test for curvature in the response function and lack of fit.

(c) What type of model seems appropriate for this purpose?

(d) Can you fit the model in part (c) using the data from this experiment?

4.9 An article in *Industrial and Engineering Chemistry* ("More on Planning Experiments to Increase Research Efficiency," 1970, pp. 60–65) uses a 2^{5-2} design to investigate the effect of A = condensation temperature, B = amount of material 1, C = solvent volume, D = condensation time, and E = amount of material 2 on yield. The results obtained are as follows:

$$e = 23.2 \quad ad = 16.9 \quad cd = 23.8 \quad bde = 16.8$$

$$ab = 15.5 \quad bc = 16.2 \quad ace = 23.4 \quad abcde = 18.1$$

(a) Verify that the design generators used were $I = ACE$ and $I = BDE$.

(b) Write down the complete defining relation and the aliases for this design.

(c) Estimate the main effects.

(d) Prepare an analysis of variance table. Verify that the AB and AD interactions are available to use as error.

(e) Plot the residuals versus the fitted values. Also construct a normal probability plot of the residuals. Comment on the results.

4.10 Reconsider the experiment in Exercise 4.9. Suppose that four center-points were run, and that the responses are as follows: 20.1, 20.9, 19.8, and 20.4.

(a) Use the center points to obtain an estimate of pure error.

(b) Test for lack of fit and curvature. What conclusions can you draw about the type of model required for yield in this process?

4.11 An industrial engineer is conducting an experiment using a Monte Carlo simulation model of an inventory system. The independent variables in her model are the order quantity (A), the reorder point (B), the setup cost (C), the backorder cost (D), and the carrying cost rate (E). The response variable is average annual cost. To conserve computer time, she decides to investigate these factors using a 2_{III}^{5-2} design with $I = ABD$ and $I = BCE$. The results she obtains are $de = 95$, $ae = 134$, $b = 158$, $abd = 190$, $cd = 92$, $ac = 187$, $bce = 155$, and $abcde = 185$.

(a) Verify that the treatment combinations given are correct. Estimate the effects, assuming three-factor and higher interaction are negligible.

(b) Suppose that a second fraction is added to the first. The runs in this new design are $ade = 136$, $e = 93$, $ab = 187$, $bd = 153$, $acd = 139$, $c = 99$, $abce = 191$, and $bcde = 150$. How was this second fraction obtained? Add these runs to the original fraction, and estimate the effects.

(c) Suppose that the fraction $abc = 189$, $ce = 96$, $bcd = 154$, $acde = 135$, $abe = 193$, $bde = 152$, $ad = 137$, and $(1) = 98$ was run. How was this fraction obtained? Add these data to the original fraction and estimate the effects.

4.12 Carbon anodes used in a smelting process are baked in a ring furnace. An experiment is run in the furnace to determine which factors influence the weight of packing material that is stuck to the anodes after baking. Six variables are of interest, each at two levels: A = pitch/fines ratio (0.45, 0.55); B = packing material type (1, 2); C = packing material temperature (ambient, 325°C); D = flue location (inside, outside); E = pit temperature (ambient, 195°C); and F = delay time before packing (zero, 24 hours). A 2^{6-3} design is run, and three replicates are obtained at each of the design points. The weight of packing material stuck to the anodes is measured in grams. The data in run order are as follows: $abd = (984, 826, 936)$; $abcdef = (1275, 976, 1457)$; $be = (1217, 1201, 890)$; $af = (1474, 1164, 1541)$; $def = (1320, 1156, 913)$; $cd = (765, 705, 821)$; $ace = (1338, 1254, 1294)$; and $bcf = (1325, 1299, 1253)$. We wish to minimize the amount of stuck packing material.

(a) Verify that the eight runs correspond to a 2_{III}^{6-3} design. What is the alias structure?

(b) Use the average weight as a response. What factors appear to be influential?

(c) Use the range of the weights as a response. What factors appear to be influential?

(d) What recommendations would you make to the process engineers?

4.13 A 16-run experiment was performed in a semiconductor manufacturing plant to study the effects of six factors on the curvature or camber of the substrate devices produced. The six variables and their levels are shown below:

Run	Lamination Temperature (°C)	Lamination Time (sec)	Lamination Pressure (ton)	Firing Temperature (°C)	Firing Cycle Time (hr)	Firing Dew Point (°C)
1	55	10	5	1580	17.5	20
2	75	10	5	1580	29	26
3	55	25	5	1580	29	20
4	75	25	5	1580	17.5	26
5	55	10	10	1580	29	26
6	75	10	10	1580	17.5	20
7	55	25	10	1580	17.5	26
8	75	25	10	1580	29	20
9	55	10	5	1620	17.5	26
10	75	10	5	1620	29	20
11	55	25	5	1620	29	26
12	75	25	5	1620	17.5	20
13	55	10	10	1620	29	20
14	75	10	10	1620	17.5	26
15	55	25	10	1620	17.5	20
16	75	25	10	1620	29	26

Each run was replicated four times, and a camber measurement was taken on the substrate. The data are shown below:

Run	Camber for Replicate (in. / in.)				Total (10^{-4} in. / in.)	Mean (10^{-4} in. / in.)	Standard Deviation
	1	2	3	4			
1	0.0167	0.0128	0.0149	0.0185	629	157.25	24.418
2	0.0062	0.0066	0.0044	0.0020	192	48.00	20.976
3	0.0041	0.0043	0.0042	0.0050	176	44.00	4.0825
4	0.0073	0.0071	0.0039	0.0030	223	55.75	25.025
5	0.0047	0.0047	0.0040	0.0089	223	55.75	22.410
6	0.0219	0.0258	0.0147	0.0296	920	230.00	63.639
7	0.0121	0.0090	0.0092	0.0086	389	97.25	16.029

	Camber for Replicate (in. / in)				Total	Mean	Standard
Run	1	2	3	4	$(10^{-4}$ in. / in.)	$(10^{-4}$ in. / in.)	Deviation
8	0.0255	0.0250	0.0226	0.0169	900	225.00	39.42
9	0.0032	0.0023	0.0077	0.0069	201	50.25	26.725
10	0.0078	0.0158	0.0060	0.0045	341	85.25	50.341
11	0.0043	0.0027	0.0028	0.0028	126	31.50	7.681
12	0.0186	0.0137	0.0158	0.0159	640	160.00	20.083
13	0.0110	0.0086	0.0101	0.0158	455	113.75	31.12
14	0.0065	0.0109	0.0126	0.0071	371	92.75	29.51
15	0.0155	0.0158	0.0145	0.0145	603	150.75	6.75
15	0.0093	0.0124	0.0110	0.0133	460	115.00	17.45

(a) What type of design did the experimenters use?

(b) What are the alias relationships in this design?

(c) Do any of the process variables affect average camber?

(d) Do any of the process variables affect the variability in camber measurements?

(e) If it is important to reduce mean camber as much as possible, what recommendations would you make?

(f) Fit an appropriate model for both mean camber and the standard deviation of camber. If we would like to reduce the variability in camber while keeping the mean camber as close as possible to 100×10^{-4} in. / in., what recommendations would you make?

4.14 A spin coater is used to apply photoresist to a bare silicon wafer. This operation usually occurs early in the semiconductor manufacturing process, and the average coating thickness and the variability in the coating thickness has an important impact on downstream manufacturing steps. Six variables are used in the experiment. The variables and their high and low levels are as follows:

Factor	Low Level	High Level
Final spin speed	7300 rpm	6650 rpm
Acceleration rate	5	20
Volume of resist applied	3 cc	5 cc
Time of spin	14 s	6 s
Resist batch variation	Batch 1	Batch 2
Exhaust pressure	Cover off	Cover on

The experimenter decides to use a 2^{6-1} design, and to make three readings on resist thickness on each test wafer. The data are shown in the table below.

(a) Verify that this is a 2^{6-1} design. Discuss the alias relationships in this design.

(b) What factors appear to affect average resist thickness?

(c) Since the volume of resist applied has little affect on average thickness, does this have any important practical implications for the process engineers?

(d) Project this design into a smaller design involving only significant factors. Graphically display the results. Does this aid in interpretation?

(e) Use the range of resist thickness as a response variable. Is there any indication that any of these factors affect the variability in resist thickness?

(f) Where would you recommend that the process engineers run the process?

Data for Problem 4.14

	A	B	C	D	E	F	Resist Thickness			
Run	Volume	Batch	Time	Speed	Acc.	Cover	Left	Center	Right	Average
1	5	Batch 2	14	7350	5	Off	4531	4531	4515	4525.7
2	5	Batch 1	6	7350	5	Off	4446	4464	4428	4446
3	3	Batch 1	6	6650	5	Off	4452	4490	4452	4464.7
4	3	Batch 2	14	7350	20	Off	4316	4328	4308	4317.3
5	3	Batch 1	14	7350	5	Off	4307	4295	4289	4297
6	5	Batch 1	6	6650	20	Off	4470	4492	4495	4485.7
7	3	Batch 1	6	7350	5	On	4496	4502	4482	4493.3
8	5	Batch 2	14	6650	20	Off	4542	4547	4538	4542.3
9	5	Batch 1	14	6650	5	Off	4621	4643	4613	4625.7
10	3	Batch 1	14	6650	5	On	4653	4670	4645	4656
11	3	Batch 2	14	6650	20	On	4480	4486	4470	4478.7
12	3	Batch 1	6	7350	20	Off	4221	4233	4217	4223.7
13	5	Batch 1	6	6650	5	On	4620	4641	4619	4626.7
14	3	Batch 1	6	6650	20	On	4455	4480	4466	4467
15	5	Batch 2	14	7350	20	On	4255	4288	4243	4262
16	5	Batch 2	6	7350	5	On	4490	4534	4523	4515.7
17	3	Batch 2	14	7350	5	On	4514	4551	4540	4535
18	3	Batch 1	14	6650	20	Off	4494	4503	4496	4497.7
19	5	Batch 2	6	7350	20	Off	4293	4306	4302	4300.3
20	3	Batch 2	6	7350	5	Off	4534	4545	4512	4530.3
21	5	Batch 1	14	6650	20	On	4460	4457	4436	4451
22	3	Batch 2	6	6650	5	On	4650	4688	4656	4664.7
23	5	Batch 1	14	7350	20	Off	4231	4244	4230	4235
24	3	Batch 2	6	7350	20	On	4225	4228	4208	4220.3
25	5	Batch 1	14	7350	5	On	4381	4391	4376	4382.7
26	3	Batch 2	6	6650	20	Off	4533	4521	4511	4521.7
27	3	Batch 1	14	7350	20	On	4194	4230	4172	4198.7

Data for Problem 4.14

	A	B	C	D	E	F	Resist Thickness			
Run	Volume	Batch	Time	Speed	Acc.	Cover	Left	Center	Right	Average
28	5	Batch 2	6	6650	5	Off	4666	4695	4672	4677.7
29	5	Batch 1	6	7350	20	On	4180	4213	4197	4196.7
30	5	Batch 2	6	6650	20	On	4465	4496	4463	4474.7
31	5	Batch 2	14	6650	5	On	4653	4685	4665	4667.7
32	3	Batch 2	14	6650	5	Off	4683	4712	4677	4690.7

4.15 Consider the leaf spring experiment in Exercise 4.7. Suppose that factor E (quench oil temperature) is very difficult to control during manufacturing. Where would you set factors A, B, C, and D to reduce variability in the free height as much as possible regardless of the quench oil temperature used?

4.16 Construct a 2^{7-2} design by choosing two four-factor interactions as the independent generators. Write down the complete alias structure for this design. Outline the analysis of variance table. What is the resolution of this design?

4.17 Consider the 2^5 design in Exercise 3.12. Suppose that only a one-half fraction could be run. Furthermore, two days were required to take the 16 observations, and it was necessary to confound the 2^{5-1} design in two blocks. Construct the design and analyze the data.

4.18 Analyze the data in the first replicate of Exercise 3.9 as if it came from a 2_{IV}^{4-1} design with $I = ABCD$. Project the design into a full factorial in the subset of the original four factors that appear to be significant.

4.19 Repeat Exercise 4.18 using $I = -ABCD$. Does use of the alternate fraction change your interpretation of the data?

4.20 Project the 2_{IV}^{4-1} design in Example 4.1 into two replicates of a 2^2 design in the factors A and B. Analyze the data and draw conclusions.

4.21 Construct a 2_{III}^{6-3} design. Determine the effects that may be estimated if a second fraction of this design is run with all signs reversed.

4.22 Consider the 2_{III}^{6-3} design in Exercise 4.21. Determine the effects that may be estimated if a second fraction of this design is run with the signs for factor A reversed.

4.23 Fold over the 2_{III}^{7-3} design in Table 4.13. Verify that the resulting design is a 2_{IV}^{8-4} design. Is this minimal design?

4.24 Fold over a 2_{III}^{5-2} design. Verify that the resulting design is a 2_{IV}^{6-2} design. Compare this design to the 2_{IV}^{6-2} design in Table 4.14.

4.25 Suppose that you have run a 2_{IV}^{4-1} design with $I = ABCD$, and that you have fit the model $y = \beta_0 + \beta_1 x_1 + \beta_2 x_2 + \beta_3 x_3 + \beta_4 x_4 = \varepsilon$.

  (a) Assume that the true model for the system is $E(y) = \beta_0 + \sum_{i=1}^{4} \beta_i x_i + \beta_{12} x_1 x_2$. Are the least squares estimates of β_0, β_1, β_2, β_3, and β_4 in the fitted model unbiased? Is the result you obtained an "intuitive" one?

  (b) Assume that the true model is $E(y) = \beta_0 + \sum_{i=1}^{4} \beta_i x_i + \beta_{12} x_1 x_2 + \beta_{11} x_1^2 + \beta_{22} x_2^2$. Are the least squares estimates of β_0, β_1, β_2, β_3, and β_4 biased estimates?

CHAPTER 5

Process Improvement with Steepest Ascent

In previous chapters we dedicated considerable attention to constructing and applying designed experiments with views toward model building. The designs discussed are very efficient for development of empirical equations which relate controllable factors to an important response. Clearly these empirical regression equations serve to provide information about the properties of the system from which the data is taken. Signs and magnitudes of coefficients and the presence or absence of interaction in the system underscore important pieces of information for the user. Often the sole utility of the model is for interpretations such as these. However, there are many other times in which a model is used for *process optimization* or process improvement. The result of a model building procedure is an equation. A statistical analyst is armed with mathematical and analytical techniques, the purpose of which is to optimize the process. This and other chapters that follow will deal in process optimization at different levels. Here we assume that the practitioner is experimenting with a system (perhaps a new system) in which the goal is not to find a *point of optimimum response*, but to search for a new region in which the process or product is improved. Perhaps the current working region for designed experiments is only based on an educated guess. It is felt as if improvement can be found. For example, in a chemical process one seeks to find a new region where improved yields will be experienced.

The experimental design, model building procedure, and sequential experimentation that is used in searching for a region of improved response comprise the **method of steepest ascent**. The type of designs used are those discussed in Chapters 2 and 3, namely, two-level factorial and fractional factorial designs. One might keep in mind that the strategy involves sequential movement from one region in the factors to another. As a result the total operation may involve more than one experiment. Thus, design economy and model simplicity may be very important. One begins by assuming that a first-order model (a planar representation) is a reasonable approximation of the system in the initial region of $x_1, x_2, \ldots, x_k$. Than the method of steepest ascent contains the following steps:

1. Fit a planar (first-order) model using an orthogonal design. Two-level designs will be quite appropriate, although center runs are often recommended.

2. Compute a path of steepest ascent if maximizing response is required. If minimum response is required one should compute the path of **steepest descent**. The path of steepest ascent is computed so that one may expect the *maximum increase* in response. Steepest decent produces a path that results in a maximum decrease in response.

3. Conduct experimental runs along the path. That is, do single or replicated runs and observe the response value. The results will normally show improving values of response. At some region along the path the improvement will decline and eventually disappear. This stems from the deterioration of the simple first-order model once one strays too far from the initial experimental region. Often the first experimental run should be taken near the design perimeter, or at coordinates corresponding to a value of 1.0 in an important variable.

4. At some location, where an approximation of the maximum (or minimum) response is located on the path, a base for a second experiment is chosen. The design should again be a first-order design. It is quite likely that center runs for testing curvature, and degrees of freedom for interaction-type lack of fit are important at this point.

5. A second experiment is conducted and another first-order model is fitted. A test for lack of fit is made. If lack of fit is not significant, a second path based on the new model is computed. Single or replicated experiments along this second path are conducted. It is quite likely that the improvement will not be as strong as that enjoyed in the first path. After improvement is diminished, one has a base for conducting a more elaborate experiment and a more sophisticated process optimization.

This step-by-step procedure is meant to be only a guideline from which to work. It is quite possible that only one stage (and hence one path) will be used. Clearly, if interaction or quadratic lack-of-fit contributions appear to be quite prominent at the second stage, the analyst will likely not conduct experiments along the path which is based on a planar model. Augmentation of the first-order design to allow higher-order models will be discussed in Chapter 6.

5.1   THE COMPUTATION OF THE PATH OF STEEPEST ASCENT

The coordinates along the path of steepest ascent depends on the nature of the regression coefficients in the fitted first-order model. Size and magnitude are both important. The following describes the movement in, say, x_j ($j = 1, 2, \ldots, k$) relative to the movement of the other factors. The reader should

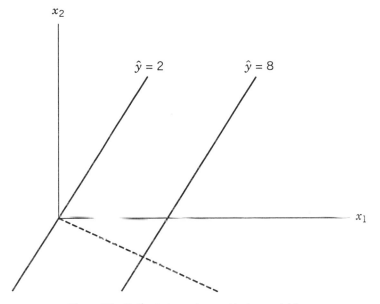

Figure 5.1 Path of steepest ascent in two variables.

bear in mind that the variables are in coded form with the center of the design being $(0, 0, \ldots, 0)$.

The movement in x_j along the path of steepest ascent is proportional to the magnitude of the regression coefficient b_j with the direction taken being the sign of the coefficient. Steepest descent requires the direction to be opposite the sign of the coefficient.

For example, if the fitted regression produces an equation $\hat{y} = 2 + 3x_1 - 1.5x_2$ the path of the steepest ascent will result in x_1 moving in a positive direction and x_2 in a negative direction. In addition, x_1 will move twice as fast; that is, x_1 moves two units for every single unit movement in x_2. Figure 5.1 indicates the nature of the path for this example. The path is indicated by the dashed line. Note that the path is perpendicular to the lines of constant response. This is clearly the quickest movement (steepest ascent) toward large response values. For $k = 3$, these lines become planes and the path of steepest ascent moves perpendicular to these planes.

While a path created by moving x_j a relative distance that is proportional to the regression coefficient b_j is a reasonable and intuitive selection for a path, the reader may enjoy a better understanding of the procedure through a mathematical development of the path. Consider the fitted first-order regression model:

$$\hat{y} = b_0 + b_1 x_1 + b_2 x_2 + \cdots + b_k x_k$$

By the path of steepest *ascent* we mean that which produces a maximum estimated response with the constraint that $\sum_{i=1}^{k} x_i^2 = r^2$. In other words, of all points that are a fixed distance from the center (distance = r) of the design, we seek that point $x_1, x_2, x_3, \ldots, x_k$ for which $\hat{y}$ is maximized. One must remember that in the metric of the coded design variables the design center is $(0, 0, \ldots, 0)$. As a result, the constraint given by $\sum_{i=1}^{k} x_i^2 = r^2$ is that of a sphere with radius r.

The maximization procedure involves the use of Lagrange multipliers. Maximization requires the partial derivatives with respect to x_j ($j = 1, 2, \ldots, k$) of

$$L = b_0 + b_1 x_1 + b_2 x_2 + \cdots + b_k x_k - \lambda \left(\sum_{i=1}^{k} x_i^2 - r^2 \right) \qquad (5.1)$$

The derivative with respect to x_j is clearly given by

$$\frac{\partial L}{\partial x_j} = b_j - 2\lambda x_j \qquad (j = 1, 2, \ldots, k)$$

Allowing $\partial L / \partial x_j = 0$ gives the following coordinate of x_j of the path of steepest ascent

$$x_j = \frac{b_j}{2\lambda} \qquad (j = 1, 2, \ldots, k)$$

Now, the constant $1/2\lambda$ may be viewed as a constant of proportionality. That is, the coordinates are given by

$$x_1 = \rho b_1, x_2 = \rho b_2, \ldots, x_k = \rho b_k \qquad (5.2)$$

where, for *steepest ascent*, the constant ρ is positive. For *steepest descent*, ρ is taken to be negative. Now, this implies that the choice of ρ, which is related to λ, merely determines the distance from the design center that the resulting point will reside. As a result, of course, the constant ρ is determined by the practitioner. One can view the path of steepest ascent as being created as indicated in Figure 5.2. Suppose again that $\hat{y} = 2 + 3x_1 - 1.5x_2$. The path is drawn again and coordinates are viewed as points that contain maximum values of $\hat{y}$ at fixed radii. Each point on the path then, is a point of maximum response. Again, note that the path involves a movement in a positive direction for x_1 and a negative direction for x_2, with a change of two units in x_1 for every single unit change in x_2. The following example will provide an illustration of the computation of the path of steepest ascent for a real-life illustration.

Example 5.1 Many manufacturing companies in the United States and abroad use molded parts as components of the process. Shrinkage is often a problem. Often a molded die for a part will be built larger than nominal to allow for part shrinkage. In the following experiment a new die is being produced, and ultimately it is important to find the proper settings to minimize shrinkage. In the following experiment, the response values are deviations from nominal—that is, shrinkage. The factor levels in natural and

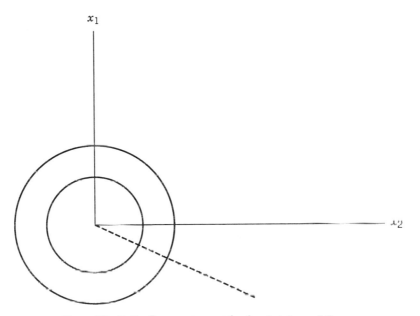

Figure 5.2   Path of steepest ascent for $\hat{y} = 2 + 3x_1 - 1.5x_2$.

design units are as follows:

		Design Units	
		−1	+1
x_1:	Injection velocity (ft/sec)	1.0	2.0
x_2:	Mold temperature (°C)	100	150
x_3:	Mold pressure (psi)	500	1000
x_4:	Back pressure (psi)	75	120

The design was a 2^4 factorial with no replicated observations. The response is in centimeters $\times 10^4$. The actual data will not be given, but the first-order model fit to the data is given by

$$\hat{y} = 80 - 5.28x_1 - 6.22x_2 - 1.21x_3 - 1.07x_4$$

This linear regression model relates predicted shrinkage to the design variables in coded design units. It appears that a movement to a region with higher injection velocity and mold temperature will be beneficial. In addition, a slight increase in mold pressure and back pressure may be beneficial. One must keep in mind that we are searching for the path of steepest decent. Then an increase proportional to the regression coefficient in each factor will define the proper path for future experiments. Another way of viewing the path of steepest ascent is as follows: If the regression coefficients are scaled to the unit vector (i.e., $\Sigma_{i=1}^{k} b_i^2 = 1.0$), the coefficients become a point on the path. This is equivalent to using $\rho = 1.0$ in Equation (5.2).

One usually determines increments along the path on the basis of a movement in one particular variable. In this case suppose we determine that a change of one coded unit (i.e., 0.5 ft/ sec) in injection velocity should determine an increment. Then at each coordinate of the path of steepest descent there will be an increase of 0.5 ft/ sec. As a result, in coded design units we will have

$$\left(\frac{6.22}{5.28}\right) \text{ design units increase in } x_2$$

$$\left(\frac{1.21}{5.28}\right) \text{ design units increase in } x_3$$

$$\left(\frac{1.07}{5.28}\right) \text{ design units increase in } x_4$$

As a result, the following is a table of computed points along the path of steepest descent:

	Coded Units				Natural Units			
	x_1	x_2	x_3	x_4	ft/ sec	°C	psi	psi
Base	0	0	0	0	1.5	125	750	97.50
Increment = Δ	1.0	1.178	0.23	0.203	0.5	29.45	57.5	4.53
Base + Δ	1.0	1.178	0.23	0.203	2.0	154.45	807.5	102.03
Base + 2Δ	2.0	2.356	0.46	0.406	2.5	183.90	865.0	106.56
Base + 3Δ	3.0	3.534	0.69	0.609	3.0	213.35	922.5	111.09
Base + 4Δ	4.0	4.712	0.92	0.812	3.5	242.80	980.0	115.62

The steepest descent procedure in this case would involve experimental runs along the path. This strategy might continue until it is clear that further experimentation would be fruitless.

5.2 CONSIDERATION OF INTERACTION AND CURVATURE

As we indicated earlier in this chapter, it is expected that no more than two steepest ascent (or descent)-type procedures will prove to be useful. This general strategy is at its best when the researcher begins experimentation far from the region of optimum conditions. Here one expects that the first-order approximation might be quite reasonable. As the experimental region moves near the region of optimum conditions, it is certainly expected that curvature would be more prevalent and, of course, interactions among the factors would cease to be negligible. As a result, the experimental design used to carry out the strategy should certainly allow estimation (and testing) of certain potentially important interactions. In later stages of the strategy, the design should allow some information regarding model curvature. The use of center runs allows a single degree of freedom estimate of curvature (see

Chapter 3). Clearly, if interaction is found to be important and/or a test for curvature finds significant quadratic terms, the researcher would suspect that the steepest ascent (descent) methodology would become effective. Augmentation of the design to allow the fitting of a complete second-order model should then be done.

When second-order terms describing interaction and model curvature $(x_1^2, x_2^2, \ldots)$ begin to dominate, then continuing the ascent exercise and experimentation will be self-defeating. However, the question arises, "What do we mean by 'dominant'?" It is possible that second-order terms can be statistically significant, yet the first-order approximation allows a reasonably successful experimental strategy. One must keep in mind that "statistical significance" only implies that the effects are real in comparison to experimental error. The second-order effects may be small in magnitude when compared to their first-order counterparts. As a result, there will certainly be situations where one should compute the path and take experimental trials even though certain second-order effects are significant.

Example 5.2 For this example, we consider an illustration of a steepest ascent experiment in which it is of interest to maximize the reaction yield. For factors, A (amount of reactant A), B (reaction time), C (amount of reactant C), and D (temperature) are being considered. The natural and coded levels are given as follows:

	Coded Levels		
	-1	$+1$	
Natural Levels	10	15	(factor A, grams)
	1	2	(factor B, minutes)
	25	35	(factor C, grams)
	75	85	(factor D, °C)

A $\frac{1}{2}$ fraction of the 2^4 was used as the design, with the yield values given as follows:

(1):	62.0	ad:	61.8
ab:	69.0	bc:	64.7
cd:	57.0	bd:	62.2
ac:	64.5	$abcd$:	66.3

The fitted linear regression model is given by

$$\hat{y} = 63.44 + 1.9625x_1 + 2.1125x_2 - 0.3125x_3 - 1.6125x_4$$

The basis for computation of points along the path was chosen to be 1 gram of reactant A. This corresponds to $1/2.5 = 0.4$ design units. As a result, the corresponding movement in the other design variables are $(2.1125/1.9625)(0.4) = 0.4306$ design units for x_2, $(0.3125/1.9625)(0.4) = 0.0637$ design units for x_3, and $(1.6125/1.9625)(0.4) = 0.3287$ design units for x_4. The

Table 5.1 Coordinates on the Path of Steepest Ascent in Natural Variables for Example 5.2

Run	x_1	x_2	x_3	x_4	y
Base	12.5	1.5	30	80	
Δ	1.0	$(0.4306)(0.5) = 0.215$	$-0.0637(5) = -0.319$	$(-0.3287)(5) = -1.643$	
Base $+ \Delta$	13.5	1.715	29.681	78.357	
Base $+ 2\Delta$	14.5	1.930	29.362	76.714	
Base $+ 3\Delta$	15.5	2.145	29.043	75.071	
9 Base $+ 4\Delta$	16.5	2.360	28.724	73.428	74.0
10 Base $+ 6\Delta$	18.5	2.790	28.086	70.142	77.0
11 Base $+ 8\Delta$	20.5	3.220	27.448	73.856	81.0
12 Base $+ 9\Delta$	21.5	3.435	27.129	65.213	78.7

movement along the path will be positive for x_1 and x_2, and negative for x_3 and x_4. Table 5.1 shows the appropriate coordinates along the path in the natural variables. At some location along the design perimeter, experimental runs should be made. In this case, runs were made at base $+ 4\Delta$ (after four increments): Runs 9, 10, 11, and 12 indicate new experimental runs. Theoretically, one expects an increase in response as runs are taken along the path. Eventually, of course, deterioration should occur due to the fact that the first order approximation is no longer valid. In this case a reduction is experienced after run 11. Any further follow-up experiment should involve an experimental design centered in the vicinity of run 11.

5.2.1 What About a Second Phase?

As we indicated earlier, one should expect that the evidence of curvature in the system and interaction among the factors will eventually necessitate an abandonment of steepest ascent. Any additional phases of the procedure beyond the first will usually not bring about the level of success enjoyed in the initial phase. In addition, one should be careful to use a design in the second phase that allows for testing lack of fit that includes interaction and curvature induced by quadratic terms. In our example a reasonable design for the second stage is a $\frac{1}{2}$ fraction of a 2^4 (or a complete 2^4) *augmented by center runs*. This allows for three degrees of freedom for testing interaction type lack of fit and one degree of freedom for testing curvature. If the lack of fit is quite significant one might expect little or no success with steepest ascent.

5.2.2 What Transpires Following Steepest Ascent?

It should be emphasized at this point that quality improvement through analysis of designed experiments, when successful, is usually a lengthy **iterative** experience. This is illustrated quite well in dealing with the strategy of steepest ascent. In our example the investigator may well invest in a 2^4

factorial with, say, five center runs for a second phase of steepest ascent. However, if curvature and interaction are found to be quite evident, the steepest ascent procedure will certainly soon be truncated. At this point the investigators will surely be interested in finding optimum conditions through the use of a fitted *second-order model*. This involves a computation of estimated optimum conditions. The first-order two-level design with center runs is nicely augmented to allow estimation of second-order terms. For example, a $\frac{1}{2}$ fraction (resolution **IV**) of a 2^4 factorial with five center runs is augmented as follows:

x_1	x_2	x_3	x_4	
-1	-1	-1	-1	
1	1	-1	-1	
1	-1	1	-1	
1	-1	-1	1	
-1	1	1	-1	
-1	1	-1	1	Initial design
-1	-1	1	1	
1	1	1	1	
0	0	0	0	
0	0	0	0	
0	0	0	0	
0	0	0	0	
0	0	0	0	
1	-1	-1	-1	
-1	1	-1	-1	
-1	-1	1	1	
-1	-1	-1	1	
1	1	1	-1	
1	1	-1	1	
-1	1	1	1	
1	-1	1	1	Augmentation
-2	0	0	0	
2	0	0	0	
0	-2	0	0	
0	2	0	0	
0	0	-2	0	
0	0	2	0	
0	0	0	-2	
0	0	0	2	

The total design allows for efficient estimation of the terms of the model

$$y = \beta_0 + \sum_{i=1}^{4} \beta_i x_i + \sum_{i=1}^{4} \beta_{ii} x_i^2 + \sum\sum_{i<j} \beta_{ij} x_i x_j + \epsilon$$

This model is called the *second-order response surface model* and is used for process optimization. The design is called the **central composite design**. Process optimization with the second-order model is discussed in Chapter 6. Designs for fitting second-order models, including the central composite design, are discussed in Chapters 7 and 8.

The point made here is that the total strategy of product improvement or optimization can very well involve both steepest ascent (region seeking) and more formal response surface optimization in an iterative procedure. The transition from one to the other can be made quite easily without any waste of experimental effort.

5.3   EFFECT OF SCALE (CHOOSING RANGE OF FACTORS)

The methodology of proceeding along the path of steepest ascent is generally a precursor to a more elaborate experimental effort and optimization involving a more sophisticated model and analysis. In Chapters 3 and 4, considerable attention was placed on two-level designs with variable screening being an important goal. It was suggested at that point that choosing ranges on the factor is a vital decision and something that the researcher should not take lightly. Clearly, variable screening and steepest ascent are early steps in the process optimization experience. Sloppy decisions with little forethought in these early stages may lead to very inefficient process optimization at a later stage. It should be clear by now to the reader that there is a connection between selection of ranges of the variables and the *choice of scale*. For example, the coding in Example 5.1 suggests a decision in which one design unit in injection velocity is 0.5 ft/ sec. An *equivalent unit* in mold temperature is 25°C. This choice of scale is a decision that is made by the practitioner. Note that the regression coefficient on back pressure (x_4) is considerably smaller in magnitude than those or x_1 and x_2. Perhaps the implication is that the choice of range (and hence, choice of scale) on back pressure was incorrect. Choosing design factor ranges certainly should improve with increased experience with the system. If variable screening has already been accomplished, then one would expect a more educated choice of scale for the hill-climbing exercise of steepest ascent. One can only use the latest information available.

A change of scale does not change the direction that a factor should move along the path of steepest ascent. However, it changes the relative magnitude of movement of the factor. Suppose now we have an ideal situation with time (x_1) and temperature (x_2) in which the true regression structure involving yield y is as follows:

$$E(y) = \beta_0 + \beta x_1 + \beta x_2$$

where β is a coefficient that corresponds to a $(+1, -1)$ scaling for ranges of temperature of 50°C and time of 1.0 hr. Suppose researcher A chooses the above ranges with $(+1, -1)$ scaling while researcher B chooses a 50°C range of temperature, but a 0.5 range of time, again with $(+1, -1)$ scaling on the true factors. The model that is relevant to researcher B is given by

$$E(y) = \beta_0 + \beta x_1 + (\beta/2)x_2$$

Thus the expected value of the time regression coefficient b_2 for researcher B is one-half that of the same coefficient for researcher A. As a result, the relative movement in x_2 compared to x_1 along the path, in design units will be half as great. As an example, suppose design levels are as follows for the researchers:

	Researcher A			
	Natural Levels		Coded Design Units	
Temperature	200°F	250°F	-1	$+1$
Time	1.0	2.0	-1	$+1$

	Researcher B			
	Natural Levels		Coded Design Units	
Temperature	200°F	250°F	-1	$+1$
Time	1.25	1.75	-1	$+1$

Suppose both researchers use 2^2 factorial experimental plans. Let us assume that the steepest ascent coordinates are to be based on a change of 25°F in temperature. The steepest ascent picture is as follows:

	A		B	
Base	225	1.5	225	1.5
Δ	25	0.5	25	0.125
Base + Δ	250	2.0	250	1.625
Base + 2Δ	275	2.5	275	1.750

Note that the actual change in time for researcher B is *one-fourth* the change incurred by researcher A. The 0.125 incremental change for researcher B results from the fact that the change per coded unit change in x_2 is only one-half that experienced by researcher A on the average. In addition, the computation of the coordinates for the natural variable must take into account that the design unit described by researcher B is only one-half that described by researcher A. As a result the reader should be able to ascertain the following general rule that reflects the distinction between steepest ascent coordinates in a k variable problem.

Suppose researcher A chooses scale factor (range) $r_1, r_2, \ldots, r_k$ *and researcher B chooses* $r'_1, r'_2, \ldots, r'_k$, *where* $a_j = r_j / r'_j$. *Refer to the relative movements along the path in the natural variables as follows:* $\Delta_1, \Delta_2, \ldots, \Delta_k$ *and* $\Delta'_1, \Delta'_2, \ldots, \Delta'_k$ *for researcher A and B, respectively; then* $\Delta_j / \Delta'_j = a_j^2$ *for* $j = 1, 2, \ldots, k$.

This does suggest that the user of steepest ascent must use whatever knowledge of the system is at his or her disposal to determine the range of the variable and hence the scale—that is, the definition of a design unit. As we indicated earlier, each experimental experience with a particular system allows for a more educated choice of intervals on the variable being studied. One can view this general procedure as one in which there are truly many paths of steepest ascent (or descent) that will lead to a region where the response is improved. The method itself can be a learning device that will allow for a better choice of ranges in future experiments, particularly those experiments in which the goal is to find optimum process conditions. One should not let that difficulty in choosing ranges prohibit the use of this region-seeking method. The methodology allows the user to be the beneficiary of more appropriate regions and ranges for future experiments.

5.4 CONFIDENCE REGION FOR DIRECTION OF STEEPEST ASCENT

It is useful to take into account sampling variation in assessing the nature of the path of steepest ascent. One must remember that the path is based on the regression coefficients and these coefficients have sampling properties characterized by standard errors. As a result, the path has sampling variation. This sampling variation can lead to a confidence region for the path itself. The value of the confidence region may be derived by plots, say in the case of two or three variables. A graphical analysis may indicate the amount of flexibility the practitioner has in experiments along the path. A tighter region gives the user confidence that the path is being estimated well.

Suppose there are k design variables and, indeed, coefficients $b_1, b_2, \ldots, b_k$ provide estimates of the relative movement of variables along the path. Assuming that the first-order model is correct, the *true* path is defined by parameters $\beta_1, \beta_2, \ldots, \beta_k$, and

$$E(b_i) = \beta_i \quad (i = 1, 2, \ldots, k)$$

and the true coefficients are proportional to the relative movement along the path [i.e., Equation (5.2)] implies

$$\beta_i = \gamma X_i \quad (i = 1, 2, \ldots, k) \tag{5.3}$$

where X_i represents the *direction cosines* of the path. In other words, $X_1, X_2, \ldots, X_k$ are the constants which, if known, could be used to compute any coordinates on the true path. We can view the relationship in Equation (5.3) as a *regression model* without an intercept. The subject of the statistical inference here will be the X_i. Though the regression structure may appear to be rather unorthodox, one should consider the model

$$b_i = \gamma X_i + \varepsilon_i \qquad (i = 1, 2, \ldots, k) \qquad (5.4)$$

The variance of the b_i is constant across all coefficients if one uses a standard two-level orthogonal design. Call the estimated variance s_b^2. A second variance, s_b^{2*}, is found from the error mean square of the regression of Equation (5.4). The quantity s_b^{2*} is given by

$$s_b^{2*} = \frac{\sum_{i=1}^{k} (b_i - \hat{\gamma} X_i)^2}{k - 1}$$

where

$$\hat{\gamma} = \frac{\sum_{i=1}^{k} b_i X_i}{\sum_{i=1}^{k} X_i^2}$$

Then, of course, $k - 1$ is the number of error degrees of freedom for the regression. As a result,

$$\frac{s_b^{2*}}{s_b^2} \sim F_{k-1, \nu_b} \qquad (5.5)$$

where ν_b is the number of error degrees of freedom associated with the estimate s_b^2 (from the steepest ascent experiment). Values of $X_1, X_2, \ldots, X_k$ that fall inside the confidence region are those for which the resulting values of s_b^{2*} / s_b^2 do not appear to refute Equation (5.5). In particular, coordinates outside the confidence region are those for which s_b^{2*} is significantly larger than s_b^2. As a result, the $100(1 - \alpha)\%$ confidence region is defined as the set of values $X_1, X_2, \ldots, X_k$ for which

$$\sum_{i=1}^{k} \frac{(b_i - \hat{\gamma} X_i)^2 / (k - 1)}{s_b^2} \leq F_{\alpha, k-1, \nu_b} \qquad (5.6)$$

where $F_{\alpha, k-1, \nu_b}$ is the upper $100(1 - \alpha)\%$ point of the F_{k-1, ν_b} distribution.

What Does the Confidence Region Mean?

One must understand that specific coordinates $X_1, X_2, \ldots, X_k$ define the direction along the path. For example, $X_1 = \sqrt{0.5}$, $X_2 = \sqrt{0.3}$, and

$X_3 = \sqrt{0.2}$ define a direction and also represent coordinates that are a unit distance away from the design origin. (Keep in mind that we remain in design unit scaling). The confidence region turns out to be a cone (or a hypercone in more than three variables) with the apex at the design origin and all points a unit distance from the origin satisfying

$$\sum_{i=1}^{k}\left[b_i - \left(\sum_{i=1}^{k} b_i X_i\right) X_i\right]^2 / (k - 1) \le s_b^2 F_{\alpha, k-1, \nu_b} \qquad (5.7)$$

Example 5.3 Consider the path of steepest ascent illustrated in Figure 5.2. The coefficients are $b_1 = 3$ and $b_2 = -1.5$. Suppose further that the estimates of the variances of the coefficients are both $\frac{1}{4}$ and there are four error degrees of freedom. The value $F_{0.05, 2, 4} = 7.71$. As a result, the 95% confidence region for the path of steepest ascent at fixed distance $X_1^2 + X_2^2 = 1.0$ is determined by solutions (X_1, X_2) to

$$9 + 2.25 - (3X_1 - 1.5X_2)^2 \le \tfrac{1}{4}(7.71)$$

or

$$(3X_1 - 1.5X_2)^2 \ge 9.3225 \qquad (5.8)$$

As a result, the total confidence region on the path is illustrated in Figure 5.3. Similar graphical approaches to the problem are reasonable for $k = 3$. See Box and Draper (1987). For $k > 3$ one cannot display a graphical picture of the confidence region. Of course, one can always substitute any value $(X_1, X_2, \ldots, X_k)$ into Equation (5.6) to determine if the point falls inside the confidence region. In general, it may be rather difficult to determine the relative size of the confidence region, and yet the user needs to have some indication of whether or not it is permissible to continue. Box and Draper describe an interesting analytic procedure for determining what percentage of the possible directions was excluded by the 95% confidence region. This then produces some impression about how "tight" the confidence region is for any specific example. In our simple $k = 2$ example, this might correspond to the determination of the angle θ in Figure 5.3. It turns out that

$$\theta = \arcsin\left\{\frac{(k - 1)s_b^2 F_{\alpha, k-1, \nu_2}}{\displaystyle\sum_{i=1}^{k} b_i^2}\right\}^{1/2} \qquad (5.9)$$

For our example, we have

$$\theta = \arcsin\left\{\left[\left(\frac{1}{4}\right)(7.71)\left(\frac{1}{11.25}\right)\right]^{1/2}\right\}$$

$$= \arcsin[0.413]$$

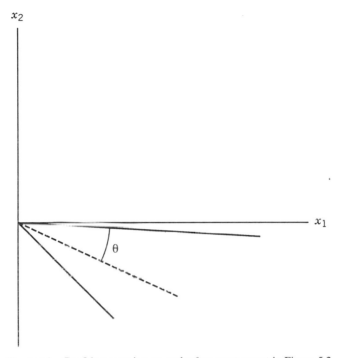

Figure 5.3   Confidence region on path of steepest ascent in Figure 5.2.

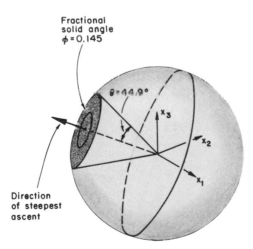

Figure 5.4   Confidence cone for direction of steepest ascent for $k = 3$ [From Box and Draper (1987), with permission.]

Thus $\theta \cong 24.4°$. As a result, the confidence region illustrated by Figure 5.3 sweeps an angle of $2(\theta) = 48.8°$. Of interest is the ratio $48.8/360 = 0.135$, suggesting that 86.5% of the possible directions taken from the design origin are excluded. Clearly, the larger the percentage the more accurate the computed path. For $k = 3$ the confidence region defines a cone and the quality of the confidence region depends on the fraction of the area of the total sphere that is taken up by the fractional angle created by the confidence cone. Consider, for example, Figure 5.4, taken from Box and Draper (1987). The dashed line represents the computed path. The ratio of the shaded area to the total area of the sphere determines the quality of the computed path.

5.5 STEEPEST ASCENT SUBJECT TO A LINEAR CONSTRAINT

Anytime a researcher encounters the task of sequential experimentation that involves considerable movement, there is always the likelihood that the path may move into an area of the design space where one or more of the variables are not in a permissible range from an engineering or scientific point of view. For example, it is quite possible that an ingredient concentration may exceed practical limits. As a result, in many situations it becomes necessary to build the path of steepest assent with a constraint imposed in the design variables. Suppose, in fact, that we view the constraint in the form of a boundary. That is, we are bounded by

$$c_0 + c_1 x_1 + c_2 x_2 + \cdots + c_k x_k = 0 \qquad (5.10)$$

This bound need not involve all k variables. For example, $x_j = c_0$ may represent a boundary on a single design variable. One must keep in mind that in practice the constraint must be formulated in terms of the natural (uncoded) variables and then written in the form of coded variables for manipulation.

Figure 5.5 illustrates, for $k = 2$, the necessary steepest ascent procedure when a linear constraint is to be applied. The procedure is as follows:

1. Proceed along the usual path of steepest ascent until contact is made with the constraining line (or plane for $k > 2$). Call this point of contact point **O**.

2. Beginning at point **O**, proceed along an adjusted or modified path.

3. Take experiments along the modified path, as usual, with the stopping rule based on the same general principles as those discussed in Sections 5.1 and 5.2.

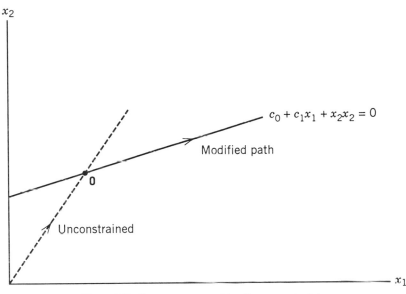

Figure 5.5   Steepest ascent with a linear constraint.

What is the Modified Path?
Figure 5.5 clearly outlines the modified path for the $k = 2$ illustration. In general, however, the proper direction vector is one that satisfies the constraint and still makes the greatest possible progress toward maximizing (or minimizing) response. (Clearly the modified path in our illustration is correct.)

It turns out (and is intuitively reasonable) that the modified path is given by the direction vector

$$b_i - dc_i \qquad (i = 1, 2, \ldots, k) \tag{5.11}$$

for which

$$\sum_{i=1}^{k} (b_i - dc_i)^2$$

is minimized. That is, the direction along the constraint line (or plane) is taken so as to be "closest" to the original path. Once again we can consider this to be a simple regression situation in which the b_i are being regressed against the c_i. As a result, the quantity d plays the role of a regression slope and minimization of the "residual" sum of squares produces the value

$$d = \frac{\sum\limits_{i=1}^{k} b_i c_i}{\sum\limits_{i=1}^{k} c_i^2} \tag{5.12}$$

Thus, the modified path begins at point **O** and proceeds using the direction vector $b_1 - dc_1, b_2 - dc_2, \ldots, b_k - dc_k$.

It remains then to determine the point **O**, that is, the point that lies on the original path but is also on the constraint plane. From Equation (5.2) we know that $x_j = \rho b_j$ for $j = 1, 2, \ldots, k$. But we also know that for point **O**, the constraint equation [Equation (5.10)] must hold. As a result, the collision between the original path of steepest ascent and the constraint plane must occur for $\rho = \rho_0$, satisfying

$$c_0 + (c_1 b_1 + c_2 b_2 + \cdots + c_k b_k)\rho_0 = 0.$$

Thus,

$$\rho_0 = \frac{-c_0}{\sum\limits_{i=1}^{k} c_i b_i} \tag{5.13}$$

As a result, the modified path starts at $x_{j,0} = \rho_0 b_j$ (for $j = 1, 2, \ldots, k$) and moves along the modified path, being defined by

$$x_{j,m} = x_{j,0} + \lambda(b_j - dc_j) \qquad (j = 1, 2, \ldots, k) \tag{5.14}$$

where d is given by Equation (5.12) and λ is merely a proportionality constant. Note that the modified path is like the standard path except it does not start at the origin and the regression coefficient b_j is replaced by $b_j - dc_j$ in order to accommodate the constraint.

Linear constraints on design variables occur frequently in practice. In the chemical and related fields, constraints are often imposed in situations in which the design variable include concentrations of components in a system. In a gasoline blending system there will certainly be constraints imposed on the components in the system. General information regarding **designs for mixture problems** will be presented in Chapters 11 and 12. In the example that follows we illustrate steepest ascent in a situation where mixture factors are involved and a linear constraint must be applied to modify the path of steepest ascent.

Example 5.4 Consider a situation in which the breaking strength of a certain type of fabric is a function of the amount of three important components in a kilogram of raw material. We will call the components ξ_1, ξ_2, and ξ_3. The levels used in the experiment are as follows:

Material	Amount -1	Grams $+1$
1	100	150
2	50	100
3	20	40

The remaining ingredients in the raw material are known to have no impact on fabric strength in grams per square inch. However, it is important that ξ_1 and ξ_2, the amounts of material 1 and 2 respectively, be constrained by the following equation:

$$\xi_1 + \xi_2 \leq 500$$

The design-level centering and scaling are, of course, given by

$$x_1 = \frac{\xi_1 - 125}{25}, \qquad \xi_1 = 25x_1 + 125$$

$$x_2 = \frac{\xi_2 - 75}{25}, \qquad \xi_2 = 25x_2 + 75$$

$$x_3 = \frac{\xi_3 - 30}{10}, \qquad \xi_3 = 10x_3 + 30$$

As a result, the constraint reduces to

$$25x_1 + 25x_2 \leq 300$$

Suppose the fitted regression function is given by

$$\hat{y} = 150 + 1.7x_1 + 0.8x_2 + 0.5x_3$$

Because it is desirable to find the condition of increased fabric strength, the unconstrained path of steepest ascent will proceed with increasing values of all three components. The increments along the path are to correspond to changes in ξ_1 of 25 grams or 1 design unit. This corresponds to changes in x_2 and x_3 of 0.47 and 0.294 units, respectively.

Now, based on the above information, the standard path of steepest ascent can be computed. But at what point does the path make contact with the constraint and with the constraint plane? From Equation (5.13) we obtain

$$\rho_0 = \frac{300}{(25)(1.7) + (25)(0.8)} = \frac{300}{62.5} = 4.8$$

As a result, the modified path starts at

$$x_{j,0} = 4.8b_j \qquad (j = 1, 2, 3)$$

which results in the coordinates $(8.16, 3.84, 2.4)$ in design units. As a result, the coordinates of the modified path are given by Equation (5.14); that is,

$$x_{j,m} = 4.8b_j + \lambda(b_j - dc_j) \qquad (j = 1, 2, 3)$$

where

$$d = \frac{(25)(1.7) + (25)(0.8)}{1,250} = 0.05$$

As a result, the modified path is given by

$$x_{1,m} = 8.16 + \lambda(0.45)$$

$$x_{2,m} = 3.84 + \lambda(-0.45)$$

$$x_{3,m} = 2.4 + \lambda(0.5)$$

Thus, the following table shows a set of coordinates on the path of steepest ascent, followed by points along the modified path.

	x_1	x_2	x_3
Base	0	0	0
Δ	1	0.47	0.294
Base $+ \Delta$	1	0.47	0.294
Base $+ 2\Delta$	2	0.94	0.598
$\vdots$	$\vdots$	$\vdots$	$\vdots$
Base $+ 5\Delta$	5	2.35	1.470
Point O	8.16	3.84	2.4
	8.61	3.39	2.9
	9.06	2.94	3.4
	9.51	2.49	3.9
	$\vdots$	$\vdots$	$\vdots$

For the modified path we are using $\lambda = 1.0$ for convenience. Note that all points on the modified path meet the requisite of the constraint.

EXERCISES

5.1 Given the fitted response function

$$\hat{y} = 72.0 + 3.6x_1 - 2.5x_2$$

which is found to fit well in a localized region of (x_1, x_2).

(a) Plot contours (lines in this case) of constant response, y, in the (x_1, x_2) plane

(b) Plot the path of steepest ascent generated from this response function.

(c) Plot the 95% confidence regions on the location of the path of steepest ascent.

5.2 In a metallurgy experiment it is desired to test the effect of four factors and their interactions on the concentration (percent by weight) of a particular phosphorous compound in costing material. The variables are: A, percent phosphorous in the refinement; B, percent remelted material; C, fluxing time, and D, holding time. The four factors are varied in a 2^4 factorial experiment with two castings taken at each factor combination. The 32 castings were made in random order.

Treatment Combination	Weight % of Phosphorous Compound		
	Replication 1	Replication 2	Total
(1)	30.3	28.6	58.9
a	28.5	31.4	59.9
b	24.5	25.6	50.1
ab	25.9	27.2	53.1
c	24.8	23.4	48.2
ac	26.9	23.8	50.7
bc	24.8	27.8	52.6
abc	22.2	24.9	47.1
d	31.7	33.5	65.2
ad	24.6	26.2	50.8
bd	27.6	30.6	58.2
abd	26.3	27.8	54.1
cd	29.9	27.7	57.6
acd	26.8	24.2	51.0
bcd	26.4	24.9	51.3
$abcd$	26.9	29.3	56.2
Total	428.1	436.9	865.0

(a) Build a first-order response function.

(b) Write out a table of the path of steepest ascent (unconstrained) in the coded design variables.

(c) It is important to constrain the percent phosphorous and percent remelted material. In fact, in the metric of the coded variables we obtain

$$x_1 + x_2 = 2.7$$

where x_1 is percent phosphorous and x_2 is percent remelted material. Redo the path of steepest ascent subject to the above constraint.

5.3 Consider the 2^2 factorial design featured in the illustration in Figure 3.1 in Chapter 3. Factor A is the concentration of a reactant, factor B is feed rate, and the response is the viscosity of the output material. As one can tell by the computed effects and the analysis of variance shown, the main effects dominate in the analysis.

(a) Fit a first-order model.

(b) Compute and plot path of steepest ascent.

(c) Suppose that in the metric of the coded variables, the concentration cannot exceed 3.5 and the feed rate cannot go below -2.7. Show the "practical" path of steepest ascent—that is, the path that accounts for these constraints.

5.4 It is stated in the text that the development of the path of steepest ascent makes use of the assumption that the model is truly first order in nature. However, even if there is a modest amount of curvature or interaction in the system, the use of steepest ascent can be extremely useful in determining a future experimental region. Suppose that in a system involving x_1 and x_2 the actual model is given by

$$E(y) = \beta_0 + \beta_1 x_1 + \beta_2 x_2 + \beta_{12} x_1 x_2$$

with

$$\beta_0 = 14$$

$$\beta_1 = 5$$

$$\beta_2 = -10$$

$$\beta_{12} = 3$$

Again, assume that x_1 and x_2 are in "coded" form.

(a) Show a plot of the path of steepest ascent (based on actual parameters) if the interaction is ignored.

(b) Show a plot of the path of steepest ascent for the model with interaction. Note that this path is not linear.

(c) Comment.

5.5 Consider Example 3.3 in Chapter 3.

(a) Using a first-order regression model plot the path of steepest ascent.

(b) Indicate expected responses in a table for various points on the path.

(c) The test for curvature indicated that no quadratic terms are significant. Suppose that, instead of having center runs, four additional runs were placed on the factorial points (one at each point). Will the steepest ascent be improved? Explain.

(d) Refer to the part (c) above. Give an argument for improvement that takes the confidence region on the path of steepest ascent into account.

5.6 Consider Example 3.4 in Chapter 3. There are two design variables and four blocks. The experiment involves complete blocks. And, of course, it is assumed that there is no interaction between blocks and the design variables.

(a) Write the linear regression model for this experiment. Assume blocks are a part of the model. Assume no interaction between the design variables.

(b) Show that the path of steepest ascent is the same whether or not blocking is involved in the experiment.

(c) Suppose the block effects are extremely important. Explain how blocking will improve the precision of the path of steepest ascent.

5.7 Consider the fractional factorial experiment illustrated in Exercise 4.9.

(a) From the analysis, write out a first-order regression model. All factors are quantitative. Assume that the levels are centered and scaled to ± 1 levels.

(b) Write out a table giving the path of steepest ascent for achieving maximum yield.

(c) Suppose, for example, that the CE interaction is extremely important but the analyst does not realize it. How does this effect the computed path in part (b)? Give a qualitative explanation.

5.8 Consider the data below. Apply the method of steepest ascent and create an appropriate path.

Natural Variables		Coded Variables		
ξ_1	ξ_2	x_1	x_2	Response
80	40	-1	-1	15
100	40	1	-1	32
80	60	-1	1	25
100	60	1	1	40
90	50	0	0	33
90	50	0	0	27
90	50	0	0	30

5.9 Suppose the model you fit in Exercise 5.8 was the following:

$$\hat{y} = 28 + 8x_1 - 4.5x_2$$

(Note that this is not necessarily the model *you* actually found when you worked this problem. It's just one that we will use in *this* question, under the assumption that it is the correct model.)

(a) Which variable would you use to define the step size along the path of steepest ascent in this model, and why?

(b) Using the variable you selected in part (a), choose a step size that is large enough to take you to the boundary of the experimental region in that particular direction. Find and show the coordinates of this point on the path of steepest ascent in the coded variables x_1 and x_2.

(c) Find and show the coordinates of the point from part (c) in terms of the natural variables.

(d) Are the points $\xi_1 = 107.4$ and $\xi_2 = 59.8$ on the path of steepest ascent?

5.10 Consider the first-order model

$$\hat{y} = 100 + 5x_1 + 8x_2 - 3x_3$$

This model was fit using a 2^3 design in the coded variables $-1 \le x_i \le +1$, $i = 1, 2, 3$. The model fit was adequate. The region of exploration on the natural variables was

$$\zeta_1 = \text{temperature } (100, 110°C), \zeta_2 = \text{time}(1, 2 \text{ hr}),$$

$$\zeta_3 = \text{pressure } (50, 75 \text{ psi})$$

(a) Which variable would you choose to define the step size along the path of steepest ascent, and why?

(b) Using the variable in part (a), choose a step size large enough to take you to the boundary of the experimental region in that particular direction. Find and show the coordinates of this point on the path of steepest ascent in the coded variables x_i.

(c) Find and show the coordinates of this point on the path of steepest ascent from part (b) using the natural variables.

(d) Find a unit vector that defines the path of steepest ascent.

(e) What step size multiplier for the unit vector in (d) above would give the same point on the path of steepest ascent you found in parts (b) and (c)?

5.11 Consider the fitted first-order model

$$\hat{y} = 15.96 + 1.02x_1 + 3.4x_2 - 2.4x_3$$

where x_i is a coded variable such that $-1 \le x_i \le 1$, and the natural variables are

Pressure	24–30 ps.g (x_1)
Time	1–2 min (x_2)
Pounds caustic	2–4 lb (x_3)

If the step size in terms of the natural variable time is 15 min, find the coordinates of a point on the path of steepest ascent.

5.12 Consider the data of Exercise 5.8. Show graphically a confidence region for the path of steepest ascent.

5.13 Consider again the data of Exercise 5.8. Do tests for interaction and curvature. From these tests, do you feel comfortable doing the steepest ascent? Explain why or why not.

CHAPTER 6

The Analysis of Response Surfaces

6.1 SECOND-ORDER RESPONSE SURFACE

In previous developments we have confined ourselves to models that are either first order in the design variables or contain first-order plus interaction terms. In Chapters 3 and 4, attention was focused on two-level designs. These designs are natural for fitting models containing first-order main effects and low-order interactions. Indeed, the material in Chapter 5 dealing with steepest ascent made use of first-order models only. Nevertheless, it has been our intention to motivate the need for fitting second-order response surface models. For example, in Chapter 3 we discussed the use of multiple center runs as an augmentation of the standard two-level design. Interest centered on the detection of model curvature, that is, the detection of terms $\beta_{11}x_1^2, \beta_{22}x_2^2, \ldots, \beta_{kk}x_k^2$. Recall that the use of center runs only to augment a standard two-level design allows only a single degree of freedom for estimation of (and thus testing of) these second-order coefficients. As a result, efficient estimates of $\beta_{11}, \beta_{22}, \ldots, \beta_{kk}$ require additional design points.

In Chapter 5, we indicated that in most attempts at product improvement through the gradient technique, called steepest ascent, the investigator will encounter situations where the lack of fit attributable to curvature (pure second-order terms) is found to be quite significant. In these cases it is likely that the model containing first-order terms and, say, two-factor interaction terms $\beta_{12}, \beta_{13}, \ldots, \beta_{k-1,k}$ is woefully inadequate. As a result, a *second-order response surface model*, discussed in the next section, is a reasonable choice.

6.2 SECOND-ORDER APPROXIMATING FUNCTION

In Chapter 1 the notion of approximating polynomial functions was discussed. In response surface methodology (RSM) work it is assumed that the true functional relationship

$$y = f(\mathbf{x}, \theta) \tag{6.1}$$

is, in fact, unknown. Here $(x_1, x_2, \ldots, x_k)$ are in centered and scaled design units. The genesis of the first-order approximating model or the model that contains first-order terms and low-order interaction terms is the notion of the Taylor series approximation of Equation (6.1). Clearly, a Taylor series expansion of Equation (6.1) through second-order terms could involve a model of the type

$$y = \beta_0 + \beta_1 x_1 + \beta_2 x_2 + \cdots + \beta_k x_k + \beta_{11} x_1^2 + \cdots + \beta_{kk} x_k^2 + \beta_{12} x_1 x_2$$
$$+ \beta_{13} x_1 x_3 + \cdots + \beta_{k-1,k} x_{k-1} x_k + \epsilon \tag{6.2}$$

The model of Equation (6.2) is the second-order response surface model. Note that it contains all model terms through order two. The ϵ term is the usual random error term. The assumptions on the random error are those discussed in Chapters 2, 3, and 4.

6.2.1   The Nature of the Second-Order Function and Second-Order Surface

The model of Equation (6.2) is an interesting and widely usable model to describe experimental data in which system curvature is readily abundant. We do not mean to imply here that all systems containing curvature are well accommodated by this model. There are often times in which a model nonlinear in the parameters is necessary. In some rare cases one may even require the use of cubic terms say $x_1^2 x_2$, $x_1 x_2^2$, and so on, to achieve an adequate fit. In other cases the curvature may be quite easily handled through the use of a log transformation on the response itself; for example, in the case where $k = 2$,

$$\ln y = \beta_0 + \beta_1 x_1 + \beta_2 x_2 + \beta_{12} x_1 x_2$$

In other situations, transformation of the design variables may lead to a satisfactory approximating function—for example,

$$y = \beta_0 + \beta_1 (\ln x_1) + \beta_2 (\ln x_2)$$

At times, of course, transformation of both response and design variables may be needed.

The second-order model described in Equation (6.2) is quite useful and is easily accommodated via the use of a wide variety of experimental designs. Families of designs for fitting the second-order model are discussed in Chapters 7 and 8. Note from the model of Equation (6.2) that the model contains $1 + 2k + k(k - 1)/2$ parameters. As a result, the experimental design used must contain at least $1 + 2k + k(k - 1)/2$ *distinct design points*. In addition, of course, the design must involve at least three levels of each design variable to estimate the pure quadratic terms.

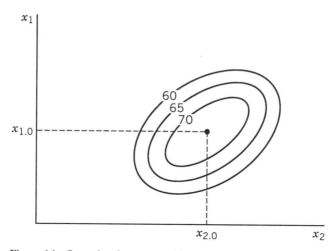

Figure 6.1   Second-order system with point of maximum response.

The geometric nature of the second-order function is displayed in Figures 6.1, 6.2, and 6.3. This will become vital in Section 6.3 as we discuss the location of estimated optimum conditions. Figures 6.1 and 6.2 show contours of constant response for a hypothetical $k = 2$ situation. In Figure 6.1 the center of the system or **stationary point** is a point of maximum response. In Figure 6.2 the stationary point is a point of minimum response. In both cases the response picture displays concentric ellipses. In Figure 6.3, a hyperbolic system of contours is displayed. Note that the center is neither a maximum nor minimum point. In this case, the stationary point is called a **saddle point**

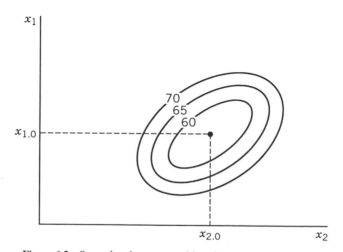

Figure 6.2   Second-order system with point of minimum response.

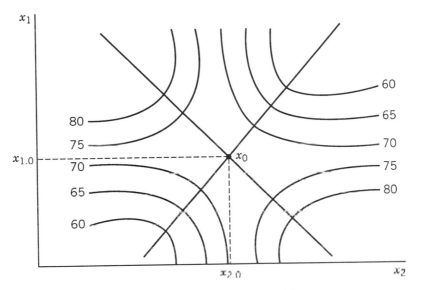

Figure 6.3   Second-order system with saddle point.

and the system of contours is called a *saddle system*. The detection of the nature of the system and the location of the stationary point is an important part of the second-order analysis. Obviously, modern computer graphics blend nicely with this type of analysis. Three dimensional graphics can be very helpful to the data analyst in the determination of the nature of a scientific system. Clearly the ellipses in Figures 6.1 and 6.2 reflect *contours* of constant response. In the $k = 3$ case these contours become *response surfaces*.

6.2.2   Illustration of Second-Order Response Surface

The nature of the response surface system (maximum, minimum, or saddle point) depends on the signs and magnitude of the coefficients in the model of Equation (6.2). The second-order coefficients (interaction and pure quadratic terms) play a vital role. One must keep in mind that the coefficients used are estimates of the β's of Equation (6.2). As a result the contours (or surfaces) represent contours of *estimated* response. Thus, even the system itself (saddle, maximum, or minimum points) is part of the estimation process. The stationary point and the general nature of the system arise as a result of a *fitted model*, not the true structure.

Consider an example in $k = 2$ variables and the fitted second-order model given by

$$\hat{y} = 80 + 2x_1 + 3x_2 - 1.5x_1^2 - 2x_2^2 - 1.0x_1x_2$$

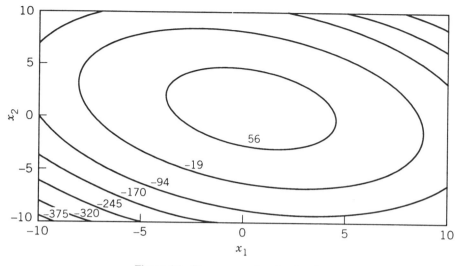

Figure 6.4   Response surface for $k = 2$.

An analysis of this response function would determine the location of the stationary point and the nature of the response system. In this case some simple graphics deals with both issues. Figure 6.4 shows the response surface with contours of constant $\hat{y}$ plotted against x_1 and x_2. It appears as if the system produces a stationary point which has maximum estimated response. Figure 6.5 shows the three-dimensional contour plot verifying that the system produces maximum response at the center of the system. The stationary point

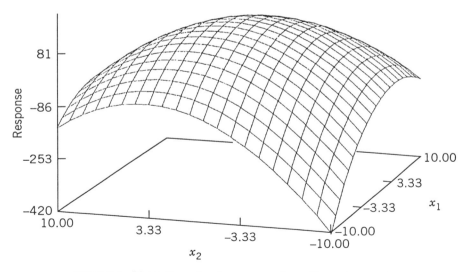

Figure 6.5   Three-dimensional response surface with maximum point.

is found by computing $(\partial \hat{y} / \partial x_1, \partial \hat{y} / \partial x_2)$ and setting to zero.

$$\frac{\partial \hat{y}}{\partial x_1} = 2 - 3x_1 - x_2 = 0$$

$$\frac{\partial \hat{y}}{\partial x_2} = 3 - x_1 - 4x_2 = 0$$

The system of equations

$$3x_1 + x_2 = 2$$

$$x_1 + 4x_2 = 3$$

gives as the solution $x_1 = 5/11$ and $x_2 = 7/11$. The estimated response at the stationary point is given by $\hat{y} = 81.409$.

As a second illustration, consider the fitted function

$$\hat{y} = 80 + 2x_1 + 3x_2 - 1.5x_1^2 + 2x_2^2 - 1.0x_1x_2$$

The three-dimensional response surface system is shown in Figure 6.6. It is clear from the figure that the system is a saddle system. Figure 6.7 reveals two-dimensional contours for the same system. The stationary point is neither a maximum nor a minimum point. Again, the stationary point is

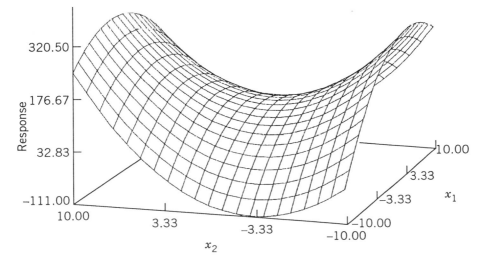

Figure 6.6 Three-dimensional plot for saddle system.

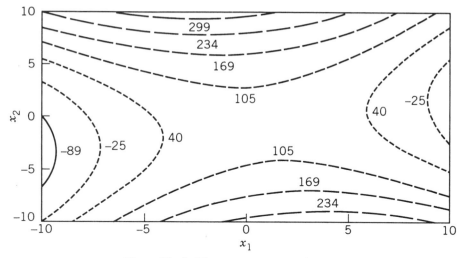

Figure 6.7   Saddle response contour plots.

found by simultaneously setting $\partial \hat{y}/\partial x_1$ and $\partial \hat{y}/\partial x_2$ to zero:

$$\frac{\partial \hat{y}}{\partial x_1} = 2 - 3x_1 - x_2 = 0$$

$$\frac{\partial \hat{y}}{\partial x_2} = 3 + 4x_2 - x_1 = 0$$

This requires the solution for (x_1, x_2) of

$$3x_1 + x_2 = 2$$

$$x_1 - 4x_2 = 3$$

This results in a stationary point given by $x_1 = 11/13$ and $x_2 = -7/13$.

In much of what follows we deal in a general way with the determination of the nature of the total response surface system and the stationary point. It turns out that the strategy that deals in quality improvements is profoundly dependent on the nature of the response system.

Example 6.1   Suppose we are interested in developing a response surface model to relate yield of reaction (%) against temperature and reaction time. The experimental design, commonly referred to as a **central composite design**, may involve the use of as many as five levels of each of the two variables. The factor levels (natural and design units) are as follows:

	-1.414	-1	0	1	1.414
temperature	110.86°C	115°C	125°C	135°C	139.14°C
time	257.58 sec	270 sec	300 sec	330 sec	342.42 sec

Obviously, then, the formulae relating design units to natural units are given by

$$x_1 = \frac{\text{temperature} - 125}{10}$$

$$x_2 = \frac{\text{time} - 300}{30}$$

The design matrix and accompanying response values are given by

x_1	x_2	y
-1	-1	88.55
-1	1	86.29
1	-1	85.80
1	1	80.44
-1.414	0	85.50
1.414	0	85.39
0	-1.414	86.22
0	1.414	85.70
0	0	90.21
0	0	90.85
0	0	91.31

The Central Composite Design. Before we embark on a discussion of the analysis, we should make a few comments about the experimental design. Notice that there are five levels in an experimental design that involves three components. They are (i) a 2^2 factorial, at levels ± 1, (ii) a one-factor-at-a-time array given by

x_1	x_2	
-1.414	0	
1.414	0	(6.3)
0	-1.414	
0	1.414	

and (iii) three center runs. A graphical depiction of this design is given by the following:

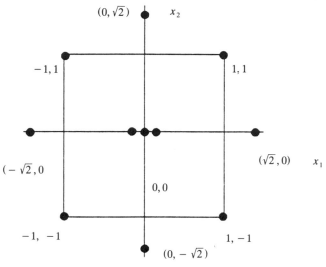

The runs are indicated by the dots. Note that the design essentially consists of eight equally spaced points on a circle of radius $\sqrt{2}$ and three runs in the design center. The set of four design points in display (6.3) are called **axial points**. This terminology stems from the fact that the points lie on the x_1- or x_2-axes. Note also that in the axial portion of the design the factors are not varying simultaneously but rather in a one factor at a time array. As a result, no information regarding the x_1x_2 interaction is attained from this portion of the design. However, the axial portion allows for efficient estimation of pure quadratic terms, that is, x_1^2 and x_2^2. In Chapter 7, we will discuss the central composite design and several other design families in a more thorough way.

Analysis. The second-order model was fitted to the data in the coded design variables for the example, giving the result

$$\hat{y} = 90.790 - 1.095x_1 - 1.045x_2 - 2.781x_1^2 - 0.775x_1x_2 - 2.524x_2^2$$

Figure 6.8 displays a set of two-dimensional contours of constant response. This figure was constructed using the Design-Expert software system. Notice from the figure that the contour system is a set of concentric ellipses with a point of maximum response in the design region. The point of maximum response resides at $\mathbf{x}_0 = (-0.1716, -0.1806)$. The estimated response at this point is $\hat{y} = 90.978\%$.

While graphical analysis plays an important role in RSM, there are times when a more formal analytical approach to the second-order surface is necessary. This is particularly true in the case where several design variables are present in the system. The formal analysis determines the nature of the stationary point and provides other information regarding the nature of the second-order system. One should keep in mind that merely finding the nature of the stationary point does not allow sufficient information for the researcher. Often the nature of the system is elucidated through further analytical techniques combined with graphical information. In addition, it is often necessary for the scientist or engineer to use constrained optimization to arrive at potential operating conditions. This is particularly true when the stationary point is a saddle point or point of maximum or minimum response that resides well outside the experimental region.

6.3   A FORMAL ANALYTICAL APPROACH TO THE SECOND-ORDER ANALYSIS

Consider again the second-order response surface model of Equation (6.2). However, let us consider the *fitted model* in matrix notation as

$$\hat{y} = b_0 + \mathbf{x'b} + \mathbf{x'\hat{B}x} \tag{6.4}$$

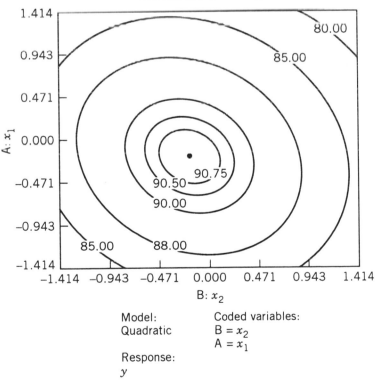

Figure 6.8   Two-dimensional contours for Example 6.1.

where b_0, $\mathbf{b}$, and $\hat{\mathbf{B}}$ contain estimates of the intercept, linear, and second-order coefficients, respectively. In fact $\mathbf{x}' = [x_1, x_2, \ldots, x_k]$, $\mathbf{b}' = [b_1, b_2, \ldots, b_k]$, and $\hat{\mathbf{B}}$ is the $k \times k$ symmetric matrix

$$\hat{\mathbf{B}} = \begin{bmatrix} b_{11} & b_{12}/2 & \cdots & b_{1k}/2 \\ & b_{22} & \cdots & b_{2k}/2 \\ & & \ddots & \vdots \\ \text{sym.} & & & b_{kk} \end{bmatrix} \tag{6.5}$$

One can differentiate $\hat{y}$ in Equation (6.4) with respect to $\mathbf{x}$ and obtain

$$\partial \hat{y} / \partial \mathbf{x} = \mathbf{b} + 2\hat{\mathbf{B}}\mathbf{x}$$

Allowing the derivative to be set to $\mathbf{0}$, one can solve for the stationary point of the system. As a result, we obtain the solution $\mathbf{x}_s$ as

$$\mathbf{x}_s = -\mathbf{B}^{-1}\mathbf{b}/2 \tag{6.6}$$

The point $\mathbf{x}_s$ is the stationary point of the system.

6.3.1   Nature of the Stationary Point (Canonical Analysis)

The nature of the stationary point is determined from the signs of the eigenvalues of the matrix $\hat{\mathbf{B}}$. It turns out that the relative magnitudes of these eigenvalues can be helpful in the total interpretation. For example, let the $k \times k$ matrix $\mathbf{P}$ be the matrix whose columns are the normalized eigenvectors associated with the eigenvalues of $\hat{\mathbf{B}}$. We know that

$$\mathbf{P}'\hat{\mathbf{B}}\mathbf{P} = \boldsymbol{\Lambda}$$

where $\boldsymbol{\Lambda}$ is a diagonal matrix containing the eigenvalues of $\hat{\mathbf{B}}$ as main diagonal elements. Now if we *translate* the model of Equation (6.4) to a new center, namely the stationary point, and *rotate* to axes corresponding to the principal axes of the contour system, we have

$$\mathbf{z} = \mathbf{x} - \mathbf{x}_s$$
$$\mathbf{w} = \mathbf{P}'\mathbf{z} \tag{6.7}$$

This translation gives

$$\begin{aligned} \hat{y} &= b_0 + (\mathbf{z} + \mathbf{x}_s)'\mathbf{b} + (\mathbf{z} + \mathbf{x}_s)'\hat{\mathbf{B}}(\mathbf{z} + \mathbf{x}_s) \\ &= \left[b_0 + \mathbf{x}_s'\mathbf{b} + \mathbf{x}_s'\hat{\mathbf{B}}\mathbf{x}_s \right] + \mathbf{z}'\mathbf{b} + \mathbf{z}'\mathbf{B}\mathbf{z} + 2\mathbf{x}_s'\mathbf{B}\mathbf{z} \\ &= \hat{y}_s + \mathbf{z}'\hat{\mathbf{B}}\mathbf{z} \end{aligned}$$

because $2\mathbf{x}'_s\mathbf{Bz} = \mathbf{z}'\mathbf{b}$ from Equation (6.6). The rotation gives

$$\hat{y} = \hat{y}_s + \mathbf{w}'\mathbf{P}'\hat{\mathbf{B}}\mathbf{P}\mathbf{w}$$

$$= \hat{y}_s + \mathbf{w}'\boldsymbol{\Lambda}\mathbf{w} \tag{6.8}$$

The w-axes are the principal axes of the contour system. Equation (6.8) can be written

$$\hat{y} = \hat{y}_s + \sum_{i=1}^{k} \lambda_i w_i^2 \tag{6.9}$$

where $\hat{y}_s$ is the estimated response at the stationary point and $\lambda_1, \lambda_2, \ldots, \lambda_k$ are the eigenvalues of $\hat{\mathbf{B}}$. The variables $w_1, w_2, \ldots, w_k$ are called **canonical variables**.

The translation and rotation described in the foregoing leads to Equation (6.9). This equation nicely describes the nature of the stationary point and the nature of the system around the stationary point. The *signs* of the λ's determine the nature of $\mathbf{x}_s$, and the relative magnitude of the eigenvalues help the user gain a better understanding of the response system:

1. If $\lambda_1, \lambda_2, \ldots, \lambda_k$ are all negative, the stationary point is a point of *maximum* response.
2. If $\lambda_1, \lambda_2, \ldots, \lambda_k$ are all positive, the stationary point is a point of *minimum* response.
3. If $\lambda_1, \lambda_2, \ldots, \lambda_k$ are mixed in sign, the stationary point is a *saddle* point.

This entire translation and rotation of axes described here is called a *canonical analysis* of the response system. Obviously, if a saddle system occurs the experimenter will often be led to an alternative type of analysis. However, even if a maximum or minimum resides at $\mathbf{x}_s$, the total canonical analysis can be helpful. This is particularly true when one is dealing with a relatively small number of design variables. For example, suppose $k = 3$ and one seeks a large value for the response. Suppose also that $\mathbf{x}_s$ is a point of maximum response (all three eigenvalues are negative) but $|\lambda_1| \cong 0$ and $|\lambda_2| \cong 0$. The implication here is that there is considerable flexibility in locating an acceptable set of operating conditions. In fact, there is essentially an entire plane rather than merely a point where the response is approximately at its maximum value. In fact, it is simple to see that the conditions described by

$$w_3 = 0$$

may in fact describe a rich set of conditions for nearly maximum response.

Now, of course, $w_3 = 0$ describes the plane

$$w_3 = \mathbf{p}'_3 \begin{bmatrix} x_1 - x_{1,s} \\ x_2 - x_{2,s} \\ x_3 - x_{3,s} \end{bmatrix} = 0$$

where the vector $\mathbf{p}_3$ is the normalized eigenvector associated with the negative eigenvalue λ_3.

6.3.2 Ridge Systems

The foregoing is only one of many situations in which the canonical form of Equation (6.9) helps to interpret the second-order surface and generally either indicates flexibility in operating conditions or provides information regarding areas for future experiments. The system described above is called a **ridge system**. If the stationary point is in the region of the experimental design, the system is a *stationary ridge system*. The stationary point is a maximum and yet there is an approximate maximum on a plane within the design region. A simple stationary ridge is described by Figure 6.9. Here we have two design variables with $\lambda_1 < 0$ and $\lambda_2 < 0$, but $|\lambda_2| \cong 0$. The figure shows the design region and illustrates the use of the canonical variables. Clearly the stationary point is a point of maximum response, but there is essentially a "line maximum." Consider locations in the design region where $w_1 = 0$; the insensitivity of the response to movement along the w_2-axes render little change in response off the value at the stationary point. The contours of a constant response are concentric ellipses that are greatly

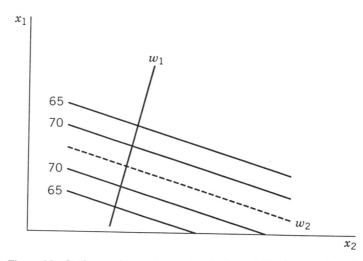

Figure 6.9   Stationary ridge system in two design variables (line maximum).

elongated along the w_2-axes. Sizable changes in response are experienced along the w_1-axis.

There are other ridge systems that signal certain moves in strategy by the analyst. While the stationary ridge suggests flexibility in choice of operating conditions, the *rising (or falling)* ridge suggests that movement outside the experimental region for additional experimentation might warrant considerations. For example, consider a $k = 2$ situation in which the stationary point is remote from the experimental region. Assume also that one needs to maximize response. Let λ_1 and λ_2 both be negative but assume λ_2 is near zero. The condition is displayed in Figure 6.10. The remoteness of the stationary point from the experimental region might suggest here that a movement (experimentally) along the w_2 axis may well result in an increase in response values. Indeed the rising (or falling) ridge often is a signal to the researcher that he or she has perhaps made a faulty or premature selection of the experimental design region. This is not at all uncommon in applications of RSM.

The rising ridge in these variables again is signaled by a stationary point that has a maximum response and is remote from the design region. In addition however, one of the eigenvalues is near zero. The limiting condition for three variables is displayed in the illustration in Figure 6.11. Here the eigenvalues for w_1, w_2, and w_3 are $(-,-,\cong 0)$. Experimentation along the w_3-axis toward the stationary point is certainly suggested.

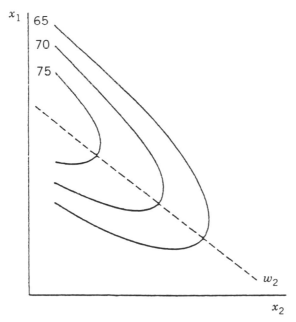

Figure 6.10   Rising ridge in two variables.

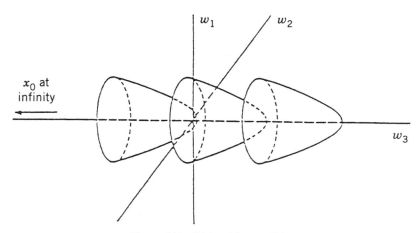

Figure 6.11   Rising ridge conditions.

Examples of ridge systems will be seen in illustrations in other portions of this text. However, it is timely at this point to discuss the role of contour plots in the practical analysis of the second-order fitted surface.

6.3.3   Role of Contour Plots

The role of contour plots or contour "maps" is one of the most revealing ways of illustrating and interpreting the response surface system. The contour plots are merely two-dimensional (or sometimes three-dimensional) graphs that show contours of constant response with the axis system being a specific pair of the design variables say x_i and x_j, while the other design variables are held constant. Modern graphics allow interesting interpretations that can be of benefit to the user. In fact, in almost all practical response surface situations, the analysis should be followed with a display of contour plotting. The reader should recall the contour plot in Example 6.1. The plots are particularly necessary when the stationary point is not of practical significance (saddle point or, say, a maximum point when one seeks a point of minimum response) or when the stationary point is remote from the design region. Clearly, ridge systems can only be interpreted when one can observe two-dimensional systems as a set of "snapshots" of the design region. Of course, if there is an excessive number of design variables, contour plotting is more difficult since many factors need to be held constant. The method of ridge analysis (Section 6.4) can often aid the user in an understanding of the system. One cautionary note regarding contour plots should certainly be made at this point. The investigator must bear in mind that the contours are only "estimates" and if repeated data sets were observed using the same design the complexion of the response system will change slightly or perhaps drastically. In other words, the *contours are not generated by deterministic equations*. Every point on a contour has a standard error. This will be illustrated in a more technical way in a subsequent section.

Example 6.2   The following data was taken from an experiment designed for estimating optimum conditions for storing bovine semen to retain maximum survival. The variables under study are percent sodium citrate, percent glycerol, and the equilibration time in hours. The response observed is percent survival of motile spermatozoa. The data with the design levels in coded design units are as follows:

Treatment Combination	Percent Sodium Citrate	Percent Glycerol	Equilibration Time, (hr)	Percent Survival
1	−1	−1	−1	57
2	1	−1	−1	40
3	−1	1	−1	19
4	1	1	−1	40
5	−1	−1	1	54
6	1	−1	1	41
7	−1	1	1	21
8	1	1	1	43
9	0	0	0	63
10	−2	0	0	28
11	2	0	0	11
12	0	−2	0	2
13	0	2	0	18
14	0	0	−2	56
15	0	0	2	46

The fitted second-order response function is given by

$$\hat{y} = 66.111 - 1.313x_1 - 2.313x_2 - 1.063x_3 - 11.264x_1^2 - 13.639x_2^2 - 3.357x_3^2$$
$$+ 9.125x_1x_2 + 0.625x_1x_3 + 0.875x_2x_3$$

The design levels relate to the natural levels in the following way:

	−2	−1	0	1	2
Sodium Citrate	1.6	2.3	3.0	3.7	4.4
Glycerol	2.0	5.0	8.0	11.0	14.0
Equilibration Time	4.0	10.0	16.0	22.0	28.0

Note that the design consists of a 2^3 factorial component, one center run, and six "one-factor-at-a-time" experimental runs in which each factor is set at level $+2$ and -2. The 15 experimental runs, involving 3 factors at five evenly spaced levels, is another example of a central composite design. From Equation (6.6) the stationary point is given by

$$\mathbf{x}_s = \begin{bmatrix} x_{1,s} \\ x_{2,s} \\ x_{3,s} \end{bmatrix} = \begin{bmatrix} -0.1158 \\ -0.1294 \\ -0.1841 \end{bmatrix}$$

with eigenvalues of the $\hat{\mathbf{B}}$ matrix given by

$$\lambda_1 = -3.327, \qquad \lambda_2 = -7.797, \qquad \lambda_3 = -17.168$$

The canonical and stationary point analysis indicates that the stationary point is a point of maximum estimated response—that is, *maximum estimated percent survival*. The estimated response at the maximum is $\hat{y} = 66.435$. As a result, the estimated condition sought by the experimenters is found inside the experimental design region. Table 6.1 shows a SAS PROC RSREG output of the response function and the canonical analysis.

Clearly, in this case, the researcher may be quite satisfied to operate at the conditions of the computed stationary point. However, because of economic considerations and scientific constraints of the problem, it is often of interest to determine how sensitive the estimated response is to movements away

Table 6.1   SAS RSREG Canonical and Stationary Point Analysis of Data of Example 6.2[a]

	Canonical Analysis of Response Surface (Based on Coded Data)	
	Critical Value	
Factor	Coded	Uncoded
x_1	-0.057896	-0.115792
x_2	-0.064709	-0.129418
x_3	-0.092074	-0.184147

Predicted value at stationary point 66.434567

	Canonical Analysis of Response Surface (Based on Coded Data)		
	Eigenvectors		
Eigenvalues	x_1	x_2	x_3
-13.308207	0.084930	0.079716	0.993193
-31.187893	0.787001	0.605961	-0.115934
-68.670567	-0.611078	0.791490	-0.011272

Stationary point is a maximum.

[a]The uncoded levels in SAS are what we have called "coded." This will be explained later in this chapter

from the stationary point. In this regard, contour plotting can be very useful. For example, in this case, the stationary point x_s corresponds to 2.9% sodium citrate, 7.61 glycerol, and 14.9 hours equilibration time. The equilibration time may be viewed as being somewhat high from a practical point of view, even though the optimum level is inside the experimental design region. Scientific curiosity suggests the need to determine how much is lost in estimated percent survival if equilibration time is set at values 14 hours, 12 hours, and 10 hours. An analytical procedure would answer the question, but one gains considerably more information from the set of pictures produced from contour plotting. Figures 6.12, 6.13, and 6.14 from the Design-Expert package reveal the two-dimensional plots giving contours of constant estimated percent survival for fixed values of equilibration times of 14 hours, 12 hours, and 10 hours, respectively. The Design-Expert contour plots in the figures show coded values of $x_3 = -0.333$, $x_3 = -0.667$, and -1.00, corresponding to the natural levels of 14, 12, and 10 hours, respectively. For 14 hours equilibration time, the maximum estimated survival time is very

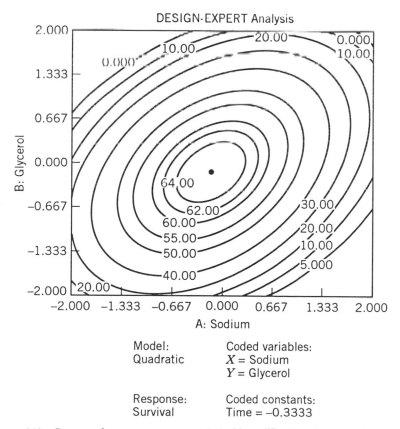

Figure 6.12 Contour of constant percent survival with equilibration time fixed at 14 hours.

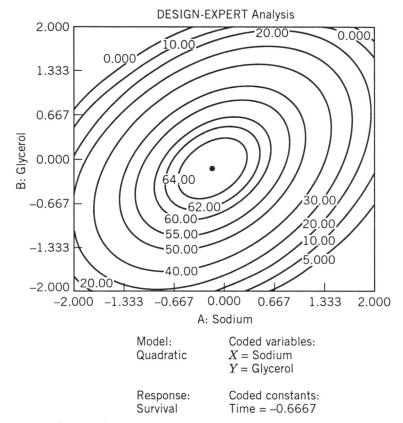

Figure 6.13   Contour of constant percent survival with equilibration time fixed at 12 hours

nearly 65%. In fact, Figure 6.13 reveals that equilibration time as low as 12 hours can result in estimated survival time that exceeds 64%. As a result, one may wish to recommend using values of equilibration time considerably below optimum and still not lose much in percent survival.

6.4 RIDGE ANALYSIS OF THE RESPONSE SURFACE

Often the analyst is confronted with a situation in which the stationary point itself is essentially of no value. For example, the stationary point may be a saddle point, which of course is neither a point of estimated maximum or minimum response. In fact, even if one encounters a stationary point that is a point of maximum or minimum response, if the point is outside the experimental region, it would not be advisable to suggest it as a candidate for operating conditions because the fitted model is not reliable outside the

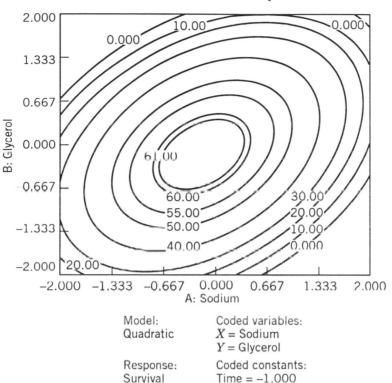

Figure 6.14 Contour of constant percent survival with equilibration time fixed at 10 hours.

region of the experiment. Certainly the use of contour plotting can be beneficial in these situations. Of course, the goal of the analysis is to determine the nature of the system inside or on the perimeter of the experimental region. As a result, a useful procedure involves a **constrained optimization** algorithm. It is particularly useful when several design variables are in the response surface model. The methodology, called **ridge analysis**, produces a locus of points, each of which is a point of maximum response, with a constraint that the point resides on a sphere of a certain radius. For example, for $k = 2$ design variables a typical ridge analysis might produce the locus of points as in Figure 6.15. The origin is $x_1 = x_2 = 0$, the design center; a given point, say $\mathbf{x}_p$, is a point of maximum (or minimum) estimated response, with the constraint being that the point be on a sphere of radius $(\mathbf{x}_p' \mathbf{x}_p)^{1/2} = r_p$. In other words, *all points are constrained stationary points.* Indeed, if the stationary point of the system is not a maximum (or minimum) inside the experimental region, then the point on the "path" at the design perimeter is the point of maximum (minimum) estimated response over all

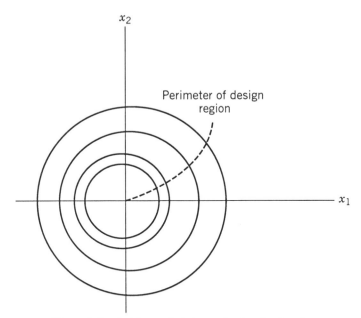

Figure 6.15 A ridge analysis locus of points for $k = 2$.

points on the path. This, then, may be viewed as a reasonable candidate for recommended operating conditions.

6.4.1   What Is the Value of Ridge Analysis?

The purpose of ridge analysis is to *anchor* the stationary point to be inside the experimental region. The output of the analysis is the set of coordinates of the maxima (or minima) along with the predicted response, $\hat{y}$, at each computed point on the path. From this, the analyst gains useful information regarding the roles of the design variables inside the experimental region. He or she is also given some candidate locations for suggested improved operating conditions. This is not to infer that additional information would not come from the contour plots. However, when one is faced with a large number of design variables, many contour plots may be required. Typical output of ridge analysis might simply be a set of two-dimensional plots, despite the value of k. For example, consider Figures 6.16 and 6.17.

In this case there are three design variables, and let us assume that the perimeter of the experimental design is at radius $\sqrt{3}$. If one seeks to maximize response, a candidate set of conditions is given by $x_1 \cong 1.5, x_2 \cong 0.4, x_3 \cong -0.7$. The estimated maximum response achieved is given by $\hat{y} \cong 60$. Another useful feature of ridge analysis is that one obtains some guidelines regarding where future experiments should be made in order to achieve

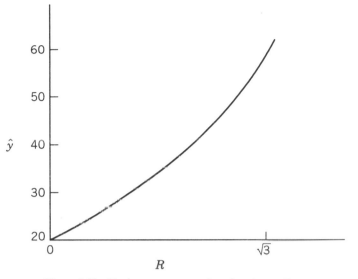

Figure 6.16 Maximum response plotted against radius.

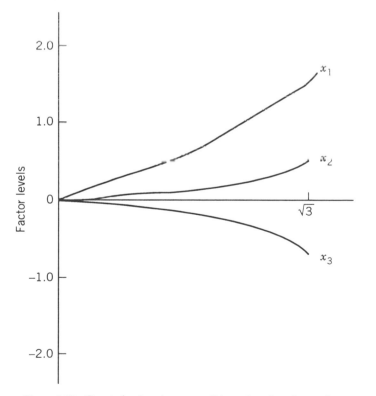

Figure 6.17 Constrained optimum conditions plotted against radius.

conditions that are more desirable than those experienced in the current experiment.

One must keep in mind that while ridge analysis as a formal tool is more ideal when the design region is spherical, often the design required is not spherical but rather cuboidal in nature. In the latter case, one can still gain important practical information about the behavior of the response system inside the experimental region, particularly when one encounters several important design variables.

6.4.2 Mathematical Development of Ridge Analysis

In this section we shall sketch the development of ridge analysis though the details involved in some of the theorems of ridge analysis will be relegated to Appendix 4. Many of the details are similar to those of steepest ascent. In fact, *ridge analysis is steepest ascent applied to second order models*. The reader should recall that the intent of steepest ascent was to provide a path to an improved product in the case of a system that had not been well-studied. A rather inexpensive first-order experiment is the usual vehicle for steepest ascent. However, ridge analysis is generally used when the practitioner feels that he or she is in or quite near the region of the optimum. Nevertheless, both are constrained optimization procedures.

Consider the fitted second-order response surface model of Equation (6.4):

$$\hat{y} = b_0 + \mathbf{x}'\mathbf{b} + \mathbf{x}'\hat{\mathbf{B}}\mathbf{x}$$

We maximize $\hat{y}$ subject to the constraint

$$\mathbf{x}'\mathbf{x} = R^2$$

where $\mathbf{x}' = [x_1, x_2, \ldots, x_k]$ and the center of the design region is taken to be $x_1 = x_2 = \cdots = x_k = 0$. Using Lagrange multipliers, we need to differentiate

$$L = b_0 + \mathbf{x}'\mathbf{b} + \mathbf{x}'\hat{\mathbf{B}}\mathbf{x} - \mu(\mathbf{x}'\mathbf{x} - R^2)$$

with respect to the vector $\mathbf{x}$. The derivative $\lambda L / \partial \mathbf{x}$ is given by

$$\frac{\partial L}{\partial \mathbf{x}} = \mathbf{b} + 2\hat{\mathbf{B}}\mathbf{x} - 2\mu\mathbf{x}$$

To determine constrained stationary points we set $\partial L / \partial \mathbf{x} = \mathbf{0}$. This results in

$$(\hat{\mathbf{B}} - \mu\mathbf{I})\mathbf{x} = -\tfrac{1}{2}\mathbf{b} \tag{6.10}$$

As a result, for a fixed μ, a solution x of Equation (6.10) is a stationary point on $R = (x'x)^{1/2}$. However, the appropriate solution x is that which results in a maximum $\hat{y}$ on R or a minimum $\hat{y}$ on R, depending on which is desired. It turns out that the appropriate choice of μ depends on the eigenvalues of the $\hat{B}$ matrix. The reader should recall the important role of these eigenvalues in the determination of the nature of the stationary point of the response system (see Section 6.3.1). The following are rules for selection of values of μ:

1. If μ exceeds the largest eigenvalue of $\hat{B}$, the solution x in Equation (6.10) will result in an absolute maximum for $\hat{y}$ on $R = (x'x)^{1/2}$.

2. If μ is smaller the smallest eigenvalue of $\hat{B}$, the solution in Equation (6.10) will result in an absolute minimum for $\hat{y}$ on $R = x'x)^{1/2}$.

Appendix 4 provides some mathematical insight into 1 and 2 above. We should also reflect at this point on the relationship between R and μ. The analyst desires to observe results on a locus of points like that depicted in Figure 6.15. As a result, of course, the radii of the solution to Equation (6.10) should fall in the interval $[0, R_b]$, where R_b is a radius approximately representing the boundary of the experimental region. The value R is actually controlled through the choice of μ. In the "working regions" of μ, namely $\mu > \lambda_k$ or $\mu < \lambda_1$, where λ_1 is the smallest eigenvalue of $\hat{B}$ and λ_k is the largest eigenvalue of $\hat{B}$, R is a monotonic function of μ. In fact, Figure 6.18 gives the relationship between R and μ throughout the spectrum of the eigenvalue.

As a result, a computer algorithm of ridge analysis involves the substitution of $\mu > \lambda_k$ (for a designed maximum response) and increases μ until radii near the design perimeter are encountered. Future increasing of μ results in coordinates that are closer to the design center. The same applies for $\mu < \lambda_1$ (for desired minimum response) with decreasing values of μ being required.

Example 6.3   A chemical process that converts 1,2-propanediol to 2,5-di-methylpiperazine is the object of an experiment to determine optimum process conditions—that is, conditions for maximum conversion [see Myers (1976)]. The following factors were studied:

$$x_1 = \frac{\text{Amount of } NH_3 - 102}{51}$$

$$x_2 = \frac{\text{Temperature} - 250}{20}$$

$$x_3 = \frac{\text{Amount of } H_2O - 300}{200}$$

$$x_4 = \frac{\text{Hydrogen pressure} - 850}{350}$$

As in previous examples, the type of design used for fitting a second-order model was a central composite design. The design and observation vector are as follows:

$$
\mathbf{D} = \begin{array}{c} \\ \begin{array}{cccc} x_1 & x_2 & x_3 & x_4 \end{array} \\ \left[\begin{array}{rrrr} -1 & -1 & -1 & -1 \\ +1 & -1 & -1 & -1 \\ -1 & +1 & -1 & -1 \\ +1 & +1 & -1 & -1 \\ -1 & -1 & +1 & -1 \\ +1 & -1 & +1 & -1 \\ -1 & +1 & +1 & -1 \\ +1 & +1 & +1 & -1 \\ -1 & -1 & -1 & +1 \\ +1 & -1 & -1 & +1 \\ -1 & +1 & -1 & +1 \\ +1 & +1 & -1 & +1 \\ -1 & -1 & +1 & +1 \\ +1 & -1 & +1 & +1 \\ -1 & +1 & +1 & +1 \\ +1 & +1 & +1 & +1 \\ 0 & 0 & 0 & 0 \\ -1.4 & 0 & 0 & 0 \\ +1.4 & 0 & 0 & 0 \\ 0 & -1.4 & 0 & 0 \\ 0 & +1.4 & 0 & 0 \\ 0 & 0 & -1.4 & 0 \\ 0 & 0 & +1.4 & 0 \\ 0 & 0 & 0 & -1.4 \\ 0 & 0 & 0 & +1.4 \end{array}\right] \end{array}, \qquad \mathbf{y} = \left[\begin{array}{r} 58.2 \\ 23.4 \\ 21.9 \\ 21.8 \\ 14.3 \\ 6.3 \\ 4.5 \\ 21.8 \\ 46.7 \\ 53.2 \\ 23.7 \\ 40.3 \\ 7.5 \\ 13.3 \\ 49.3 \\ 20.1 \\ 32.8 \\ 31.1 \\ 28.1 \\ 17.5 \\ 49.7 \\ 49.9 \\ 34.2 \\ 31.1 \\ 43.1 \end{array}\right]
$$

The fitted second-order function is given by

$$
\hat{y} = 40.198 - 1.511x_1 + 1.284x_2 - 8.739x_3 + 4.955x_4 - 6.332x_1^2
$$

$$
+ 2.194x_1x_2 - 4.292x_2^2 - 0.144x_1x_3 + 8.006x_2x_3 + 0.0196x_3^2
$$

$$
+ 1.581x_1x_4 + 2.806x_2x_4 + 0.294x_3x_4 - 2.506x_4^2
$$

SAS RSREG was used to analyze the data. Table 6.2 shows computer printout that reveals the ridge analysis of maximum response created by the ridge analysis. The stationary point (in design units) is given by $x_{1,s} = 0.265$, $x_{2,s} = 1.034$, $x_{3,s} = 0.291$, and $x_{4,s} = 1.668$. The response at the stationary point in $\hat{y}_s = 43.52$ but the eigenvalues of the $\hat{\mathbf{B}}$ matrix are -7.55, -6.01, -2.16, and $+2.60$, indicating a saddle point. They indicate that a ridge

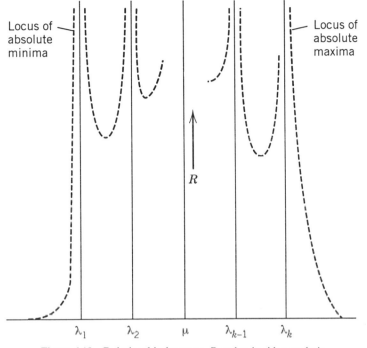

Figure 6.18 Relationship between R and μ in ridge analysis.

analysis might reveal reasonable candidates for operating conditions, with the implied constraint that the candidates lie inside or on the boundary of the experimental design. In this case the factorial points are at a distance of two units from the design center. Note that among the reasonable candidates for coordinates, we have

$$x_1 = -0.1308, \qquad x_2 = -0.7861, \qquad x_3 = -1.8281, \qquad x_4 = 0.1514$$

at a radius of two units. The estimated response is given by $\hat{y}_s = 64.61$. Note also that the estimated standard error of prediction at this location is given by

$$s_{\hat{y}} = 16.543$$

This, of course, reflects the variability in $\hat{y}$ conditioned on the given set of coordinates, that is, given that prediction is desired at the location indicated in x_1, x_2, x_3, and x_4. Note how this standard error begins to grow rather rapidly after a radius of 1.4. Recall that the axial distance in the design is 1.4, and the factorial points are at a distance of 2.0. It turns out that a better choice of axial distance would have been a value of 2.0. This would have prevented the quick acceleration of the standard error of prediction as one

Table 6.2   Annotated SAS RSREG for Ridge Analysis of Data in Example 6.3

Radius	Estimated Response	Standard Error of Prediction
0	40.198215	8.321708
0.100000	41.207095	8.304643
0.200000	42.195254	8.254609
0.300000	43.175752	8.175379
0.400000	44.159990	8.073806
0.500000	45.157604	7.960229
0.600000	46.176477	7.848751
0.700000	47.222896	7.757356
0.800000	48.301790	7.707668
0.900000	49.416989	7.724092
1.000000	50.571468	7.832082
1.100000	51.767542	8.055561
1.200000	53.007026	8.414079
1.300000	54.291354	8.920649
1.400000	55.621668	9.581090
1.500000	56.998888	10.394958
1.600000	58.423759	11.357423
1.700000	59.896893	12.461286
1.800000	61.418794	13.698565
1.900000	62.989882	15.061512
2.000000	64.610510	16.543132

| Radius | Factor Values | | | |
	x_1	x_2	x_3	x_4
0	0	0	0	0
0.100000	−0.012558	0.006391	−0.087085	0.047091
0.200000	−0.021700	0.001210	−0.177274	0.090009
0.300000	−0.028715	−0.014124	−0.269867	0.127073
0.400000	−0.034566	−0.037947	−0.364067	0.157545
0.500000	−0.039887	−0.068576	−0.459104	0.181466
0.600000	−0.045054	−0.104485	−0.554349	0.199375
0.700000	−0.050267	−0.144399	−0.649356	0.212036
0.800000	−0.055621	−0.187319	−0.743845	0.220255
0.900000	−0.061150	−0.232486	−0.837668	0.224773
1.000000	−0.066860	−0.279337	−0.930765	0.226226
1.100000	−0.072739	−0.327459	−1.023128	0.225144
1.200000	−0.078773	−0.376547	−1.114783	0.221959
1.300000	−0.084944	−0.426379	−1.205773	0.217020
1.400000	−0.091237	−0.476786	−1.296146	0.210608
1.500000	−0.097636	−0.527644	−1.385954	0.202951
1.600000	−0.104128	−0.578859	−1.475246	0.194235
1.700000	−0.110702	−0.630359	−1.564069	0.184612
1.800000	−0.117348	−0.682087	−1.652464	0.174207
1.900000	−0.124058	−0.734001	−1.740472	0.163123
2.000000	−0.130824	−0.786066	−1.828127	0.151447

approaches the design perimeter. In fact, in this case one might feel obliged to recommend conditions at radius of 1.4 or 1.5 where the predicted response is smaller but the standard error is considerably smaller. We will shed more light on the standard error of prediction in the next section and in future chapters. This criterion plays a huge role in discussions that relate to the choice of experimental design and comparisons among experimental designs.

For further discussion and developments related to ridge analysis the reader is referred to Hoerl (1959, 1964, 1985), who originated the concept, and Draper (1963), who developed the mathematical ideas behind the concept.

6.5 SAMPLING PROPERTIES OF RESPONSE SURFACE RESULTS

In response surface analysis there is considerable attention given to the use of contour plots and the estimation of optimum conditions—either absolute optimum conditions or constrained optimum conditions as in the case of ridge analysis. It is always a temptation to offer interpretations that treat these optimum conditions or observed contours of constant response as if the values were scientifically exact. One must bear in mind that in any RSM analysis the computed stationary point is only an estimate and any point on any contour (and, indeed, the contour itself) possesses sampling variability. Thus, standard errors and, at times, confidence regions afford the analyst a more realistic assessment of the quality of the point estimate. In addition, it forces the researcher to properly "tone down" or temper interpretations in many cases. Interpretations that are made as if the results are deterministic can be erroneous and highly misleading. Quite frankly, it is our impression that too often in practice the random nature and hence "noise" associated with an RSM result is completely ignored.

The first type of sampling property we shall introduce deals with the standard error of predicted response, or standard error of estimated mean response. This concept was discussed briefly in the development of standard regression analysis in Chapter 2. It was also discussed in the previous section.

6.5.1 Standard Error of Predicted Response

The most fundamental sampling property in any model building exercise is the quantity $s_{\hat{y}(x)}$, where in the most general framework

$$\hat{y}(x) = x^{(m)'}b$$

where $b = (X'X)^{-1}X'y$. Here, of course, the X matrix is as discussed in Chapter 2. The vector $x^{(m)}$ is a function of the location at which one is predicting response; the (m) notation indicates that $x^{(m)}$ has been expanded to "model space"; that is, the vector reflects the form of the model as $\mathbf{X}$ does. For example, for $k = 2$ design variables and a second-order model we have

$$\mathbf{x}^{(m)'} = \left[1, x_1, x_2, x_1^2, x_2^2, x_1 x_2\right]$$

$$\mathbf{b}' = \left[b_0, b_1, b_2, b_{11}, b_{22}, b_{12}\right]$$

Under the i.i.d. error assumptions of the first- and second-order models discussed in Chapters 5 and 6, and, of course, with the assumption of constant error variance σ^2 we have

$$\text{Var } \hat{y}(\mathbf{x}) = \mathbf{x}^{(m)'}(\mathbf{X}'\mathbf{X})^{-1}\mathbf{x}^{(m)} \cdot \sigma^2$$

As a result, an **estimated standard error** of $\hat{y}(\mathbf{x})$ is given by

$$s_{\hat{y}(\mathbf{x})} = s\sqrt{\mathbf{x}^{(m)'}(\mathbf{X}'\mathbf{X})^{-1}\mathbf{x}^{(m)}} \tag{6.11}$$

where s is the root mean squared error of the fitted response surface—that is, for a fitted model with p parameters

$$s = \sqrt{\sum_{i=1}^{k}(y_i - \hat{y}_i)^2 / (n - p)}$$

The standard error, $s_{\hat{y}(\mathbf{x})}$ is used in attaching confidence limits around a predicted response. That is, for $\hat{y}(\mathbf{x}) = \mathbf{x}^{(m)'}\mathbf{b}$, a prediction at some location $\mathbf{x}$, the $100(1 - \alpha)\%$ confidence interval on the mean response $E(y|\mathbf{x})$ is given by

$$\hat{y}(\mathbf{x}) \pm t_{\alpha/2, n-p} s\sqrt{\mathbf{x}^{(m)'}(\mathbf{X}'\mathbf{X})^{-1}\mathbf{x}^{(m)}}$$

This standard error of predicted values, $s_{\hat{y}(\mathbf{x})}$, and the corresponding confidence intervals are the simplest form of sampling properties associated with an RSM analysis. The standard error can provide the user a rough idea about the relative quality of predicted response values in various locations in the design region. For example, the reader should refer to Table 6.2, which displays the ridge analysis for Example 6.3. Note, for example, how prediction with this second-order response surface becomes worse as one gets near the design perimeter. This may very well lead, in some cases, to the conclusion that a reasonable choice of recommended operating conditions might be further inside the perimeter when the predicted value at the optimum on the boundary has a relatively large standard error. The standard error of prediction should be computed at any point that the investigator considers a potentially useful location in the design region.

Standard Error of Prediction as a Function of Design

Because of the prominent role of $(X'X)^{-1}$ in Equation (6.11), it should be apparent that $s_{\hat{y}(x)}$ is very much dependent on the experimental design. In Table 6.2, the larger standard error that is experienced close to the design perimeter is a result of the choice of design. In the case of the central composite design, the choice of the "axial distance" (± 1.4 in this case) and number of center runs have an important influence on the quality of $\hat{y}(x)$ as a predictor. In Chapter 7 we deal with properties of first- and second-order designs. Much attention is devoted to Var $\hat{y}(x)/\sigma^2 = x^{(m)'}(X'X)^{-1}x^{(m)}$. In fact, designs are compared on the basis of relative distribution of values of $x^{(m)'}(X'X)^{-1}x^{(m)}$ in the design space.

What Are the Restrictions on the Standard Error of Prediction?

The user of RSM can learn a great deal about relative quality of prediction using the fitted response surface in various locations in the design space. The prediction standard error, a function of the model, the design, and the location x, is quite fundamental. However, it is important for the reader to understand what the standard error of prediction *is not* as well as what it is.

The value $s_{\hat{y}(x)}$ in Equation (6.11) is an estimate of the standard deviation of $\hat{y}(x)$ over repeated experiments in which the same design and model are employed, and thus repeated values of $\hat{y}$ are calculated at the same location x.

As a result, if the analyst calculates a stationary point x_s (a point of maximum estimated response) and determines the standard error of prediction at x_s, it cannot be said that the resulting confidence interval is a proper confidence interval on the maximum response. One must keep in mind that repeated experiments would not produce x_s coordinates at the same location. Indeed, repeated experiments may not all result in stationary points which are maxima. The computed confidence interval here is a confidence interval **conditional** on prediction at the point x_s, which only happens to be a stationary point in the present experiment. The confidence region on the location of the stationary point and the confidence interval on the maximum response are considerably more complicated.

6.5.2 Confidence Region on the Location of the Stationary Point

There are instances when the computed stationary point is of considerable interest to the user. It may be the location in the design variables that represents recommended operating conditions for the process. However, there remains the obvious question, "How good are the estimates of these coordinates?" Perhaps there is sufficient noise in the estimator that a secondary, less costly set of conditions results in a mean response that is not significantly different from that produced by the location of the stationary point. These issues can be dealt with nicely through the construction of a confidence region on the location of the stationary point.

Suppose we again consider the fitted second-order response surface model

$$\hat{y} = b_0 + \sum_{i=1}^{k} b_i x_i + \sum_{i=1}^{k} b_{ii} x_i^2 + \sum_{i<j}^{k} \sum b_{ij} x_i x_j$$

$$= b_0 + \mathbf{x'b} + \mathbf{x\hat{B}x}$$

One should recall from material in Section 6.3 that the stationary point is computed from setting the derivatives $\partial \hat{y}(x)/\partial \mathbf{x}$ to zero. We have

$$\partial \hat{y}(\mathbf{x})/\partial \mathbf{x} = \mathbf{b} + 2\mathbf{\hat{B}x}$$

The jth derivative, $d_j(\mathbf{x})$, can be written $d_j(\mathbf{x}) = b_j + 2\mathbf{\hat{B}'_j x}$, where the vector $\mathbf{\hat{B}'_j}$ is the jth row of $\hat{B}$, the matrix of quadratic coefficients described by Equation (6.5). These derivatives are simple linear functions of $x_1, x_2, \ldots, x_k$. We denote the vector of derivatives as the k-dimensional vector $\mathbf{d}(\mathbf{x})$. Now, suppose we consider the derivatives evaluated at $\mathbf{t}$, where the coordinates of $\mathbf{t}$ are the coordinates of the *true stationary point of the system* (which of course are unknown). If the errors around the model of Equation (6.2) are i.i.d. normal, $N(0, \sigma)$, then

$$\mathbf{d(t)} \sim N\big(\mathbf{0}, \mathrm{Var}\,\mathbf{d(t)}\big)$$

where $\mathrm{Var}\,\mathbf{d(t)}$ is the variance-covariance matrix of $\mathbf{d(t)}$. As a result [see Graybill (1976), Myers and Milton (1991)],

$$\frac{\mathbf{d'(t)}\big[\widehat{\mathrm{Var}\,\mathbf{d(t)}}\big]^{-1}\mathbf{d(t)}}{k} \sim F_{k,\,n-p} \qquad (6.12)$$

Here, of course, $F_{k,n-p}$ is the familiar F-distribution with k and $n - p$ degrees of freedom. It is important to note that $\mathrm{Var}\,\mathbf{d(t)}$ in a $k \times k$ matrix that contains the error variance σ^2 as a multiplier. In addition, $\mathrm{Var}\,\mathbf{d(t)}$ is clearly a function of $\mathbf{t}$. For example, in the case of two design variables,

$$d_1(\mathbf{t}) = b_1 + 2\left(b_{11}t_1 + \frac{b_{12}}{2}t_2\right)$$

$$d_2(\mathbf{t}) = b_2 + 2\left(\frac{b_{12}}{2}t_1 + b_{22}t_2\right)$$

and

$$\mathrm{Var}\,\mathbf{d(t)} = \begin{bmatrix} \mathrm{Var}\,d_1(\mathbf{t}) & \mathrm{Cov}\big[d_1(\mathbf{t}), d_2(\mathbf{t})\big] \\ \mathrm{Cov}\big[d_1(\mathbf{t}), d_2(\mathbf{t})\big] & \mathrm{Var}\,d_2(\mathbf{t}) \end{bmatrix}$$

One can easily observe that elements in this matrix came from $(\mathbf{X'X})^{-1}\sigma^2$, the variance–covariance matrix of regression coefficients. The role of t_1 and t_2 should also be apparent. In display (6.12) the "$\wedge$" in $[\widehat{\text{Var}}\,d(\mathbf{t})]$ merely implies that σ^2 is replaced by the familiar s^2, the model error mean square. Now, based on display (6.13) we have

$$\Pr\left\{\mathbf{d'(t)}\left[\widehat{\text{Var}}\,\mathbf{d(t)}\right]^{-1}\mathbf{d(t)} \le kF_{\alpha,k,n-p}\right\} = 1 - \alpha \qquad (6.13)$$

Here, of course, $F_{\alpha,k,n-p}$ is the upper αth percent point of the F-distribution for k and $n - p$ degrees of freedom. Note that while $\mathbf{t}$ is truly unknown, all other quantities in Equation (6.12) are known. As a result, values $t_1, t_2, \ldots, t_k$ that fall inside the $100(1 - \alpha)\%$ confidence region for the stationary point are those that satisfy the following inequality:

$$\mathbf{d'(t)}\left[\widehat{\text{Var}}\,\mathbf{d(t)}\right]^{-1}\mathbf{d(t)} \le kF_{\alpha,k,n-p} \qquad (6.14)$$

This useful and important approach was developed by Box and Hunter (1954).

6.5.3 Use and Computation of the Confidence Region on the Location of the Stationary Point

As one can easily observe from Equation (6.14), the computation of the confidence region is quite simple. However, the display of the confidence region carries with it a "clumsiness" that is similar to that of the response surface display, particularly when several design variables are involved. For two or three design variables a graphical approach can be very informative. The general size (or volume) of the confidence region can be visualized and the analyst can gain some insight regarding how much flexibility is available in the recommendation of "optimum" conditions.

Example 6.4 Consider the data of Example 6.1. The design variables are reaction time and temperature, and yield was the response. A point of maximum response is sought. A central composite design with factorial points scaled to ± 1 and an axial distance of $\sqrt{2}$ was used. The stationary point is a point of maximum response and is located at $x_1 = -0.1716$, $x_2 = -0.1806$. The major components of the inequality of Equation (6.14) are quite simple. There are $p = 6$ parameters in the second-order system and

$$d_1(\mathbf{t}) = -1.095 - 5.562t_1 - 0.775t_2$$

$$d_2(\mathbf{t}) = -1.045 - 0.775t_1 - 5.048t_2$$

The variances and covariances are obtained from the $(\mathbf{X'X})^{-1}$ matrix with

$$\mathbf{X'X} = \begin{array}{c} \quad\quad\quad b_1 \;\; b_2 \;\; b_{11} \;\; b_{22} \;\; b_{12} \\ \begin{bmatrix} 11 & 0 & 0 & 8 & 8 & 0 \\ & 8 & 0 & 0 & 0 & 0 \\ & & 8 & 0 & 0 & 0 \\ & & & 12 & 4 & 0 \\ & & & & 12 & 0 \\ \text{sym.} & & & & & 4 \end{bmatrix} \end{array}$$

and

$$(\mathbf{X'X})^{-1} = \begin{bmatrix} \frac{1}{3} & 0 & 0 & -\frac{1}{6} & -\frac{1}{6} & 0 \\ & \frac{1}{8} & 0 & 0 & 0 & 0 \\ & & \frac{1}{8} & 0 & 0 & 0 \\ & & & 0.1772 & 0.0521 & 0 \\ & & & & 0.1772 & 0 \\ \text{sym.} & & & & & \frac{1}{4} \end{bmatrix}$$

The error mean square from the fitted response surface is $s^2 = 3.1635$ (5 degrees of freedom). Apart from s^2, the variances and covariances of the coefficients of the fitted response surface came by $(\mathbf{X'X})^{-1}$ above. As a result, elements of $\widehat{\mathrm{Var}}\,\mathbf{d}(t)$ are easily determined. For example,

$$\widehat{\mathrm{Var}}\,d_1(\mathbf{t}) = \frac{s^2}{\sigma^2}\{\mathrm{Var}\,b_1 + 4t_1^2\,\mathrm{Var}\,b_{11} + t_2^2\,\mathrm{Var}\,b_{12}\}$$

Note that the covariances involving b_1 are zero and $\mathrm{Cov}(b_{11}, b_{12}) = \mathrm{Cov}(b_{12}, b_{22}) = 0$. Thus we have

$$\widehat{\mathrm{Var}}\,d_2(\mathbf{t}) = \frac{s^2}{\sigma^2}\{\mathrm{Var}\,b_2 + t_1^2\,\mathrm{Var}\,b_{12} + 4t_2^2\,\mathrm{Var}\,b_{22}\}$$

$$\widehat{\mathrm{Cov}}[d_1(\mathbf{t}), d_2(\mathbf{t})] = \frac{s^2}{\sigma^2}\{4t_1t_2\,\mathrm{Cov}(b_{11}, b_{22}) + t_1t_2\,\mathrm{Var}\,b_{12}\}$$

As a result, we have

$$\widehat{\mathrm{Var}}\,d_1(\mathbf{t}) = 3.1635\{\tfrac{1}{8} + 4t_1^2(0.1772) + \tfrac{1}{4}t_2^2\}$$

$$\widehat{\mathrm{Var}}\,d_2(\mathbf{t}) = 3.1635\{\tfrac{1}{8} + \tfrac{1}{4}t_1^2 + 4t_2^2(0.1772)\}$$

$$\widehat{\mathrm{Cov}}[d_1(\mathbf{t}), d_2(\mathbf{t})] = 3.1635\{4(0.0521)t_1t_2 + \tfrac{1}{4}t_1t_2\}$$

Equation (6.14) can now be used to plot the confidence region. The reader should note how the design has a profound impact on the important elements of $(\mathbf{X'X})^{-1}$. The confidence region is displayed in Figure 6.19.

The display in Figure 6.19 presents a rather bleak picture concerning the quality of estimation of the stationary point. The dotted contour encloses the 95% confidence region on the location of the stationary point, while the solid contour depicts the 90% confidence region. One must not forget the interpretation, which is that a true response surface maximum at any location inside the 95% contour could readily have produced the data that was observed. The 95% interval does not close inside the experimental design region.

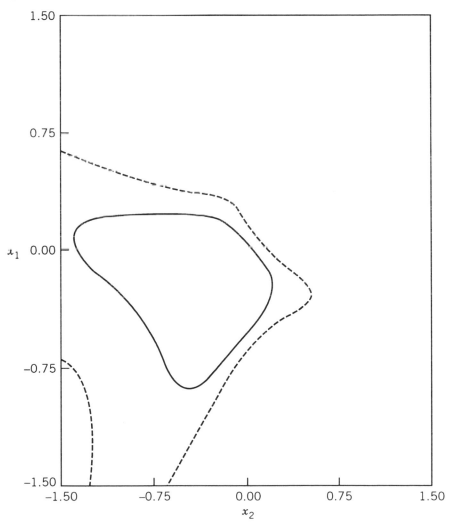

Figure 6.19 Confidence regions on location of stationary point for data of Example 6.4.

Table 6.3   Data for Second Example Showing Confidence Region on Location of Stationary Point

x_1	x_2	y
-1	-1	87.6
-1	1	85.7
1	-1	86.5
1	1	86.9
$-\sqrt{2}$	0	86.7
$\sqrt{2}$	0	86.8
0	$-\sqrt{2}$	87.4
0	$\sqrt{2}$	86.7
0	0	90.3
0	0	91.0
0	0	90.8

Confidence regions on the location of optima like those shown in Figure 6.19 are not unusual. The quality of the confidence region depends a great deal on the nature of the design and the fit of the model to the data. Recall that R^2 is approximately 82%. One can note the role of the variances and covariances of coefficients, and, of course, the error mean square in the analysis. The appearance of the confidence region shown in Figure 6.19 should serve as a lesson regarding the emphasis that is put on a "point estimate" of optimum conditions. As we have indicated (and will continue to expound), process improvement should be regarded as a continuing, iterative process, and any optimum found in any given experiment should perhaps not be considered as important as what one clearly gains in process improvement and what one learns about the process in general.

Consider now a second data set with the same design as appears in the previous example. However, one will notice that the fit is considerably better. The data appear in Table 6.3. The second-order coefficients are given by

$$\hat{y} = 90.7000 + 0.0302x_1 - 0.3112x_2 - 2.0316x_1^2 + 0.5750x_1x_2 - 1.8816x_2^2$$

The value of R^2 is 0.9891.

The stationary point (maximum) is at $(-0.004373, -0.08339)$ with $\hat{y}_s = 90.7130$. One would certainly expect the confidence region to be considerably tighter in this case. Figure 6.20 shows the confidence region around the estimated stationary point. Once again, both the 90% and 95% regions are shown. We show the graph using the scale as in the previous example in order to emphasize the contrast. Clearly, here, the stationary point is esti-

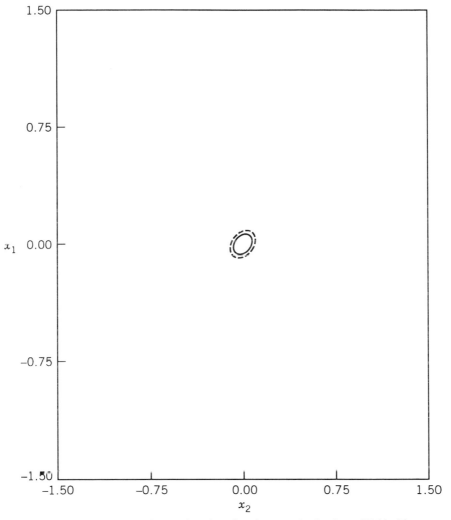

Figure 6.20 Confidence regions on location of stationary point for data of Table 6.3.

mated very well. It should be emphasized that a large confidence region is not necessarily bad news. A model that fits quite well yet generates a large confidence region on the stationary point may produce a "flat" surface which implies flexibility in choosing the optimum. This is clearly of interest to the engineer or scientist.

The two examples here illustrate contrasting situations. One should not get the feeling that the RSM in the first example will not produce important and interesting results about the process. However, in many RSM situations, *estimation of optimum conditions is very difficult*.

6.6 RESPONSE SURFACE ANALYSIS WITH MULTIPLE RESPONSES

Up to this point we have emphasized the use of response surface analyses for improvement of quality of products and processes. We have focused on modeling a measured response or a function of design variables and letting the analysis indicate areas in the design region where the process is likely to give *desirable results*, the term "desirable" being a function of the predicted response. However, in many instances the term "desirable" is a function of more than one response. In a chemical process there are almost always several properties of the product output that must be considered in the definition of "desirable." In all consumer products (food, tobacco), the scientist must deal with taste as a response but also must consider undesirable byproducts. In the pharmaceutical or biomedical area, the clinician is primarily concerned with the efficacy of the drug or remedy but must not ignore the possibility of serious side effects. In a tool life problem, cutting speed (x_1) and depth of cut (x_2) influence the primary response, the life of the tool. However, a secondary response, rate of metal removed, may also be important in the study.

Example 6.5   In many practical situations, multiple responses are observed and there is a need to reach some type of compromise as far as optimum conditions are concerned. While there have been many creative approaches put forth in the statistics literature for analysis of multiple responses, there are many times when a simple overlaying of contour plots will enlighten the scientist or engineer about the process. Often, then, a reasonable compromise will be apparent. Obviously, the methodology becomes more difficult as the number of responses and factors become large.

In a process designed to purify an antibiotic product [see Lind et al. (1960)], a response surface approach was used in the solvent extraction portion of the process. The yield of the product and cost of the operation represent very critical responses. The operation involved extracting the antibiotic into an organic solvent. Chemicals A and B, the reagents, were added to form material that is soluble in the solvent. Concentration of A and B and the pH of the extraction environment were chosen as the design variables. In the past, it was standard practice to use excessive amounts of reagents—a practice that seemed to produce high yield. However, it also resulted in a relatively high cost for the operation.

Based on previous experience, a range was chosen so that the maximum yield would be included in the design region. However, a cost response was considered for the first time. It was decided that fitted functions should contain

$$x_1 = \frac{\%A - 1.5}{0.5}, \qquad x_2 = \frac{\%B - 1.5}{0.5}, \qquad x_3 = \frac{pH - 5.0}{0.5}$$

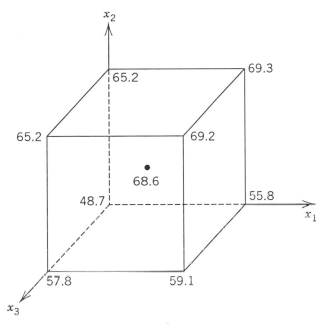

Figure 6.21   Results of the 2^3 factorial in yield response.

The study was done in phases, using a 2^3 factorial with center runs to test initially for quadratic effects in the yield response. The design layout appears in Figure 6.21. The response values on the figure indicate averages. Two runs were made at factorial points and four runs were made in the center. The first-order response function is given by

$$\hat{y} = 62.7 + 4.11x_1 + 11.86x_2 + 30.63x_3$$

This model would, indeed, imply that increasing both reagents results in a corresponding increase in yield. However, the use of center runs allows a test for "curvature"—that is, a one-degree-of-freedom test on the sum of the terms $\beta_{11}x_1^2$, $\beta_{22}x_2^2$, and $\beta_{33}x_3^2$, though the terms individually are not estimable. The reader should recall this test from material in Chapter 3.

Table 6.4 outlines the analysis of variance table in which the test for curvature is made. Clearly, curvature is significant and thus a complete second-order model must be fit. The reader should recall that in Chapter 5 the use of the central composite design was motivated from the sequential augmentation of a two-level factorial (or fraction) with additional design points that allow the fitting of a second-order response surface. This augmentation is well served by the addition of axial points. Six axial points are added with $\alpha = 1.0$ chosen as the axial distance. Figure 6.22 reflects the design and the average of the response observations. In the end, four observations were

Table 6.4 Analysis of Variance Table for 2^3 Factorial in Example 6.5

Source of Variation	Sum of Squares	Degrees of Freedom	Mean Square	F
First-order terms (x_1, x_2, x_3)	668.05	3	222.68	Significant
Lack of fit	226.69	5		
Interactions	56.22	4 $\big\}$	14.06	
Pure quadratic terms	170.47	1	170.47	Significant
Error	293.53	11	26.69	

taken at each design point. The fitted response function is given by

$$\hat{y} = 65.05 + 1.63x_1 + 3.28x_2 + 0.93x_3 - 2.93x_1^2 - 2.02x_2^2$$
$$- 1.07x_3^2 - 0.53x_1x_2 - 0.68x_1x_3 - 1.44x_2x_3$$

The canonical analysis results in a stationary point given by

$$x_{1,s} = 0.2256, \qquad x_{2,s} = 0.8589, \qquad x_{3,s} = 0.2150$$

The eigenvalues are $\lambda_1 = -0.6687$, $\lambda_2 = -3.1515$, and $\lambda_3 = -2.199$. Thus, the stationary point is a point of maximum response. The estimated yield at the stationary point is $\hat{y} = 66.542$. These optimum conditions correspond to %A = 1.6128, %B = 1.9294, and pH = 4.8925. Figure 6.23 shows a plot of yield contours with pH = 5.0, where the yield is nearly optimal.

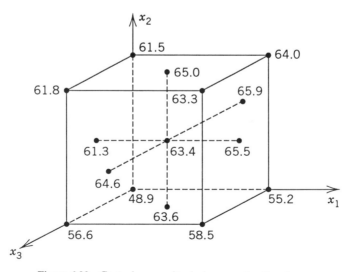

Figure 6.22 Central composite design on extraction data.

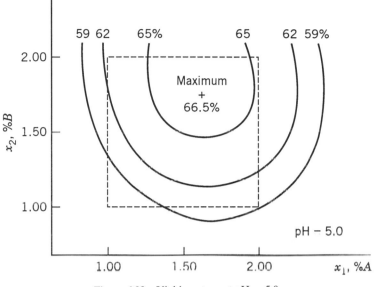

Figure 6.23   Yield contour at pH = 5.0.

As we indicated earlier, the need to include cost as a second response prompted the fitting of a second-order response function for cost of the extraction operation. Figure 6.24 displays two-dimensional cost contours at fixed pH levels. Figure 6.25 displays an overlaying of the yield and cost contour for fixed pH = 5.0. Shown is the point representing conditions of the product that was used before this study was undertaken. Certain things become apparent from the contours. High yield is not obtained from using excessive amounts of reagents A and B. In addition, the current product appears to be too costly with yield that is far from what may be achieved at the estimated maximum. Finally, while the product with maximum yield is not identical to the product with minimum cost, it appears that a reasonable compromise can be achieved. For example, the analyst may wish to seek coordinates that minimize cost at a prefixed satisfactory yield. Or, in fact, the roles of cost and yield may be reversed. On the other hand, a simple inspection of the contours would suggest that if a yield of say 65% appears to produce a good product, a pH fixed at 5.0, reagent A = 1.3%, and reagent B = 1.65% result in minimum cost at that particular yield value. In this case the estimated yield is 4% higher than previous products and a reduction in cost of almost $6/ kg product will be enjoyed.

6.6.1   The Use of Many Responses: The Desirability Function

The overlaying of contour plots along with separate response surface analyses often give the user workable solutions for product improvement as long as the number of responses is not too great. However, when the problem involves, say, four or more responses or design variables, then contour overlay methodology becomes unruly. Derringer and Such (1980) developed an interesting procedure which can be very useful when several responses are involved. The method makes use of a desirability function in which the researchers own priorities and desires on the response values are built into the optimization procedure.

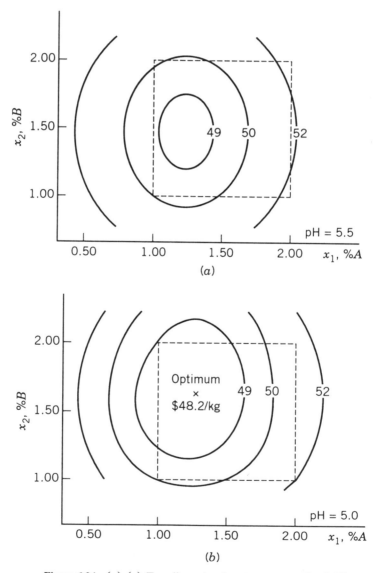

Figure 6.24 (*a*)–(*c*). Two-dimensional cost contours at fixed pH.

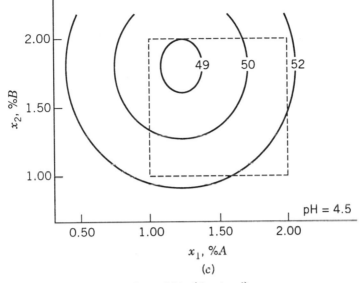

Figure 6.24   (*Continued*).

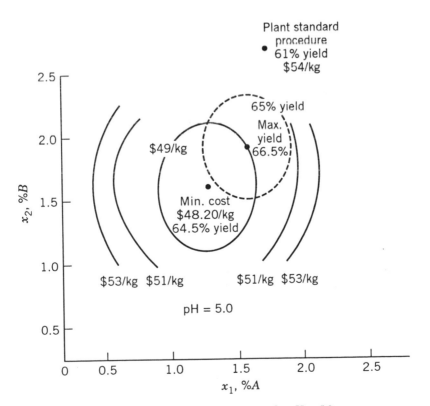

Figure 6.25   Yield and cost contour for pH = 5.0.

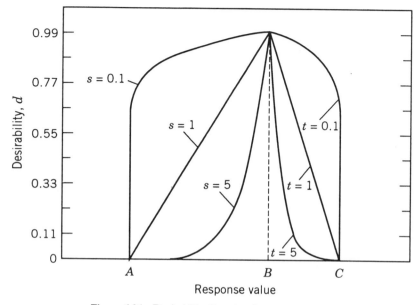

Figure 6.26   Desirability function for target value B.

One- and two-sided desirability functions are used depending on whether the response is to be maximized or minimized or has an assigned target value. For a response with a target value level, A, B, and C are assigned so that $A \leq B \leq C$. A product is considered unacceptable if $y < A$ or $y > C$. The value B is the "most desirable value" (target). The quantity d, the desirability, is defined as

$$d = \begin{cases} \left(\dfrac{\hat{y} - A}{B - A} \right)^{s}, & A \leq \hat{y} \leq B \\[3mm] \left(\dfrac{\hat{y} - C}{B - C} \right)^{t}, & B \leq \hat{y} \leq C \end{cases}$$

with d being 0 if $\hat{y} > C$ or $\hat{y} < A$. One can then use Figure 6.26 to allocate power values s and t according to one's subjective impression about the role of this response in the total desirability of the product.

In the case where a response should be maximized, one chooses a value B such that $d = 1$ for any $\hat{y} > B$. In other words, the researcher is quite happy if $\hat{y} \geq B$. However, we assume that the product is unacceptable if $\hat{y} < A$. Here, of course, $B = C$. Then the desirability function is given by

$$d = \left\{ \frac{\hat{y} - A}{B - A} \right\}^{s}, \qquad A \leq \hat{y} \leq B$$

In the case where a response is to be minimized, a value of B is chosen such that $\hat{y} \leq B$ is quite desirable and produces a $d = 1$. A value for $y > C$ is considered unacceptable and therefore results in a $d = 0$. In this case, $A = B$. The desirability function is given by

$$d = \left\{ \frac{\hat{y} - C}{B - C} \right\}^s, \qquad B \leq \hat{y} \leq C$$

In most applications, values of A, B, and C are chosen according to the researcher's priorities. Choices for s and t may be more difficult. However, there are plots that can be useful. Consider Figure 6.26. One can easily see that choices of s and t (two-sided case) are determined by how important it is for $\hat{y}$ to be close to the target B. Using $s = t = 0.1$ essentially does not require the response to be close to target. But a choice of s and t as large as, say, a value of 5 implies that the desirability value is very low unless $\hat{y}$ gets very close to target. Figure 6.27 shows the appearance of the desirability function for the maximum $\hat{y}$ case. The minimum $\hat{y}$ case looks similar. Again, of course, $s \ll 1$ implies that a product may not be close to the desirable value of $\hat{y} = B$ and yet still be quite acceptable. But $s = 8$, say, implies that the product is nearly unacceptable unless $\hat{y}$ is close to B.

The procedure involves creating a desirability function for each of the m responses. Then a *single composite response* is developed which is the *geometric mean* of the desirabilities of the individual responses. In other

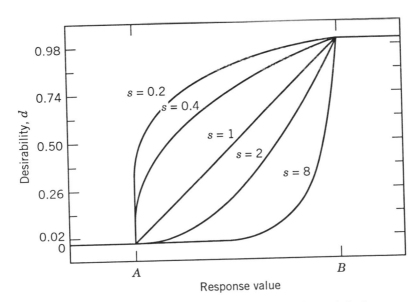

Figure 6.27 Desirability function for a response to be maximized.

words, a response is defined as

$$D = \{d_1 \cdot d_2 \cdot d_3 \cdots d_m\}^{1/m}$$

Now, taking into account all m responses, clearly one wishes to choose the conditions $\mathbf{x}$ on the design variables that **maximize** D. A value of D close to 1.0 implies that all responses are in a desirable range simultaneously. One should keep in mind that the formulation of D given here is only one possibility. Other functions of the d_i may work as well. Also, there are many ways of implementing this Derringer–Suich methodology. The following represent two possibilities:

1. Fit a separate response surface for each of the m responses. Create $d_1, d_2, \ldots, d_m$ and then $D = (\prod_{s=1}^{m} d_i)^{1/m}$ at each design point according to values of the fitted response, the $\hat{y}_i$. Then build a response surface with

$$D = f(x_1, x_2, \ldots, x_k)$$

and find conditions on $\mathbf{x}$ that maximizes D. Make use of canonical analysis, possibly ridge analysis, and so on.

2. Compute $d_1, d_2, \ldots, d_m$ and D at each design point according to the individual observations—that is, the values $y_1, y_2, \ldots, y_m$. Then build a response surface with the computed response D and use appropriate methodology for finding $\mathbf{x}$ that maximizes $\hat{D}$.

Method 1 above has its own variations. One may replace responses in the design with single univariate numerical values, namely the D values, which are generated from the $m\hat{y}$ numbers. Then a second-order response surface is conducted. On the other hand, the m fitted response surface functions may be used to create a functional form for D which is nonlinear in $x_1, x_2, \ldots, x_k$. Then "optimum" conditions are found through a nonlinear optimization algorithm.

Caution needs to be used with any multiple response procedure. For the Derringer–Suich method, when the "optimum" or desirable condition (i.e., large D) on $\mathbf{x}$ has been determined, the researcher should do confirmatory runs at that condition to be sure that he or she is satisfied that all responses are in a satisfactory region.

Example 6.6   Consider the following example in three design variables: reaction time, temperature, and percent catalyst. The responses are percent conversion, and thermal activity. The design is a central composite design with six center runs:

$A (x_1)$ Time	$B (x_2)$ Temperature	$C (x_3)$ Catalyst	y_1 Conversion	y_2 Activity
-1.000	-1.000	-1.000	74.00	53.20
1.000	-1.000	-1.000	51.00	62.90
-1.000	1.000	-1.000	88.00	53.40
1.000	1.000	-1.000	70.00	62.60
-1.000	-1.000	1.000	71.00	57.30
1.000	-1.000	1.000	90.00	67.90
-1.000	1.000	1.000	66.00	59.80
1.000	1.000	1.000	97.00	67.80
-1.682	0.000	0.000	76.00	59.10
1.682	0.000	0.000	79.00	65.90
0.000	-1.682	0.000	85.00	60.00
0.000	1.682	0.000	97.00	60.70
0.000	0.000	-1.682	55.00	57.40
0.000	0.000	1.682	81.00	63.20
0.000	0.000	0.000	81.00	59.20
0.000	0.000	0.000	75.00	60.40
0.000	0.000	0.000	76.00	59.10
0.000	0.000	0.000	83.00	60.60
0.000	0.000	0.000	80.00	60.80
0.000	0.000	0.000	91.00	58.90

It is important to maximize y_1 while y_2 is held between 55 and 60. Indeed, the target on activity is 57.5. The Design-Expert software accommodates the Derringer–Suich procedure. The fitted second-order function for conversion is given by

$$\hat{y} = 81.09 + 1.0284x_1 + 4.043x_2 + 6.2037x_3 - 1.8336x_1^2 + 2.9382x_2^2$$

$$-5.1915x_3^2 + 2.1250x_1x_2 + 11.375x_1x_3 - 3.8750x_2x_3$$

The response function for activity is given by

$$\hat{y}_2 = 59.85 + 3.5830x_1 + 0.25460x_2 + 2.2298x_3 + 0.83479x_1^2$$

$$+ 0.07484x_2^2 + 0.05716x_3^2 - 0.38750x_1x_2 - 0.3750x_1x_3$$

$$+ 0.3125x_2x_3$$

Table 6.5 shows the results of the Derringer–Suich procedure. For the conversion response, $A = 80$ and $B = 97$ describe the desires of the experimenter. The circled portion in the table gives the weights in the desirability

Table 6.5 Derringer–Suich Results for Conversion and Activity Example

Res	Response	Units	Observations	Minimum	Maximum	Transformation	Model
Y_1	Conversion	%	20	51.00	97.00	None	Quad
Y_2	Activity		20	53.20	67.90	None	Quad

Numerical Optimization Results, Cycle = 1

Factor	Minimum	Maximum	Start	Finish
A	−1.682	1.682	0.000	−0.5486
B	−1.682	1.682	0.000	1.682
C	−1.682	1.682	0.000	−0.5955

Response	Observed Minimum	Observed Maximum	Goal	Low	High	First	Second	Result
Conversion	51.00	97.00	Max	80	97	0.10		95.18
Activity	53.20	67.90	57.5	55	60	0.10	0.10	57.50

Desirability function = 0.9944

Res	Response	Units	Observations	Minimum	Maximum	Transformation	Model
Y_1	Conversion	%	20	51.00	97.00	None	Quad
Y_2	Activity		20	53.20	67.90	None	Quad

Numerical Optimization Results, Cycle = 1

Factor	Minimum	Maximum	Start	Finish
A	−1.682	1.682	0.000	−0.5458
B	−1.682	1.682	0.000	1.682
C	−1.682	1.682	0.000	−0.5976

Response	Observed Minimum	Observed Maximum	Goal	Low	High	First	Second	Result
Conversion	51.00	97.00	Max	80	97	1.00		95.18
Activity	53.20	67.90	57.5	55	60	1.00	1.00	57.50

Desirability function = 0.9451

Table 6.5(*Continued*)

Res	Response	Units	Observations	Minimum	Maximum	Transformation	Model
Y_1	Conversion	%	20	51.00	97.00	None	Quad
Y_2	Activity		20	53.20	67.90	None	Quad

Numerical Optimization Results, Cycle = 1

Factor	Minimum	Maximum	Start	Finish
A	−1.682	1.682	0.000	−0.5456
B	−1.682	1.682	0.000	1.682
C	−1.682	1.682	0.000	−0.5978

	Observed		Optimization Parameters			Weights		
Response	Minimum	Maximum	Goal	Low	High	First	Second	Result
Conversion	51.00	97.00	max	80	97	10.00		95.18
Activity	53.20	67.90	57.5	55	60	1.00	1.00	57.50
Desirability function = 0.5686								

Res	Response	Units	Observations	Minimum	Maximum	Transformation	Model
Y_1	Conversion	%	20	51.00	97.00	None	Quad
Y_2	Activity		20	53.20	67.90	None	Quad

Numerical Optimization Results, Cycle = 1

Factor	Minimum	Maximum	Start	Finish
A	−1.682	1.682	0.000	−0.5444
B	−1.682	1.682	0.000	1.682
C	−1.682	1.682	0.000	−0.5986

	Observed		Optimization Parameters			Weights		
Response	Minimum	Maximum	Goal	Low	High	First	Second	Result
Conversion	51.00	97.00	Maximum	80	97	10.00		95.18
Activity	53.20	67.90	57.5	55	60	10.00	10.00	57.50
Desirability function = 0.5686								

Table 6.5 (*Continued*)

Res	Response	Units	Observations	Minimum	Maximum	Transformation	Model
Y_1	Conversion	%	20	51.00	97.00	None	Quad
Y_2	Activity		20	53.20	67.90	None	Quad

Numerical Optimization Results, Cycle = 1

Factor	Minimum	Maximum	Start	Finish
A	−1.682	1.682	0.000	−0.5499
B	−1.682	1.682	0.000	1.682
C	−1.682	1.682	0.000	−0.5945

Response	Observed Minimum	Maximum	Optimization Parameters Goal	Low	High	Weights First	Second	Result
Conversion	51.00	97.00	Maximum	80	97	1.00		95.18
Activity	53.20	67.90	57.5	55	60	10.00	10.00	57.50
Desirability function = 0.9451								

Res	Response	Units	Observations	Minimum	Maximum	Transformation	Model
Y_1	Conversion	%	20	51.00	97.00	None	Quad
Y_2	Activity		20	53.20	67.90	None	Quad

Numerical Optimization Results, Cycle = 1

Factor	Minimum	Maximum	Start	Finish
A	−1.682	1.682	0.000	−0.8552
B	−1.682	1.682	0.000	1.682
C	−1.682	1.682	0.000	−0.3955

Response	Observed Minimum	Maximum	Optimization Parameters Goal	Low	High	Weights First	Second	Result
Conversion	51.00	97.00	max	80	97	0.10		94.08
Activity	53.20	67.90	57.5	55	60	10.00	10.00	57.50
Desirability function = 0.9906								

Table 6.5 (*Continued*)

Res	Response	Units	Observations	Minimum	Maximum	Transformation	Model
Y_1	Conversion	%	20	51.00	97.00	None	Quad
Y_2	Activity		20	53.20	67.90	None	Quad

	Numerical Optimization Results, Cycle = 1			
Factor	Minimum	Maximum	Start	Finish
A	−1.682	1.682	0.000	−0.5549
B	−1.682	1.682	0.000	1.682
C	−1.682	1.682	0.000	−0.5908

	Observed		Optimization Parameters			Weights		
Response	Minimum	Maximum	Goal	Low	High	First	Second	Result
Conversion	51.00	97.00	Maximum	80	97	0.10		95.18
Activity	53.20	67.90	57.5	55	60	1.00	1.00	57.50
Desirability function = 0.9944								

Res	Response	Units	Observations	Minimum	Maximum	Transformation	Model
Y_1	Conversion	%	20	51.00	97.00	None	Quad
Y_2	Activity		20	53.20	67.90	None	Quad

	Numerical Optimization Results, Cycle = 1			
Factor	Minimum	Maximum	Start	Finish
A	−1.682	1.682	0.000	−0.5447
B	−1.682	1.682	0.000	1.682
C	−1.682	1.682	0.000	−0.5985

	Observed		Optimization Parameters			Weights		
Response	Minimum	Maximum	Goal	Low	High	First	Second	Result
Conversion	51.00	97.00	max	80	97	1.00		95.18
Activity	53.20	67.90	57.5	55	60	0.10	0.10	57.50
Desirability function = 0.9451								

Table 6.5 (*Continued*)

Res	Response	Units	Observations	Minimum	Maximum	Transformation	Model
Y_1	Conversion	%	20	51.00	97.00	None	Quad
Y_2	Activity		20	53.20	67.90	None	Quad

Numerical Optimization Results, Cycle = 1

Factor	Minimum	Maximum	Start	Finish
A	−1.682	1.682	0.000	−2.42E − 02
B	−1.682	1.682	0.000	1.682
C	−1.682	1.682	0.000	−0.2375

Response	Observed Minimum	Observed Maximum	Optimization Parameters Goal	Low	High	Weights First	Second	Result
Conversion	51.00	97.00	Maximum	80	97	10.00		96.93
Activity	53.20	67.90	57.5	55	60	0.10	0.10	59.77

function—that is, s and t for the activity response and s for the conversion response. Note that in seven out of the nine situations (combination of weights), similar results are obtained, with the optimum conditions being

$$x_1: \quad -0.5486; \qquad x_2: \quad 1.682; \qquad x_3: \quad -0.5955$$

giving conversion 95.18% and activity 57.50. Note, however, that when a weight of $s = 10.0$ is used on conversion with $s = t = 0.1$ on activity, the optimum conditions are different and the result is 95.93% on conversion and 59.77 on activity. In addition, when $s = 0.10$ for conversion and $s = t = 10.0$ for activity, optimum conditions are different and the result is 94.08% on conversion and 57.5% on activity. As one can readily see, different choices on weights lead to different results because of varying priorities or goals for the responses. Several choices for weights can produce results from which the experimenter may choose.

6.6.2 Other Multiple Response Techniques

As we indicated earlier, multiple response procedures will always involve compromise between important responses. There have been several creative ideas put forth in the literature, and some have enjoyed considerable use in industry. In fact, because logical solutions to multiple response problems involve maximizing (or minimizing) a response with constraints on other responses, the use of nonlinear programming techniques can be very useful.

An example which illustrates this is given in Montgomery (1992). There are other noteworthy approaches that are listed in the following subsections.

Dual Response Approach

In the case of two responses, a useful approach termed *dual response approach* was introduced by Myers and Carter (1973). It is assumed that the two responses can be categorized as *primary* and *secondary*. The goal then involves the determination of conditions x on the design variables that produces max $\hat{y}_p(x)$ [or min $\hat{y}(x)$] subject to a constraint on $\hat{y}_s(x)$. Here $\hat{y}_p(x)$ is the primary response function and $\hat{y}_s(x)$ is the secondary response function. The methodology, which is not unlike ridge analysis discussed earlier in this chapter, produces a locus of coordinates in which various values of $\hat{y}_s(x)$ are considered. For example, in the yield–cost example (Example 6.5), one might let the secondary response be yield and determine conditions on x that minimize cost for fixed satisfactory levels of yield. Biles (1975) discussed the extension of this concept to more than two responses. Nonlinear optimization procedures can be very useful for multiple response RSM. In an example that follows, one of these methods is illustrated. Del Gastillo and Montgomery (1993) discuss these approaches.

Khuri–Conlon Approach

Khuri and Conlon (1981) introduced a procedure that is based on a distance function that computes the overall closeness in which the response functions are achieving their respective optima at the same set of conditions. That is, it gives a measure of closeness to the *ideal optimum*. One then finds conditions on x that minimize this distance function over all x in the experimental region.

Example 6.7   There are many nonlinear optimization algorithms that are available for producing "optimum" conditions. The Design-Expert system allows for inequality constraints on secondary responses. This procedure is illustrated with the conversion and activity example in the previous section. Table 6.6 gives the result. The "goal optimization" procedure is designed to maximize $\hat{Y}_1$ (conv.) subject to $55 < \hat{Y}_2 < 60$. The procedure was allowed to go through 10 cycles, and the most attractive conditions for accomplishing the constrained optimization are given by

$$x_1: \ 0.0257; \qquad x_2: \ 1.682; \qquad x_3: \ -0.2062$$

with a conversion of 96.10% and activity of 60. Nonlinear optimization procedures are very useful in multiple response situations. However, the Derringer–Such method can be particularly beneficial when many responses are involved. The use of multiple choices of s and t allows the user to select from among many alternative compromises.

Table 6.6 Nonlinear Optimization Results for Conversion and Activity Example

Numerical Optimization Results, Cycle = 9

Factor	Minimum	Maximum	Start	Finish
A	−1.682	1.682	−0.7774	2.57E − 02
B	−1.682	1.682	−0.1858	1.682
C	−1.682	1.682	0.3345	−0.2062

Response	Observed Minimum	Observed Maximum	Optimization Parameters Goal	Optimization Parameters Low	Optimization Parameters High	Weights First	Weights Second	Result
Conversion	51.00	97.00	max					96.10
Activity	53.20	67.90		55	60			60.00

6.7 CHOICE OF METRIC FOR THE RESPONSE

The use of the second-order response function provides curvature for the response surface system. However, there are situations in which additional data transformation allows for a better fit to the system. In situations where the range on the response is fairly large, a change in metric on the response will often be successful (e.g., the use of $\ln y$, $y^{1/2}$, etc.). The methodology often used is that of the Box–Cox procedure [see Box and Cox (1964)]. This is discussed in depth in Myers (1990) and Montgomery and Peck (1992). The procedure essentially makes use of the so-called **power transformation** on the response—that is, the implementation of y^λ, where λ varies from say −2 to 2. The spectrum on λ includes $\ln y$ as an important candidate model. As a result, formally, one should view the response as taking the form

$$w = \frac{y^\lambda - 1}{\lambda} \tag{6.15}$$

Because $\lim_{\lambda \to 0}[(y^\lambda - 1)/\lambda] = \ln y$, a choice of $\lambda = 0$ results in the use of $\ln y$.

Choice of λ

Several computer packages allow the analyst to use the Box–Cox transformation, though the proper metric generally needs to be determined prior to accessing the methodology for RSM. The estimate of λ is produced via maximum likelihood estimation assuming normal errors around the transformed model. The choice of λ allows the natural log ($\lambda = 0$), no transformation ($\lambda = 1.0$), square root ($\lambda = 0.5$), reciprocal ($\lambda = -1$), and, of course, others. While the procedure itself allows for a continuous spectrum in the interval $[-2, 2]$ or even $[-3, 3]$, "natural" choices such as those mentioned above should be adopted in practice. For example, if $\lambda = 0.15$ appears to be the "optimal" value, the use of ln y would usually allow for simplicity. One must keep in mind that the choice of λ comes from the data and is hence a random variable.

A choice of λ for the power transformation that results in *minimum residual sum of squares* in the transformed model is equivalent to maximum likelihood estimation of λ. However, in the execution of the procedure it is essential to adjust for scale. For example, it should be clear to the reader that the error sum of squares for a model in linear units in y cannot be compared to the error sum of squares for a model in which the response is ln y. As a result, the value of w in Equation (6.15) must be adjusted for scale *solely for the purpose of computing* λ. A useful adjustment involves the computation of

$$z = \frac{y^{\lambda} - 1}{\lambda(\tilde{y})^{\lambda-1}} = \frac{w}{(\tilde{y})^{\lambda-1}}$$

where $\tilde{y} = \{\prod_{i=1}^{n} y_i\}^{1/n}$, the geometric mean of the response observations. The procedure reduces to the following:

1. Fit the response surface model with z as the response.
2. Choose λ which produces minimum residual sum squares.

As we indicated earlier, the potential for improvement with a transformation is greatest when the range on the response is large, for example, when

$$[y_{max}/y_{min}] > 3$$

6.7.1 Other Reasons for Data Transformation

Much attention has been paid in this text to model assumptions—that is, normal i.i.d. errors. Normality and the constant variance assumption are often the subject of diagnostic information, residual plots, and so on [see Montgomery (1991a), Myers (1990), Montgomery and Peck (1992)]. Transforming the response variable can be used for reasons other than to improve the fit of the model. A transformation on the response may be an effective tool when an assumption has been violated—for example, the assumption of homogeneous error variance. In RSM situations, both the log and square root transformations are often effective *variance stabilizing transformations*. In fact, it is often the case that when the range on the response is large, the appropriate Box–Cox power transformation also aids in variance stabilizing if nonhomogeneous variance exists. The result is an improved fit as well as a more attractive set of residual plots.

Simplicity Derived from Data Transformation
In addition to improved fit to the data and error assumptions that are more palatable, a transformed response may also result in a simpler model that is easier to interpret. This is particularly important in RSM work where we deal so often in empirical models. For example, we often depend on the second-order function to approximate the system. If the system clearly has curvature, the system curvature may be nicely depicted through a square or log transformation. If this is the case, there is often no need for the higher-order terms; perhaps even the interactions disappear. This certainly makes for a simpler analysis. Certainly the roles of the design variables in the quest for optimum conditions are much better defined. The example that follows provides a nice illustration.

Example 6.8 The data in this example involve three design variables in a 3^3, or $3 \times 3 \times 3$, factorial array. It deals with the behavior of worsted yarn under cycles of repeated loading. The design variables are as follows:

$$\text{Length of test specimen (mm):} \quad x_1 = \frac{\text{Length} - 300}{50}$$

$$\text{Amplitude of load cycle (mm):} \quad x_2 = \text{Length} - 9$$

$$\text{Load (grams):} \quad x_3 = \frac{\text{Length} - 45}{5}$$

The response is the number of cycles to failure. The data, taken from Box and Draper (1987), are given in Table 6.7. Note the sizable range in response for this example. Also note that this is the first example of a second-order design that does not fall into the class of central composite designs. The 3^3 factorial provides three levels of each factor with 27 design points—certainly more than ample for a fitted second-order function. The resulting second-order model is given by

$$\hat{y} = 550.7 + 660x_1 - 535.9x_2 - 310.8x_3 + 238.7x_1^2 + 275.7x_2^2$$
$$- 48.3x_3^2 - 456.5x_1x_2 - 235.7x_1x_3 + 143.0x_2x_3$$

The R^2 value is 0.975. An analysis of variance is given below:

Source	SS($\times 10^{-3}$)	df	MS($\times 10^{-3}$)	F
First–order terms	14,748.5	3	4,916.2	70
Added second-order terms	4,224.3	6	704.1	9.5
Residual	1,256.6	17	73.9	
Total	20,229.4	26		

Table 6.7 Table of Data for Worsted Yarn

Run Number	Length, x_1	Amplitude, x_2	Load, x_3	Cycles to failure, Y	$y = \log_{10} Y$
1	−1	−1	−1	674	2.83
2	0	−1	−1	1414	3.15
3	1	−1	−1	3636	3.56
4	−1	0	−1	338	2.53
5	0	0	−1	1022	3.01
6	1	0	−1	1368	3.19
7	−1	1	−1	170	2.23
8	0	1	1	442	2.65
9	1	1	−1	1140	3.06
10	−1	−1	0	370	2.57
11	0	−1	0	1198	3.08
12	1	−1	0	3184	3.50
13	−1	0	0	266	2.42
14	0	0	0	620	2.79
15	1	0	0	1070	3.03
16	−1	1	0	118	2.07
17	0	1	0	332	2.52
18	1	1	0	884	2.95
19	−1	−1	1	292	2.47
20	0	−1	1	634	2.80
21	1	−1	1	2000	3.30
22	−1	0	1	210	2.32
23	0	0	1	438	2.64
24	1	0	1	566	2.75
25	−1	1	1	90	1.95
26	0	1	1	220	2.34
27	1	1	1	360	2.56

The fit appears to be fairly good. However, both first- and second-order terms appear to be necessary.

Now consider the same data with $\log_{10} y$ as the response. Here, a second-order polynomial was fit. However, *no second-degree term was significant*. In fact, the fitted first-order model, given by

$$\widehat{\log_{10}} \, y = 2.751 + 0.362x_1 - 0.274x_2 - 0.171x_3$$

resulted in an $R^2 > 0.99$, with only three model terms (apart from the intercept). Here, as in most modeling exercises, simplicity is of vital importance. The elimination of the quadratic terms and interaction terms with the change in response metric not only allows a better fit than the second-order model with the natural metric, but the roles of the design variables are clear.

To maximize cycles to failure, a simple steepest ascent path with an increase in x_1, decrease in x_2, and decrease in x_3 (see Chapter 5) may well be worthwhile.

6.7.2 Additional Remarks Concerning Transformation of Response

The data analyst cannot expect that every example will result in a solution so neat and clean as the one in Example 6.8. However, in many cases, alternative metric for the response should be considered, particularly when the range in the response is large. In RSM we view the second-order model as providing the curvature that the system exhibits. This approximating function is obviously not the true model. But if the curvature can be better accommodated by an alternative approximating function that is simple and easier to interpret and analyze, then the alternative function should be used. Often the analyst is reluctant to forego usage of what he or she perceives to be the natural metric of the response. But what does natural really mean? If the true function is, in some sense, an exponential function, then log y is the metric that will lead to the simplest form and best fit to the data. What is more natural than this? Similar arguments may well apply to other members of the power transformation family.

6.8 FURTHER COMMENTS CONCERNING RESPONSE SURFACE ANALYSIS

There are several important items that deal with RSM that only become apparent after one gains a certain level of experience. In a textbook presentation of RSM, a great deal is taken for granted—the model, the experimental region, the proper choice of response, and, of course, the goal of the study. All, however, need to be handled with care, and each one needs to be a topic of discussion between statistician and subject matter and scientist.

The choice of experimental region needs to be made very carefully. Choice of range in the natural levels cannot be taken lightly. Ranges that are too small may result in an important factor becoming insignificant in the analysis. This is particularly crucial in the phase of the analysis when variable screening is being done. Often the natural sequential nature of RSM allows the user to make intelligent choices of variable ranges after preliminary phases of the study have been analyzed. For example, following variable screening, the user should have better choices of ranges for a steepest ascent procedure. After steepest ascent is accomplished, choice of variable ranges to be used for a second-order analysis should be easier to make.

Let us consider briefly the issue of the goal of the experiment. The textbook develops and emphasizes the need to find optimum conditions.

However, one should keep in mind that estimated optimum conditions from one experiment is just that—an estimate. The next experiment may well find the estimated optimum conditions to be at a different set of coordinates. Also, process technology changes from time to time, and a computed set of optimum conditions may well be a fleeting concept. What is often more important is for the analysis to reveal important information about the process and information about the roles of the variables. The computation of a stationary point, a canonical analysis, or a ridge analysis may well lead to important information about the process, and this in the long run will often be more valuable than a single set of coordinates representing an estimate of optimum conditions.

EXERCISES

6.1 In a study to determine the nature of a response system which relates dry modulus of rupture (psi) in a certain ceramic material with three important independent variables, the following quadratic regression equation was determined [see Hackney and Jones (1969)]:

$$\hat{y} \times 10^{-2} = 6.88 - 0.1466x_1^2 + 0.1875x_1x_2 + 0.2050x_1x_3$$

$$+ 0.0325x_1 - 0.0053x_2^2 - 0.1450x_2x_3 + 0.2588x_2$$

$$+ 0.1359x_3^2 - 0.1363x_3$$

The independent variables represent ratios of concentration of various ingredients in the material.
(a) Determine the stationary point.
(b) Put the response surface into canonical form and determine the nature of the stationary point.
(c) Find the appropriate expressions relating the canonical variables to the independent variables, x_1, x_2, and x_3.
(d) Generate two-dimensional graphs showing contours of constant estimated modulus of rupture. Use $x_3 = -1, 0, 1$.

6.2 In a chemical engineering experiment dealing with heat transfer in a shallow fluidized bed, data are collected on the following four regressor variables: fluidizing gas flow rate, lb/hr (x_1); supernatant gas flow rate, lb/hr (x_2); supernatant gas inlet nozzle opening, millimeters

(x_3); supernatant gas inlet temperature, °F (x_4). The responses measured are heat transfer efficiency (y_1) and thermal efficiency (y_2). The data are as follows:

Observation	y_1	y_2	x_1	x_2	x_3	x_4
1	41.852	38.75	69.69	170.83	45	219.74
2	155.329	51.87	113.46	230.06	25	181.22
3	99.628	53.79	113.54	228.19	65	179.06
4	49.409	53.84	118.75	117.73	65	281.30
5	72.958	49.17	119.72	117.69	25	282.20
6	107.702	47.61	168.38	173.46	45	216.14
7	97.239	64.19	169.85	169.85	45	223.88
8	105.856	52.73	169.85	170.86	45	222.80
9	99.438	51.00	170.89	173.92	80	218.84
10	111.907	47.37	171.31	173.34	25	218.12
11	100.008	43.18	171.43	171.43	45	219.20
12	175.380	71.23	171.59	263.49	45	168.62
13	117.800	49.30	171.63	171.63	45	217.58
14	217.409	50.87	171.93	170.91	10	219.92
15	41.725	54.44	173.92	71.73	45	296.60
16	151.139	47.93	221.44	217.39	65	189.14
17	220.630	42.91	222.74	221.73	25	186.08
18	131.666	66.60	228.90	114.40	25	285.80
19	80.537	64.94	231.19	113.52	65	286.34
20	152.966	43.18	236.84	167.77	45	221.72

The above data set is a good example of what often happens in practice. An attempt was made to use a particular second-order design. However, errors in controlling the variables produced a design which is only an approximation of the standard design.

(a) Center and scale the design variables. That is, create the design matrix in coded form.

(b) Fit a second-order model for both responses.

(c) In the case of transfer efficiency (y_1), do a canonical analysis and determine the nature of the stationary point. Do the same for thermal efficiency (y_2).

(d) For the case of transfer efficiency, what levels (natural levels) of the design variables would you recommend if maximum transfer efficiency is sought.

(e) Do the same for thermal efficiency; that is, find levels of the design variables that you recommend for maximization of thermal efficiency.

(f) The chemical engineer requires that y_2 exceeds a value of 60. Find levels (natural) of x_1, x_2, and x_3 that maximize y_1 subject to the constraint $y_2 > 60$.

6.3 Rayon whiteness is an important factor for scientists dealing in fabric quality. Whiteness is affected by pulp quality and other processing variables. Some of the variables include: acid bath temperature °C (x_1); cascade acid concentration, % (x_2); water temperature, °C (x_3); sulfide concentration, % (x_4); amount of chlorine bleach, lb/min (x_5). The following is an experimental design involving these five design variables. The response is a measure of whiteness

x_1	x_2	x_3	x_4	x_5	y
−1	−1	−1	−1	−1	71.5
1	1	−1	−1	−1	76.0
1	−1	1	−1	−1	79.9
1	−1	−1	1	−1	83.5
1	−1	−1	−1	1	89.5
−1	1	1	−1	−1	84.2
−1	1	−1	1	−1	85.7
−1	1	−1	−1	1	94.5
−1	−1	1	1	−1	89.4
−1	−1	1	−1	1	97.5
−1	−1	−1	1	1	103.2
1	1	1	1	−1	108.7
1	1	1	−1	1	115.2
1	1	−1	1	1	111.5
1	−1	1	1	1	102.3
−1	1	1	1	1	108.1
−2	0	0	0	0	80.2
2	0	0	0	0	84.1
0	−2	0	0	0	77.2
0	2	0	0	0	85.1
0	0	−2	0	0	71.5
0	0	2	0	0	84.5
0	0	0	−2	0	77.5
0	0	0	2	0	79.2
0	0	0	0	−2	71.0
0	0	0	0	2	90.2
0	0	0	0	0	72.1
0	0	0	0	0	72.0
0	0	0	0	0	72.4
0	0	0	0	0	71.7
0	0	0	0	0	72.8

The coding of the design variables is as follows:

$$x_1 = \frac{\text{Temp.} - 45}{10} ; \quad x_2 = \frac{\text{Conc.} - 0.5}{0.2} ; \quad x_3 = \frac{\text{Temp.} - 85}{3} ;$$

$$x_4 = \frac{\text{Conc.} - 0.25}{0.05} ; \quad x_5 = \frac{\text{Amt.} - 0.4}{0.1}$$

(a) Fit an appropriate second-order model in the metric of the coded variables.

(b) Find the stationary point and give the nature of the stationary point.

(c) Give a recommendation for operating conditions on the design variables. It is important to minimize whiteness.

(d) Compute the standard error of prediction at the following design locations:

   (i) Design center $(0, 0, 0, 0, 0)$

   (ii) All factorial points

   (iii) All axial points

Comment on the relative stability of the standard error of prediction.

6.4 A client from the Department of Mechanical Engineering at Virginia Polytechnic Institute and State University asked for help in analyzing an experiment dealing with gas turbine engines. Voltage output of engines was measured at various combinations of blade speed and voltage measuring sensor extension. The data are as follows:

y (volts)	Speed x_1 (in. / sec)	Extension x_2 (in.)
1.23	5300	0.000
3.13	8300	0.000
1.22	5300	0.012
1.92	8300	0.012
2.02	6800	0.000
1.51	6800	0.012
1.32	5300	0.006
2.62	8300	0.006
1.65	6800	0.006
1.62	6800	0.006
1.59	6800	0.006

(a) Write the design matrix in coded form.

(b) What type of design do we have here?

(c) Fit an appropriate model.

(d) Find the stationary point and interpret.

(e) It is important to determine what value of speed and extension results in a response that exceeds 2.8 volts. Describe where in the design region this voltage can be achieved. Use a two-dimensional contour plot to answer the question.

6.5 In Exercise 3.9 we discussed an application dealing with effects on cracking of a titanium alloy. An additional study was made in which the same factors were studied except the "heat treatment" method. The "high" or "+" method was used and held constant. However, in this case, a second-order design was used so the analyst could fit a second-order model involving curvature. The three factors are pouring temperature x_1, titanium content x_2, and amount of grain refiner x_3. The following gives the design points and the response data. The response is the length of crack (in millimeters) induced in a sample of the alloy.

x_1	x_2	x_3	y
-1	-1	-1	1.32
$+1$	-1	-1	1.25
-1	$+1$	-1	1.45
$+1$	$+1$	-1	1.30
-1	-1	$+1$	1.86
$+1$	-1	$+1$	1.30
-1	$+1$	$+1$	1.51
$+1$	$+1$	$+1$	1.37
0	0	0	1.27
0	0	0	1.22
0	0	0	1.20
0	0	0	1.21
-1.732	0	0	1.88
1.732	0	0	1.06
0	-1.732	0	1.75
0	1.732	0	1.18
0	0	-1.732	1.15
0	0	1.732	1.68

(a) Fit an appropriate model with the data of the above central composite design.

(b) From the fitted model in (a), estimate the conditions that give rise to minimum shrinkage. Be sure that these estimated conditions are not remote from the current experimental design.

(c) Compute the predicted value and its standard error at your recommended set of conditions.

(d) Does this experiment give rise to suggestions for future experimental runs? Explain!

6.6 Consider the reaction yield data of Example 6.1. This was an illustration of the use of a central composite design in two design variables. Suppose four center runs were used rather than only three. Suppose, in addition, that the experiment was done during a 2-day period and that there definitely is a day-to-day effect. As a result, a blocking effect must be used in the model. In fact, the experiment and response data are given as follows:

x_1	x_2	Block	y
-1	-1	1	88.55
-1	1	1	86.29
1	-1	1	85.80
1	1	1	80.40
0	0	1	90.21
0	0	1	90.85
-1.414	0	2	85.50
1.414	0	2	85.39
0	-1.414	2	86.22
0	1.414	2	85.70
0	0	2	91.31
0	0	2	91.94

(a) Analyze this data with the "block effect" in the model. In other words,

$$y_u = \beta_0 + \beta_1 x_{1u} + \beta_2 x_{2u} + \beta_{11} x_{1u}^2 + \beta_{22} x_{2u}^2$$
$$+ \beta_{12} x_{1u} x_{2u} + \delta z_u + \epsilon_u$$

where $u = 1, 2, \ldots, 12$; δ is the coefficient of a blocking variable (two levels). You may allow $z_u = -1$ if the uth observation is in block 1 and $z_u = +1$ if the uth observation is in block 2.

(b) Fit a second-order model and comment on the regression coefficients in the model compared to those shown in Example 6.1.

(c) Did the block effect have any influence on the fitted model or standard error of coefficients? Explain.

(d) Will the estimated set of optimum conditions be influenced by the blocking variable? Explain.

6.7 In the computer industry, it is important to use experimental design and response surface methods to analyze performance data from integrated circuit / packet-switched computer networks. Here, the design variables are network input variables which deal with network size, traffic load, link capacity, and so on. The response measures deal with the nature of the network performance. In this experiment

[Schmidt and Launsby (1990)] the design variables are:

CS: Circuit switch arrival rate (voice calls/ min)
PS: Packet switch arrival rate (packets/ sec)
SERV: Voice call service rate (sec)
SLOTS: Number of time slots per link (a capacity indicator)

Two responses were measured. They were:

ALU: Average link utilization
BLK: Fraction of voice calls blocked

The data are as follows:

OBS	CS	PS	SERV	SLOTS	ALU	BLK
1	2	150	120	34	0.282	0.000
2	2	150	120	46	0.194	0.000
3	2	150	240	34	0.372	0.000
4	2	150	240	46	0.275	0.000
5	2	450	120	34	0.553	0.016
6	2	450	120	46	0.396	0.000
7	2	450	240	34	0.676	0.031
8	2	450	240	46	0.481	0.003
9	6	150	120	34	0.520	0.012
10	6	150	120	46	0.386	0.000
11	6	450	120	46	0.594	0.016
12	6	300	180	40	0.231	0.000
13	4	300	180	40	0.325	0.000
14	4	300	180	52	0.427	0.000
15	4	300	180	40	0.561	0.016
16	4	300	180	40	0.571	0.005
17	4	300	180	40	0.561	0.011
18	4	300	180	40	0.581	0.020
19	4	300	180	40	0.562	0.017
20	4	300	180	40	0.608	0.036
21	4	300	180	40	0.560	0.007
22	5	400	210	43	0.739	0.070
23	5	400	210	43	0.746	0.082
24	5	400	150	37	0.725	0.067
25	5	400	150	37	0.725	0.084
26	5	400	150	43	0.615	0.027
27	5	400	210	40	0.785	0.106
28	5	400	210	40	0.782	0.125
29	4	400	210	37	0.747	0.084

30	4	400	210	37	0.745	0.101
31	4	400	210	40	0.695	0.063
32	4	400	210	40	0.667	0.057
33	4	400	180	40	0.648	0.035
34	5	400	180	40	0.722	0.074
35	5	400	180	40	0.731	0.077
36	5	400	180	37	0.775	0.110
37	5	400	180	37	0.779	0.139
38	4	400	180	37	0.712	0.057
39	4	400	180	37	0.701	0.065
40	3	400	210	37	0.652	0.017
41	3	400	210	43	0.541	0.003
42	3	400	150	37	0.570	0.003
43	3	400	150	43	0.476	0.005

(a) Write the design in coded form. Use the "coding"

$$x_1 = (CS - 4)/2$$

$$x_2 = (PS - 300)/150$$

$$x_3 = (SERV - 180)/60$$

$$x_4 = (SLOTS - 40)/6$$

(b) Fit two second-order models for the two responses.

(c) Consider the response surface for the BLK response. Can the model be improved by a power transformation. Discuss!

(d) Use the two models above to determine optimum conditions for the two responses *separately*. We need to maximize ALU and minimize BLK. Use ridge analysis or any other optimization algorithm.

(e) Use appropriate software to find conditions that

$$\max \widehat{ALU|BLK} < 0.005$$

(f) Use the Derringer–Suich procedure to find optimum conditions. Use the conditions that ALU cannot be less than 0.5 and BLK cannot exceed 0.10.

6.8 In Schmidt and Launsby (1990), a simulation program was revealed that simulates an "auto bumper" plating process using thickness as the response with time, temperature, and pH as the design variables. An experiment was conducted in order that a response surface optimization could be accomplished. The coding of the design variables are given by

$$x_1 = \frac{Time - 8}{4}$$

$$x_2 = \frac{Temp - 24}{8}$$

$$x_3 = \frac{Nickel - 14}{4}$$

The design in the natural units is given below along with the thickness values.

Run Number	Time	Temperature	Nickel	Thickness
1	4.00	16.00	10.00	113
2	12.00	16.00	10.00	756
3	4.00	32.00	10.00	78
4	12.00	32.00	10.00	686
5	4.00	16.00	18.00	87
6	12.00	16.00	18.00	788
7	4.00	32.00	18.00	115
8	12.00	32.00	18.00	696
9	4.00	16.00	10.00	99
10	12.00	16.00	10.00	739
11	4.00	32.00	10.00	10
12	12.00	32.00	10.00	712
13	4.00	16.00	18.00	159
14	12.00	16.00	18.00	776
15	4.00	32.00	18.00	162
16	12.00	32.00	18.00	759
17	8.00	24.00	14.00	351
18	8.00	24.00	14.00	373
19	8.00	24.00	14.00	353
20	8.00	24.00	14.00	321
21	12.00	24.00	14.00	736.1
22	4.00	24.00	14.00	96.0
23	8.00	32.00	14.00	328.9
24	8.00	16.00	14.00	303.5
25	8.00	24.00	18.00	358.2
26	8.00	24.00	10.00	347.7

(a) Write the design in coded form. Name the type of design.

(b) Fit a complete second-order model in the coded metric.

(c) Edit the model by eliminating obviously insignificant terms.

(d) Check model assumptions by plotting residuals against the design variables separately.

(e) The purpose of this experiment was not to merely find optimum conditions (conditions that maximize thickness) but to gain an impression about the role of the three design variables. Show contour plots, fixing levels of nickel at 10, 14, and 18.

(f) Use the plots to produce a recommended set of conditions that maximize thickness.

(g) Compute the standard error of prediction at the location of maximum thickness.

6.9 Refer to Exercise 6.8. The same simulation model was used to construct another response surface using a different experimental design. This second design, in natural units, is as follows, along with the thickness data.

Run Number	Time	Temperature	Nickel	Thickness
1	4	16	14	122
2	12	16	14	790
3	4	32	14	100
4	12	32	14	695
5	4	24	10	50
6	12	24	10	720
7	4	24	18	118
8	12	24	18	650
9	8	16	10	330
10	8	32	10	314
11	8	16	18	302
12	8	32	18	340
13	8	24	14	364
14	8	24	14	342
15	8	24	14	404

(a) Using the same coding as in Exercise 6.8, write out the design matrix in coded form.

(b) What is the design type?

(c) Fit and edit a second order response surface model. That is, fit and eliminate insignificant terms.

(d) Find conditions that maximize thickness, with the constraint that the condition falls inside the design region. Use an appropriate software package. In addition, compute the standard error of prediction at the location of optimum conditions.

6.10 Please refer to Exercises 6.8 and 6.9. In addition to the designs in Exercises 6.8 and 6.9, a *D-optimal* design was used to generate data on thickness from the simulation model. The notion of *D*-optimality will be discussed at length in Chapter 8. The design and data are as follows:

Run Number	Time	Temperature	Nickel	Thickness
1	12	32	10	717
2	4	16	10	136
3	12	16	18	787
4	4	32	18	78
5	4	16	18	87
6	4	32	10	80
7	12	16	10	760
8	8	24	10	282
9	8	32	14	318
10	4	24	14	96
11	12	32	18	682
12	12	24	14	747
13	12	24	14	764
14	12	24	14	742

(a) Using the same coding scheme as in Exercises 6.8 and 6.9, write the design in coded units.

(b) Fit a second-order model and edit the model. That is, eliminate insignificant terms.

(c) Find conditions of estimated maximum thickness. Can this be done with a two-dimensional contour plot? Explain. Compute the standard error of prediction at the set of optimum conditions.

6.11 Exercises 6.8, 6.9, and 6.10 illustrate the use of three different types of designs for modeling the same system. All three designs are used because of their capability to efficiently fit a *complete* second-order model, even if the edited model contains no interactions. Another type of design that allows for estimation of *linear and pure quadratic effects* is the *Taguchi L_{27}*. The Taguchi approach to quality improvement will be discussed in detail in Chapter 10. The L_{27} was also used to generate data on thickness as in the previous exercises. The design and response data are given by

Run Number	Time	Temperature	Nickel	Thickness
1	4	16	10	113
2	4	16	14	152
3	4	16	18	147
4	4	24	10	65
5	4	24	14	53
6	4	24	18	130
7	4	32	10	83
8	4	32	14	40
9	4	32	18	94
10	8	16	10	339
11	8	16	14	303
12	8	16	18	381
13	8	24	10	348
14	8	24	14	342
15	8	24	18	420
16	8	32	10	327
17	8	32	14	255
18	8	32	18	322
19	12	16	10	745
20	12	16	14	780
21	12	16	18	740
22	12	24	10	772
23	12	24	14	769
24	12	24	18	755
25	12	32	10	735
26	12	32	14	726
27	12	32	18	757

(a) Again, code the design as in previous exercises.

(b) Form the complete **X** matrix for the L_{27} (using coded design levels).

(c) Comment on whether or not this design is adequate for a *complete* second-order model (including interactions).

6.12 In this exercise we attempt to make comparisons among the central composite, Box–Behnken, and *D*-optimal design used in Exercises 6.8, 6.9, and 6.10. We will not deal with any particular edited model, but, rather, a complete second-order model. Compute the *scaled* standard error of prediction

$$\sqrt{\frac{N \operatorname{Var} \hat{y}(\mathbf{x})}{\sigma^2}} = \sqrt{N\left[\mathbf{x}^{(2)\prime}(\mathbf{X}'\mathbf{X})^{-1}\mathbf{x}^{(2)}\right]}$$

at the following locations in scaled design units:

x_1	x_2	x_3
-1	-1	-1
1	-1	-1
-1	1	-1
-1	-1	1
1	1	-1
1	-1	1
-1	1	1
1	1	1
-1	0	0
1	0	0
0	-1	0
0	1	0
0	0	-1
0	0	1
0	0	0

Comment on comparisons among the three designs.

6.13 In their paper, Derringer and Suich (1980), illustrate their multiple response procedure with an interesting data set. The following central composite design shows data obtained in the development of a tire tread compound on four responses: PICO Abrasion Index, y_1; 200% modulus, y_2; elongation at break, y_3; hardness, y_4. Each column was taken at the 20 sets of conditions shown, where x_1, x_2, x_3 are coded levels of the variables x_1 = hydrated silica level, x_2 = silane coupling agent level, and x_3 = sulfur. The following inequalities represent desirable conditions on the responses:

$$y_1 > 120$$

$$y_2 > 1{,}000$$

$$400 < y_3 < 600$$

$$60 < y_4 < 75$$

Compound number	x_1	x_2	x_3	y_1	y_2	y_3	y_4
1	−1	−1	1	102	900	470	67.5
2	1	−1	−1	120	860	410	65
3	−1	1	−1	117	800	570	77.5
4	1	1	1	198	2294	240	74.5
5	−1	−1	−1	103	490	640	62.5
6	1	−1	1	132	1289	270	67
7	−1	1	1	132	1270	410	78
8	1	1	−1	139	1090	380	70
9	−1.633	0	0	102	770	590	76
10	1.633	0	0	154	1690	260	70
11	0	−1.633	0	96	700	520	63
12	0	1.633	0	163	1540	380	75
13	0	0	−1.633	116	2184	520	65
14	0	0	1.633	153	1784	290	71
15	0	0	0	133	1300	380	70
16	0	0	0	133	1300	380	68.5
17	0	0	0	140	1145	430	68
18	0	0	0	142	1090	430	68
19	0	0	0	145	1260	390	69
20	0	0	0	142	1344	390	70

(a) Fit a second-order model for all responses.

(b) Develop two-dimensional contours (at $x_3 = -1.6, -1, 0, 1.0, 1.6$) of constant response for all four responses. Can you determine sets of conditions on x_1, x_2, and x_3 that meet the above requirements? If so, list them.

(c) Use the Derringer–Such procedure to determine other competing conditions.

6.14 Researchers are interested in studying the effect of three design variables on the yield of a chemical process involving precipitating stoichiometric calcium hydrogen orthophosphate. The three design variables and their selected levels are as follows:

$$x_1 = \frac{\text{Mole ratio of NH}_3 \text{ to CaCl}_2 - 0.85}{0.09}$$

$$x_2 = \frac{\text{Additional time (minutes)} - 50}{24}$$

$$x_3 = \frac{\text{Initial pH of solution} - 3.5}{0.9}$$

Write out the design matrix in natural variables for a central composite plan that has coded axial distance $\sqrt{3}$.

CHAPTER 7

Experimental Designs for Fitting Response Surfaces—I

Many of our examples in Chapter 6 displayed the fitting of second-order response surfaces with data taken from real-life applications of response surface methodology (RSM). In many of those the experimental design possesses properties required to fit a second-order response surface—that is,

(i)   at least three levels of each design variable
(ii) at least $1 + 2k + k(k - 1)/2$ distinct design points

The above represent minimum conditions for the fitting of a second-order model. The reader has also noticed the reference to the *central composite design* in Chapter 6. This second-order design class was represented in almost all of the illustrations. The composite design is by far the most popular second-order design in use by practitioners. As a result, it will receive considerable attention in subsections that follow. First, however, we feel as if some general commentary should be presented regarding the choice of designs for response surface analysis.

7.1 DESIRABLE PROPERTIES OF RESPONSE SURFACE DESIGNS

There are many classes of experimental designs in the literature, and there are many criteria on which experimental designs are based. Indeed, there are many computer packages that give optimal designs based on special criteria and input from the user. Special design criteria and important issues associated with computer-generated design of experiments will be discussed in a later section and in Chapter 8. However, it is first important for the reader to review a set of properties that should be taken into account when the choice of a response surface design is made. Some of the important characteristics are as follows:

1. Result in a good fit of the model to the data.
2. Give sufficient information to allow a test for lack of fit.
3. Allow models of increasing order to be constructed sequentially.
4. Provide an estimate of "pure" experimental error.
5. Be insensitive (robust) to the presence of outliers in the data.
6. Be robust to errors in control of design levels.
7. Be cost effective.
8. Allow for experiments to be done in blocks.
9. Provide a check on the homogeneous variance assumption.
10. Provide a good distribution of Var $\hat{y}(\mathbf{x})/\sigma^2$.

As one can readily see, not all of the above properties are required in every RSM experience. However, most of them must be given serious consideration on each occasion in which one designs experiments. Most of the properties are self-explanatory. We have already made reference to item 3. Item 5 is important if one expects outliers to occur. Though we have not discussed blocking in the context of fitting second-order models, we will consider the concept in detail later in this chapter. Number 10 is very important! The reader has been exposed to the notion of prediction variance in Chapters 2 and 6. In later sections we shall discuss the importance of stability of prediction variance.

There are several purposes behind the introduction of the list of 10 characteristics at this point in the text. Of primary importance is a reminder to the reader that designing an experiment is not necessarily so easy. Indeed, a few of the 10 items may be important and yet the researcher may not be aware of the magnitude of their importance. Some items do conflict with each other. As a result, there are tradeoffs that almost always exist when one chooses an appropriate design. An excellent discourse on the many considerations that one must consider in choosing a response surface design is given by Box and Draper (1975).

7.2 OPERABILITY REGION, REGION OF INTEREST, AND MODEL INADEQUACY

At the end of the previous chapter we discussed the choice of ranges on the design variables; this naturally leads to an introduction of the notion of the **region of interest** in the design variables. As we indicated in Chapter 6, we are assuming that the true function relating y to the design variables is unknown. In fact,

$$E(y) = f(\mathbf{x}, \boldsymbol{\theta})$$

and we are assuming that in the region of interest, $R(\mathbf{x})$, f is well approximated by a low-order polynomial. This, of course, brings on the notion of the first order or second-order response functions. They are approximations (justified by a Taylor series expansion of f) in the confined region R.

It is important for the reader to understand that while the experimental design may be confined to R, the region R may change from experiment to experiment (say, for example, in a steepest ascent experiment). However, there is a secondary region called the **region of operability** $O(\mathbf{x})$. This is the region in which the equipment, electronic system, chemical process, drug, and so on, works; and it is theoretically possible to do the experiment and observe response values. In some cases, data points in the design may be taken outside $R(\mathbf{x})$. Obviously, if one takes data too far outside $R(\mathbf{x})$, then the adequacy of our response surface representation (polynomial model) may be in question. Though the framework of the region of interest/operability is important for the reader to understand, knowledge of $R(\mathbf{x})$ or $O(\mathbf{x})$ in a given situation may simply be too idealistic. The region R changes and at times O is not truly known until much is known about the process.

Figure 7.1 gives the reader a pictorial depiction of O and R. In this case with two different experiments we have R and R', both being considerably more confined than O. The region R is the current "best guess" at where the optimum is. It is also the region in which we feel that $\hat{y}(\mathbf{x})$ should be a good predictor or, say, a good estimator of $f(\mathbf{x}, \boldsymbol{\theta})$. However, the user should constantly be aware of model inadequacy in cases where $\hat{y}(\mathbf{x})$ is not a good approximation of $f(\mathbf{x}, \boldsymbol{\theta})$.

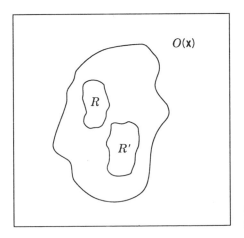

Figure 7.1 Region of operability and region of interest.

Model Inadequacy and Model Bias

If we consider the vector of fitted values as $\hat{\mathbf{y}}(\mathbf{x})$, then

$$\hat{\mathbf{y}}(\mathbf{x}) = \mathbf{Xb} \tag{7.1}$$

where

$$\mathbf{b} = (\mathbf{X'X})^{-1}\mathbf{X'y}$$

is the least squares estimator of $\boldsymbol{\beta}$. We devoted considerable discussion in Chapter 6 to the meaning of lack of fit. If a "better" approximation of f is

$$\mathbf{f} \cong \mathbf{X}\boldsymbol{\beta} + \mathbf{Z}\boldsymbol{\gamma} \tag{7.2}$$

where Z contains, say, terms of higher order than those in $\mathbf{X}$, then the expected value of $\mathbf{b}$ is given by

$$E(\mathbf{b}) = (\mathbf{X'X})^{-1}\mathbf{X'}[\mathbf{X}\boldsymbol{\beta} + \mathbf{Z}\boldsymbol{\gamma} + \boldsymbol{\varepsilon}]$$

$$= \boldsymbol{\beta} + (\mathbf{X'X})^{-1}\mathbf{X'Z}\boldsymbol{\gamma}$$

$$= \boldsymbol{\beta} + \mathbf{A}\boldsymbol{\gamma}$$

Here, of course, $\boldsymbol{\varepsilon}$ is the usual random error and $\mathbf{A} = (\mathbf{X'X})^{-1}\mathbf{X'Z}$ is called the *alias matrix*. The alias matrix essentially transcribes **bias errors** to $\mathbf{b}$, the estimator of $\boldsymbol{\beta}$. As in the case where there is appreciable lack of fit, bias errors appear in predicted values and fitted values. In fact,

$$E[\hat{\mathbf{y}}(\mathbf{x})] = \mathbf{X}(\boldsymbol{\beta} + \mathbf{A}\boldsymbol{\gamma})$$

Thus, the bias associated with the vector of predicted values is given by

$$E[\hat{\mathbf{y}}(\mathbf{x})] - \mathbf{f} \cong (\mathbf{X}\boldsymbol{\beta} + \mathbf{X}\mathbf{A}\boldsymbol{\gamma}) - \mathbf{X}\boldsymbol{\beta} + \mathbf{Z}\boldsymbol{\gamma}$$

$$= (\mathbf{XA} - \mathbf{Z})\boldsymbol{\gamma} \tag{7.3}$$

Thus, when the "true" model (or truly the best approximation) is given by Equation (7.2), the bias is transmitted to fitted values.

Equation (7.3) is important in that it plays a role in the lack of fit test discussed in Chapter 6. In fact, if the model is misspecified as we have described here, the residual mean square in the response surface analysis of variance has as its expectation

$$E(s^2) = \sigma^2 + \frac{1}{p}\{\boldsymbol{\gamma'}[\mathbf{XA} - \mathbf{Z}]'[\mathbf{XA} - \mathbf{Z}]\boldsymbol{\gamma}\}$$

Thus, in Equation (7.3) if we call $\mathbf{\Delta} = (\mathbf{XZ} - \mathbf{Z})$ with the vector of bias in $\hat{y}(\mathbf{x})$ being $\mathbf{\Delta\gamma}$, then

$$E(s^2) = \sigma^2 + \frac{1}{p}\{\mathbf{\gamma'\Delta'\Delta\gamma}\} \tag{7.4}$$

(see Appendix 5). Indeed, the lack-of-fit test is detecting, when significant, the presense of $\mathbf{\gamma'\Delta'\Delta\gamma}$, the latter being nonzero when $\mathbf{\gamma} \neq 0$—that is, when f is not well represented by $\mathbf{X\beta}$ [see Equation (7.2)].

In addition to lack of fit, the concept described by Equation (7.3) plays an important role in the choice of an experimental design. In fact, the choice of design is based on control of two types of error, **variance error** and **bias error**. If we consider this in the context of a particular fitted value, say $\hat{y}(\mathbf{x})$, we have a loss-type criterion:

$$
\begin{aligned}
E\left[\hat{y}(\mathbf{x}_0) - f(\mathbf{x}_0, \mathbf{\theta})\right]^2 &= E\left[\hat{y}(\mathbf{x}_0) - E\hat{y}(\mathbf{x}_0) + E\hat{y}(\mathbf{x}_0) - f(\mathbf{x}_0, \mathbf{\theta})\right]^2 \\
&= E\left[\hat{y}(\mathbf{x}_0) - E\hat{y}(\mathbf{x}_0)\right]^2 + \left[E\hat{y}(\mathbf{x}_0) - f(\mathbf{x}_0, \mathbf{\theta})\right] \\
&= \text{Var } \hat{y}(\mathbf{x}_0) + \left[\text{Bias } \hat{y}(\mathbf{x}_0)\right]^2 \tag{7.5}
\end{aligned}
$$

The term Var $\hat{y}(\mathbf{x}_0)$ is clear. It has been discussed in Chapter 6 and will be discussed further in this chapter. Suppose we call Bias $\hat{y}(\mathbf{x}_i)$ the bias evaluated at the ith data point. The term [Bias $\hat{y}(\mathbf{x}_i)]^2$ is a single element in $\mathbf{\gamma'\Delta'\Delta\gamma}$ in Equation (7.4). In fact,

$$\mathbf{\gamma'\Delta'\Delta\gamma} = \sum_{i=1}^{N} \left(\text{Bias } \hat{y}(\mathbf{x}_i)\right)^2$$

The statistics literature contains an enormous amount of information on experimental design for regression or response surface modeling. However, the majority of this technical or theoretical information deals with variance oriented design criteria—that is, the type of design criteria that ignore model misspecification. In other words, these criteria essentially assume that the model that is specified by the user is correct. In what follows, we shall initially focus on variance-type criteria. However, the errors associated with model misspecification will be revisited in Chapter 9. In what follows, we begin with experimental designs for first-order response models.

7.3 DESIGN OF EXPERIMENTS FOR FIRST-ORDER MODELS

Consider a situation in which N experimental runs are conducted on k design variables $x_1, x_2, \ldots, x_k$ and a single response y. A model is postulated of the types

$$y_i = \beta_0 + \beta_1 x_{i1} + \beta_2 x_{i2} + \cdots + \beta_k x_{ik} + \varepsilon_i \qquad (i = 1, 2, \ldots, N)$$

As a result a fitted model is given by

$$\hat{y}_i = b_0 + b_1 x_{i1} + b_2 x_{i2} + \cdots + b_k x_{ik}$$

Here, of course, the b_j are found by the method of least squares. Obviously, the most intuitive variance type of criterion involves the variance of the regression coefficients—that is, $\text{Var}(b_i)$ ($i = 0, 1, 2, \ldots, k$). Let us assume that each design variable is in coded design units and that the ranges for the design levels are such that $x_j \in [-1, +1]$. The key ingredient in choosing a design that minimizes the variances is the concept of **orthogonality**.

7.3.1 The First-Order Orthogonal Design

The terms **orthogonal design**, *orthogonal array*, and *orthogonal main effect plan* received an unusual amount of attention in the 1950s and 1960s. A first order orthogonal design is one for which $\mathbf{X'X}$ is a diagonal matrix. The above definition implies, of course, that the columns of the $\mathbf{X}$ matrix are mutually orthogonal. In other words, if we write

$$\mathbf{X} = [\mathbf{1}, \mathbf{x}_1, \mathbf{x}_2, \ldots, \mathbf{x}_k]$$

where $\mathbf{x}_j$ is the jth column of $\mathbf{X}$, then a first-order orthogonal design is such that $\mathbf{x}_i' \mathbf{x}_j = 0$ for all $i \neq j$, and, of course, $\mathbf{1'x}_j = 0$ for $j = 1, 2, \ldots, k$. It should be clear that if two columns are orthogonal the levels of the two corresponding variables are linearly independent. The implication is that the roles of the two variables are being assessed independent of each other. This underscores the virtues of orthogonality. In the first-order situations, an orthogonal design in which the ranges on these variables are taken to extremes results in minimizing $\text{Var}(b_i)$ on a per observation basis. In other words,

For the first-order model and a fixed sample N, if $x_j \in [-1, +1]$ for $j = 1, 2, \ldots, k$, then $\text{Var}(b_i / \sigma^2)$ for $i = 1, 2, \ldots, k$ is minimized if the design is orthogonal and all x_i levels in the design are ± 1 for $i = 1, 2, \ldots, k$.

Thus, the elements on the diagonals of $(\mathbf{X'X})^{-1}$ are minimized [recall that $\text{Var } \mathbf{b} = \sigma^2 (\mathbf{X'X})^{-1}$] by making off-diagonals of $\mathbf{X'X}$ zero and by forcing the diagonals of $\mathbf{X'X}$ to be as large as possible. The notion of *linear independence* and *variable levels at ± 1 extremes* as a desirable set of conditions should certainly be intuitive to the reader.

Example 7.1 The two-level factorial plans and fractions of resolution $\geq$ III do, in fact, minimize the variances of all coefficients (variances scaled by σ^2) on a per observation basis. To illustrate this, consider the following $\frac{1}{2}$ fraction of a 2^4 factorial (resolution IV) with defining relation $ABCD = I$

The **X** matrix is given by

$$
\mathbf{X} = \begin{array}{c}
\begin{array}{cccc} x_1 & x_2 & x_3 & x_4 \end{array} \\
\begin{bmatrix}
1 & -1 & -1 & -1 & -1 \\
1 & 1 & 1 & -1 & -1 \\
1 & 1 & -1 & 1 & -1 \\
1 & 1 & -1 & -1 & 1 \\
1 & -1 & 1 & 1 & -1 \\
1 & -1 & 1 & -1 & 1 \\
1 & -1 & -1 & 1 & 1 \\
1 & 1 & 1 & 1 & 1
\end{bmatrix}
\end{array}
$$

One can easily determine that $\mathbf{X'X} = 8\mathbf{I}_5$ and $(\mathbf{X'X})^{-1} = \frac{1}{8}\mathbf{I}_5$. Indeed, one can say that *no other design with 8 experimental runs can result in variances smaller than* $\sigma^2/8$. It should be apparent to the reader that a full 2^4 factorial results in all coefficient variances being $\sigma^2/16$. In other words, if the size of the design is doubled, the variances of coefficients are cut in half. However, both are considered optimal designs on a per observation basis.

Example 7.2   Consider a situation in which there are three design variables and the user wishes to use eight experimental runs but also desires to have design replication. A $\frac{1}{2}$ fraction of a 2^3 factorial (resolution III) is used with each design point replicated. The **X** matrix for the design is as follows (defining relation $ABC = I$):

$$
\mathbf{X} = \begin{array}{c}
\begin{array}{ccc} x_1 & x_2 & x_3 \end{array} \\
\begin{bmatrix}
1 & 1 & -1 & -1 \\
1 & 1 & -1 & -1 \\
1 & -1 & 1 & -1 \\
1 & -1 & 1 & -1 \\
1 & -1 & -1 & 1 \\
1 & -1 & -1 & 1 \\
1 & 1 & 1 & 1 \\
1 & 1 & 1 & 1
\end{bmatrix}
\end{array}
$$

Note that the columns are orthogonal and all levels are at ± 1 extremes. Thus the design is "variance optimal" for the model

$$\hat{y} = b_0 + b_1 x_1 + b_2 x_2 + b_3 x_3$$

That is, all coefficients in the above model have minimum variance over all designs with sample size $N = 8$. In fact, the variance covariance matrix is given by

$$\text{Var } \mathbf{b} = \left(\sigma^2/8\right)\mathbf{I}_4$$

It is interesting that for the situation described in the previous example a full 2^3 factorial results in the same variance–covariance matrix. Though the two designs are quite different, from a variance point of view they are equivalent for the first order model in three design variables. Obviously in the case of the $\frac{1}{2}$ fraction (replicated) there are no degrees of freedom for lack of fit, whereas the 2^3 factorial enjoys four lack of fit degrees of freedom that are attributable to x_1x_2, x_1x_3, x_2x_3, and $x_1x_2x_3$. On the other hand, the replicated $\frac{1}{2}$ fraction allows 4 degrees of freedom for replication error (pure error), and, of course, the full 2^3 possesses no degrees of freedom for replication error. As a result, even though they are "variance equivalent," the two orthogonal designs would find use in somewhat different circumstances.

It should be apparent to the reader that for the use of two-level factorials or fractions for a first-order model, orthogonality (and hence variance optimality) is obtained with resolution at least III. In the case of resolution III, the $\mathbf{x}_i$ columns in the $\mathbf{X}$ matrix will be aliased with the $\mathbf{x}_j\mathbf{x}_k$ columns. However, no $\mathbf{x}_i$ column will be aliased with any $\mathbf{x}_j$ column. In the case of 2^k factorials or *regular fractions, any two columns in the* $\mathbf{X}$ *matrix that are not aliased are, indeed, orthogonal.*

The notion of orthogonality and its relationship to design resolution clearly suggests that variance optimal designs should also be available if the fitted response model contains first-order terms and one or more interaction terms. The following section addresses this topic.

7.3.2 Orthogonal Designs for Models Containing Interaction

As we indicated earlier, it is not uncommon for the process or system to require one or more interaction terms to accompany the linear main effects terms in a first-order model. The two-level factorials and fractions are quite appropriate. As in the case of the first-order model, the orthogonal design is variance optimal—the sense again being that variances of all coefficients are minimized on a per observation basis. Based on the discussion earlier in Section 7.3, it should be apparent that the two-level full factorial is orthogonal for models containing main effects and any (or all) interactions involving cross products of linear terms. If a fraction is to be used, one must have sufficient resolution to ensure that no model terms are aliased. For example, if two factor interaction terms appear in the model, a resolution of at least V is appropriate.

As an illustration suppose for $k = 4$ the analyst postulates a model

$$y = \beta_0 + \sum_{i=1}^{4} \beta_1 x_i + \sum_{i<j} \sum \beta_{ij} x_i x_j + \varepsilon$$

then clearly no fractional factorial will be orthogonal because a resolution IV fraction will result in aliasing two factor interactions with each other. In fact, a $\frac{1}{2}$ fraction in this case would involve only eight design points and the above model contains 11 model terms. A full 2^4 factorial is an orthogonal design for the above model.

Consider a second illustration in which five design variables are varied in a study in which the model can be written as

$$y = \beta_0 + \sum_{i=1}^{5} \beta_i x_i + \sum \sum_{i<j} \beta_{ij} x_i x_j + \varepsilon$$

In this case a complete 2^5 or a resolution V 2^{5-1} serves as an orthogonal design. In this case, linear main effects are orthogonal to each other and orthogonal to two-factor interactions while two-factor interactions are orthogonal to each other. The following X matrix illustrates the concept of orthogonality in the case of the $\frac{1}{2}$ fraction.

$$X = \begin{array}{c c c c c c c c c c c c c c c}
x_1 & x_2 & x_3 & x_4 & x_5 & x_1x_2 & x_1x_3 & x_1x_4 & x_1x_5 & x_2x_3 & x_2x_4 & x_2x_5 & x_3x_4 & x_3x_5 & x_4x_5 \\
+ & + & - & - & - & - & - & - & - & + & + & + & + & + & + \\
+ & - & + & - & - & - & + & + & + & - & - & - & + & + & + \\
+ & - & - & + & - & + & - & + & + & - & + & + & - & - & + \\
+ & - & - & - & + & - & + & + & - & + & + & - & + & - & - \\
+ & - & - & - & - & + & + & + & + & - & + & + & - & + & - \\
+ & + & + & + & - & - & + & + & - & - & + & - & - & - & + \\
+ & + & + & - & + & - & + & - & + & - & - & + & - & + & - \\
+ & + & + & - & - & + & + & - & - & + & - & + & + & - & - \\
+ & + & - & + & + & - & - & + & + & - & - & + & + & - & - \\
+ & + & - & + & - & + & - & + & - & + & - & - & + & + & - \\
+ & + & - & - & + & + & - & - & + & + & + & - & - & - & + \\
+ & - & + & + & + & - & - & - & - & + & + & + & - & - & - \\
+ & - & + & + & - & + & - & - & + & - & + & - & + & - & - \\
+ & - & - & + & + & + & + & - & - & - & - & - & + & + & + \\
+ & - & + & - & + & + & - & + & - & - & + & + & - & - & + \\
+ & + & + & + & + & + & + & + & + & + & + & + & + & + & +
\end{array}$$

The defining contrast $ABCDE = I$ was used to construct the $\frac{1}{2}$ fraction. It is clear that the columns of the X matrix are orthogonal to each other and that

$$X'X = 16I_{16}$$

and thus the variances of the coefficients are given by

$$\mathrm{Var}(b_0) = \mathrm{Var}(b_i) = \mathrm{Var}(b_{ij}) = \frac{\sigma^2}{16} \qquad (i = 1,2,3,4,5; i \neq j)$$

The Saturated Design

The reader should note that in the previous illustration there are 16 design points and 16 model terms. The result is that there are *zero residual degrees of freedom*. This type of scenario deserves special attention. The absence of any lack-of-fit degrees of freedom implies, of course, that no test can be made for lack of fit; that is, there is no information to test for model inadequacy. In this situation the design is said to be **saturated**. While this type of situation should be avoided, there certainly are practical situations in which cost constraints only allow a saturated design. The saturated design will continue to receive attention in future chapters. The discussion of the saturated design invites a question regarding what form the analysis takes. Can any inferences be made regarding model terms? The reader should recall the discussion in Chapter 3 regarding use of normal probability plots for effects. For an orthogonal design with a first-order model or a first-order-with-interaction model the model coefficients are simple functions of the *computed effects* discussed in Chapter 3. As a result, the use of normal probability plots for detection of "active effects" can be used to detect "active regression coefficients." However, the reader should understand that when at all possible, saturated designs should be avoided in favor of designs that contain degrees of freedom for pure error and degrees of freedom for lack of fit.

Effect of Center Runs

In Chapters 3 and 4 we discussed the utility of center runs as an important augmentation of the two-level factorial and fraction. This was discussed again and illustrated in Chapter 5. Obviously, the addition of center runs to an orthogonal design does not alter the orthogonality property. However, the design is no longer a variance optimal design; that is, the variance of regression coefficients are no longer minimized on a per observation basis. Obviously, the requirement that all levels be at ± 1 extremes is not met. Indeed, the center runs (runs at the zero coordinates in design units) add nothing to information about linear effects and interactions. However, let us not forget the importance of the detection of quadratic effects (see Chapter 3), a task which is handled very effectively and efficiently with the use of center runs.

What Design Should Be Used?

The discussion of center runs and saturated designs in the context of orthogonal designs underscores a very important point. The design that is used in a given situation depends a great deal on the goal of the experiment and the assumptions that one is willing to make. It should also be clear that an optimal design (i.e., variance optimal) is not always appropriate. Consider an example in which one is interested in fitting the model:

$$y_i = \beta_0 + \beta_1 x_{i1} + \beta_2 x_{i2} + \beta_3 x_{i3} + \varepsilon_i \qquad (i = 1, 2, 3, \ldots, 8)$$

that is, a first-order model in three design variables. Among the candidate designs are three orthogonal designs that we will attempt to discuss and compare. All three of these designs involve eight experimental runs. Consider the following:

Design 1: 2^3 factorial

Design 2: Resolution III fraction of 2^3 augmented with four center runs

Design 3: Resolution III fraction of 2^3 with replicate runs at each design point

Designs 1 and 3 are both variance optimal designs. The variances of regression coefficients will be $\sigma^2/8$ for Designs 1 and 3. For Design 2, the variance of the intercept term will be $\sigma^2/8$ but the variances of the linear coefficients will be $\sigma^2/4$. The appropriate comments concerning all three designs follow:

Design 1. It does not allow for estimation of replication error (pure error). There are four lack-of-fit degrees of freedom, but no error degrees of freedom are available to test lack of fit. The design is appropriate if there is strong prior information that there is no model inadequacy.

Design 2. Pure error degrees of freedom (3 df) are available for testing linear effects. One can also test for curvature (one df for pure quadratic terms). It should be used when one accepts the possibility of quadratic terms but can absolutely rule out interaction (the presence of pure quadratic terms with no interaction is not particularly likely).

Design 3. The design contains four pure error degrees of freedom but no lack-of-fit degrees of freedom associated with interaction terms or pure quadratic terms. Tests of significance can be made on model coefficients for the first-order model, but the design should not be used if either quadratic effects or interaction may be present. There is no model adequacy check of any kind.

From the above description it would seem that all these candidates possess disadvantages. However, one must keep in mind that with a sample size of eight and four model terms there is "not much room" for an efficient check for model inadequacy because both lack of fit and pure error information is needed. Incidentally, a full 2^3 augmented with, say, four center runs would provide three pure error degrees of freedom that can be used to test pure quadratic curvature and the presence of interaction.

7.3.3 Other First-Order Orthogonal Designs—The Simplex Design

The majority of applications of orthogonal designs for first-order models should suggest the use of the two-level designs that we have discussed here and in Chapters 3 and 4. However, it is important for the reader to be aware of a special class of first-order orthogonal order designs that are saturated. This class is called the class of *simplex designs*. Let it be clear that these designs are saturated for a *first-order model*—a fact that implies that no information is available at all to study interactions.

The main feature of the simplex design is that it requires $N = k + 1$ observations to fit a first-order model in k variables. Geometrically, the design points represent the vertices of a regular sided figure. That is, the design points, given by the rows of the design matrix

$$
\mathbf{D} = \begin{bmatrix}
x_{11} & x_{21} & \cdots & x_{k1} \\
x_{12} & x_{22} & & x_{k2} \\
\vdots & \vdots & \vdots & \vdots \\
x_{1,N} & x_{2,N} & \cdots & x_{k,N}
\end{bmatrix}
$$

are points in k dimensions such that the angle that any two points make with the origin is θ, where

$$
\cos(\theta) = -\frac{1}{(N-1)} = -\frac{1}{k}
$$

For $k = 2$, $N = 3$, $\cos(\theta) = -1/2$, and thus $\theta = 120°$. Thus, for this case, the points are coordinates of an *equilateral triangle*. For $k = 3$ and $N = 4$, the design points are the vertices of a tetrahedron. For $k = 2$, the design matrix is given by

$$
\mathbf{D} = \begin{bmatrix}
\sqrt{3/2} & -1/\sqrt{2} \\
-\sqrt{3/2} & -1/\sqrt{2} \\
0 & 2/\sqrt{2}
\end{bmatrix} \qquad (7.6)
$$

Figure 7.2 reveals that the design for $k = 2$ contains points on the vertices of an equilateral triangle. The three rows in the design matrix correspond to points 1, 2, and 3, respectively. The **X** matrix is the design matrix augmented by a column of ones. Thus

$$
\mathbf{X} = \begin{bmatrix}
\sqrt{3/2} & -1/\sqrt{2} \\
-\sqrt{3/2} & -1/\sqrt{2} \\
0 & 2/\sqrt{2}
\end{bmatrix} \qquad (7.7)
$$

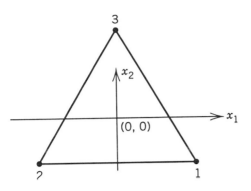

Figure 7.2 Design points for the simplex in two variables.

As a result, one can readily observe the orthogonality from

$$\mathbf{X'X} = \begin{bmatrix} 3 & 0 & 0 \\ 0 & 3 & 0 \\ 0 & 0 & 3 \end{bmatrix}$$

For $k = 3$, the design involves the use of $N = 4$ points and any two points make an angle θ with the origin, where $\cos \theta = -\frac{1}{3}$. The appropriate design is given by

$$\mathbf{D} = \begin{array}{c} \begin{array}{ccc} x_1 & x_2 & x_3 \end{array} \\ \begin{bmatrix} 1 & -1 & -1 \\ -1 & 1 & -1 \\ -1 & -1 & 1 \\ 1 & 1 & 1 \end{bmatrix} \end{array} \qquad (7.8)$$

If we use $\mathbf{X} = [\mathbf{1}|\mathbf{D}]$ it is clear that

$$\mathbf{X'X} = 4\mathbf{I}_4$$

and hence the design is orthogonal for fitting the first-order model. Figure 7.3 shows the coordinates displayed in three dimensions. Note that any two points form the same angle with the origin $(0, 0, 0)$. Also note that in this special case the simplex is a $\frac{1}{2}$ fraction of a 2^3. The one we have chosen to display is the resolution III fraction with $ABC = I$ as the defining relation. Indeed, the fraction formed from $ABC = -I$ also is a simplex.

Rotation and Construction of the Simplex

For a specific value of k there are numerous simplex designs that can be constructed. In fact, if one were to rotate the rows of the design matrix through an angle, say ϕ, the resulting design is still a simplex and thus still possesses the property of being first-order orthogonal. For example, a change in the orientation of the equilateral triangle of the design in Equation (7.6)

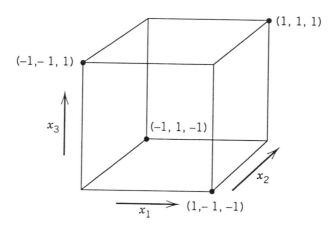

Figure 7.3 Simplex in three design variables.

will result in a second simplex. For $k = 3$, a design that is an alternative to that shown in Equation (7.8) is given by

$$
\mathbf{D} =
\begin{array}{ccc}
x_1 & x_2 & x_3 \\
\end{array}
\begin{bmatrix}
0 & \sqrt{2} & -1 \\
-\sqrt{2} & 0 & 1 \\
0 & -\sqrt{2} & -1 \\
\sqrt{2} & 0 & 1
\end{bmatrix}
\tag{7.9}
$$

It is easily seen that this design is also first order orthogonal, although from a pragmatic point of view the design in Equation (7.8) is preferable.

The construction of the $\mathbf{X}$ matrix for the k-dimensional simplex is quite simple. One merely starts with an orthogonal matrix $\mathbf{O}$ of order $N \times N$ with the elements of the first column being equal. The $\mathbf{X}$ matrix is then

$$
\mathbf{X} = \boldsymbol{\sigma} \cdot N^{1/2}
$$

Thus, it is easily seen that $\mathbf{X'X} = N(\mathbf{O'O}) = N\mathbf{I}_N$. For example, the $k = 2$ simplex in Equation (7.7) can be constructed from the orthogonal matrix

$$
\mathbf{O} =
\begin{bmatrix}
1 & 1 & -1 \\
1 & -1 & -1 \\
1 & 0 & 2 \\
1/\sqrt{3} & 1/\sqrt{2} & 1/\sqrt{6}
\end{bmatrix}
$$

The values at the bottom of each column are to be multiplied by each element in the column.

$$\mathbf{X} = \mathbf{O} \cdot \sqrt{3} = \begin{bmatrix} 1 & \sqrt{3/2} & -1/\sqrt{2} \\ 1 & -\sqrt{3/2} & -1/\sqrt{2} \\ 1 & 0 & 2/\sqrt{2} \end{bmatrix}$$

which is the **X** matrix in Equation (7.7). The **X** matrix given in the design of Equation (7.8) can be constructed from the simple orthogonal matrix

$$\mathbf{O} = \begin{bmatrix} 1 & 1 & -1 & -1 \\ 1 & -1 & 1 & -1 \\ 1 & -1 & -1 & 1 \\ 1 & 1 & 1 & 1 \\ 1/2 & 1/2 & 1/2 & 1/2 \end{bmatrix}$$

The reader should keep in mind that the simplex design is first-order saturated, which implies that there is no information available for lack of fit. If the analyst expects system interaction or pure quadratic type of curvature at all, the simplex is not an appropriate design.

Example 7.3   A simplex design is used in a laboratory experiment designed to build a first-order relationship between growth (y) of a particular organism and the percentage of glucose (x_1), concentration of yeast extract (x_2), and the time in hours (x_3) allowed for organism growth. In coded form the design variables are

$$x_1 = \frac{\% \text{ glucose} - 30}{1.0}$$

$$x_2 = \frac{\% \text{ yeast} - 0.5}{0.10}$$

$$x_3 = \frac{\text{hr} - 45}{15}$$

Table 7.1 gives the design in terms of the original and coded variables and indicates the observed response.
The fitted equation is given by

$$\hat{y} = 9.54 + 0.97x_1 + 0.40x_2 + 1.59x_3$$

Table 7.1   Experimental Data for Organism Growth Experiment Using Simplex Design

Run Number	Percent Glucose, x_1 Uncoded	Coded	Percent Yeast, x_2 Uncoded	Coded	Time, x_3 (hr) Uncoded	Coded	Growth, (g/liter)
1	3.0	0	0.641	$\sqrt{2}$	30	-1	8.52
2	1.586	$-\sqrt{2}$	0.500	0	60	1	9.76
3	3.0	0	0.359	$-\sqrt{2}$	30	-1	7.38
4	4.414	$\sqrt{2}$	0.500	0	60	1	12.50

7.3.6   Another Variance Type Property—Prediction Variance

We earlier indicated that two-level orthogonal first-order designs with levels at ± 1 extremes result in variances of coefficients that are minimized on a per observation basis. This provided the inspiration for the "variance optimal" terminology that we used consistently as we discussed examples and applied the concept to models that involve interaction terms. Criteria involving variances of coefficients are certainly important. In fact, in Chapter 8 we explain the notion of special norms on the $(X'X)^{-1}$ as we more formally deal with design optimality. However, we should also be attentive to the criterion of **prediction variance**. This concept was discussed in Chapter 2 in dealing with confidence intervals around $\hat{y}(x)$ in regression problems, and again in Chapter 6 as a part of the total RSM analysis. However, the prediction variance concept is a very important aspect in the study of experimental design. It is important for the reader to become familiar with it in what is the simplest scenerio—namely, the first-order model.

Prediction Variance for the First-Order Model

Recall that the prediction variance, or variance of a predicted value, is given by

$$\text{Var}[\,\hat{y}(x)\,] = x^{(m)'}(X'X)^{-1}x^{(m)} \cdot \sigma^2$$

where $x^{(m)}$ is a function of the location in the design variables at which one predicts and also a function of the model. In fact, the (m) in $x^{(m)}$ reflects the model. In the case of a strictly first-order model, we obtain

$$x^{(m)'} = [1, x_1, x_2, \ldots, x_k]$$

For a $k = 2$ model containing x_1, x_2 and their interaction, we obtain

$$x^{(m)'} = [1, x_1, x_2, x_1 x_2]$$

One can easily see that $\text{Var}[\,\hat{y}(x)\,]$ varies from location to location in the design space. In addition, the presence of $(X'X)^{-1}$ attests to the fact that the criterion is very much a function of the experimental design. The reader should view the prediction variance as a reflection of how well one predicts with the model.

In studies that are done to compare designs, it is often convenient to *scale* the prediction variance, that is, work with

$$\frac{N \operatorname{Var} \hat{y}(x)}{\sigma^2} = N\mathbf{x}^{(m)'}(\mathbf{X'X})^{-1}\mathbf{x}^{(m)} \tag{7.10}$$

The division by σ^2 makes the quantity scale-free, and the multiplication by N allows the quantity to reflect variance on a *per observation basis*. That is, if two designs are being compared, the scaling by N automatically "punishes" the design with the larger sample size. It forces a premium on efficiency.

The scaled prediction variance in Equation (7.10) is quite simple for the case of a variance optimal first-order orthogonal design. In fact, because $(\mathbf{X'X})^{-1} = (1/N)\mathbf{I}_p$, we have

$$\frac{N \operatorname{Var} \hat{y}(\mathbf{x})}{\sigma^2} = [1, x_1, x_2, \ldots, x_k] \begin{bmatrix} 1 \\ x_1 \\ x_2 \\ \vdots \\ x_k \end{bmatrix} = 1 + \sum_{i=1}^{k} x_i^2$$

$$= 1 + \rho_{\mathbf{x}}^2 \tag{7.11}$$

where $\rho_{\mathbf{x}}$ is the distance that the point $\mathbf{x'} = [x_1, x_2, \ldots, x_k]$ is away from the design origin. As a result, $N \operatorname{Var} \hat{y}/\sigma^2 = 1$ at the design center and becomes larger as one moves toward the design perimeter. The student who recalls confidence intervals on $E[y(x)]$ in linear regression should not be surprised at this result. As an illustration, for a 2^3 factorial, the following are values of $N \operatorname{Var} \hat{y}(\mathbf{x})/\sigma^2$ at different locations where one may wish to predict:

Location	$\dfrac{N \operatorname{Var} \hat{y}(\mathbf{x})}{\sigma^2}$
$0, 0, 0$	1
$\frac{1}{2}, \frac{1}{2}, \frac{1}{2}$	$1\frac{3}{4}$
$\frac{1}{2}, \frac{1}{2}, 1$	$2\frac{1}{2}$
$\frac{1}{2}, 1, 1$	$3\frac{1}{4}$
$1, 1, 1$	4

Thus, the scaled prediction variance increases *fourfold* as one moves from the design center to the design perimeter. For, say, a 2^5 factorial at the perimeter [i.e., $(\pm 1, \pm 1, \pm 1, \pm 1, \pm 1)$], $N \operatorname{Var} \hat{y}(\mathbf{x})/\sigma^2 = 6$ and hence the variance increases *sixfold*.

In the case of a lesser design, (i.e., one that is not variance optimal, or even orthogonal), this instability of prediction variance is even more pronounced. For example, consider a $\frac{1}{2}$ fraction (resolution III) of a 2^3 factorial with four center runs as an alternative to a 2^3. We know that this design is orthogonal but not variance optimal. Recall this design as Design 2 in the comparisons we made in Section 7.3.2. The 2^3 was Design 1. For Design 2 we have

$$\mathbf{X} = \begin{bmatrix} 1 & 1 & -1 & -1 \\ 1 & -1 & 1 & -1 \\ 1 & -1 & -1 & 1 \\ 1 & 1 & 1 & 1 \\ 1 & 0 & 0 & 0 \\ 1 & 0 & 0 & 0 \\ 1 & 0 & 0 & 0 \\ 1 & 0 & 0 & 0 \end{bmatrix} \qquad \mathbf{X'X} = \begin{bmatrix} 8 & 0 & 0 & 0 \\ & 4 & 0 & 0 \\ & & 4 & 0 \\ & & & 4 \end{bmatrix}$$

with the variance–covariance matrix (apart from σ^2) being

$$(\mathbf{X'X})^{-1} = \begin{bmatrix} \frac{1}{8} & 0 & 0 & 0 \\ 0 & \frac{1}{4} & 0 & 0 \\ 0 & 0 & \frac{1}{4} & 0 \\ 0 & 0 & 0 & \frac{1}{4} \end{bmatrix}$$

As a result,

$$\frac{N \operatorname{Var} \hat{y}(\mathbf{x})}{\sigma^2} = 8[1, x_1, x_2, x_3] \begin{bmatrix} \frac{1}{8} & 0 & 0 & 0 \\ 0 & \frac{1}{4} & 0 & 0 \\ 0 & 0 & \frac{1}{4} & 0 \\ 0 & 0 & 0 & \frac{1}{4} \end{bmatrix} \begin{bmatrix} 1 \\ x_1 \\ x_2 \\ x_3 \end{bmatrix}$$

$$= 1 + 2(x_1^2 + x_2^2 + x_3^2)$$

$$= 1 + 2\rho_x^2$$

As a result, it is apparent that $N \operatorname{Var} \hat{y} / \sigma^2$ for Design 2 increases much faster as one approaches the design perimeter than does Design 1. In fact, at $(\pm 1, \pm 1, \pm 1)$, $N \operatorname{Var} \hat{y}(\mathbf{x}) / \sigma^2 = 4$ for Design 1 but the value is 7 for Design 2. Incidentally, Design 3, which is a resolution III fraction of a 2^3 with duplicate runs, has prediction variance properties identical to that of Design 1; they are both variance optimal designs.

The reader should view prediction variance as another "variance-type" criterion from which comparisons among design can be made. It is different than merely comparing variances of individual coefficients. Much more attention will be given to $N \operatorname{Var} \hat{y}(\mathbf{x}) / \sigma^2$ later in this chapter and in future chapters.

7.4 DESIGNS FOR FITTING SECOND-ORDER MODELS

A Review of the Phases of RSM

Before we embark on the huge topic of experimental designs for second-order models, we should review for the reader some minimum design requirements and RSM philosophy that motivates one or two of the classes of designs that we will subsequently present in detail. Variable screening is an essential phase of RSM that was presented at length in Chapter 1. Here, of course, the two-level factorial designs and fractions play a major role. Any sequential movement (region seeking) that is necessary is also accomplished with a first-order design. Of course, there may be instances where region-seeking via steepest ascent (or descent) and/or variable screening are not required. However, the possibility of either or both should be included in the total sequential plan. At some point the researcher will be interested in fitting a second-order response surface in the design variables $x_1, x_2, \ldots, x_k$. This response surface analysis may involve optimization as discussed in Chapter 6. It may lead to even more sequential movement through the use of canonical or ridge analysis. But, despite the form of the analysis, the purpose of the experimental design is one that should allow the user to fit the second-order model

$$y = \beta_0 + \sum_{i=1}^{k} \beta_i x_i + \sum_{i=1}^{k} \beta_{ii} x_i^2 + \sum_{i<j} \sum \beta_{ij} x_i x_j + \varepsilon \qquad (7.12)$$

The model of Equation (7.12) contains $1 + 2k + k(k-1)/2$ parameters. There must be at least this number of distinct design points and at least three levels of each design variable. Now, of course, these represent minimum conditions and one should keep in mind the 10 desirable design characteristics listed at the beginning of this chapter. In what follows in this chapter and in Chapter 8 we discuss important design properties for the second-order model and specific classes of second order designs. The reader will recall from the previous section that in the case of first-order designs (or first-order-with-interaction designs), the dominant property is orthogonality. In the case of second-order designs, orthogonality ceases to be such an important issue; and estimation of individual coefficients, while still important, becomes secondary to the scaled prediction variance $N \operatorname{Var} \hat{y}(\mathbf{x})/\sigma^2$. This stems from the fact that there is often less concern with what variables belong in the model, but more emphasis placed on the quality of $\hat{y}(\mathbf{x})$ as a prediction or, rather, an estimator for $E[y(\mathbf{x})]$. We now introduce, formally, the class of central composite designs.

7.4.1 The Class of Central Composite Designs

The class of **central composite designs** (CCDs) were introduced in an informal way in Chapter 6. Some of the examples used for illustrative purposes involved the use of the CCD. The CCD is without a doubt the most popular class of second-order designs. It was introduced by Box and Wilson (1951). Much of the motivation of the CCD evolves from its use in *sequential experimentation*. It involves the use of a *two-level factorial or fraction* (resolution V) combined with the following $2k$ *axial* or *star* points:

$$
\begin{array}{cccc}
x_1 & x_2 & \cdots & x_k \\
-\alpha & 0 & \cdots & 0 \\
\alpha & 0 & \cdots & 0 \\
0 & -\alpha & \cdots & 0 \\
0 & \alpha & \cdots & 0 \\
\vdots & \vdots & \cdots & \vdots \\
0 & 0 & \cdots & -\alpha \\
0 & 0 & \cdots & \alpha
\end{array}
$$

As a result, the design involves, say, F factorial points, $2k$ axial points, and n_c center runs. The sequential nature of the design becomes very obvious. The factorial points represent a variance optimal design for a first-order model or a first-order + two-factor interaction type model. Center runs clearly provide information about the existence of curvature in the system. If curvature is found in the system, the addition of axial points allow for efficient estimation of the pure quadratic terms.

While the genesis of this design is derived from sequential experimentation, the CCD is a very efficient design in situations that call for a nonsequential batch response surface experiment. In effect, the three components of the design play important and somewhat different roles.

1. The resolution V fraction contributes in a major way in estimation of linear terms and two-factor interactions. It is variance-optimal for these terms. The factorial points are the only points that contribute to the estimation of the interaction terms.

2. The axial points contribute in a large way to estimation of quadratic terms. Without the axial points, only the sum of the quadratic terms, $\sum_{i=1}^{k} \beta_{ii}$ can be estimated. The axial points do not contribute to the estimation of interaction terms.

3. The center runs provide an internal estimate of error (pure error) and contribute toward the estimation of quadratic terms.

The areas of flexibility in the use of the central composite design resides in the selection of α, the axial distance, and n_c, the number of center runs. The choice of these two parameters can be very important. The choice of α depends to a great extent on the region of operability and region of interest. The choice of n_c often has an impact on the distribution of N Var $\hat{y}(\mathbf{x})/\sigma^2$ in the region of interest. More will be said about the choice of α and n_c in subsequent sections. Figures 7.4 and 7.5 show the CCD for $k = 2$ and $k = 3$. For the $k = 2$ case the value of α, the axial distance is $\sqrt{2}$. For $k = 3$, the value of α is $\sqrt{3}$. Note that for $k = 3$ the axial points come through the six faces to a distance $\sqrt{3}$ from the origin. Note that for $k = 2$ the design

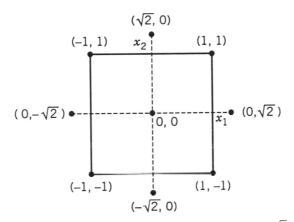

Figure 7.4 Central composite design for $k = 2$ and $\alpha = \sqrt{2}$.

represents eight points equally spaced on a circle, plus the center runs. For $k = 3$ the design represents 14 points all on a common sphere, plus center runs.

The values of the axial distance generally varies from 1.0 to $\sqrt{k}$, the former placing all axial points on the face of the cube or hypercube, the latter resulting in all points being placed on a common sphere. There are times when two or more center runs are needed and times when only one or two will suffice. We will allocate considerable space in Chapter 8 to comparison of the CCD with other types of designs. We will discuss the choice of α and n_c following a numerical example of the use of a CCD.

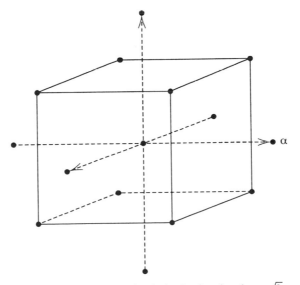

Figure 7.5 Central composite design for $k = 3$ and $\alpha = \sqrt{3}$.

Example 7.4   An experiment was conducted to study the response surface relating the strength of breadwrapper stock in grams per square inch to sealing temperature (x_1), cooling bar temperature, (x_2) and percent polyethylene additive (x_3). The definition of the design levels are [see Myers (1976)]:

$$x_1 = \frac{°F - 255}{30}$$

$$x_2 = \frac{°F - 55}{9}$$

$$x_3 = \frac{\% \text{ polyethylene} - 1.1}{0.6}$$

Five levels of each factor are involved in this design. The coded and natural levels are given by the following:

	− 1.682	− 1.000	0.000	+ 1.000	1.682
x_1	204.5	225	255	285	305.5
x_2	39.9	46	55	64	70.1
x_3	0.09	0.5	1.1	1.7	2.11

The design matrix and the **y** vector of responses are

$$
\mathbf{D} =
\begin{array}{ccc}
x_1 & x_2 & x_3 \\
\end{array}
\begin{bmatrix}
-1 & -1 & -1 \\
1 & -1 & -1 \\
-1 & 1 & -1 \\
1 & 1 & -1 \\
-1 & -1 & 1 \\
1 & -1 & 1 \\
-1 & 1 & 1 \\
1 & 1 & 1 \\
-1.682 & 0 & 0 \\
1.682 & 0 & 0 \\
0 & -1.682 & 0 \\
0 & 1.682 & 0 \\
0 & 0 & -1.682 \\
0 & 0 & 1.682 \\
0 & 0 & 0 \\
0 & 0 & 0 \\
0 & 0 & 0 \\
0 & 0 & 0 \\
0 & 0 & 0 \\
0 & 0 & 0 \\
\end{bmatrix},
\quad \mathbf{y} =
\begin{bmatrix}
6.6 \\
6.9 \\
7.9 \\
6.1 \\
9.2 \\
6.8 \\
10.4 \\
7.3 \\
9.8 \\
5.0 \\
6.9 \\
6.3 \\
4.0 \\
8.6 \\
10.1 \\
9.9 \\
12.2 \\
9.7 \\
9.7 \\
9.6 \\
\end{bmatrix}
$$

Note that the $k = 3$ CCD contains $\alpha = 1.682$ and $n_c = 6$ center runs. Before we present the RSM analysis of the data, it may be instructive to observe the $\mathbf{X}$ and $\mathbf{X'X}$ matrix and thus gain more insight into the properties of the

$$
\mathbf{X} = \begin{bmatrix}
 & x_1 & x_2 & x_3 & x_1^2 & x_2^2 & x_3^2 & x_1x_2 & x_1x_3 & x_2x_3 \\
1 & & & & 1 & 1 & 1 & 1 & 1 & 1 \\
1 & & & & 1 & 1 & 1 & -1 & -1 & 1 \\
1 & & & & 1 & 1 & 1 & -1 & 1 & -1 \\
1 & & & & 1 & 1 & 1 & 1 & -1 & -1 \\
1 & & & & 1 & 1 & 1 & 1 & -1 & -1 \\
1 & & & & 1 & 1 & 1 & -1 & 1 & -1 \\
1 & & & & 1 & 1 & 1 & -1 & -1 & 1 \\
1 & & & & 1 & 1 & 1 & 1 & 1 & 1 \\
1 & & D & & 2.828 & 0 & 0 & 0 & 0 & 0 \\
1 & & & & 2.828 & 0 & 0 & 0 & 0 & 0 \\
1 & & & & 0 & 2.828 & 0 & 0 & 0 & 0 \\
1 & & & & 0 & 2.828 & 0 & 0 & 0 & 0 \\
1 & & & & 0 & 0 & 2.828 & 0 & 0 & 0 \\
1 & & & & 0 & 0 & 2.828 & 0 & 0 & 0 \\
1 & & & & 0 & 0 & 0 & 0 & 0 & 0 \\
1 & & & & 0 & 0 & 0 & 0 & 0 & 0 \\
1 & & & & 0 & 0 & 0 & 0 & 0 & 0 \\
1 & & & & 0 & 0 & 0 & 0 & 0 & 0 \\
1 & & & & 0 & 0 & 0 & 0 & 0 & 0 \\
1 & & & & 0 & 0 & 0 & 0 & 0 & 0
\end{bmatrix}
$$

$$(7.12)$$

and

$$
\mathbf{X'X} = \begin{bmatrix}
b_0 & b_1 & b_2 & b_3 & b_{11} & b_{22} & b_{33} & b_{12} & b_{13} & b_{23} \\
10 & 0 & 0 & 0 & 13.658 & 13.658 & 13.658 & 0 & 0 & 0 \\
 & 13.658 & 0 & 0 & 0 & 0 & 0 & 0 & 0 & 0 \\
 & & 13.658 & 0 & 0 & 0 & 0 & 0 & 0 & 0 \\
 & & & 13.658 & 0 & 0 & 0 & 0 & 0 & 0 \\
 & & & & 24 & 8 & 8 & 0 & 0 & 0 \\
 & & & & & 24 & 8 & 0 & 0 & 0 \\
 & & & & & & 24 & 0 & 0 & 0 \\
 & & & & & & & 8 & 0 & 0 \\
 & & & & & & & & 8 & 0 \\
 & \text{symmetric} & & & & & & & & 8
\end{bmatrix}
$$

$$(7.13)$$

Note the many zeros in the $\mathbf{X'X}$ matrix. The linear columns are orthogonal to each other and all other columns. The same is true with the interaction columns. The quadratic columns, of course, are not orthogonal and are not

Table 7.2 Analysis of Variance for Breadwrapper Stock Data

Source	Sum of Squares	Degrees of Freedom	Mean Square	F
Regression (linear and quadratic)	70.3056	9	7.8117	7.87
Lack of fit	6.9044	5	1.3809	1.39
Error	4.9600	5	0.9920	
Total	82.1700	19		

orthogonal to the initial columns of ones (though orthogonality between the x_i^2 and the column of ones can be displayed if the x_i^2 are centered).

The least squares estimation gives the second-order response function

$$\hat{y} = 10.165 - 1.1036x_1 + 0.0872x_2 + 1.020x_3 - 0.760x_1^2$$
$$- 1.042x_2^2 - 1.148x_3^2 - 0.350x_1x_2 - 0.500x_1x_3$$
$$+ 0.150x_2x_3$$

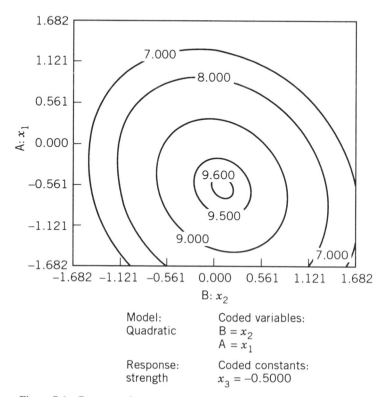

DESIGN-EXPERT Analysis

Model: Quadratic

Coded variables:
$B = x_2$
$A = x_1$

Response: strength

Coded constants:
$x_3 = -0.5000$

Figure 7.6 Contours of constant strength at percent polyethylene $= -0.50$.

An analysis of variance is shown in Table 7.2. Note that there are five degrees of freedom for lack of fit, representing contribution from cubic-type terms. The F-test for lack of fit is not significant.

The stationary point is computed based on the methodology discussed in Chapter 6:

$$\mathbf{x}_s = -\tfrac{1}{2}\hat{\mathbf{B}}^{-1}\mathbf{b}$$

$$= \begin{bmatrix} -1.011 \\ 0.260 \\ 0.681 \end{bmatrix}$$

with $\hat{y}(\mathbf{x}_s) = 11.08$.

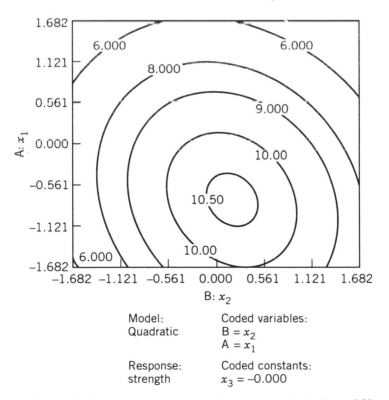

DESIGN-EXPERT Analysis

Model: Coded variables:
Quadratic $B = x_2$
                 $A = x_1$

Response: Coded constants:
strength $x_3 = -0.000$

Figure 7.7 Contours of constant strength at percent polyethylene $= 0.00$.

The eigenvalues of the $\hat{\mathbf{B}}$ matrix are found to be

$$\lambda_1 = -0.562; \quad \lambda_2 = -1.271; \quad \lambda_3 = -1.117$$

Thus the canonical form is given by

$$\hat{y} = 11.08 - 0.562w_1^2 - 1.271w_2^2 - 1.117w_3^2$$

As a result, the stationary point is a point of estimated maximum mean strength of the breadwrapper. Figures 7.6, 7.7, 7.8, and 7.9 reveal interesting contour graphs that are designed to show the flexibility available around the estimated optimum. Contours of constant response are shown for x_3 (percent

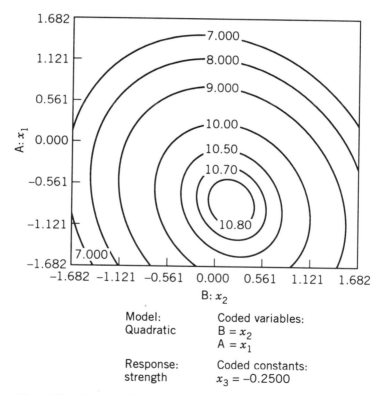

Figure 7.8 Contours of constant strength at percent polyethylene = 0.25.

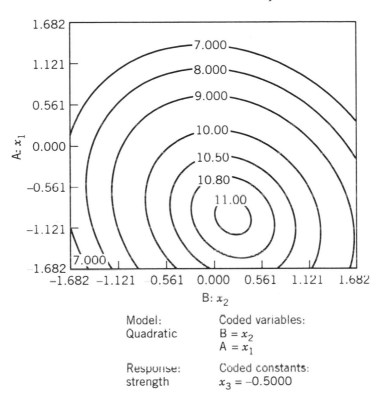

DESIGN-EXPERT Analysis

Figure 7.9 Contours of constant strength at percent polyethylene = 0.50.

polyethylene) values of -0.5, 0.00, 0.25, and 0.5. The purpose is to determine how much strength is lost by moving percent polyethylene off the optimum value of 0.6808 (coded). From the contour plots it becomes apparent that even if the polyethylene content is reduced to a value as low as 0.25 (coded), the estimated maximum strength is reduced to only slightly less than 10.9 psi. A reduction to 0.500 in polyethylene will still produce a maximum that exceeds 11.0 psi.

The purpose of this section is to formally introduce the user to the CCD and to allow more insight into its properties. However, the CCD is such an important part of the heritage and practical use of RSM that it will be revisited frequently in the text. In the next section we reintroduce the notion of prediction variance—that is, $N \operatorname{Var} \hat{y}(\mathbf{x}) / \sigma^2$ in the case of second-order models. Simultaneously, we discuss the notion of the property of rotatability. This will allow the discussion of the choice of α and n_c to be brought to focus.

7.4.2 Property of Rotatability

It is important for a second-order design to possess a reasonably stable distribution of $N \operatorname{Var} \hat{y}(\mathbf{x}) / \sigma^2$ throughout the experimental design region. It must be clearly understood that the experimenter does not know at the outset where in the design space he or she might wish to predict, or where in the design space the optimum may lie. Thus, a reasonably stable $N \operatorname{Var} \hat{y}(\mathbf{x}) / \sigma^2$ provides insurance that the quality of the $\hat{y}(\mathbf{x})$ as a prediction of future response values is roughly the same throughout the region of interest. To this end, Box and Hunter (1957) developed the notion of **design rotatability**.

A *rotatable* design is one for which $N \operatorname{Var} \hat{y}(\mathbf{x}) / \sigma^2$ has the same value at any two locations that are the same distance from the design center. In other words, $N \operatorname{Var} \hat{y}(\mathbf{x}) / \sigma^2$ is constant on spheres.

The purpose of the Box–Hunter idea of design rotatability was to, in part, impose a type of stability of $N \operatorname{Var} \hat{y}(\mathbf{x}) / \sigma^2$. The rationale of rotatability is that at two locations in the design space $\mathbf{x}_1$ and $\mathbf{x}_2$ for which the distances from the origin are the same [i.e., $(\mathbf{x}_1' \mathbf{x}_1)^{1/2} = (\mathbf{x}_2' \mathbf{x}_2)^{1/2}$], the predicted values $\hat{y}(\mathbf{x}_1)$ and $\hat{y}(\mathbf{x}_2)$ should be equally good–that is, have equal variance. While rotatability itself does not ensure stability or even near-stability throughout the design region, in many cases it provides some useful guidelines for the choice of design parameters—for example, the choice of α and n_c in the CCD. The importance of rotatability as a design property depends on many things, not the least of which is the nature of the region of interest and region of operability. It is important to note that rotatability or near-rotatability is often very easy to achieve without the sacrifice of other important design properties. In what follows, we present the foundation that will allow the determination of necessary and sufficient conditions for design rotatability.

Concept of Design Moments

Many properties of experimental designs are quantified through the choice of values of **design moments**. The concept is very much like moments of distributions that we deal with in probability theory. Given the design matrix

$$
\mathbf{D} = \begin{bmatrix}
x_{11} & x_{21} & \cdots & x_{k1} \\
x_{12} & x_{22} & \cdots & x_{k2} \\
\vdots & \vdots & \vdots & \vdots \\
x_{1n} & x_{2n} & \cdots & x_{kn}
\end{bmatrix}
$$

we say that relevant design moments are

$$[i] = \frac{1}{N} \sum_{u=1}^{N} x_{iu}$$

$$[ii] = \frac{1}{N} \sum_{u=1}^{N} x_{iu}^2$$

$$[ij] = \frac{1}{N} \sum_{u=1}^{N} x_{iu} x_{ju}$$

$$[iii] = \frac{1}{N} \sum_{u=1}^{N} x_{iu}^3$$

$$[iijj] = \frac{1}{N} \sum_{u=1}^{N} x_{iu}^2 x_{ju}^2$$

$$[iiii] = \frac{1}{N} \sum_{u=1}^{N} x_{iu}^4$$

etc.

These moments characterize the fashion in which the design points are *distributed* in k-dimensional design space. In a sense, the design moments help in characterizing the geometry of the design, which is very important. For example, the design moments come from $N^{-1}(\mathbf{X'X})$, the so-called moment matrix of the design; and from Equations (7.12) and (7.13) it is apparent that for the CCD, all *odd moments through order four* are zero. That is $[i]$, $[ij]$, $[iij]$, $[iii]$, $[ijk]$, $[ijjj]$, and $[iijk]$ are zero (all $i \neq j \neq k$). By an odd moment, we mean any moments for which at least one design variable possesses a power that is odd. Zero odd moments suggest symmetry in the design which is quite apparent when one observes the design matrix in Figures 7.2 and 7.3. In fact, the only nonzero moments for the $k = 3$ CCD are $[ii]$, $[iijj]$, and $[iiii]$ for all $i \neq j$.

The idea of design moments can nicely simplify the important property of orthogonality for first order designs. We know that for first order designs, a design is said to be orthogonal if $\mathbf{X'X}$ is a diagonal matrix. That is,

$$N^{-1}(\mathbf{X'X}) = \begin{bmatrix} 1 & 0 & \cdots & \cdots & 0 \\ 0 & [11] & 0 & \cdots & 0 \\ 0 & 0 & [22] & \cdots & 0 \\ \vdots & \vdots & \vdots & \ddots & \vdots \\ 0 & 0 & & & [kk] \end{bmatrix}$$

The moment matrix contains moments through order 2, and the design is orthogonal when $[ij] = 0$ and $[i] = 0$ for all $i \neq j$. Now, if the first-order design is also variance-optimal (i.e., variances of coefficients all minimized), all points are set to ± 1 extremes, and thus

$$N^{-1}(\mathbf{X'X}) = \bar{\mathbf{I}}_{k+1}$$

The unity values as the second *pure* moments are fixed from the ± 1 scaling. If scaling is different, then the $[ii]$ would be different.

The important variance properties of an experimental design are determined by the nature of the moment matrix. This should be evident because the matrix $(\mathbf{X'X})^{-1}$ and hence $(\mathbf{X'X})$ are so important in characterizing variance properties—that is, variances and covariances of regression coefficients as well as prediction variance.

Moment Matrix for a Rotatable Design (First and Second Order)

In this section we give the moment matrix for a rotatable design for both a second-order model and a first-order model. Additional theoretical development regarding the general case is given in Appendix 5. For the case of a first-order model, a design is rotatable if and only if *odd moments through order two are zero and the pure second moments are all equal.* In other words,

$$[i] = 0 \qquad (i = 1, 2, \ldots, k)$$

$$[ij] = 0 \qquad (i \neq j, i, j = 1, 2, \ldots, k)$$

$$[ii] = \lambda_2 \qquad (i = 1, 2, \ldots, k)$$

The quantity λ_2 is determined by the scaling of the design. As a result in the first-order case, the moment conditions for a rotatable design are equivalent to the moment conditions for a variance optimal design. Indeed, with scaling that allows levels at ± 1 in a two-level design of resolution $\geq$ III, we have

$$[i] = 0, \qquad [ij] = 0, \qquad [ii] = 1.0 \qquad (i = 1, 2, \ldots, k)$$

$$(i \neq j)$$

In other words, λ_2 is set at 1.0 due to the standard ± 1 scaling. The equivalence of rotatability to variance-optimality in the first-order case should not be surprising. The reader should recall the result of Equation (7.11). We showed quite simply that

$$\frac{N \operatorname{Var} \hat{y}(\mathbf{x})}{\sigma^2} = 1 + \rho_{\mathbf{x}}^2$$

for a first-order variance optimal design. Thus, $N \operatorname{Var} \hat{y}(\mathbf{x})/\sigma^2$ is a function of x only through ρ_x, the distance that x is away from the design origin. This implies that $N \, \text{Var} \, \hat{y}(x) / \sigma^2$ is the same at any two locations that are the same distance from the origin.

In the case of a second-order model, the moments that impact rotatability (or any variance property) are moments through order 4. Necessary and sufficient conditions for rotatability are as follows:

1. All odd moments through order four are zero.
2. $[iiii]/[iijj] = 3 \quad (i \neq j)$. $\qquad\qquad\qquad\qquad\qquad\qquad$ (7.14)

The conditions are not only simple but they are relatively easy to achieve, particularly with a CCD.

7.4.3 Rotatability and the CCD

The conditions given in Equation (7.14) are achieved or at least approximately achieved by several classes of designs. In the case of the CCD, rotatability is achieved by making a proper choice of α, the axial distance. Condition (1) above will hold as long as the factorial portion is a full 2^k or a fraction with resolution V or higher. The balance between $+1$ and -1 in the factorial columns and the orthogonality among certain columns in the $\mathbf{X}$ matrix for the CCD will result in all odd moments being zero. For condition (2), one merely seeks α for which

$$\frac{[iiii]}{[iijj]} = \frac{F + 2\alpha^4}{F} = 3$$

which results in

$$\alpha = \sqrt[4]{F} \qquad\qquad\qquad\qquad\qquad\qquad (7.15)$$

where, of course, F is the number of factorial points ($F = 2^k$ if it is a full factorial). It is important to note that rotatability is achieved by merely using α as in Equation (7.15) *despite the number of center runs*. Table 7.3 gives value of α for a rotatable design for various values of the number of design variables.

Note that for $k = 2$ and $k = 4$ the rotatable CCD contains 8 and 24 points (apart from center runs), respectively, that are equidistant from the design center. For $k = 3$, the $\alpha = 1.682$ corresponds to the CCD used in the breadwrapper example in Section 7.4.1. For $k = 2$, 3, and 4 the rotatable CCD is either exactly or very nearly a spherical design; that is, all points (apart from center runs) are exactly (or approximately for $k = 3$) a distance $\sqrt{k}$ from the design center.

Center Runs for the Rotatable CCD

The property of rotatability is an attempt at producing a certain sense of stability of $N \, \text{Var} \, \hat{y} / \sigma^2$. The "sense" is, of course, constant $N \, \text{Var} \, \hat{y}(x) / \sigma^2$ on spheres. However, the presence of a rotatable design does not imply

Table 7.3   Values of α for a Rotatable Central Composite Design

k	F	N	α
2	4	$8 + n_c$	1.414
3	8	$14 + n_c$	1.682
4	16	$2 + n_c$	2.000
5	32	$42 + n_c$	2.378
$5(\frac{1}{2}$ rep)	16	$26 + n_c$	2.000
6	64	$76 + n_c$	2.828
$6(\frac{1}{2}$ rep)	32	$44 + n_c$	2.378
7	128	$142 + n_c$	3.364
$7(\frac{1}{2}$ rep)	64	$78 + n_c$	2.828

stability throughout the design region. In fact it turns out that a spherical design (all points on a common radius) used for fitting a second-order model has an infinite N Var $\hat{y}(\mathbf{x})/\sigma^2$ since the design is *singular*, that is, $(\mathbf{X'X})$ is a singular matrix [see Box and Hunter (1957), Box and Draper (1985) and Myers (1976)]. The use of center runs does provide reasonable stability of N Var $\hat{y}(\mathbf{x})/\sigma^2$ in the design region; as a result, some center runs for a rotatable CCD are very beneficial. The use of a rotatable or near-rotatable CCD with only a small number of center runs is not a good practice. An illustration of this point is given through Figures 7.10 and 7.11. Figure 7.10 shows contours of N Var $\hat{y}(\mathbf{x})/\sigma^2$ for a $k = 2$ CCD ($\alpha = \sqrt{2}$) and one center run. (For zero center runs the design is singular.) Figure 7.11 gives the

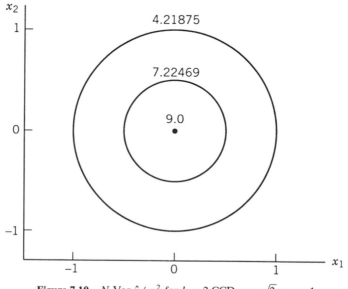

Figure 7.10 N Var $\hat{y}/\sigma^2$ for $k = 2$ CCD, $\alpha = \sqrt{2}$, $n_c = 1$.

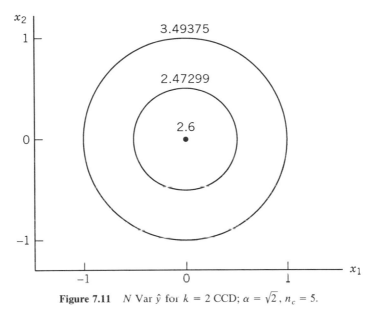

Figure 7.11 N Var $\hat{y}$ for $k = 2$ CCD; $\alpha = \sqrt{2}$, $n_c = 5$.

contours with $\alpha = \sqrt{2}$ and five center runs. Note that the design in Figure 7.11 is preferable. Please also note that the criterion involves weighting by N, which means that the design in Figure 7.11 has a larger weight. In spite of this, the $n_c = 1$ design has a scaled prediction variance at the design center that is 3.5 times as large as that of the $n_c = 5$ design. Though guidelines regarding the use of center runs with the CCD will be given in the next section, suffice it to say at this point that spherical or nearly spherical designs require 3–5 center runs in order to avoid a severe inbalance in N Var $\hat{y}(\mathbf{x})/\sigma^2$ through the design region. The message communicated by Figures 7.10 and 7.11 extends to larger values of k. Draper (1982), Giovanitti-Jensen and Myers (1989), and Myers et al. (1992b) supply information regarding center runs in the use of the CCD.

How Important Is Rotatability?
The rotatable CCD plays an important role in RSM, from both a historical and operational point of view. However, it is important for the analyst to understand that it is not necessary to have exact rotatability in a second-order design. In fact, *if the desired region of the design is spherical, the CCD that is most effective from a variance point of view is to use $\alpha = \sqrt{k}$ and 3–5 center runs.* This design is not necessarily rotatable but is near-rotatable. The recommendation is based on stability and size of N Var $\hat{y}(\mathbf{x})/\sigma^2$ in the spherical design region. For example, for $k = 3$, $\alpha = \sqrt{3}$ does not produce a rotatable design. However, the loss in rotatability is actually trivial and the larger value of α, compared to the rotatable value of $\alpha = 1.682$, results in a

design that is slightly preferable. Numerical evidence regarding this recommendation will be given later in this chapter and in Chapter 8. The reader is also referred to Box and Draper (1987), Khuri and Cornell (1987), Lucas (1976), Giovannitti-Jensen and Myers (1989), and Myers et al. (1992b) for information regarding the practical use of the CCD.

7.4.4 The Cuboidal Region and the Face Center Cube

There are many practical situations in which the scientist or engineer specifies ranges on the design variables, and these ranges are strict. That is, the region of interest and the region of operability are the same, and the obvious region for the design is a cube. For example, in an experimental study designed to study organism growth, the design variables and their ranges are percent glucose [2%, 4%], percent yeast [0.4, 0.6], and time, hr [30, 60]. Suppose it is of interest to build a second-order response surface model and the biologist is interested in predicting growth of the organism inside and on the perimeter of the cuboidal region produced by the cube. In addition, for biological reasons, one cannot experiment outside the cube, though experimentation at the extremes in the region is permissible and, in fact, desirable. This scenario, which occurs frequently in many scientific areas, suggests a central composite design in which the eight corners of the cube are centered and scaled to $(\pm 1, \pm 1, \pm 1)$ and α cannot exceed 1.0. The final design is given (in coded form) by

$$
\mathbf{D} = \begin{bmatrix}
x_1 & x_2 & x_3 \\
-1 & -1 & -1 \\
1 & -1 & -1 \\
-1 & 1 & -1 \\
-1 & -1 & 1 \\
1 & 1 & -1 \\
1 & -1 & 1 \\
-1 & 1 & 1 \\
1 & 1 & 1 \\
-1 & 0 & 0 \\
1 & 0 & 0 \\
0 & -1 & 0 \\
0 & 1 & 0 \\
0 & 0 & -1 \\
0 & 0 & 1 \\
\mathbf{0} & \mathbf{0} & \mathbf{0}
\end{bmatrix}
$$

where the $(\mathbf{0}, \mathbf{0}, \mathbf{0})$ at the design center indicates a vector of center runs.

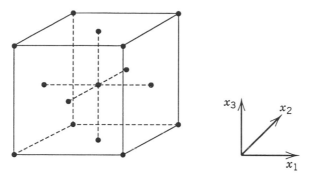

Figure 7.12 Face center cube (CCD with $\alpha = 1.0$) for $k = 3$.

Figure 7.12 shows the design, often called the *face center cube* because the axial points occur at the center of the faces, rather than outside the faces as in the case of a spherical region.

In the case of a cuboidal design region, the face center cube is an effective second-order design. When one encounters what is a natural cuboidal region, it is important that the points be "pushed to the extreme" of the experimental region. This results in the most attractive distribution of N Var $\hat{y}(\mathbf{x})/\sigma^2$. It is important for the region to be "covered" in a symmetric fashion. The face center cube accomplishes this. Of course, the design is not rotatable. However, rotatability or near-rotatability is not an important priority when the region of interest is clearly cuboidal. Rotatability (or near-rotatability) is a useful option that comes from spherical or near-spherical designs; these designs are certainly appropriate for spherical regions of interest or regions of operability, and they are less appropriate with cuboidal regions.

The face center cube is a useful design for any number of design variables. Again, a resolution V fraction is used for the factorial portion. The recommendation for center runs is quite different from that of the spherical designs. In the case of spherical designs, center runs are a necessity in order to achieve a reasonable distribution of N Var $\hat{y}(\mathbf{x})/\sigma^2$, with $n_c = 3\text{--}5$ giving good results. In the cuboidal case (i.e., with $\alpha = 1.0$), 1 or 2 center runs are sufficient to produce a reasonable stability of N Var $\hat{y}(\mathbf{x})/\sigma^2$. The sensitivity of N Var $\hat{y}(\mathbf{x})/\sigma^2$ to the number of center runs for the spherical design is seen in Figure 7.10 and 7.11. The *insensitivity* to center runs for the face center cube is illustrated in Figures 7.13, 7.14, and 7.15. Contours of values of N Var $\hat{y}(\mathbf{x})/\sigma^2$ are given for $n_c = 0$, $n_c = 1$, and $n_c = 2$ for the $k = 2$ face center cube. It is clear that many center runs are not needed to stabilize prediction variance. In fact, one center run is quite sufficient for stability though $n_c = 2$ is slightly preferable. It turns out that no further improvement is achieved beyond $n_c = 2$.

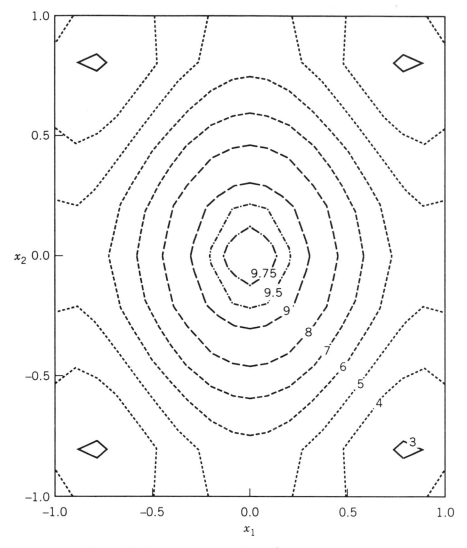

Figure 7.13   Contours of N Var $\hat{y}(\mathbf{x})/\sigma^2$; $\alpha = 1.0$, $n_c = 0$.

Though much has been said here about the impact of center runs on N Var $\hat{y}(\mathbf{x})/\sigma^2$ for both spherical and cuboidal designs, it should be noted that multiple center runs or replication of exterior points may, in many cases, be required in order to have substantial degrees of freedom for pure error.

7.4.5   When Is the Design Region Spherical?

As we have indicated throughout this section, the cuboidal CCD is appropriate when a cuboidal region of interest and cuboidal region of operability

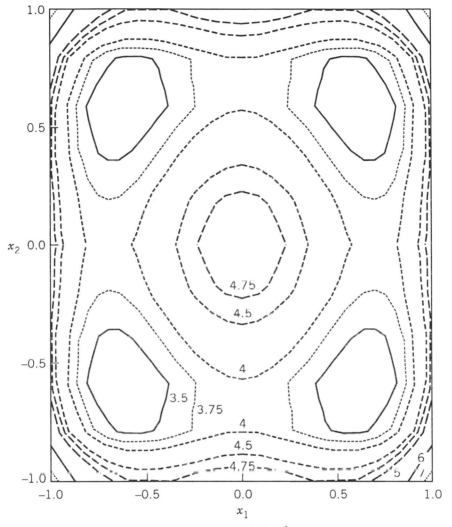

Figure 7.14 Contours of $N \, \mathrm{Var} \, \hat{y}(\mathbf{x})/\sigma^2$; $\alpha = 1.0$, $n_c = 1$.

are apparent to the practitioner. Clearly the problem may suggest ranges on the factors which may define the "corners" where the factorial points should reside. Then the question regarding design region should hinge on whether axial points outside the ranges are scientifically permissible and should be included in the region of interest. For example, consider Figure 7.16. The shaded area forms the cube, but after some deliberation it is determined that the nonshaded area is not only within the region of operability but also the researcher is interested in predicting response in the unshaded area as well as the shaded area. The region of interest is a sphere which is circumscribed around a cube. A spherical design (CCD with $\alpha = \sqrt{k}$) is certainly appropriate.

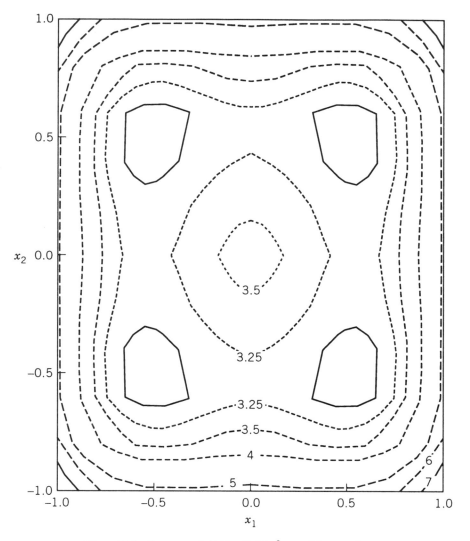

Figure 7.15   Contours of N Var $\hat{y}(\mathbf{x})/\sigma^2$; $\alpha = 1.0$, $n_c = 2$.

A second situation can occur very much like that in Figure 7.17. Ranges are chosen on design variable, but as the planning of the experiment evolves, it is determined that several (and perhaps all) of the vertices on the cube defined by the ranges are not scientifically permissible; that is, they are outside both regions of operability and the region of interest (e.g., in the case of a food product, high levels of flour, shortening, and baking time) is known to produce an unacceptable product. As a result, the corners, shaded in Figure 7.17, are "shaved off" and the design region is formed from the unshaded region. Here, the sphere is inscribed inside the cuboidal region formed from the selection of ranges.

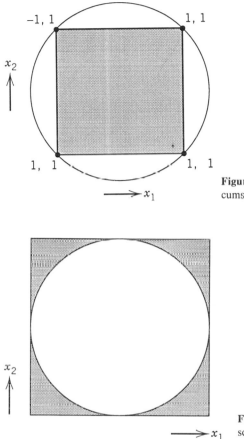

Figure 7.16 Spherical region (sphere circumscribed).

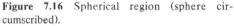

Figure 7.17 Spherical region (sphere inscribed).

It is important for the novice in experimental design to understand that in many situations the region of interest (or perhaps even the region of operability) is not clear cut. It is often difficult enough to get a commitment from a scientist or engineer on what are the interesting and permissible ranges on the factors. Mistakes are often made and adjustments adopted in future experiments. *Confusion regarding type of design should never be an excuse for not using designed experiments.* Using a spherical region when it is more naturally cuboidal, for example, will still provide important information that will, among other things, lead to more educated selection of regions for future experiments.

7.4.6 Summary Statements Regarding CCD

The CCD is an efficient design that is ideal for sequential experimentation and allows a reasonable amount of information for testing lack of fit while not involving an unusually large number of design points. The design accommodates a spherical region with five levels of each factor and a choice of $\alpha = \sqrt{k}$. The design can be a three-level design to accommodate a cuboidal region with the choice of $\alpha = 1.0$. In the spherical case, the design is either rotatable or very near rotatable, and 3–5 center runs should be used. In the cuboidal case, 1–2 center runs will suffice. Though the cuboidal and spherical designs are natural choices depending on the region of interest, the reader should not get the impression that the use of the CCD must be confined to those choices of α. There will be practical situations for which $\alpha = 1.0$ or $\alpha = \sqrt{k}$ cannot be used. The CCD should not be ruled out in these situations.

The natural competitor for the CCD is, of course, the three-level factorial, that is, the 3^k factorial. The 3^2 design is, in fact, the face center cube and is thus a CCD. But when k becomes moderately large, the 3^k factorial design involves an excessive number of design points. In the case of three factors and a natural cuboidal region, the 3^3 factorial is an efficient design if the researcher can afford to use the required 27 design points. However, for $k > 3$ the number of design points for the 3^k is usually considered impracticable for most applications. A comparison of the 3^3 with the face center cube and other designs via the consideration of prediction variance will be revealed in Chapter 8.

7.4.7   The Box–Behnken Design

Box and Behnken (1960) developed a family of efficient *three-level* designs for fitting second-order response surfaces. The methodology for design construction is interesting and quite creative. The class of designs is based on the construction of balanced incomplete block designs. For example, a balanced incomplete block design with three treatments and three blocks is given by

	Treatment		
	1	2	3
Block 1	X	X	
Block 2	X		X
Block 3		X	X

The pairing together of treatments 1 and 2 symbolically implies, in the response surface setting, that design variables x_1 and x_2 are paired together in a 2^2 factorial (scaling ± 1) while x_3 remains fixed at the center ($x_3 = 0$). The same applies for blocks 2 and 3, with a 2^2 factorial being represented by each pair of treatments while the third factor remains fixed at 0. As a result, the $k = 3$ Box–Behnken design is given by

$$
D = \begin{array}{ccc}
\begin{array}{ccc} x_1 & x_2 & x_3 \end{array} \\
\left[\begin{array}{ccc}
-1 & -1 & 0 \\
-1 & 1 & 0 \\
1 & -1 & 0 \\
1 & 1 & 0 \\
-1 & 0 & -1 \\
-1 & 0 & 1 \\
1 & 0 & -1 \\
1 & 0 & 1 \\
0 & -1 & -1 \\
0 & -1 & 1 \\
0 & 1 & -1 \\
0 & 1 & 1 \\
0 & 0 & 0
\end{array}\right]
\end{array}
$$

The last row in the design matrix implies a *vector* of center runs. In the case of $k = 4$ the same methodology applies. Each pair of factors are linked in a 2^2 factorial. The design matrix is given by

$$
D = \begin{array}{cccc}
\begin{array}{cccc} x_1 & x_2 & x_3 & x_4 \end{array} \\
\left[\begin{array}{cccc}
-1 & -1 & 0 & 0 \\
-1 & 1 & 0 & 0 \\
1 & -1 & 0 & 0 \\
1 & 1 & 0 & 0 \\
-1 & 0 & -1 & 0 \\
-1 & 0 & 1 & 0 \\
1 & 0 & -1 & 0 \\
1 & 0 & 1 & 0 \\
-1 & 0 & 0 & -1 \\
-1 & 0 & 0 & 1 \\
1 & 0 & 0 & -1 \\
1 & 0 & 0 & 1 \\
0 & -1 & -1 & 0 \\
0 & -1 & 1 & 0 \\
0 & 1 & -1 & 0 \\
0 & 1 & 1 & 0 \\
0 & -1 & 0 & -1 \\
0 & -1 & 0 & 1 \\
0 & 1 & 0 & -1 \\
0 & 1 & 0 & 1 \\
0 & 0 & -1 & -1 \\
0 & 0 & -1 & 1 \\
0 & 0 & 1 & -1 \\
0 & 0 & 1 & 1 \\
0 & 0 & 0 & 0
\end{array}\right]
\end{array} \qquad (7.16)
$$

Note that the Box–Behnken design (BBD) is quite comparable in number of design points to the CCD for $k = 3$ and $k = 4$. (There is no BBD for $k = 2$.) For $k = 3$, the CCD contains $14 + n_c$ runs while the BBD contains $12 + n_c$ runs. For $k = 4$ the CCD and BBD both contain $24 + n_c$ design points. For $k = 5$ we have the following $40 + n_c$ experimental runs:

$$
\mathbf{D} = \begin{bmatrix}
x_1 & x_2 & x_3 & x_4 & x_5 \\
\pm 1 & \pm 1 & 0 & 0 & 0 \\
\pm 1 & 0 & \pm 1 & 0 & 0 \\
\pm 1 & 0 & 0 & \pm 1 & 0 \\
\pm 1 & 0 & 0 & 0 & \pm 1 \\
0 & \pm 1 & \pm 1 & 0 & 0 \\
0 & \pm 1 & 0 & \pm 1 & 0 \\
0 & \pm 1 & 0 & 0 & \pm 1 \\
0 & 0 & \pm 1 & \pm 1 & 0 \\
0 & 0 & \pm 1 & 0 & \pm 1 \\
0 & 0 & 0 & \pm 1 & \pm 1 \\
0 & 0 & 0 & 0 & 0
\end{bmatrix}
\tag{7.17}
$$

The CCD for $k = 5$ involves $26 + n_c$ center runs when the $\frac{1}{2}$ fraction is used in the factorial portion. When the full factorial is used, the CCD makes use of $42 + n_c$ center runs. Each row in the BBD symbolizes four design points with $(\pm 1, \pm 1)$ representing a 2^2 factorial.

For $k = 6$ the construction of the design is based on partially balanced incomplete block designs. Thus, each treatment does not occur with every other treatment the same number of times. In the Box–Behnken construction, this means that each factor does not occur in a two-level factorial structure the same number of times with every factor. The design matrix, involving $N = 48 + n_c$ runs, is given by

$$
\mathbf{D} = \begin{bmatrix}
x_1 & x_2 & x_3 & x_4 & x_5 & x_6 \\
\pm 1 & \pm 1 & 0 & \pm 1 & 0 & 0 \\
0 & \pm 1 & \pm 1 & 0 & \pm 1 & 0 \\
0 & 0 & \pm 1 & \pm 1 & 0 & \pm 1 \\
\pm 1 & 0 & 0 & \pm 1 & \pm 1 & 0 \\
0 & \pm 1 & 0 & 0 & \pm 1 & \pm 1 \\
\pm 1 & 0 & \pm 1 & 0 & 0 & \pm 1 \\
0 & 0 & 0 & 0 & 0 & 0
\end{bmatrix}
\tag{7.18}
$$

The CCD involves the use of $44 + n_c$ runs when the $\frac{1}{2}$ fraction is used in the factorial portion.

Unlike the $k = 3$, 4, and 5 cases, the factorial structures in the BBD for $k = 6$ are 2^3 factorials involving three factors. Thus each "row" of the above matrix involves eight design points.

For $k = 7$, the factorial structures again involve combinations of three design factors. The $N = 56 + n_c$ design runs are given by the following design matrix:

$$
D = \begin{bmatrix}
x_1 & x_2 & x_3 & x_4 & x_5 & x_6 & x_7 \\
0 & 0 & 0 & \pm 1 & \pm 1 & \pm 1 & 0 \\
\pm 1 & 0 & 0 & 0 & 0 & \pm 1 & \pm 1 \\
0 & \pm 1 & 0 & 0 & \pm 1 & 0 & \pm 1 \\
\pm 1 & \pm 1 & 0 & \pm 1 & 0 & 0 & 0 \\
0 & 0 & \pm 1 & \pm 1 & 0 & 0 & \pm 1 \\
\pm 1 & 0 & \pm 1 & 0 & \pm 1 & 0 & 0 \\
0 & \pm 1 & \pm 1 & 0 & 0 & \pm 1 & 0 \\
0 & 0 & 0 & 0 & 0 & 0 & 0
\end{bmatrix}
\tag{7.19}
$$

Characteristics of the Box–Behnken Design

In many scientific studies that require RSM, researchers are inclined to require three evenly spaced levels. Thus, the Box Behnken design is an efficient option and indeed an important alternative to the central composite design. As we can observe from the sample sizes, there is sufficient information available for testing lack of fit. For example, for $k = 6$ the use of, say, $n_c = 5$ would allow four degrees of freedom for pure error and 21 degrees of freedom for lack of fit. It turns out that the Box–Behnken design does not substantially deviate from rotatability, and, in fact, for $k = 4$ and $k = 7$ the design is exactly rotatable. Verification of rotatability in both cases should be quite simple for the reader. From the design matrix in Equation (7.16), it is easy to see that all odd moments are zero. This is a result of the prominence of the sets of 2^2 factorial arrays. For the second condition required for rotatability, note from Equation (7.16) that $(iiii) = \frac{12}{24}$ and $(iijj) = \frac{4}{24}$, for $i \neq j$. As a result, $(iiii)/(iijj) = 3$. An investigation of equation (7.19) suggests that the rotatability conditions hold for the $k = 7$ BBD as well.

Another important characteristic of the BBD is that it is a *spherical design*. Note, for example, in the $k = 3$ case that all of the points are so-called "edge points" (i.e., points that are on the edges of the cube); in this case, all edge points are a distance $\sqrt{2}$ from the design center. There are no factorial points or face points. Figure 7.18 displays the BBD for $k = 3$. Contrast this design to that depicted in Figure 7.12 with the face center cube. The latter displays face points and factorial points that are not the same distance from the design center though they all reside on the cube and provide good "coverage" of the cube. The BBD involves all edge points, but the entire cube is not covered. In fact, there are no points on the corner of the cube or even a distance $\sqrt{3}$ from the design center. The analyst should not view the lack of coverage of the cube a reason not to use the BBD. It is not meant to be a cuboidal design. However, the use of the BBD should be confined to situations in which one is **not interested in predicting response at the extremes**, that is, at the corners of the cube. A quick inspection of the design

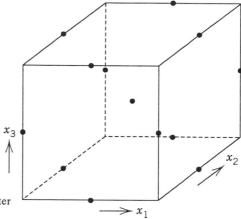

Figure 7.18 The $k = 3$ BBD with a center point.

matrices in Equations (7.16)–(7.19) further underscores the fact that the BBD is not a cuboidal design. For example, for $k = 7$, the design points are at a radius of $\sqrt{3}$, which is considerably smaller than the radius of $\sqrt{7}$ exhibited by the corner of the cube. If three levels are required and coverage of the cube is necessary, one should use a face center cube rather than the BBD.

The spherical nature of the BBD, combined with the fact that the designs are rotatable or near-rotatable, suggests that ample center runs should be

Table 7.4   Data for Example 7.5 Using Box–Behnken Design

Level	Temperature	Agitation	Rate	x_1	x_2	x_3
High	200	10.0	25	+1	+1	+1
Center	175	7.5	20	0	0	0
Low	150	5.0	15	−1	−1	−1

Standard Order	x_1	x_2	x_3	y_1
1	−1	−1	0	53
2	+1	−1	0	58
3	−1	+1	0	59
4	+1	+1	0	56
5	−1	0	−1	64
6	+1	0	−1	45
7	−1	0	+1	35
8	+1	0	+1	60
9	0	−1	−1	59
10	0	+1	−1	64
11	0	−1	+1	53
12	0	+1	+1	65
13	0	0	0	65
14	0	0	0	59
15	0	0	0	62

used. In fact, for $k = 4$ and 7, center runs are necessary to avoid singularity. The use of 3–5 center runs are recommended for the BBD.

Example 7.5   One step in the production of a particular polyamide resin is the addition of amines. It was felt that the manner of addition has a profound effect on the molecular weight distribution of the resin. Three variables are thought to play a major role: temperature at the time of addition (x_1, °C), agitation (x_2, RPM), and rate of addition (x_3, 1/ min). Because it was difficult to physically set the levels of addition and agitation, three levels were chosen and a BBD was used. The viscosity of the resin was recorded as an indirect measure of molecular weight. The data, including natural levels, design, and response values, are shown in Table 7.4. Table 7.5 shows an analysis using the package Design-Expert. Three contour plots of constant viscosity (shown in Figure 7.19) are given in order to illustrate what combination of the factors produce high- and low-molecular-

Table 7.5   Analysis of Box–Behnken Example Using Design-Expert

Factor	Factor	Units	−1 Level	+1 Level
A	x_1	°C	−1.000	1.000
B	x_2	RPM	−1.000	1.000
C	x_3	min^{-1}	−1.000	1.000

ANOVA for Quadratic Model

Source	Sum of Squares	Degrees of Freedom	Mean Squares	F-Value	Probability > F
Model	882.5	9	98.05	9.566	0.0114
Residual	51.3	5	10.25		
Lack of fit	(33.3)	3	11.08	1.231	0.4774
Pure error	(18.0)	2	9.00		
Corrected total	933.8	14			
Root mean squared error	3.20		R-squared	0.9451	
Dependent mean	57.13		Adjusted R-squared	0.8463	
Coefficient of variation	5.60%				

Final Equation in Terms of Coded Factors

$$
\begin{aligned}
Resin = \ & 62.00 \\
& + 1.0000 * x_1 \\
& + 2.6250 * x_2 \\
& - 2.3750 * x_3 \\
& - 7.3750 * x_1^2 \\
& + 1.8750 * x_2^2 \\
& - 3.6250 * x_3^2 \\
& - 2.0000 * x_1{}^* x_2 \\
& + 11.000 * x_1{}^* x_3 \\
& + 1.7500 * x_2{}^* x_3
\end{aligned}
$$

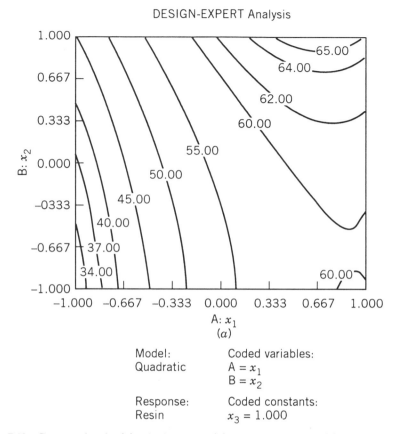

Figure 7.19   Contour plots for (a) agitation = 1.0, (b) agitation = 0.0, and (c) agitation = −1.0.

weight resins. Agitation was fixed at the low, medium, and high levels. It is clear that high-molecular-weight resins are produced with low values of agitation, whereas low-molecular-weight resins are potentially available when one uses high values of agitation. In fact, the extremes of low rates, low temperature, and high agitation clearly produces low-molecular-weight resins, whereas low agitation, low temperature, and high rate results in high-molecular-weight resins.

7.4.8   Other Spherical RSM Designs; Equiradial Designs

There are some special and interesting *two-factor designs* that serve as alternatives to the central composite design. They are the class of equiradial designs, beginning with a pentagon, five equally spaced points on a circle; the pentagon, hexagon, heptagon, and so on, do require center runs because they are designs on a common sphere and, as we shall show, are rotatable.

DESIGN-EXPERT Analysis

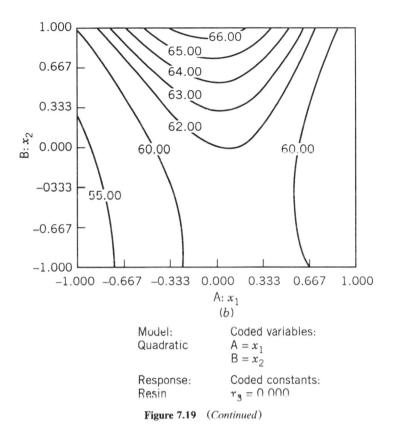

Figure 7.19 (*Continued*)

The design matrix for the equiradial design can be written

$$\left\{\rho\cos[\theta + 2\pi u/n_1], \rho\sin[\theta + 2\pi u/n_1]\right\}, \qquad u = 0, 1, 2, \ldots, n_1 - 1$$

$$(7.20)$$

where ρ is the design radius. Here, n_1 is the number of points on the sphere, say, 5, 6, 7, 8, and so on. In addition to the n_1 points indicated by (7.20), we will assume n_c center runs. The value that one chooses for ρ merely determines the nature of the scaling. It turns out that the choice of θ has no impact on $\mathbf{X'X}$; that is, it has no impact on the design as far as the variance structure is concerned. As a result, all of the equiradial second-order designs (i.e., using $n_1 = 5, 6, 7, 8$, etc.) are such that the $\mathbf{X'X}$ matrix is *invariant to design rotation*.

DESIGN-EXPERT Analysis

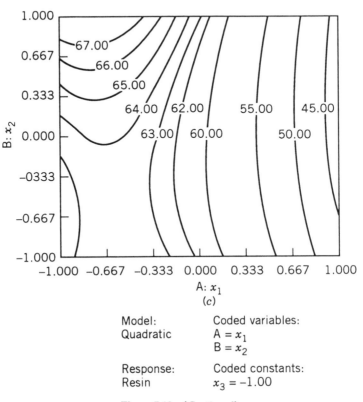

(c)

Model: Coded variables:
Quadratic   A = x_1
                    B = x_2

Response: Coded constants:
Resin $x_3 = -1.00$

Figure 7.19   (*Continued*)

A very useful and interesting special case is the **hexagon**—that is, six equally spaced points on a circle. Thus $n_1 = 6$; setting $\theta = 0$ and $\rho = 1$, we have the following design matrix using $n_c = 3$:

$$\mathbf{D} = \begin{bmatrix} x_1 & x_2 \\ 1 & 0 \\ 0.5 & \sqrt{0.75} \\ -0.5 & \sqrt{0.75} \\ -1 & 0 \\ -0.5 & -\sqrt{0.75} \\ 0.5 & -\sqrt{0.75} \\ 0 & 0 \\ 0 & 0 \\ 0 & 0 \end{bmatrix} \qquad (7.21)$$

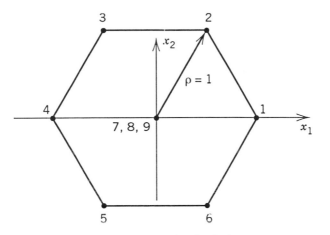

Figure 7.20 Design points for the hexagon.

Note that x_1 is at five levels and x_2 is at three levels for this particular rotation. Figure 7.20 shows the design matrix. Note the six evenly spaced points.

It is of interest to consider the important design moments for the hexagon —that is, the moments through order four. From Equation (7.21) we have, for the hexagon,

$$[i] = [ij] = [iij] = [iii] = [iiij] = 0, \quad i \neq j = 1, 2 \text{ (odd moments zero)}$$

$$[iiii] = \frac{9/4}{N}; [iijj] = \frac{3/4}{N}, \quad i \neq j = 1, 2$$

Because all design moments through order four are zero and $[iiii]/[iijj] = 3$, the hexagon is indeed rotatable. The reader should also note that these conditions are independent of ρ and θ.

It turns out that all equiradial designs for $n_1 > 5$ are rotatable. Appendix 6 shows a development of the moment conditions. The relevant moments for the equiradial designs are as follows:

1. All odd moments through order four are zero.

2.

$$[ii] = \frac{\rho^2 n_1}{2N}, \quad i = 1, 2.$$

$$[iiii] = \frac{3\rho^4 n_1}{8N}, \quad i = 1, 2.$$

$$[iijj] = \frac{\rho^4 n_1}{8N}, \quad i \neq j = 1, 2.$$

Of course, these conditions imply that all equiradial designs for $n_1 \geq 5$ are rotatable.

Special Case of The Equiradial Design—The CCD for $k = 2$

For $n_1 = 8$, the equiradial design is the **octagon**. As one might expect, the use of Equation (7.20) for $n_1 = 8$ produces (using $\rho = \sqrt{2}$)

$$
\begin{array}{cc}
x_1 & x_2 \\
\left[\begin{array}{cc}
-1 & -1 \\
-1 & 1 \\
1 & -1 \\
1 & 1 \\
\sqrt{2} & 0 \\
-\sqrt{2} & 0 \\
0 & -\sqrt{2} \\
0 & \sqrt{2} \\
\mathbf{0} & \mathbf{0}
\end{array}\right]
\end{array}
$$

which, of course, is the $k = 2$ rotatable CCD.

Summary Comments on the Equiradial Designs

The equiradial designs are an interesting family of designs that enjoy some usage in the case where only two design variables are involved. One should use the CCD (the octagon) whenever possible. However, there are situations when cost constraints do not allow the use of the octagon. The pentagon (plus center runs) is a saturated design and thus should not be used unless absolutely necessary. The hexagon is a nice design that allows one degree of freedom for lack of fit. The heptagon also has its place in some applications. The octagon, hexagon, and nonagon possess the added property of *orthogonal blocking* which will be discussed in the next section.

We have already indicated that the equiradial designs require the use of center runs. Reasonable stability of $N \text{ Var } \hat{y} / \sigma^2$ is achieved with 2–4 center runs.

7.4.9 Orthogonal Blocking in Second-Order Designs

In many RSM situations the study is too large to allow all runs to be made under homogeneous conditions. As a result, it is important and interesting to consider second-order designs that facilitate *blocking*—that is, the inclusion of block effects. It is important that the assignment of the design points to block be done so as to minimize the impact on the model coefficients. The property of the experimental design that we seek is that of **orthogonal blocking**. To say that a design admits orthogonal blocking implies that the block effects in the model are orthogonal to model coefficients. A simple first-order example serves as an illustration. Consider a 2^2 factorial to be used for the fitting of a first-order model. Suppose we use $AB = I$ as the defining relation to form two $\frac{1}{2}$ fractions which are placed in separate blocks. As a result, we have the model

$$y_i = \beta_0 + \beta_1 x_{i1} + \beta_2 x_{i2} + \delta_1 z_{i1} + \delta_2 z_{i2} + \varepsilon_i \tag{7.22}$$

where δ_1 and δ_2 are "block effect" coefficients. Here z_{i1} and z_{i2} are "dummy" or indicator variables; that is, $z_{i1} = 1$ if y_i is in block 1 and $z_{i1} = 0$ otherwise. The variable $z_{i2} = 1$ if y_i is in block 2 and $z_{i2} = 0$ otherwise. The X matrix is then

$$\mathbf{X} = \begin{matrix} & x_1 & x_2 & z_1 & z_2 \\ \begin{bmatrix} 1 & -1 & -1 & 1 & 0 \\ 1 & 1 & 1 & 1 & 0 \\ \hdashline 1 & -1 & 1 & 0 & 1 \\ 1 & 1 & -1 & 0 & 1 \end{bmatrix} & \begin{matrix} \\ \left.\rule{0pt}{1.2em}\right\} \text{Block 1} \\ \\ \left.\rule{0pt}{1.2em}\right\} \text{Block 2} \end{matrix} \end{matrix} \tag{7.23}$$

Note that block 1 contains the design points $\{(1), ab\}$, whereas block 2 contains $\{a, b\}$. Now, in Equation (7.23), the X matrix is singular; the last two columns add to the first column. This underscores the fact that the parameter δ_1 and δ_2 are not estimable. As a result, one of the indicator variables should not be included. We shall eliminate the z_2 column and center z_1. As a result, the model of Equation (7.22) can be rewritten

$$y_i = \beta_0^* + \beta_1 x_{i1} + \beta_2 x_{i2} + \delta_1 (z_{i1} - \bar{z}_1) + \varepsilon_i \tag{7.24}$$

and thus the X matrix is written

$$\mathbf{X} = \begin{matrix} & x_1 & x_2 & \delta_1 \\ \begin{bmatrix} 1 & -1 & -1 & 1/2 \\ 1 & 1 & 1 & 1/2 \\ \hdashline 1 & -1 & 1 & -1/2 \\ 1 & 1 & -1 & -1/2 \end{bmatrix} & \begin{matrix} \left.\rule{0pt}{1.2em}\right\} \text{Block 1} \\ \\ \left.\rule{0pt}{1.2em}\right\} \text{Block 2} \end{matrix} \end{matrix} \tag{7.25}$$

Now, the concept of orthogonal blocking is nicely illustrated through the use of Equations (7.24) and (7.25). In this example the block effect is orthogonal to the regression coefficients if

$$\sum_{u=1}^{4} (z_{u1} - \bar{z}_1) x_{1u} = 0$$

$$\sum_{u=1}^{4} (z_{u2} - \bar{z}_1) x_{2u} = 0 \tag{7.26}$$

Thus, as the reader might have expected, the assignment of design points to blocks in this example provides an array in which the block factor has no effect on the regression coefficients. In fact, if one considers the least squares estimator $\mathbf{b} = (\mathbf{X'X})^{-1}\mathbf{X'y}$, the regression coefficients β_0^*, β_1, and β_2 are estimated *as if the blocks were not in the experiment*. The variances of the estimates of β_1 and β_2 in this variance optimal design are not changed in the presence of blocks.

The above example serves two purposes. First, it reinforces the material in Chapter 3 regarding blocking in two-level designs. In the above illustration, the $x_1 x_2$ interaction is "confounded with blocks." If one were to augment the $\mathbf{X}$ matrix with the $x_1 x_2$ column, this confounding becomes evident. The linear effects, or linear coefficients, are independent of block effects, and this is reinforced from the fact that Equations (7.26) hold. Secondly, the example allows us a foundation for setting up conditions for orthogonal blocking for second-order designs. Though we now move on to the second-order case, there are several exercises at the end of this chapter that deal with first-order models and first-order-plus-interaction models.

Conditions for Orthogonal Blocking in Second-Order Designs

Consider a second-order model with k design variables and b blocks. The model, then, can be written

$$y_u = \beta_0^* + \sum_{i=1}^{k} \beta_i x_{ui} + \sum_{i=1}^{k} \beta_{ii} x_{ui}^2 + \sum_{i<j} \sum \beta_{ij} x_{ui} x_{uj} + \sum_{m=1}^{b} \delta_m (z_{um} - \bar{z}_m),$$

$$u = 1, 2, \ldots, N \qquad (7.27)$$

Here we have used the indicator again, with $z_{mu} = 1$ if the uth observation is in the mth block. We have centered the indicator variables as in our earlier example. In addition, we have included all of the block effects in the model even though it results in a singular model. This will have no impact on the conditions for orthogonal blocking. From Equation (7.27), block effects are orthogonal to regression coefficients if

$$\sum_{u=1}^{N} x_{ui}(z_{um} - \bar{z}_m) = 0, \qquad i = 1, 2, \ldots, k; \, m = 1, 2, \ldots, b \quad (7.28)$$

$$\sum_{u=1}^{N} x_{ui} x_{uj}(z_{um} - \bar{z}_m) = 0, \qquad i \neq j; \, m = 1, 2, \ldots, b \qquad (7.29)$$

$$\sum_{u=1}^{N} x_{ui}^2(z_{um} - \bar{z}_m) = 0, \qquad i = 1, 2, \ldots, k; \, m = 1, 2, \ldots, b \quad (7.30)$$

In what follows we will assume that we are dealing with designs in which $[i] = 0$ and $[ij] = 0$ for all $i \neq j$, $i, j = 1, 2, \ldots, k$. That is, the first moments and mixed second moments are all zero. Recall that these conditions hold for all central composite designs, Box–Behnken designs, and two-level factorial designs. Now, if we consider Equations (7.28) and (7.29) for a specific value of m (i.e., for a specific block), we have

$$\sum_{\text{block } m} x_{ui} = \sum_{\text{block } m} x_{ui} x_{uj} = 0, \qquad i, j = 1, 2, \ldots, k; i \neq j \quad (7.31)$$

But, of course, Equation (7.30) must hold for all blocks. Equation (7.31) implies, then, that in each block the first and second mixed moments are zero, implying that *each block must itself be a first-order orthogonal design* (condition 1).

Now consider Equation (7.31). For say the mth block, this equation implies that $\sum_{u=1}^{N} x_{ui}^2 z_{um} = \bar{z}_m \sum_{u=1}^{N} x_{ui}^2$ for $i = 1, 2, \ldots, k$. The quantity $\bar{z}_m$ is the fraction of the total runs that are in the mth block. As a result, Equation (7.30) implies that *for each design variable, the sum of squares contribution from each block is proportional to the block size* (condition 2).

Designs that fulfill conditions 1 and 2 above are not difficult to find. Once again, the central composite design is prominent in that respect due to flexibility involved in the choice of α and n_c.

Orthogonal Blocking in the CCD

It may be instructive to discuss an example CCD that does block orthogonally. Consider a $k = 2$ rotatable CCD in two blocks with two center per block. Consider the assignment of design points to blocks as follows:

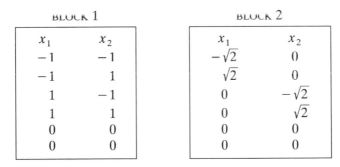

BLOCK 1		BLOCK 2	
x_1	x_2	x_1	x_2
-1	-1	$-\sqrt{2}$	0
-1	1	$\sqrt{2}$	0
1	-1	0	$-\sqrt{2}$
1	1	0	$\sqrt{2}$
0	0	0	0
0	0	0	0

The above design does block orthogonally. Each block is a first-order orthogonal design; the first moments and mixed second moments are both zero for the "design" in each block, and thus condition 1 holds. Also, the sum of squares for each design variable is 4.0, and the block sizes are equal. Hence condition 2 regarding the proportionality relationship also holds. In the example, notice that any even number of center runs with half assigned to block 1 and the other half to block 2 would have also resulted in orthogonal blocking. In the example, the factorial design points are assigned to one block and the axial points to the other block. In this case, condition 1 will automatically hold. The values for α and n_c are chosen to achieve condition 2.

As a second example let us consider, again, a CCD but assume $k = 3$ and suppose that three blocks are required. The design is as follows:

BLOCK 1			BLOCK 2			BLOCK 3		
x_1	x_2	x_3	x_1	x_2	x_3	x_1	x_2	x_3
-1	-1	-1	1	-1	-1	$-\alpha$	0	0
1	1	-1	-1	1	-1	α	0	0
1	-1	1	-1	-1	1	0	$-\alpha$	0
-1	1	1	1	1	1	0	α	0
0	0	0	0	0	0	0	0	$-\alpha$
0	0	0	0	0	0	0	0	α
						0	0	0
						0	0	0
						0	0	0

Here we have divided the $2^3 = 8$ factorial points into two blocks with $ABC = I$ the defining relation. The reader can again verify that condition 1 for orthogonal blocking holds. We can merely solve for α that allows condition 2 to hold.

$$\frac{2\alpha^2}{9} = \frac{4}{6}$$

which result in $\alpha = \sqrt{3}$. As a result, we have a spherical CCD with $\alpha = \sqrt{3}$ (recommended choice for α for Section 7.4.3) which blocks orthogonally in three blocks.

As the reader studies these two examples and the use of conditions 1 and 2, two principles obviously emerge: When one requires two blocks, there should be a factorial block and an axial block. In the case of three blocks, the factorial portion is divided into two blocks and the axial portion remains a single block. The partitioning into blocks of the factorial portion must result in a guaranteed orthogonality between blocks and the two factor interaction terms and, of course, orthogonality between blocks and linear coefficients. The number of blocks with the CCD are $2, 3, 5, 9, 17, \ldots$. *There is never a subdivision of the axial block.* The axial points are always a single block. The flexibility in the choice of center runs and the value of α allows us to achieve orthogonal blocking and near-orthogonal blocking in many situations. At this point, we will deal briefly in the special case of two blocks to give the reader further insight into the choice of α and the assignment of center runs to blocks.

Orthogonal Blocking in Two Blocks with the CCD

Suppose we call F_0 the number of center runs in the factorial block and a_0 the number of center runs in the axial block. Condition 1 holds when we assign the factorial and axial points as the two blocks. Condition 2 holds if

$$\frac{\sum\limits_{\text{ax. bl.}} x_{iu}^2}{\sum\limits_{\text{fac. bl.}} x_{iu}^2} = \frac{2k + a_0}{F + F_0}, \qquad i = 1, 2, \ldots, k \tag{7.32}$$

where, of course, $F = 2^k$ or $F = 2^{k-1}$ (resolution V fraction). The value for α that results in orthogonal blocking is developed from Equation (7.32). From Equation (7.32) we have

$$\frac{2\alpha^2}{F} = \frac{2k + a_0}{F + F_0}$$

and the solution for α is given by

$$\alpha = \sqrt{\frac{F(2k + a_0)}{2(F + F_0)}} \tag{7.33}$$

General Recommendations for Blocking with CCD

There are many instances in which the practitioner using RSM can achieve orthogonal blocking and still have designs that contain design parameters near those recommended in Section 7.4.3 when the region is spherical. We have already seen this to be the case with the $k = 2$ and $k = 3$ CCD. A value of $\alpha = \sqrt{2}$ for the case of two blocks produces orthogonal blocking as long as center runs are evenly divided among the two blocks. We will soon produce a table that gives the recommended designs. However, the case of the cuboidal region disserves special attention. In order to achieve orthogonal blocking for say a *face center cube*, one must resort to badly disproportionate block size which may often be a problem. For example, the following face center cube blocks orthogonally for $k = 2$:

$$\mathbf{D} = \begin{array}{c} \begin{array}{cc} x_1 & x_2 \end{array} \\ \left[\begin{array}{cc} -1 & -1 \\ -1 & 1 \\ 1 & -1 \\ 1 & 1 \\ 0 & 0 \\ 0 & 0 \\ 0 & 0 \\ 0 & 0 \\ \hdashline -1 & 0 \\ 1 & 0 \\ 0 & -1 \\ 0 & 1 \end{array} \right] \begin{array}{l} \\ \\ \\ \left.\rule{0pt}{3.2em}\right\} \text{Block 1} \\ \\ \\ \\ \left.\rule{0pt}{2.2em}\right\} \text{Block 2} \end{array} \end{array}$$

For three variables, if one requires say three blocks, the two factorial blocks require *eight center runs apiece* if one requires that $\alpha = 1.0$ in a cuboidal region. As a result, when ranges for the experiment seem to suggest the need for a cuboidal design region and orthogonal blocking is required, one might need to consider a spherical design in order that blocking be better accommodated.

In the case of spherical designs, many very practical designs exist. Table 7.6 shows a working table of useful second-order designs that block orthogonally. For designs in which "$\frac{1}{2}$ rep" is indicated, the factorial portion is a $\frac{1}{2}$ fraction of the 2^k. In all cases in which there are multiple blocks from the factorial portion, defining contrasts are chosen so that three factor interactions or higher are confounded with blocks.

Blocking with the Box–Behnken Design

The flexibility of the CCD makes it the most reasonable alternative when blocking is required in an RSM analysis. However, there are other second order designs that do block orthogonally. Included among these is the class of Box–Behnken designs for $k = 4$ and $k = 5$. Consider first the design matrix in Equation (7.16) for the $k = 4$ BBD. Using ± 1 notation to indicate the 2^2 factorial structure among pairs of variables, we have

$$
\mathbf{D} = \begin{bmatrix}
x_1 & x_2 & x_3 & x_4 \\
\pm 1 & \pm 1 & 0 & 0 \\
0 & 0 & \pm 1 & \pm 1 \\
0 & 0 & 0 & 0 \\
\hline
\pm 1 & 0 & 0 & \pm 1 \\
0 & \pm 1 & \pm 1 & 0 \\
0 & 0 & 0 & 0 \\
\hline
\pm 1 & 0 & \pm 1 & 0 \\
0 & \pm 1 & 0 & \pm 1 \\
0 & 0 & 0 & 0
\end{bmatrix}
\begin{array}{l}
\left.\rule{0pt}{20pt}\right\} \text{Block 1} \\
\left.\rule{0pt}{20pt}\right\} \text{Block 2} \\
\left.\rule{0pt}{20pt}\right\} \text{Block 3}
\end{array}
$$

The requirement on the vectors of center runs is that each block contains the same number of center runs. This, of course, results in equal block sizes, a requirement that results from equal sum of squares for each factor in each block.

The $k = 5$ BBD blocks orthogonally in two blocks. Consider the design of Equation (7.17). This design can be partitioned into two parts so that conditions for orthogonal blocking hold. Again, the center runs should be even in number, and half of them should be assigned to each block. The reader can easily verify that conditions 1 and 2 for orthogonal blocking both hold. See Exercise 7.27.

Table 7.6 Design Parameters for Some Useful Designs that Block Orthogonally

k	2	3	4	5	5($\frac{1}{2}$ rep)	6	6($\frac{1}{2}$ rep)	7	7($\frac{1}{2}$ rep)
Factorial block									
F: Number of points in factorial portion	4	8	16	32	16	64	32	128	64
Number of blocks in factorial portion	1	2	2	4	1	8	2	16	8
Number of points in each block from factorial portion	4	4	8	8	16	8	16	8	8
Number of added center points in each block	3	2	2	2	6	1	4	1	1
Total number of points in each block	7	6	10	10	22	9	20	9	9
Axial block									
Number of axial points	4	6	8	10	10	12	12	14	14
Number of added center points	3	2	2	4	1	6	2	11	4
Total number of points in block	7	8	10	14	11	18	14	25	18
Grand total of points in the design	14	20	30	54	33	90	54	169	90
Value of α for orthogonal blocking	1.4142	1.6330	2.0000	2.3664	2.0000	2.8284	2.3664	3.3636	2.8284
Value of α for rotatability	1.4142	1.6818	2.0000	2.3784	2.0000	2.8284	2.3784	3.3333	2.8284

Source: G. E. P. Box and J. S. Hunter (1957).

Orthogonal Blocking with Equiradial Designs

It has already been established that the CCD for $k = 2$ and $\alpha = \sqrt{2}$ involves eight points equally spaced on a circle—that is, an octagon. Much has already been said about orthogonal blocking with this design. The use of an even value for n_c and $n_c/2$ in the factorial block and $n_c/2$ assigned to the axial block results in a very efficient design for $k = 2$ that blocks orthogonality. But other equiradial designs block orthogonally too. For example, consider the hexagon. Figure 7.20 shows the hexagonal design. The hexagon is a combination of two equilateral triangles. For example, in the figure, points 1, 3, and 5 form one equilateral triangle and points 2, 4, and 6 form another. The reader should recall that an equilateral triangle is a simplex design which is first-order orthogonal; thus if one assigns points 1, 3, and 5 to one block and points 2, 4, and 6 to a second block, condition 1 for orthogonal blocking will hold. For condition 2 please recall the design matrix of the simplex from Equation (7.21). The order of the points are in the matrix are 1 through 6. Note

$$\sum_{\text{block 1}} x_{ui}^2 = \sum_{\text{block 2}} x_{ui}^2 = 1.5, \qquad i = 1, 2$$

and thus the assignment of points 1, 3, and 5 to block 1 and 2, 4, and 6 to block 2 with n_c even and $n_c/2$ assigned to each block will result in a design which blocks orthogonally in two blocks.

The nonagon (nine equally spaced points on a circle) blocks orthogonally in three blocks. The nonagon combines *three* equilateral triangles, and thus $n_c \geq 3$, n_c is divisible by three, and $n_c/3$ assigned to each block gives a design that blocks orthogonally. We will not discuss the details. Actually, there are other equiradial designs that block orthogonally but the plans described here are the more practical ones.

Form of Analysis with a Blocked Experiment

The first and foremost topic of discussion in this area should deal with the model. It is important to understand that we are assuming that blocks enter the model as additive effects. This was outlined in the model form in Equation (7.27). It is assumed that there is no interaction between blocks and other model terms. This concept should be thoroughly discussed among the practitioners involved. If it is clear that blocks interact with the factors, then it may mean that the analysis should involve the fitting of a separate response surface for each block. This often results in difficulty regarding interpretation of results. Another important issue centers around what is truly meant by orthogonal blocking. As we indicated earlier, orthogonal means that "block effects" are orthogonal to model coefficients. However, the final fitted model will obviously include estimates of block effects. The fitted model is not at all unlike an analysis of covariance model where blocks play the role of "treatments" and one is assuming that for each block the model coefficients are the same (quite like assuming equal slopes in ANACOVA). But, of course, the presence of block effects imply that we have multiple intercepts, one for each block. Thus, for, say, three blocks we have

$$\hat{y}_{\text{block 1}} = \hat{\beta}_{0,1} + f(\mathbf{x})$$

$$\hat{y}_{\text{block 2}} = \hat{\beta}_{0,2} + f(\mathbf{x})$$

$$\hat{y}_{\text{block 3}} = \hat{\beta}_{0,3} + f(\mathbf{x})$$

where $f(\mathbf{x})$ is a second-order function containing linear quadratic, and interaction terms, all of which are the same for each block. Now, the coefficients (apart from the intercepts) can be computed as if there were no blocking. As a result, the stationary point, canonical analysis, and ridge analysis path are computed, say, from SAS RSREG by ignoring blocking because none of these computations involves the intercepts. Then for the purpose of computing predicted response, separate intercepts are merely found from the overall intercept and block effects. One must understand that the essence of the RSM analysis is intended to be a description of the system *in spite of blocking*; and blocks are assumed to have no impact on the nature and shape of the response surface. The following example should be beneficial.

Example 7.6 In a chemical process it is important to fit a response surface and find conditions on time (x_1) and temperature (x_2) which give a high yield. It is important that a yield of at least 80% be achieved. A list of

Table 7.7 Natural and Design Levels for Chemical Reaction Experiment

Natural Variables		Design Units		Block	Yield Response
Time	Temperature	x_1	x_2		
80	170	−1	−1	1	80.5
80	180	−1	1	1	81.5
90	170	1	−1	1	82.0
90	180	1	1	1	83.5
85	175	0	0	1	83.9
85	175	0	0	1	84.3
85	175	0	0	1	84.0
85	175	0	0	2	79.7
85	175	0	0	2	79.8
85	175	0	0	2	79.5
92.070	175	1.414	0	2	78.4
77.930	175	−1.414	0	2	75.6
85	182.07	0	1.414	2	78.5
85	167.93	0	−1.414	2	77.0

Table 7.8 Canonical Analysis Using SAS

Canonical Analysis of Response Surface
(Based on Coded Data)

	Critical Value	
Factor	Coded	Uncoded
x_1	0.263376	0.372414
x_2	0.236583	0.334529
Predicted value at stationary point	82.136490	

Canonical Analysis of Response Surface
(Based on Coded Data)

	Eigenvectors	
Eigenvalues	x_1	x_2
-1.845262	0.160138	0.987095
-2.635808	0.987095	-0.160138

Stationary point is a maximum.

Parameter	Degrees of Freedom	Parameter Estimate	Standard Error	t for H_O: Parameter $= 0$	Probability $> |t|$
Intercept	1	81.866214	1.205283	67.923	0.0000
x_1	1	0.932541	1.043884	0.893	0.3978
x_2	1	0.577712	1.043884	0.553	0.5951
$x_1 * x_1$	1	-1.308163	1.086667	-1.204	0.2631
$x_2 * x_1$	1	0.125000	1.476164	0.0847	0.9346
$x_2 * x_2$	1	-0.933049	1.086667	-0.859	0.4155

natural and coded variables appears in Table 7.7. The design requires 14 experimental runs. However, two batches of raw materials must be used so seven experimental runs were made for each batch. The experiment was run in two blocks and the order of the runs were made randomly in each block. The factorial runs plus three center runs were made in one block (batch), and the axial runs plus three center runs were made in the second block (batch). The resulting design does block orthogonally. A second-order analysis was performed. Table 7.8 shows the canonical analysis produced by SAS. The stationary point is a maximum. The "uncoded" form is the design unit metric indicated in Table 7.7. (The reader recalls that SAS will code further to put the most distant point at unit radius from the center.) The parameter estimates in our design unit produced the fitted model

$$\hat{y} = 81.866 + 0.933x_1 + 0.578x_2 - 1.308x_1^2 + 0.125x_1x_2 - 0.933x_2^2$$

Please note that p-values indicate that model terms are insignificant. However, the data were entered into SAS as *if no blocking was done*. This renders coefficients and canonical analysis correct, but standard errors are inflated because of variability due to blocks. The appropriate location of maximum response is given by

$$x_{1,s} = 0.372$$

$$x_{2,s} = 0.335$$

in our design units. The predicted value at the stationary point is 82.136. Figure 7.21 shows the contour plot from Design-Expert.

Further analysis using Design-Expert can sort out the block effects. Table 7.9 gives a printout. Note that while 81.867 is the overall intercept,

DESIGN-EXPERT Analysis

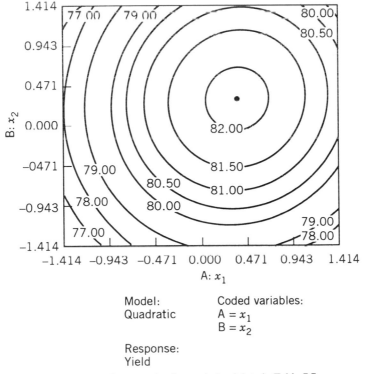

Model:
Quadratic

Coded variables:
A = x_1
B = x_2

Response:
Yield

Figure 7.21 Contour plot for analysis of data in Table 7.7.

Table 7.9   Design Expert Analysis for Blocking Analysis

Factor	Factor	Units	−1 Level	+1 Level
A	x_1		80.000	90.000
B	x_2		170.000	180.000

ANOVA for Quadratic Model

Source	Sum of Squares	Degrees of Freedom	Mean Square	F-Value	Probability > F
Blocks	69.53	1			
Model	27.48	5	5.496	206.4	0.0001
Residual	0.19	7	0.027		
Lack of fit	(0.05)	3	0.018	0.5307	0.6851
Pure error	(0.13)	4	0.033		
Corrected total	97.20	13			

Root mean squared error	0.163	R-squared	0.9933
Dependent mean	80.586	Adjusted R-squared	0.9884
Coefficient of variation	0.20%		

Predicted residual sum of squares (PRESS) = 0.76

Independent Variable	Coefficient Estimate	Degrees of Freedom	Standard Error	t for H_O Coefficient = 0	Probability > |t|
Intercept	81.867	1	0.067	1229	
Block 1	2.229				
Block 2	−2.229				
$A = x_1$	0.933	1	0.058	16.16	0.0001
$B = x_2$	0.578	1	0.058	10.01	0.0001
$A^2 = x_1^2$	−1.309	1	0.060	−21.79	0.0001
$B^2 = x_2^2$	−0.933	1	0.060	−15.54	0.0001
$AB = x_1 x_2$	0.125	1	0.082	1.532	0.1694

Final Equation in Terms of Design Units

$$\text{Yield} = 81.867$$
$$+0.933 * x_1$$
$$+0.578 * x_2$$
$$-1.309 * x_1^2$$
$$-0.933 * x_2^2$$
$$+0.125 * x_1 x_2$$

estimates of block effects are 2.229 and -2.229 for block 1 and block 2, respectively. As a result, the two separate intercepts are $81.867 \pm 2.229 =$ 84.096 and 79.638, respectively.

EXERCISES

7.1 Consider the design of an experiment to fit the first-order response model

$$y_i = \beta_0 + \beta_1 x_{1i} + \beta_2 x_{2i} + \beta_3 x_{3i} + \beta_4 x_{4i} + \beta_5 x_{4i}$$

The design used is a $\frac{1}{4}$ fraction of a 2^5 with defining relations:

$$x_1 x_2 x_3 = I$$

$$x_3 x_4 x_5 - I$$

The factors are quantitative, and thus we use the notation $x_1, x_2, x_3, \ldots$ rather than $A, B, C, \ldots$.

(a) Construct the **X** matrix for the first-order model using the design described above.

(b) Is the design orthogonal? Explain why or why not.

(c) Is the design a variance-optimal design—that is, one that results in minimum values of $N \operatorname{Var} b_i / \sigma^2$. Note that the "weight" of N is used in order to take sample size into account.

7.2 Consider the situation of Exercise 7.1. Suppose we use four center runs to augment the $\frac{1}{4}$ fraction of the 2^5 design constructed.

(a) What is the advantage of using the center runs?

(b) Will the resulting 12-run design be orthogonal? Explain.

(c) Will the resulting design be variance optimal?

(d) If your answer to (c) is no, give a 12-run first-order design with values of $\operatorname{Var} b_i / \sigma^2$ smaller than the 12-run design with center runs.

(e) Give the variances of regression coefficients for both designs—that is, the 2_{III}^{5-2} with four center runs and your design in (d).

7.3 Consider again the design constructed in Exercise 7.1. Suppose the fitted model is first order in nature and the analysis reveals a rather large effect for the interaction $x_1 x_4 = x_2 x_5$. As a result, a "second phase" would suggest an augmentation of the design. Suppose a second phase is a "fold-over" of the design in Exercise 7.1. Is the resulting design orthogonal and thus variance optimal for the model

$$y_u = \beta_0 + \sum_{i=1}^{5} \beta_i x_{ui} + \sum_{i<j} \sum \beta_{ij} x_{ui} x_{uj} + \varepsilon_u, \qquad u = 1, 2, 3, \ldots, 16$$

Explain why or why not.

7.4 Consider the L_{27} design (in coded from) discussed in Exercise 6.11. Suppose we consider this design for fitting the first-order model

$$y_u = \beta_0 + \sum_{i=1}^{3} \beta_i x_{ui} + \varepsilon_u, \qquad u = 1, 2, \ldots, 27$$

(a) Is the design first-order orthogonal? Explain.
(b) Is the design variance optimal? Explain.
(c) Compare the $N \operatorname{Var} b_i / \sigma^2$ using the L_{27} with, say, a 2^3. Compare with a Plackett–Burman 12-run design. Compare with a Plackett–Burman 28-run design.
(d) Your answer to (c) above should be that the Plackett–Burman 12-run design and 2^3 design for a first-order model are *variance equivalent on a per observation basis*. What are extra advantages enjoyed by the 2^3 over that of the Plackett–Burman 12?

7.5 Consider a situation involving seven design variables when, in fact, a first-order model is required but only 8 runs can be used. Design a saturated eight-run design that is variance optimal.

7.6 Consider a situation in which 5 factors are to be studied and a first-order model is postulated though it is not quite clear that this is the correct model. Sixteen design runs are to be used. The designs to be considered are as follows:
(i) 2_{III}^{5-2} duplicated
(ii) 2_V^{5-1}
(iii) A Plackett–Burman 12-run design augmented with four center runs.
Discuss the advantages and disadvantages of the 3 designs. Use the terms *variance optimal* and *model misspecification* in your discussion.

7.7 Consider the "model inadequacy" material in Section 7.2. The material can be used to nicely illustrate the aliasing ideas discussed in Chapter 4. Aliasing plays an important role when one deals with fractional factorials for fitting first-order and first-order-plus-interaction response surface models. Suppose one fits a first-order model using a 2_{III}^{3-1} with defining relation

$$I = x_1 x_2 x_3$$

However, suppose the true model is

$$E(y) = \beta_0 + \beta_1 x_1 + \beta_2 x_2 + \beta_3 x_3 + \beta_{12} x_1 x_2 + \beta_{13} x_1 x_3 + \beta_{23} x_2 x_3$$

Using Equation (7.2), we obtain

$$\mathbf{X} = \begin{array}{c} \begin{array}{ccc} x_1 & x_2 & x_3 \end{array} \\ \begin{bmatrix} 1 & -1 & 1 & 1 \\ 1 & 1 & -1 & 1 \\ 1 & 1 & 1 & -1 \\ 1 & -1 & -1 & -1 \end{bmatrix} \end{array}$$

$$\mathbf{Z} = \begin{array}{c} \begin{array}{ccc} x_1 x_2 & x_1 x_3 & x_2 x_3 \end{array} \\ \begin{bmatrix} -1 & -1 & 1 \\ -1 & 1 & -1 \\ 1 & -1 & -1 \\ 1 & 1 & 1 \end{bmatrix} \end{array}$$

$$\mathbf{b} = \begin{bmatrix} b_0 \\ b_1 \\ b_2 \\ b_3 \end{bmatrix}$$

$$\mathbf{\gamma} = \begin{bmatrix} \beta_{12} \\ \beta_{13} \\ \beta_{23} \end{bmatrix}$$

Use the expression

$$E(\mathbf{b}) = \mathbf{\beta} + (\mathbf{X'X})^{-1}\mathbf{X'Z\gamma}$$

$$= \mathbf{\beta} + \mathbf{A\gamma}$$

to verify the following expected values:

$$E(b_0) = \beta_0$$
$$E(b_1) = \beta_1 - \beta_{23}$$
$$E(b_2) = \beta_2 - \beta_{13}$$
$$E(b_3) = \beta_3 - \beta_{13}$$

7.8   In Exercise 7.7 if the "true" model is given by

$$E(y) = \beta_0 + \beta_1 x_1 + \beta_2 x_2 + \beta_3 x_3 + \beta_{12} x_1 x_2 + \beta_{13} x_1 x_3$$
$$+ \beta_{23} x_2 x_3 + \beta_{123} x_1 x_2 x_3$$

Show that the expected values are given by

$$E(b_0) = \beta_0 - \beta_{123}$$
$$E(b_1), E(b_2), E(b_3) \text{ are as in Exercise 7.7.}$$

Exercises 7.7 and 7.8 illustrate the effect of aliasing in the sense of a regression model. Thus, in the case of these two exercises, the aliasing of x_1 with $x_2 x_3$ implies that the coefficient b_1 is not an unbiased estimator of β_1 but is biased by β_{23}, the coefficient of $x_2 x_3$. Indeed, the coefficient of the interaction described by the defining contrast, namely β_{123}, biases the estimated intercept β_0.

7.9   Consider the same situation as in Exercise 7.7. However, the *other fraction* given by $x_1 x_2 x_3 = I$ is used. If the fitted first-order model is incorrect and the true model is as stated in Exercise 7.8, show that the biases in the coefficients are indicated by

$$E(b_0) = \beta_0 + \beta_{123}$$
$$E(b_1) = \beta_1 + \beta_{23}$$
$$E(b_2) = \beta_2 + \beta_{13}$$
$$E(b_3) = \beta_3 + \beta_{12}$$

7.10   Consider a $\frac{1}{4}$ fraction of a 2^5 factorial with defining relations

$$x_1 x_2 x_3 = I$$
$$x_3 x_4 x_5 = I$$

Suppose a first-order model is fit to the data and the "true" model includes, additionally, all two-factor interactions. Use the expression

$$E(\mathbf{b}) = \boldsymbol{\beta} + \mathbf{A}\boldsymbol{\gamma}$$

with, of course, $\boldsymbol{\gamma}$ containing all two-factor interactions to develop $E(b_1)$, $E(b_2)$, $E(b_3)$, $E(b_4)$, and $E(b_5)$ where the fitted first-order model is

$$\hat{y} = b_0 + b_1 x_1 + b_2 x_2 + b_3 x_3 + b_4 x_4 + b_5 x_5$$

7.11 Consider Exercise 4.2 in Chapter 4. The strong effects are those for the factors, A, B, C, and AB, and the fitted regression is given by

$$\hat{y} = 60.65625 + 11.60625x_1 + 34.03125x_2 + 10.45625x_3 + 6.58125x_1x_2$$

(a) Verify the above coefficients by considering this five-parameter model in the form

$$y = X\beta + \varepsilon$$

and by using

$$b = (X'X)^{-1}X'y$$

(b) Is the design in Table 4.5 orthogonal for the model that was fit? If so, verify it. Is the design variance optimal?

(c) Is the design orthogonal for a model containing x_1, x_2, x_3, x_4, x_5, x_1x_2, x_1x_3, x_1x_4, x_1x_5, x_2x_3, x_2x_4, x_2x_5, x_3x_4, x_3x_5, and x_4x_5?

7.12 Consider the design listed in Table 4.9 with the injection molding data. Answer the following questions.

(a) For the model that is fit, namely

$$\hat{y} = b_0 + b_1x_1 + b_2x_2 + b_{12}x_{12}$$

give the variance of each coefficient assuming common error variance σ^2.

(b) Using the error mean square from Table 4.10, give the estimated standard errors of all four coefficients.

(c) In light of the plot in Figure 4.14, why are the results in (b) somewhat precarious?

(d) Is this design orthogonal for a model containing all linear terms and the two-factor interactions x_1x_2, x_1x_3, x_1x_4, x_1x_5, and x_1x_6? If so, justify it; if not, explain why not.

7.13 In Section 4.5 of Chapter 4 the 2_{III}^{7-4} design is discussed. Here we may be interested in building a regression model for studying seven factors. The design is given in Table 4.14. Suppose one fits a first-order model but, in fact, all two factor interactions are in the true model. Write out $E(b_1)$, $E(b_2)$, $E(b_3)$, $E(b_4)$, $E(b_5)$, $E(b_6)$, and $E(b_7)$. Explain all your terms.

7.14 Consider Exercise 4.7 in Chapter 4.

(a) Is the design first order orthogonal?

(b) From your answer in Exercise 4.7(b), write a response surface model.

(c) Give estimated standard errors of all coefficients.

(d) From your model in Exercise 7.14(b) estimate the standard error of prediction at all design locations.

(e) Compute the estimated standard error of prediction at the design center.

7.15 Consider Exercise 4.8 in which four center runs were added to the design in Exercise 4.7. Answer the questions asked in Exercise 7.14(d) and 7.14(e) again. Use the replication error variability (3 df) in computing your mean square error.

7.16 (a) Consider Exercise 4.11 in Chapter 4. For the data in Exercise 4.11(a) develop a first-order regression model. Estimate the standard error of prediction at each design point.

(b) Do all the work in (a) again after the second fraction in Exercise 4.11(b) is added. Comment on what was gained with the addition of the second fraction.

7.17 Suppose we are interested in fitting a response surface model in five design variables. The analyst is quite sure that the form of the model should be

$$y_i = \beta_0 + \beta_1 x_{i1} + \beta_2 x_{i2} + \beta_3 x_{i3} + \beta_4 x_{i4} + \beta_5 x_{i5}$$
$$+ \beta_{12} x_{i1} x_{i2} + \beta_{13} x_{i1} x_{i2} + \varepsilon_i$$

The analyst can form 12 experimental runs in the design. Discuss and compare the following candidate designs. Use the variances of individual coefficients in your discussion. However, do not let your discussion be confined to this criterion. The candidate designs are as follows:

D_1: 2_{III}^{5-2} with four center runs
D_2: Plackett–Burman with 12 runs

7.18 Equation (7.11) is a very important expression. It gives the scaled prediction variance for a first-order orthogonal design in which all design points are at the ± 1 extremes. It also implies that this class of design gives equal prediction variance at any two locations that are on the same sphere. This implies that this class of designs is rotatable. Does the same property hold for orthogonal designs used to accommodate a model containing interaction? Illustrate with a 2^2 factorial with the model

$$y_i = \beta_0 + \beta_1 x_{i1} + \beta_2 x_{i2} + \beta_{12} x_{i1} x_{i2}$$

Compute $N \, \mathrm{Var} \, \hat{y}/\sigma^2$ at the following locations on the same sphere:

(i)	$1, 1$	**(v)**	$\sqrt{2}, 0$
(ii)	$-1, 1$	**(vi)**	$-\sqrt{2}, 0$
(iii)	$1, -1$	**(vii)**	$0, -\sqrt{2}$
(iv)	$1, 1$	**(viii)**	$0, \sqrt{2}$

Comment on your results.

7.19 A first-order orthogonal design with all points at ± 1 extremes remains rotatable for a first-order model if all points are duplicated the same number of times. True or false? Justify your answer.

7.20 Consider the model in Exercise (7.17) along with designs $\mathbf{D}_1$ and $\mathbf{D}_2$. Which design is superior using the scaled prediction variance criterion. Use as illustrations the following locations:

$$(1, 1, 1, 1, 1)$$
$$(-1, -1, -1, -1, -1)$$
$$(1, 1, -1, -1, -1)$$
$$(1, -1, -1, 1, 1)$$
$$(1, -1, -1, -1, -1)$$
$$(-1, 1, 1, 1, 1)$$

7.21 Consider a central composite design with $\alpha = \sqrt{F}$. Use the result in Equation (7.14) to show that the number of center runs has no impact on the rotatability property.

7.22 Consider a situation in which there are two design variables and one is interested in fitting a second-order model. A curious and interesting comparison surfaces when one considers the central composite design (octagon) and the hexagon. Both are equiradial designs. The beauty of the hexagon (plus center runs) or even the pentagon (plus center runs) lies in the high relative efficiency in spite of their relatively small design size. One should always choose the octagon unless it is too costly. However, it is interesting to compare the two *on a per observation basis*. Let us consider a comparison between a hexagon and a rotatable CCD. To make a valid comparison, one should scale the designs so they are comparable, that is, reside on the same radius (keep in mind that scaling is really arbitrary). The factorial points on the CCD are at radius $\sqrt{2}$. As scaled in Equation (7.21), the hexagon lies on a pick of a radius 1. As a result, for proper comparison the hexagon should be multiplied by $\sqrt{2}$. Thus we have

$$
\mathbf{D}_1 = \begin{bmatrix}
\sqrt{2} & 0 \\
\sqrt{2}/2 & \sqrt{3/2} \\
-\sqrt{2}/2 & \sqrt{3/2} \\
-\sqrt{2} & 0 \\
-\sqrt{2}/2 & -\sqrt{3/2} \\
\sqrt{2}/2 & -\sqrt{3/2} \\
0 & 0 \\
0 & 0 \\
0 & 0
\end{bmatrix}
\quad \text{(hexagon)}
$$

with column headers $x_1 \quad x_2$.

And, of course,

$$
\mathbf{D}_2 = \begin{bmatrix}
-1 & -1 \\
-1 & 1 \\
1 & -1 \\
1 & 1 \\
-\sqrt{2} & 0 \\
\sqrt{2} & 0 \\
0 & -\sqrt{2} \\
0 & \sqrt{2} \\
0 & 0 \\
0 & 0 \\
0 & 0
\end{bmatrix}
\quad \text{(CCD)}
$$

with column headers $x_1 \quad x_2$.

Use $\mathbf{D}_1$ and $\mathbf{D}_2$ to construct scaled prediction variances. Both designs are rotatable, so a rather complete comparison can be made by filling in the following table:

ρ	$N \, \mathrm{Var}\, \hat{y}/\sigma^2$ for D_1	$N \, \mathrm{Var}\, \hat{y}/\sigma^2$ for D_2
0		
0.5		
1.0		
1.5		
$\sqrt{3}$		

Comment on your study.

7.23 It is stated in Section 7.4.9 that for second-order models, orthogonal blocking is much more easily accommodated with the use of the spherical CCD than with use of the face center cube (cuboidal CCD). However, orthogonal blocking is accommodated with the face center cube when *axial points are replicated*. Show that the following design blocks orthogonally in two blocks.

$$
D = \begin{array}{c} \\ \\ \\ \\ \\ \\ \\ \\ \\ \\ \\ \\ \\ \\ \\ \end{array}
\begin{array}{cc}
x_1 & x_2 \\
\left[\begin{array}{rr}
-1 & -1 \\
-1 & 1 \\
1 & -1 \\
1 & 1 \\
0 & 0 \\
0 & 0 \\
0 & 0 \\
0 & 0 \\
-1 & 0 \\
-1 & 0 \\
1 & 0 \\
1 & 0 \\
0 & -1 \\
0 & -1 \\
0 & 1 \\
0 & 1
\end{array}\right]
\end{array}
\begin{array}{l}
\\
\\
\\
\text{Block 1} \\
\\
\\
\\
\\
\\
\\
\\
\text{Block 2} \\
\\
\\
\\
\end{array}
$$

7.24 Consider Table 7.6. The designs suggested in the table are those that block orthogonally and are rotatable or near rotatable. For example, consider the design under the column headed $k = 5(\frac{1}{2}$ rep). Construct the design completely and verify that it meets both conditions for orthogonal blocking.

7.25 Answer the same question as in Exercise 7.24 but use the $k = 5$ line with 4 blocks from the factorial portion and one block from the axial portion.

7.26 Those designs given in Table 7.6 are not the only ones that block orthogonally and are practical. The values for α and n_c can be altered in a very flexible manner. Suppose, for example, that for $k = 4$, the runs size does not allow a design that is shown in the table. Rather, we must use

$$
\mathbf{D} = \begin{bmatrix}
 & x_1 & x_2 & x_3 & x_4 \\
-1 & -1 & -1 & -1 \\
1 & 1 & -1 & -1 \\
1 & -1 & 1 & -1 \\
1 & -1 & -1 & 1 \\
-1 & 1 & 1 & -1 \\
-1 & 1 & -1 & 1 \\
-1 & -1 & 1 & 1 \\
1 & 1 & 1 & 1 \\
0 & 0 & 0 & 0 \\
\hdashline
1 & -1 & -1 & -1 \\
-1 & 1 & -1 & -1 \\
-1 & -1 & 1 & -1 \\
-1 & -1 & -1 & 1 \\
1 & 1 & 1 & -1 \\
1 & 1 & -1 & 1 \\
1 & -1 & 1 & 1 \\
-1 & 1 & 1 & 1 \\
0 & 0 & 0 & 0 \\
\hdashline
-\alpha & 0 & 0 & 0 \\
\alpha & 0 & 0 & 0 \\
0 & -\alpha & 0 & 0 \\
0 & \alpha & 0 & 0 \\
0 & 0 & -\alpha & 0 \\
0 & 0 & \alpha & 0 \\
0 & 0 & 0 & -\alpha \\
0 & 0 & 0 & \alpha
\end{bmatrix}
$$

This design can be made to block orthogonally. What value of α results in orthogonal blocking?

7.27 Construct a Box-Behnken design in 5 variables that blocks orthogonally for a second order model. Show the assignment of design points to blocks.

CHAPTER 8

Experimental Designs for Fitting Response Surfaces II

In Chapter 7 we dealt with standard second-order response surface designs. The central composite designs (CCDs) and Box–Behnken designs (BBDs) are extremely popular with practitioners and some of the equiradial designs enjoy usage for the special case of two design variables. The popularity of these standard designs is linked to the list of 10 important properties of response surface designs listed in Section 7.1. The CCD and BBD rate quite well in the 10 properties listed, particularly when they are augmented with center runs as recommended. One of the most important features associated with these two designs deals with run size. The run size is large enough to provide a comfortable margin for lack of fit but not so large as to reveal "wasted" degrees of freedom or experimental extravagance.

Despite the importance of the CCD and the BBD, there are instances in which the researcher cannot afford the required number of runs. As a result, there is certainly a need for design classes that are either *saturated* or *near-saturated*, in other words, second-order designs that contain close to (but not less than) p design points where

$$p = 1 + 2k + \frac{k(k-1)}{2} \tag{8.1}$$

There are several design classes that allow saturated or near-saturated second-order designs. These are designs that should not be used unless cost prohibits the use of one of the standard designs. In a later section we will make some comparisons that will shed light on this point. In the section that follows we will introduce a few of these design classes and make reference to others.

8.1 DESIGNS THAT REQUIRE A RELATIVELY SMALL RUN SIZE

8.1.1 The Small Composite Design

The small composite design garners its name from the ideas of the central composite, but the factorial portion is neither a complete 2^k nor a resolution

V fraction, but, rather, a special resolution III fraction in which no four-letter word is among the defining relations. This type of fraction is often called *resolution III.** As a result, the total run size is reduced from that of the CCD —hence the term **small** composite design. For $k = 3$, we have

$$
\mathbf{D} =
\begin{array}{ccc}
x_1 & x_2 & x_3 \\
\left[\begin{array}{ccc}
-1 & -1 & -1 \\
1 & 1 & -1 \\
1 & -1 & 1 \\
-1 & 1 & 1 \\
-\alpha & 0 & 0 \\
\alpha & 0 & 0 \\
0 & -\alpha & 0 \\
0 & \alpha & 0 \\
0 & 0 & -\alpha \\
0 & 0 & \alpha \\
\mathbf{0} & \mathbf{0} & \mathbf{0}
\end{array}\right]
\end{array}
\tag{8.2}
$$

The factorial portion is the fraction generated using $ABC = -I$. Obviously, the alternative fraction would be equally satisfactory in the construction of the small composite design. In this case, p, the number of second-order parameters, is 10 and thus the design is one degree of freedom above saturation. Multiple center runs allow degrees of freedom for pure error but if the fitted model is second order, there will be one degree of freedom for lack of fit.

If we focus on the example design in Equation (8.2), it becomes obvious that in the factorial portion linear main effect terms are aliased with two factor interaction terms. Hartley (1959) observed that in spite of this, all coefficients in the second-order model are estimable because the linear coefficients benefit from the axial points, though axial points provide no information in the estimation of interaction coefficients. One may gain some additional insight into this by observing the $\mathbf{X'X}$ matrix for the $k = 3$ small composite design (SCD) contrased with that of the $k = 3$ full central composite (assume a single center run for both designs).

$$
\mathbf{X'X}_{\text{SCD}} =
\begin{array}{cccccccccc}
& x_1 & x_2 & x_3 & x_1^2 & x_2^2 & x_3^2 & x_1x_2 & x_1x_3 & x_2x_3 \\
\left[\begin{array}{cccccccccc}
11 & 0 & 0 & 0 & 4+2\alpha^2 & 4+2\alpha^2 & 4+2\alpha^2 & 0 & 0 & 0 \\
 & 4+2\alpha^2 & 0 & 0 & 0 & 0 & 0 & 0 & 0 & -4 \\
 & & 4+2\alpha^2 & 0 & 0 & 0 & 0 & 0 & -4 & 0 \\
 & & & 4+2\alpha^2 & 0 & 0 & 0 & -4 & 0 & 0 \\
 & & & & 4+2\alpha^4 & 4 & 4 & 0 & 0 & 0 \\
 & & & & & 4+2\alpha^4 & 4 & 0 & 0 & 0 \\
 & & & & & & 4+2\alpha^4 & 0 & 0 & 0 \\
 & & & & & & & 4 & 0 & 0 \\
 & & & & & & & & 4 & 0 \\
 & & & & & & & & & 4
\end{array}\right]
\end{array}
$$

$$\tag{8.3}$$

$$
\mathbf{X'X}_{\text{CCD}} =
\begin{array}{c}
\begin{array}{ccccccccc}
x_1 & x_2 & x_3 & x_1^2 & x_2^2 & x_3^2 & x_1x_2 & x_1x_3 & x_2x_3
\end{array}\\
\left[
\begin{array}{ccccccccc}
15 & 0 & 0 & 8+2\alpha^2 & 8+2\alpha^2 & 8+2\alpha^2 & 0 & 0 & 0\\
 & 8+2\alpha^2 & 0 & 0 & 0 & 0 & 0 & 0 & 0\\
 & & 8+2\alpha^2 & 0 & 0 & 0 & 0 & 0 & 0\\
 & & & 8+2\alpha^2 & 0 & 0 & 0 & 0 & 0\\
 & & & & 8+2\alpha^4 & 8 & 8 & 0 & 0 & 0\\
 & & & & & 8+2\alpha^4 & 8 & 0 & 0 & 0\\
 & & & & & & 8+2\alpha^4 & 0 & 0 & 0\\
 & & & & & & & 8 & 0 & 0\\
 & & & & & & & & 8 & 0\\
 & & & & & & & & & 8
\end{array}
\right]
\end{array}
$$

$$(8.4)$$

The distinction between the two designs stems from the fact that in the CCD all linear main effects and two factor interactions are mutually orthogonal, whereas in the SCD the -4 values reflect the nonorthogonality of x_1 with x_2x_3, x_2 with x_1x_3, and x_3 with x_1x_2. The correlation among these model terms has a strong impact on the variance of the regression coefficients of x_1, x_2, x_3, x_1x_2, x_1x_3, and x_2x_3. In fact, a comparison can be made by creating scaled variances for the two designs, that is, $N \operatorname{Var} b_i/\sigma^2$ for $i = 1, 2, 3$ and $N \operatorname{Var} b_{ij}/\sigma^2$ for $i, j = 1, 2, 3; \; i \neq j$. Table 8.1 shows these values taken from appropriate diagonals of the matrix $N(\mathbf{X'X})^{-1}$. In our illustration, α was taken to be $\sqrt{3}$ and $n_c = 3$ for each design. Notice that the designs are nearly identical for estimation of pure quadratic coefficients. However, the small composite design suffers considerably in efficiency for estimation of linear and interaction coefficients.

The efficiencies of model coefficients certainly suggest that the small composite design should not be an alternative to the CCD unless cost constraints prohibit the latter. However, coefficient by coefficient comparisons of variances is certainly not the only nor necessarily the best method for comparing designs. Graphical methods for comparing designs in terms of scaled prediction variance (i.e., $N \operatorname{Var} \hat{y}(\mathbf{x})/\sigma^2$) can be used. These methods will be presented later in this chapter, and the comparison between the CCD and SCD will be displayed prominently.

Small Composite Design for k = 2
For $k = 2$ there is a small composite design in which the factorial portion is the fraction of a 2^{2-1}. The design is given by

$$
\mathbf{D} =
\begin{array}{c}
\begin{array}{cc}
x_1 & x_2
\end{array}\\
\left[
\begin{array}{cc}
-1 & -1\\
1 & 1\\
-\alpha & 0\\
\alpha & 0\\
0 & -\alpha\\
0 & \alpha\\
0 & 0
\end{array}
\right]
\end{array}
$$

Table 8.1 Scaled Variances of Model Coefficients for CCD and SCD

	b_0	b_i	b_{ii}	b_{ij}
CCD ($N = 17$)	5.6666	1.2143	1.3942	2.125
SCD ($N = 13$)	4.3333	2.1666	1.1074	5.4166

The defining relation for the factorial portion is given by $AB = I$. With six parameters to estimate, this design affords one degree of freedom for lack of fit. For $k = 2$ the hexagon is a more efficient design. This SCD design may find some use if the design region is cuboidal, in which case α must be 1.0.

Small Composite Designs For $k > 3$

The general SCD is a resolution III* fraction of a 2^k augmented with axial points and runs in the center of the design. This structure applies for all $k > 2$. For example, for $k = 4$ a SCD is given by

$$
\begin{array}{cccc}
x_1 & x_2 & x_3 & x_4 \\
\left[\begin{array}{cccc}
-1 & -1 & -1 & +1 \\
1 & -1 & -1 & -1 \\
-1 & 1 & -1 & -1 \\
-1 & -1 & 1 & 1 \\
1 & 1 & -1 & 1 \\
1 & -1 & 1 & -1 \\
-1 & 1 & 1 & -1 \\
1 & 1 & 1 & 1 \\
-\alpha & 0 & 0 & 0 \\
\alpha & 0 & 0 & 0 \\
0 & -\alpha & 0 & 0 \\
0 & \alpha & 0 & 0 \\
0 & 0 & -\alpha & 0 \\
0 & 0 & \alpha & 0 \\
0 & 0 & 0 & -\alpha \\
0 & 0 & 0 & \alpha \\
\mathbf{0} & \mathbf{0} & \mathbf{0} & \mathbf{0}
\end{array}\right]
\end{array}
\qquad (8.5)
$$

The factorial portion of the design in Equation (8.5) is a $\frac{1}{2}$ fraction of a 2^4 with $ABD = I$ as the defining relation. The design points yield 17 degrees of freedom for parameter estimation. Thus, the design is not saturated but results in only two degrees of freedom for lack of fit. In comparison, the CCD for $k = 4$ allows 10 degrees of freedom for lack of fit. Comparisons between scaled variances of coefficients of the linear and interaction coefficients of the SCD with that of the CCD will essentially tell the same story as in the $k = 3$ case. The superiority of the CCD is quite apparent.

An additional important point should be made regarding the $k = 4$ case. One is tempted to use a resolution IV fraction rather than resolution III* for

Table 8.2 Sample Size for the Small Composite Design Constructed for Regular Resolution III* Fraction Plus Axials (Center Point Not Included)

k	2	3	4	5	6	7	8	9
p	6	10	15	21	28	36	45	55
SCD	6	10	16	$*26^a$	28	46	48	82

[a] The asterisk indicates $16 + 10$ points where 16 points are from the 2^{5-1}. However, a Resolution V fraction should be used in this case.

$k = 4$ and possibly for larger values of k where higher fractions might be exploited. For example, for $k = 6$ a resolution IV 2^{6-2} fraction is available. This allows 16 factorial points plust 12 axial points plus center runs, and the result is *one degree of freedom over saturation*. However, the design *cannot* involve a resolution IV fraction! The result is a complete aliasing among two-factor interactions. One must be mindful of the fact that unlike linear or pure quadratic coefficients, all information on interaction coefficients is derived from the factorial portion. Thus, aliasing among two-factor interaction terms in the factorial portion results in aliasing of two-factor interactions for the entire design. Of course, a resolution III* fraction in the $k = 6$ case produces a useful SCD.

Table 8.2 supplies sample sizes for small composite designs for $k = 2$ through 9 constructed as we have indicated. Please note also the listing of the number of parameters required for the fitting of a second-order model.

Use of the Plackett–Burman Design with the SCD

There are situations in which the resolution III* fraction for the factorial portion of the SCD can be nicely supplied by the Plackett–Burman family. Draper (1985) and Draper and Lin (1990) have show that many small composite designs exist. The formation of these designs are as follows:

1. For the factorial portion use k columns of a Plackett–Burman design as outlined in Chapter 4. If duplicate runs exist in the PBD one may wish to reduce the sample size by removing one of the duplicates.
2. Add $2k$ axial runs.
3. If $\alpha = \sqrt{k}$, center runs are suggested to avoid singularity or near singularity.

In some cases, the use of Plackett–Burman designs (PBDs) can produce SCDs that are smaller in run size than those produced by using regular resolution III fractions. For example, for $k = 5$ the use of a 12-run PBD plus 10 axial points gives $N = 22$ plus center runs, which is a savings compared to the $N = 26$ run design listed in Table 8.2. For $k = 7$ the use of the 24-run PBD plus 14 axial points results in $N = 38$ plus center runs, which, again, is a smaller design than the one listed in Table 8.2. On the other hand, often the Plackett–Burman based design can serve as a nice compromise in run size between the SCD in Table 8.2 and the CCD. For example, for $k = 4$ the SCD in Table 8.2 requires 16 plus center runs and the standard CCD requires $N = 24$ plus center runs. But, a composite design containing 12 points from the PBD plus 8 axial points affords $N = 20$ plus center runs. This design allows 6 degrees of freedom for lack of fit. As another example, consider the case of $k = 6$. The SCD is saturated (see Table 8.2) and thus one may wish to avoid it if possible. On the other hand, the basic CCD involving the resolution V 2^{6-1} fraction plus axial points involves $N = 44$ plus center runs, which may not be practicable. But, a design consisting of a 24-point PBD augmented with 12 axial points produces a nice $N = 36$ plus center run compromise, a design that carries nine degrees of freedom for lack of fit.

Example 8.1 The reader should recall that the construction of a Plackett–Burman design begins with the "basic line" and involves cyclic permutation in forming the columns of the design matrix. A row of negative signs are used to complete the formulation of the design. For $k = 5$ a small composite is given by

$$
\mathbf{D} =
\begin{bmatrix}
x_1 & x_2 & x_3 & x_4 & x_5 \\
+ & - & + & - & - \\
+ & + & - & + & - \\
- & + & + & - & + \\
+ & - & + & + & - \\
+ & + & - & + & + \\
+ & + & + & - & + \\
- & + & + & + & - \\
- & - & + & + & + \\
- & - & - & + & + \\
+ & - & - & - & + \\
- & + & - & - & - \\
- & - & - & - & - \\
-\alpha & 0 & 0 & 0 & 0 \\
\alpha & 0 & 0 & 0 & 0 \\
0 & -\alpha & 0 & 0 & 0 \\
0 & \alpha & 0 & 0 & 0 \\
0 & 0 & -\alpha & 0 & 0 \\
0 & 0 & \alpha & 0 & 0 \\
0 & 0 & 0 & -\alpha & 0 \\
0 & 0 & 0 & \alpha & 0 \\
0 & 0 & 0 & 0 & -\alpha \\
0 & 0 & 0 & 0 & \alpha \\
0 & 0 & 0 & 0 & 0
\end{bmatrix}
$$

The result is a 23-point design (including the center point) that allows two degrees of freedom for lack of fit. Pure error degrees of freedom are based on center replication.

Further Comments on the Small Composite Design

As in the case of the CCD, the value of α depends on whether or not the natural region of the experiment is cuboidal or spherical. In the case where constraints require only three levels and a natural cuboidal region, a small composite form of the face center cube requires $\alpha = 1.0$. Otherwise, the use of $\alpha = \sqrt{k}$ is suggested with 3–5 center runs if the cost allows.

The small composite designs derivable from regular fractions do suffer in efficiency due to the aliasing in the factorial portion. Cost constraints will often necessitate the use of near-saturated or saturated second-order designs although, as we shall demonstrate, there are small run designs more efficient than the SCD. Perhaps one of the most compelling virtues of the SCD arises in sequential experimentation. Most of the discussion regarding sequential experimentation has centered on the transition from a first-order fitted model to a second-order model. One should regard response surface methodology (RSM) as nearly always involving iterative experimentation and iterative decision making. A first-stage experiment with an SCD may suggest that the current region is, indeed, the proper one. That is, there are no ridge conditions or no need to move to an alternative region. Then a second-stage augmentation of the factorial portion is conducted; the result is now either a full factorial or resolution V fraction that accompanies the axial points. As a result, the second stage is more efficient than what was used in the first stage. Further information about the SCD and comparisons with other designs will be given later in the chapter.

8.1.2 Koshal Design

Another class of design requiring a small run size is the family of Koshal designs. Koshal (1933) introduced this type of design for use in an effort to solve a set of likelihood equations. The family of designs are saturated for modeling of any response surface of order d $(d = 1, 2, \dots)$. We will not supply general details but will list Koshal designs for $d = 1$ and 2.

Koshal Design for First-Order Model

For the first-order model the Koshal design is simply the *one-factor-at-a-time-design*. For k design variables we simply have

$$\mathbf{D} = \begin{array}{c} \begin{array}{cccc} x_1 & x_2 & \cdots & x_k \end{array} \\ \left[\begin{array}{cccc} 0 & 0 & \cdots & 0 \\ 1 & 0 & \cdots & 0 \\ 0 & 1 & \cdots & 0 \\ \vdots & \vdots & \vdots & \vdots \\ 0 & 0 & \cdots & 0 \\ 0 & 0 & \cdots & 1 \end{array} \right] \end{array}$$

Thus, for the special case in which three variables are of interest we have

$$\mathbf{D} = \begin{array}{c} \begin{array}{ccc} x_1 & x_2 & x_3 \end{array} \\ \left[\begin{array}{ccc} \mathbf{0} & \mathbf{0} & \mathbf{0} \\ 1 & 0 & 0 \\ 0 & 1 & 0 \\ 0 & 0 & 1 \end{array} \right] \end{array}$$

As a result, four coefficients can be estimated in the model

$$y_i = \beta_0 + \beta_1 x_{1i} + \beta_2 x_{2i} + \beta_3 x_{3i} + \varepsilon_i \qquad (i = 1, 2, 3, 4)$$

It is important to note that the Koshal design for the first-order model does not allow estimation of interaction.

First Order Plus Interaction

The Koshal family can be nicely extended to accommodate the first-order model with iteration. One simply augments the first-order Koshal design with "interaction rows." For example, the appropriate design for $k = 3$ is given by

$$\mathbf{D} = \begin{array}{c} \begin{array}{ccc} x_1 & x_2 & x_3 \end{array} \\ \left[\begin{array}{ccc} \mathbf{0} & \mathbf{0} & \mathbf{0} \\ 1 & 0 & 0 \\ 0 & 1 & 0 \\ 0 & 0 & 1 \\ 1 & 1 & 0 \\ 1 & 0 & 1 \\ 0 & 1 & 1 \end{array} \right] \end{array}$$

Second-Order Koshal Design

The Koshal design for fitting a second-order model must, of course, include at least three levels. The one-factor-at-a-time idea remains the basis for the design. For $k = 3$, the design is given by

$$\mathbf{D} = \begin{array}{c} \begin{array}{ccc} x_1 & x_2 & x_3 \end{array} \\ \left[\begin{array}{ccc} \mathbf{0} & \mathbf{0} & \mathbf{0} \\ 1 & 0 & 0 \\ 0 & 1 & 0 \\ 0 & 0 & 1 \\ 2 & 0 & 0 \\ 0 & 2 & 0 \\ 0 & 0 & 2 \\ 1 & 1 & 0 \\ 1 & 0 & 1 \\ 0 & 1 & 1 \end{array} \right] \end{array}$$

The design contains three levels and 10 design points. Another form of the second order Koshal design is given by

$$
\mathbf{D} = \begin{array}{c} \begin{array}{ccc} x_1 & x_2 & x_3 \end{array} \\ \begin{bmatrix}
0 & 0 & 0 \\
1 & 0 & 0 \\
0 & 1 & 0 \\
0 & 0 & 1 \\
-1 & 0 & 0 \\
0 & -1 & 0 \\
0 & 0 & -1 \\
1 & 1 & 0 \\
1 & 0 & 1 \\
0 & 1 & 1
\end{bmatrix} \end{array}
$$

It should be quite clear to the reader how to extend the Koshal designs to more than three variables.

8.1.3 Hybrid Designs

Roquemore (1976) developed a set of saturated or near-saturated second-order designs called **hybrid designs.** The hybrid designs are very efficient. They were created via an imaginative idea that involves the use of a central composite design for $k - 1$ variables, and the levels of the kth variable are supplied in such a way as to create certain symmetries in the design. The result is a class of designs that are economical and either rotatable or near-rotatable for $k = 3, 4, 6,$ and 7. For example, for $k = 3$ we have the hybrid 310 with design matrix

$$
\mathbf{D}_{310} = \begin{array}{c} \begin{array}{ccc} x_1 & x_2 & x_3 \end{array} \\ \begin{bmatrix}
0 & 0 & 1.2906 \\
0 & 0 & -0.1360 \\
-1 & -1 & 0.6386 \\
1 & -1 & 0.6386 \\
-1 & 1 & 0.6386 \\
1 & 1 & 0.6386 \\
1.736 & 0 & -0.9273 \\
-1.736 & 0 & -0.9273 \\
0 & 1.736 & -0.9273 \\
0 & -1.736 & -0.9273
\end{bmatrix} \end{array}
$$

It is important to note that the efficiency of the hybrid is based to a large extent on the fact that eight of the 10 design points and two center runs on x_1 and x_2 are a central composite in these two variables. Then four levels are chosen on the remaining variable, x_3, with these levels chosen to *make odd moments zero, all second pure moments equal, and a near-rotatable design*. The design name, 310, comes from $k = 3$ and 10 distinct design points.

Roquemore developed two additional $k = 3$ hybrid designs. The design matrices are given by

$$
\mathbf{D}_{311A} = \begin{bmatrix}
x_1 & x_2 & x_3 \\
0 & 0 & \sqrt{2} \\
0 & 0 & -\sqrt{2} \\
-1 & -1 & 1/\sqrt{2} \\
1 & -1 & 1/\sqrt{2} \\
-1 & 1 & 1/\sqrt{2} \\
1 & 1 & 1/\sqrt{2} \\
\sqrt{2} & 0 & -1/\sqrt{2} \\
-\sqrt{2} & 0 & -1/\sqrt{2} \\
0 & \sqrt{2} & -1/\sqrt{2} \\
0 & -\sqrt{2} & -1/\sqrt{2} \\
\mathbf{0} & \mathbf{0} & \mathbf{0}
\end{bmatrix}
$$

$$
\mathbf{D}_{311B} = \begin{bmatrix}
x_1 & x_2 & x_3 \\
0 & 0 & \sqrt{6} \\
0 & 0 & -\sqrt{6} \\
-0.7507 & 2.1063 & 1 \\
2.1063 & 0.7507 & 1 \\
0.7507 & -2.1063 & 1 \\
-2.1063 & -0.7507 & 1 \\
0.7507 & 2.1063 & -1 \\
2.1063 & -0.7507 & -1 \\
-0.7507 & -2.1063 & -1 \\
-2.1063 & 0.7507 & -1 \\
\mathbf{0} & \mathbf{0} & \mathbf{0}
\end{bmatrix}
$$

Both the 311A and 311B are near-rotatable. Eleven design points include a center run to avoid near-singularity.

There are three hybrid designs for $k = 4$. They are given by

$$
\mathbf{D}_{416A} =
\begin{array}{cccc}
x_1 & x_2 & x_3 & x_4 \\
\left[\begin{array}{cccc}
0 & 0 & 0 & 1.7844 \\
0 & 0 & 0 & -1.4945 \\
-1 & -1 & -1 & 0.6444 \\
1 & -1 & -1 & 0.6444 \\
-1 & 1 & -1 & 0.6444 \\
1 & 1 & 1 & 0.6444 \\
-1 & -1 & 1 & 0.6444 \\
1 & 1 & 1 & 0.6444 \\
-1 & 1 & 1 & 0.6444 \\
1 & 1 & 1 & 0.6444 \\
1.6853 & 0 & 0 & -0.9075 \\
-1.6853 & 0 & 0 & -0.9075 \\
0 & 1.6853 & 0 & -0.9075 \\
0 & -1.6553 & 0 & -0.9075 \\
0 & 0 & 1.6853 & -0.9075 \\
0 & 0 & -1.6853 & -0.9075
\end{array}\right]
\end{array}
$$

$$
\mathbf{D}_{416C} =
\begin{array}{cccc}
x_1 & x_2 & x_3 & x_4 \\
\left[\begin{array}{cccc}
0 & 0 & 0 & 1.7658 \\
0 & 0 & 0 & 0 \\
\pm 1 & \pm 1 & \pm 1 & 0.5675 \\
\pm 1.4697 & 0 & 0 & -1.0509 \\
0 & \pm 1.4697 & 0 & -1.0509 \\
0 & 0 & \pm 1.4697 & -1.0509
\end{array}\right]
\end{array}
$$

$$
\mathbf{D}_{416B} =
\begin{array}{cccc}
x_1 & x_2 & x_3 & x_4 \\
\left[\begin{array}{cccc}
0 & 0 & 0 & 1.7317 \\
0 & 0 & 0 & -0.2692 \\
\pm 1 & \pm 1 & \pm 1 & 0.6045 \\
\pm 1.5177 & 0 & 0 & -1.0498 \\
0 & \pm 1.5177 & 0 & -1.0498 \\
0 & 0 & \pm 1.5177 & -1.0498
\end{array}\right]
\end{array}
$$

We have condensed the notation for 416C and 416B. The ± 1 notation indicates a 2^3 factorial in x_1, x_2, and x_3 with x_4 fixed. The design points following the 2^3 factorial are axial points in x_1, x_2, and x_3 with x_4 fixed. A center run is used in 416C to avoid near-simularity. The ± 1.4697 or ± 1.5177 notation implies two axial points.

There are two hybrid designs for six design variables. Both include 28 runs and thus are both saturated. The 628A design is as follows:

$$
\mathbf{D}_{628A} =
\begin{bmatrix}
x_1 & x_2 & x_3 & x_4 & x_5 & x_6 \\
0 & 0 & 0 & 0 & 0 & 4/\sqrt{3} \\
\pm1 & \pm1 & \pm1 & \pm1 & \pm1 & 1/\sqrt{3} \\
\pm2 & 0 & 0 & 0 & 0 & -2/\sqrt{3} \\
0 & \pm2 & 0 & 0 & 0 & -2/\sqrt{3} \\
0 & 0 & \pm2 & 0 & 0 & -2/\sqrt{3} \\
0 & 0 & 0 & \pm2 & 0 & -2/\sqrt{3} \\
0 & 0 & 0 & 0 & \pm2 & -2/\sqrt{3} \\
0 & 0 & 0 & 0 & 0 & 0
\end{bmatrix}
$$

The ±1 notation indicates a resolution V 2^{5-1} in x_1, x_2, x_3, x_4, and x_5 while x_6 is held at $1/\sqrt{3}$. The ±2 implies two axial points. A second six-variable design, the 628B, is constructed in a manner similar to that of 628A. The components of the design are as follows:

1. Resolution V 2^{5-1} in $x_1 - x_5$ with $x_6 = 0.6096$
2. Axial points in $x_1 - x_5$ with $\alpha = 2.1749$ and $x_6 = -1.0310$
3. Two additional center points with $x_6 = 2.397$ and -1.8110

General Comments Concerning the Hybrid Designs

As we indicated earlier, the hybrid designs represent a very efficient class of saturated or near-saturated second-order designs. Unlike the small composite designs, the hybrid designs are quite competitive with central composite designs *when the criteria for comparison takes design size into account* —for example, when one makes comparisons using N Var $\hat{y}(\mathbf{x})/\sigma^2$. Further discussion of design comparisons are in Sections 8.2 and 8.3.

It has been our experience that hybrid designs are not used as much in industrial applications as they should. In general, the use of saturated or near-saturated response surface designs should be avoided. However, it is inevitable that they will find use in applications where cost constraints are major obstacles. In this case the hybrid design is a good choice. It is quite likely that many practitioners are not aware of the class of hybrid designs. It is equally likely that potential users are reluctant to use the hybrid designs because of the "messy levels" required of the extra design variable—that is, the variable not involved in the central composite structure. One must remember that the levels reported in the design matrices in this text are those solved for by Roquemore in order to achieve certain ideal design conditions. An efficient design will result if one merely approximates these levels. Minor "errors in control" do not alter the efficiencies of these designs. One should assign this extra design variable to a factor that is easiest to alter and thus accommodates these levels. Of course, many practitioners are attracted to designs that involve three evenly spaced levels or levels that provide little difficulty when the experiment is conducted.

8.1.4 Some Saturated or Near-Saturated Cuboidal Designs

During the 1970s and early 1980s, considerable attention was given to the development of efficient second-order designs on a cube. This movement settled down considerably in the 1980s, and attention shifted to developments that led to general computer-generated experimental designs. Computer-generated design will be discussed in general in Section 8.2. However, there remain several interesting design classes for a cube that have potential use. Designs developed by Notz (1982), Hoke (1974), and Box and Draper (1974) are particularly noteworthy. As an example, the Notz three-variable design involves seven design points from a 2^3 factorial plus three one-factor-at-a-time axial points. Specifically, this design is given by

$$D = \begin{bmatrix} x_1 & x_2 & x_3 \\ -1 & -1 & 1 \\ 1 & -1 & -1 \\ -1 & 1 & -1 \\ 1 & 1 & 1 \\ 1 & 1 & -1 \\ 1 & -1 & 1 \\ -1 & 1 & 1 \\ 1 & 0 & 0 \\ 0 & 1 & 0 \\ 0 & 0 & 1 \end{bmatrix}$$

A design consisting of one additional point, namely $(+1, +1, +1)$ to complete the 2^3 factorial, resulting in 11 design points, is a nice augmentation of the three factor Notz design. The four-variable Notz design is constructed in a similar manner, with 11 points from the 2^4 factorial plus four axial points.

The Hoke designs represent combinations of factorial, axial, and edge points that create efficient second-order arrays for $k = 3, 4, 5,$ and 6. For specific design matrices, one should refer to the references listed above.

8.2 GENERAL CRITERIA FOR CONSTRUCTING, EVALUATING, AND COMPARING EXPERIMENTAL DESIGNS

As we indicated in the previous section, the 1980s ushered in an experimental design era in which considerable attention was drawn to the use of the computer for construction of designs for the user. This period brought many new users into the experimental design arena, simply because it became easier to find an "appropriate" design. While the virtues of **computer-generated designs** through expert systems are considerable in number, there are also negative aspects. Often users treat the computer package like a black box without really understanding what criteria are being used for constructing the design. There certainly are practical situations in which the computer is a vital tool for design construction in spite of the fact that many of the standard RSM designs discussed in this chapter and Chapter 7 are extremely useful designs.

Much of what is available in evaluation and comparison of RSM designs as well as computer-generated design were results of the work of Kiefer (1959, 1961) and Kiefer and Wolfowitz (1959) in **optimal design theory**. Their work is couched in a measure theoretic approach in which an experimental design is viewed in terms of design measure. Design optimality moved into the practical arena in the 1970s and 1980s as designs were put forth as being efficient in terms of criteria inspired by Kiefer et al. Computer algorithms were developed that allowed "best" designs to be generated by a computer package based on the practitioner's choice of sample size, model, ranges on variables, and other constraints. In the next section, we will review some practical notions of design optimality. This will lead naturally to criteria for comparison and evaluation of designs as well as computer-generated designs.

8.2.1 Practical Design Optimality

Design optimality criteria are characterized by letters of the alphabet and, as a result, are often called **alphabetic optimality criteria.** The best known and most often used criterion is D-optimality.

D-Optimality and D-Efficiency
D-Optimality is based on the notion that the experimental design should be chosen so as to achieve certain properties in the *moment matrix*

$$\mathbf{M} = \frac{\mathbf{X'X}}{N} \tag{8.6}$$

The reader recalls the importance of the elements of the moment matrix in the determination of rotatability. Also, the inverse of $\mathbf{M}$, namely

$$\mathbf{M}^{-1} = N(\mathbf{X'X})^{-1} \tag{8.7}$$

(the *scaled dispersion matrix*), contains variances and covariances of the regression coefficients, scaled by N/σ^2. As a result, control of the moment matrix by design implies control of the variances and covariances.

It turns out that an important norm on the moment matrix is the determinant; that is,

$$|\mathbf{M}| = \frac{|\mathbf{X}'\mathbf{X}|}{N^p} \qquad (8.8)$$

where p is the number of parameters in the model. Under the assumption of independent normal model errors with constant variance, the *determinant of* $\mathbf{X}'\mathbf{X}$ *is inversely proportional to the square of the volume of the confidence region on the regression coefficients.* The volume of the confidence region is relevant because it reflects how well the set of coefficients are estimated. A small $|\mathbf{X}'\mathbf{X}|$ and hence large $|(\mathbf{X}'\mathbf{X})^{-1}| = 1/|\mathbf{X}'\mathbf{X}|$ implies poor estimation of β in the model. A development demonstrating the relationship between $|\mathbf{X}'\mathbf{X}|$ and the volume of a joint confidence region on β is given in Appendix 7. *A D-optimal design is one in which* $|\mathbf{M}| = |\mathbf{X}'\mathbf{X}|/N^p$ *is maximized*; that is,

$$\underset{\zeta}{\text{Max}}\,|\mathbf{M}(\zeta)| \qquad (8.9)$$

where Max implies that the maximum is taken over all designs. As a result, it is natural to define the *D*-efficiency of a design ζ^* as

$$D_{\text{eff}} = \left(|\mathbf{M}(\zeta^*)| / \underset{\zeta}{\text{Max}}\,|\mathbf{M}(\zeta)| \right)^{1/p} \qquad (8.10)$$

Here, the $1/p$ power accounts for the p parameter estimates being assessed when one computes the determinant of the variance covariance matrix. The definition of *D*-efficiency in Equation (8.10) allows for comparing designs that have different sample sizes by comparing *D*-efficiencies.

There have been many studies that have been conducted to compare standard experimental designs through the use of *D*-efficiency. Readers are referred to Lucas (1976) for comparisons among some of the designs that we have discussed in this chapter and in Chapter 7. A paper by St. John and Draper (1975) gives an excellent review of *D*-optimality and provides much insight into pragmatic use of design optimality. One should also read Box and Draper (1971), Myers et al. (1989), and a recent text by Puhkelsheim (1995).

Recall that in Chapter 7 we discussed the notion of variance optimal designs in the first-order and first-order-plus-interaction case. While our main objective in this chapter is to continue dealing with second-order designs, it is natural to consider *D*-optimality and *D*-efficiency in the case of simpler models. Consider the variance-optimal design, namely the orthogonal design with all levels at the ± 1 *extremes of the experimental region*. The moment matrix is given by

$$\mathbf{M} = \frac{\mathbf{X'X}}{N} = \mathbf{I}_p$$

It is not difficult to show that for the first-order and first-order-plus-interaction cases, optimality extends to the determinant, that is,

$$\underset{\zeta}{\text{Max }} \mathbf{M}(\zeta) = 1,$$

and thus for these simpler models, the so-called "variance optimal" design is also optimal in the determinant sense—that is, D-optimal.

Before we embark on some comparisons among standard designs and discussion of how D-optimality impacts practical tools in the use of computer-generated design, we offer a few other important optimality criteria and methods of comparison.

A-Optimality

The concept of **A-optimality** deals with the individual variances of the regression coefficients. Unlike D-optimality, it does not make use of covariances among coefficients; recall that the variances of regression coefficients appear on the diagonals of $(\mathbf{X'X})^{-1}$. A-optimality is defined as

$$\underset{\zeta}{\text{Min }} \text{tr}\big(\mathbf{M}(\zeta)\big)^{-1} \tag{8.11}$$

where tr represents trace, that is, the sum of the variances of the coefficients (weighted by N). Some computer-generated design packages make use of A-optimality.

Criteria Dealing in the Predictions Variance (G- and Q-Optimality)

Throughout Chapter 7, we consistently made reference to the use of the scaled prediction variance $N \text{ Var } \hat{y}(\mathbf{x})/\sigma^2 = v(\mathbf{x})$ as an important measure of performance. Our feeling is that practical designers of experiments don't make use of this measure as they should. One clear disadvantage, of course, is that, unlike $\text{tr}(\mathbf{M}^{-1})$ or $|\mathbf{M}|$, $v(\mathbf{x})$, is not a single number but rather depends on the location $\mathbf{x}$ at which one is predicting. In fact, recall that

$$v(\mathbf{x}) = \frac{N \text{ Var } \hat{y}(\mathbf{x})}{\sigma^2} = N\mathbf{x}'^{(m)}(\mathbf{X'X})^{-1}\mathbf{x}^m$$

where $\mathbf{x}^m$ reflects location in the design space as well as the nature of the model. In fact, in previous discussions in Chapter 7, we focused on attempts by proper design (choice of center runs) to stabilize $v(\mathbf{x})$.

One interesting design optimality criterion that focuses on $v(\mathbf{x})$ is G-optimality. **G-optimality** and the corresponding G-efficiency emphasizes the use of designs for which the *maximum $v(\mathbf{x})$ in the region of the design is not too large*. A G-optimal design ζ is one in which we have

$$\underset{\zeta}{\text{Min}} \left[\underset{\mathbf{x} \in R}{\text{Max}} v(\mathbf{x}) \right]$$

Also, note that this is equivalent to

$$\underset{\zeta}{\text{Min}} \left[\underset{\mathbf{x} \in R}{\text{Max}} \left\{ \mathbf{x}'^{(m)} [\mathbf{M}(\zeta)]^{-1} \mathbf{x}^{(m)} \right\} \right]$$

because $N \operatorname{Var} \hat{y}(\mathbf{x}) / \sigma^2$ is merely a quadratic form in $[\mathbf{M}(\zeta)]^{-1}$. Of course, the natural selection of regions for R are the cube and sphere. The resulting G-efficiency is conceptually quite simple to grasp. It turns out (see Appendix 8) that under the standard independence and homogeneous variance on the model errors,

$$\underset{\mathbf{x} \in R}{\text{Max}} \left[v(\mathbf{x}) \right] \geq p \qquad (8.12)$$

The result in Equation (8.12) is very important and usually very surprising to practitioners. One must keep in mind that $v(\mathbf{x}) = N \operatorname{Var} \hat{y}(\mathbf{x}) / \sigma^2$ is a scale-free quantity. A G-efficiency is very easily determined for a design, say ζ, as

$$G_{\text{eff}} = \frac{p}{\underset{\mathbf{x} \in R}{\text{Max}} v(\mathbf{x})} \qquad (8.13)$$

It is always instructive to investigate efficiencies of our two-level designs of Chapters 4 and 6 with regard to first-order and first-order-plus-interaction models. Consider, for example, a 2^2 factorial in the case of the first-order model

$$y = \beta_0 + \beta_1 x_1 + \beta_2 x_2 + \varepsilon$$

with the region R being described by $-1 \leq x_1 \leq 1$ and $-1 \leq x_2 \leq 1$. We know that

$$\mathbf{M} = \mathbf{X}'\mathbf{X} / N = \mathbf{I}_3$$

and

$$v(\mathbf{x}) = [1, x_1, x_2] \mathbf{I}_3 \begin{bmatrix} 1 \\ x_1 \\ x_2 \end{bmatrix}$$

$$= 1 + x_1^2 + x_2^2$$

As a result, the maximum of $v(\mathbf{x})$ occurs at the extremes of the cube, namely at $x_1 = \pm 1$, $x_2 = \pm 1$. In addition, it is obvious that for the 2^2 factorial,

$$\underset{\mathbf{x} \in R}{\text{Max}} [v(\mathbf{x})] = 3 = p$$

Thus, the G-efficiency of the 2^2 design is 1.0. As a result, the design is G-optimal. In fact, it is simple to show that for the first-order model in k design variables for the cuboidal region, a two level resolution $\geq$ III design with levels at ± 1 results in

$$\underset{\mathbf{x} \in R}{\text{Max}} [v(\mathbf{x})] = p$$

Thus these designs are G-optimal for the first-order model.

Another design optimality criterion that addresses prediction variance is Q-optimality, or IV optimality. The attempt here is to generate a single number. This is done through an averaging process; that is, $v(\mathbf{x})$ is averaged over some region of interest R. The averaging is accomplished via the integration over R. The corresponding division by the volume of R produces an average. The Q-optimal design is given by

$$\underset{\zeta}{\text{Min}} \frac{1}{K} \int_R v(\mathbf{x}) \, d\mathbf{x} = \underset{\zeta}{\text{Min}} Q(\zeta) \tag{8.14}$$

where $K = \int_R d\mathbf{x}$. Notice, then, that we can write the criterion for a design ζ as

$$\underset{\zeta}{\text{Min}} \left\{ \frac{1}{K} \int_R \mathbf{x}'^{(m)} [\mathbf{M}(\zeta)]^{-1} \mathbf{x}^{(m)} \, d\mathbf{x} \right\} = \underset{\zeta}{\text{min}} Q(\zeta) \tag{8.15}$$

In this way the Q-optimal, or IV optimal, design is that in which the *average scaled prediction variance* is minimized. Strictly speaking, the Q-criterion given in Equation (8.15) is a special case of a more general criterion in which the scaled prediction variance is multiplied by a weight function $w(\mathbf{x})$. The utility of the weight function is to weight more heavily certain parts of R. Of course, in most practical situations $w(\mathbf{x}) = 1.0$.

The Q-optimality criterion is conceptually a very reasonable device to use for choosing experimental designs. The concept in Equation (8.15) in a different form does enjoy some use in computer generated designs. However, neither G nor Q is used as much as D-optimality. We will shed more light on this in the next section. The Q-criterion leads nicely to a Q-efficiency for design ζ^* as

$$Q_{\text{eff}} = \underset{\zeta}{\text{Min}} [Q(\zeta)] / Q(\zeta^*)$$

As in previous cases, it is instructive to apply the Q-criterion to a simple case, namely a first-order model. Consider, again, the model

$$y = \beta_0 + \sum_{j=1}^{k} \beta_j x_j + \varepsilon$$

Let us assume that the region is once again given by the cuboidal region $-1 \le x_j \le 1$. The Q-criterion can be written

$$Q(\zeta) = \frac{1}{K} \int_R v(\mathbf{x}) \, d\mathbf{x}$$

$$= \frac{N}{K} \int_R \mathbf{x}^{(m)'} (\mathbf{X}'\mathbf{X})^{-1} \mathbf{x}^{(m)} \, d\mathbf{x} \qquad (8.16)$$

where $\mathbf{x}^{(m)'} - [1, x_1, x_2, \ldots, x_k]$. Here, of course, $\int_R$ implies $\int_{-1}^{1} \int_{-1}^{1} \cdots \int_{-1}^{1}$. Because $\mathbf{x}^{(m)'}(\mathbf{X}'\mathbf{X})^{-1}\mathbf{x}^{(m)}$ is a scalar, we can write

$$Q(\zeta) = \frac{N}{K} \int_R \mathrm{tr}\left[\mathbf{x}^{(m)'}(\mathbf{X}'\mathbf{X})^{-1}\mathbf{x}^{(m)}\right] d\mathbf{x}$$

$$= \frac{N}{K} \int_R \mathrm{tr}\, \mathbf{x}^{(m)}\mathbf{x}^{(m)'}(\mathbf{X}'\mathbf{X})^{-1} \, d\mathbf{x}$$

$$= \mathrm{tr}\left[N(\mathbf{X}'\mathbf{X})^{-1}\right]\left[\frac{1}{K}\int_R \mathbf{x}^{(m)}\mathbf{x}^{(m)'} \, d\mathbf{x}\right] \qquad (8.17)$$

Now, the matrix $(1/K)\int_R \mathbf{x}^{(m)}\mathbf{x}^{(m)'} \, d\mathbf{x}$ is a $(k+1) \times (k+1)$ matrix of **region moments**, that is, the diagonal elements (apart from the initial diagonal element which is 1.0) are the $(1/K)\int_R x_i^2 \, d\mathbf{x}$. For the off-diagonals, we have the $(1/K)\int_R x_i \, d\mathbf{x}$ in the first row and first column and $(1/K)\int_R x_i x_j \, d\mathbf{x}$ for $i \ne j$ as the other off-diagonal elements. Because the region R is symmetric, $\int_R x_i \, d\mathbf{x}$ and $\int_R x_i x_j \, d\mathbf{x}$ are both zero. As a result, Equation (8.17) becomes

$$Q(\zeta) = \mathrm{tr}[\mathbf{M}(\zeta)]^{-1}[\mathbf{D}_R] \qquad (8.18)$$

when $\mathbf{D}_R$ is a diagonal matrix given by

$$\mathbf{D}_R = \begin{bmatrix} 1 & 0 & \cdots & 0 \\ 0 & & & \\ \vdots & & \frac{1}{3}I_k & \\ 0 & & & \end{bmatrix}$$

because $(1/K)\int_R x_i^2 = (1/2^k)(2^k)(\frac{1}{3})$.

Now, Equation (8.18) can be written as a simple function of the variances of the least squares estimators $b_0, b_1, b_2, \ldots, b_k$ of the coefficients in the first-order model. In fact, we have,

$$Q(\zeta) = N\left[\text{Var}(b_0) + \frac{1}{3}\sum_{i=1}^{k}\text{Var}(b_i)\right] \qquad (8.19)$$

The value of $Q(\zeta)$, the scaled prediction variance, averaged over the cuboidal region, is minimized by a design for which the variances of all coefficients are simultaneously minimized. This, as we learned in Chapter 7, is accomplished by the "variance-optimal" design, namely the two-level first-order orthogonal design (resolution $\geq$ III) with levels set at the ± 1 extremes.

If we change the region R from the cuboidal region to a spherical region, the two level resolution $\geq$ III design is both D- and Q-optimal. Again, the requirement remains that the design be first-order orthogonal with two levels set at the perimeter of the experimental region (see Exercise 8.8).

8.2.2   Use of Design Efficiencies for Comparison of Standard Second-Order Designs

Design optimality is an extremely interesting and useful concept. Considerable interest has been drawn to design of experiments and RSM because of the existence of computer packages that allow the user to generate designs with the use of the notion of alphabetic optimality. D-Optimality in particular has received much usage and attention, primarily because it lends itself to relatively simple computation and because algorithms were developed early to accommodate it. However, the reader should bear in mind the 10 or more desirable properties of an RSM experimental design. They suggest that what is required in practice, if they can be obtained, are designs that reflect several important properties, not designs that are optimal with respect to a single criterion. They suggest that what one desires are **good designs**—that is, designs that will produce reliable results under a wide variety of possible circumstances, not designs that are optimal under a fairly restrictive and idealized set of conditions and assumptions. In the case of first-order models and first-order-plus-interaction models, standard designs are optimal designs. We have seen some evidence of this in this chapter; but in the second-order case, standard RSM designs that we discussed in Chapter 7 are rarely optimal designs, and yet we know that they can be constructed to achieve many desirable properties, properties listed among the 10 given in Chapter 7.

For purposes of an RSM analysis, the standard central composite or Box–Behnken designs should be used whenever possible. When economical designs are needed (near-saturated), the hybrid designs should be used whenever possible. *This does not imply that optimal designs through computer-generated designs have no value.* It can be extremely useful and perhaps necessary in many circumstances. There are many conditions encountered by the user under which the use of a CCD or BBD is impossible. In particular, we shall see in Chapter 12 that the use of *mixture designs* often necessitates the use of computer-generated designs due to restrictions on the factor levels and design sizes. Even in nonmixture situations, constraints will often rule out the use of a standard design. We will focus on optimal designs in the context of computer-generated designs later in this chapter. But, it seems reasonable that the reader may gain some further insight into the virtues of standard designs if we discuss the performance of these designs in the context of design efficiencies. In what follows, we investigate the CCD, BBD, hybrid, and certain other saturated or near-saturated designs.

It should be emphasized that because the essentials of design optimality as introduced by Kiefer considers an experimental design as a probability measure, most finite designs (i.e., N point designs) will naturally have design efficiencies less than 1.0. As a result, it is natural that designs with small design size will have small efficiencies. The information in Table 8.3 give D- and G-efficiencies for various CCD, BBD, and hybrid designs for a spherical region. The tables are for k values from 2 through 5. For the CCD, $\alpha - \sqrt{k}$. (Further information regarding these design efficiencies can be found in Lucas (1974, 1976) and in Nalimov et al. (1970).) Several items are noteworthy. The **central composite and Box–Behnken designs are quite efficient**. Note that while the design efficiencies are highest for $n_c - 2$, one should not view these results as an indication that there is no advantage in using additional center runs. The use of 3–5 center runs, according to our recommendation in Chapter 7, brings stability in $v(\mathbf{x})$ and produces important pure error degrees of freedom. Note that for $k = 4$, the CCD and BBD have identical efficiencies. This stems from the fact that the BBD is merely a rotation of the CCD for $k - 4$. The CCD is rotatable and the variance properties do not change with rotation of the design.

The lack of G-efficiency in the CCD for a single center run should not be surprising. Recall that for $n_c = 0$ and a spherical region, $\mathbf{X}'\mathbf{X}$ is either singular or near singular. The use of only one center run results in a large value of $v(\mathbf{x})$ at the design center. As additional center runs are added, values of $v(\mathbf{x})$ grow nearer to a condition of stability; it is the nature of the G- and D-criteria to react adversely to the use of center runs in most cases.

It is natural to expect saturated or near-saturated designs to have low efficiencies. As a result, the results for some of the hybrid designs are quite impressive, especially in the case of the 311B and 416A, 416B, and 416C. Though we only included results for k from 2 through 5, it should be noted that the D-efficiencies and G-efficiencies for the 628A are both 100%, while the corresponding efficiencies for the 628B with a single center run are 90.6% and 84.4%. Further comparisons of hybrid designs with other economical designs such as the small composite design will be illustrated later. While we will not show tables of design efficiencies for the case of the cuboidal region, the following represent a synopsis of results for the CCD,

Table 8.3   *D*- and *G*-Efficiencies for Some Standard Second-Order Designs for a Spherical Region

Number of Factors	Sample Size	Design	D_{eff} (%)	G_{eff} (%)
2	9	CCD($n_c = 1$)	98.62	66.67
2	10	CCD($n_c = 2$)	99.64	96.0
2	11	CCD($n_c = 3$)	96.91	87.27
3	15	CCD($n_c = 1$)	99.14	66.67
3	16	CCD($n_c = 2$)	99.61	94.59
3	17	CCD($n_c = 3$)	97.63	89.03
3	13	BBD($n_c = 1$)	97.0	76.92
3	16	BBD($n_c = 2$)	96.53	71.43
3	15	BBD($n_c = 3$)	93.82	66.67
3	10	310	72.8	47.4
3	11	311A	79.1	78.6
3	11	311B	83.1	90.9
4	25	CCD($n_c = 1$)	99.23	60.0
4	26	CCD($n_c = 2$)	99.92	98.90
4	27	CCD($n_c = 3$)	98.86	95.24
4	25	BBD($n_c = 1$)	99.23	60.0
4	26	BBD($n_c = 2$)	99.92	98.70
4	27	BBD($n_c = 3$)	98.86	95.24
4	16	416A	81.7	8.8
4	17	416A($n_c = 1$)	90.6	74.3
4	16	416B	96.3	63.1
4	17	416B($n_c = 1$)	95.1	70.1
4	16	416C	96.9	78.0
5	43	CCD($n_c = 1$)	98.6	48.84
5	44	CCD($n_c = 2$)	99.60	87.63
5	45	CCD($n_c = 3$)	99.28	85.56
5	27	CCD($\frac{1}{2}$ rep)($n_c = 1$)	98.43	77.78
5	28	CCD($\frac{1}{2}$ rep)($n_c = 2$)	98.10	87.64
5	29	CCD($\frac{1}{2}$ rep)($n_c = 3$)	96.57	84.62
5	41	BBD($n_c = 1$)	97.93	51.22
5	42	BBD($n_c = 2$)	98.83	90.91
5	43	BBD($n_c = 3$)	98.41	88.79

Hoke, and Box–Draper designs. Again, for specific construction of the Hoke or Box–Draper design, one should consult Box and Draper (1974) and Hoke (1974).

1. The CCD for $\alpha = 1.0$ is quite efficient, particularly in the D sense. For $k = 3$, 4, and 5, this design has D-efficiency from approximately 87% to 97% *without center runs*. Center runs reduce D-efficiency and change G-efficiency very little.

2. The Hoke D_2 and D_6 designs have D-efficiencies that range from 80% to 97%. This is quite good given the small design size. The

G-efficiencies are considerably lower, but this is not unexpected with small designs. In general, the Hoke designs perform better than the Box and Draper designs.

In the next section we focus once again on design comparison. We make use of graphical methods in order to gain some perspective about the distribution of $v(\mathbf{x})$ in the design region. Our purpose is to compare standard designs and also point the reader toward computational methodology that can be used to evaluate designs in practice.

8.2.3 Graphical Procedure for Evaluating Prediction Capability of an RSM Design

Design efficiencies are single numbers that purport to evaluate the capabilities of an experimental design. In cases of interest here, the designs are RSM designs. Design efficiencies can be beneficial. Indeed, they were of benefit to us in the previous section. They enabled the reader to understand that standard RSM designs are often fairly close to the "best possible" design in the *D*-optimality or *G*-optimality sense. One must remember that in almost all cases (hybrid 628A being an exception) the best finite sample size designs have efficiencies less than 100%.

When one is interested in stability of $v(\mathbf{x})$ over the design region, a single number type of criterion such as a design efficiency may not capture the true design capability, which is actually a multidimensional concept. There is very useful information in a methodology that allows the user to know values of $v(\mathbf{x})$ throughout the design region. Obviously for $k = 2$, two-dimensional plots of $v(\mathbf{x})$ can easily be constructed. This type of plot can be used to compare designs. In addition, for a specific RSM analysis, if data have been collected, a plot showing contours of constant Vâr $\hat{y}(\mathbf{x})$ or $\sqrt{\text{Vâr } \hat{y}(\mathbf{x})}$ can aid the practitioner in visualizing confidence limits on mean response at interesting locations in the design region. The following example provides an illustration.

Example 8.2 Consider the $k = 2$ illustration of Example 6.1. The design used is a second-order rotatable CCD. Contours of constant response are shown in Figure 6.2. Figure 8.1 is a two-dimensional plot showing the estimated prediction standard deviation or estimated standard error of prediction

$$\sqrt{\hat{\text{V}}\text{ar } \hat{y}(\mathbf{x})} = s\sqrt{\mathbf{x}^{(m)\prime}(\mathbf{X}'\mathbf{X})^{-1}\mathbf{x}^{(m)}}$$

where s is the root mean square error. Note the stability on spheres. Note also that the prediction begins to break down as one moves outside the design region.

DESIGN-EXPERT Analysis

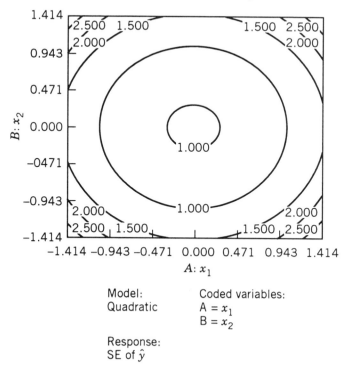

Model: Coded variables:
Quadratic   A = x_1
                B = x_2

Response:
SE of $\hat{y}$

Figure 8.1 Contours of constant standard error of prediction.

For values of $k > 2$, contours of prediction variance are, of course, considerably more unruly. However, there are other graphical procedures that allow comparison of the distribution of $v(\mathbf{x})$ among RSM designs for any number of design variables. These graphical procedures make use of **variance dispersion graphs**. The following sections define variance dispersion graphs and illustrate their use.

Variance Dispersion Graphs (VDGs)

Giovannitti-Jensen and Myers (1984) and Myers et al. (1992b) developed the notion of the variance dispersion graphs and produced case studies that allowed for interesting comparisons between standard designs. Rozum (1990) and Rozum and Myers (1991) extended the work from spherical to cuboidal regions.

A variance dispersion graph for an RSM design displays a "snapshot" that allows the practitioner to gain an impression regarding the stability of $v(\mathbf{x}) = N \, \text{Var} \, \hat{y}(\mathbf{x})/\sigma^2$ and how the design compares to an "ideal." Let us first assume a spherical design. The VDG contains three graphical components:

1. A plot of the **spherical variance** V^r against the radius r. The spherical variance is essentially $v(\mathbf{x})$ averaged *over the surface* of a sphere of radius r, namely,

$$V^r = \frac{N\psi}{\sigma^2} \int_{U_r} \text{Var } \hat{y}(\mathbf{x}) \, d\mathbf{x} \qquad (8.20)$$

where U_r implies integration over the surface of a sphere of radius r and $\psi = (\int_{U_r} d\mathbf{x})^{-1}$.

2. A plot of the maximum $v(\mathbf{x})$ on a radius r against r. That is,

$$\max_{\mathbf{x} \in U_r} \left[\frac{N \text{ Var } \hat{y}(\mathbf{x})}{\sigma^2} \right] = \max_{\mathbf{x} \in U_r} \left[v(\mathbf{x}) \right]$$

3. A plot of the minimum $v(\mathbf{x})$ on a radius r against r.

In addition to the above three plots which offer information about $v(\mathbf{x})$ for the design being evaluated, the result of Equation (8.12) is used to provide some sense of closeness to ideal. A horizontal line at $v(\mathbf{x}) = p$ is added. Obviously if the design is rotatable, the entire VDG will involve only the spherical variance and the horizontal line.

It is of interest to discuss briefly the computations associated with the spherical variance. It turns out that the integral in Equation (8.20) can be written (see Appendix 9)

$$V^r = \frac{N\psi}{\sigma^2} \int_{U_r} \mathbf{x}'^{(m)} (\mathbf{X'X})^{-1} \mathbf{x}^{(m)} \, d\mathbf{x}$$

$$= \text{tr}(\mathbf{X'X})^{-1} \mathbf{S} \qquad (8.21)$$

where $\mathbf{S}$ is the $p \times p$ matrix of region moments, with the region being the hypersphere given by U_r.

In the following section we illustrate the use of the VDG in a study to determine the utility of center runs with a rotatable CCD for $k = 5$.

Example 8.3 Consider a central composite design in five design variables with the factorial portion being the resolution V fraction of a 2^5. The value of α is 2.0 and hence the design is rotatable. As a result, $v(\mathbf{x})$ is constant on spheres and hence $\max_{\mathbf{x} \in R} [v(\mathbf{x})] = \min_{\mathbf{x} \in R} [v(\mathbf{x})] = V^r$. The plot in Figure 8.2 shows five VDGs for the number of center runs $n_c = 1$ through $n_c = 5$. The horizontal line is at $p = 21$. Recall that the maximum value of $v(\mathbf{x})$ in the design region can be no smaller than p.

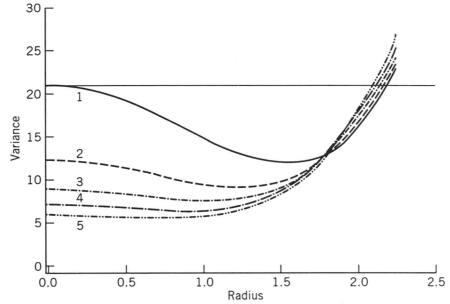

Figure 8.2   VDG showing the effect of center runs for a $k = 5$ CCD with factorial portion being $\frac{1}{2}$ fraction.

The plot in Figure 8.2 reflects values in the region $0 \le r \le \sqrt{5}$ because the factorial points are at radius $\sqrt{5}$. From the plot one can see that the design becomes more efficient as the number of center runs get larger. The effect is greatest near the design center. Note also that the improvement in $v(\mathbf{x})$ near the center becomes marginal after $n_c = 4$.

Example 8.4   In this illustration we use the variance dispersion graph to show a comparison of the distribution of $v(\mathbf{x})$ in the design region between a CCD and a BBD. The reader should recall that both of these very popular designs are spherical designs. Both designs involve three design variables with the CCD having $\alpha = \sqrt{3}$ and $n_e = 3$, while the BBD also contains three center runs. One must keep in mind that the total sample sizes are 17 and 15 for the CCD and BBD, respectively. The plot with both VDGs is found in Figure 8.3. Note that both designs contain three plots in the VDG; the min $v(\mathbf{x})$, average $v(\mathbf{x})$, and max $v(\mathbf{x})$ are shown because neither are rotatable. The plots stop at radius $\sqrt{3}$. *Both designs* have been scaled so that points are at a radius $\sqrt{3}$ from the design center.

The following represent obvious conclusions from the VDGs:

1. The $\alpha = \sqrt{3}$ CCD is nearly rotatable. This is not surprising because $\alpha = 1.682$ results in exact rotatability. The slight deviation from rotatability occurs close to the design perimeter. The BBD shows a more

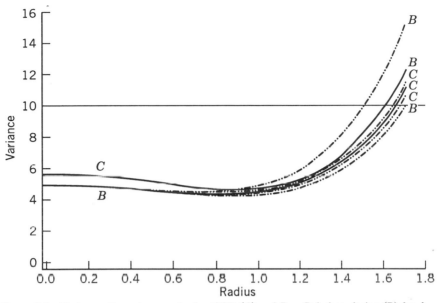

Figure 8.3 Variance dispersion graphs for CCD (C) and Box–Behnken design (B) for $k = 3$. Both designs contain three center runs.

pronounced deviation from rotatability that becomes greater as one nears the perimeter.

2. The values of $v(\mathbf{x})$ are very close for the two designs near the design center. In fact, the difference at the center is accounted for by the difference in sample sizes.

3. For prediction from radius 1.0 to $\sqrt{3}$ the CCD appears to be the better design. The stability in $v(\mathbf{x})$ is greater with the CCD, and max $v(\mathbf{x})$ is smaller than that of the BBD.

4. The comparison with the ideal design is readily seen for both designs. The maximum $v(\mathbf{x})$ in the design region is roughly 15 for the BBD and roughly 11.2 for the CCD. Thus the G-efficiencies for the two designs are $(10/11.2) \times 100 = 89\%$ for the CCD and $(10/15) \times 100 = 67\%$ for the BBD. These numbers correspond to those given in Table 8.3.

The illustration in this example nicely underscores the usefulness of the VDG for comparing competing designs. It also illustrates how a single number efficiency, while useful, cannot capture design performance. For example, the clear distinction between the two designs in terms of prediction errors is not captured in the relative comparison of D-efficiencies (97.0% and 93.82) from Table 8.3. In addition, the G-efficiencies accurately reflect the differences that exist in $v(\mathbf{x})$, *but only at the design perimeter*. Almost without exception, the max $v(\mathbf{x})$ in a design region for a second-order design (com-

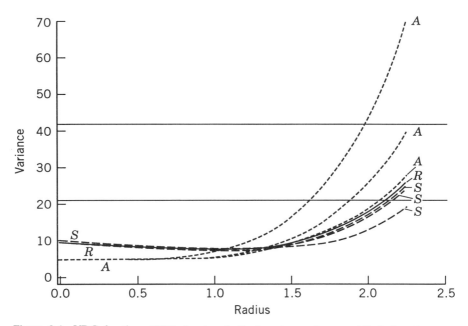

Figure 8.4 VDG for three CCDs for $k = 5$. Design A contains $\alpha = 1.5$, design R contains $\alpha = 2.0$, and design S contains $\alpha = \sqrt{5}$. Each contains a 2^{5-1} and three center runs.

pletely without exception in first-order case) occurs at the design perimeter. Thus G-efficiency is reflecting what is occurring at the design perimeter.

In Figure 8.4 we show the VDGs for three CCDs with $k = 5$. The α-values are 1.5, 2.0, and $\sqrt{5}$. In each case the factorial portion is a resolution V fraction. Each design contains three center runs, and thus the total design size is $N = 29$. The $\alpha = 1.5$ design is not expected to predict as well near the perimeter because the axial points are not placed at or near the region perimeter. The $\alpha = 2.0$ design is rotatable because $\alpha = \sqrt[4]{16} = \sqrt[4]{F} = 2.0$.

The following are conclusions drawn from Figure 8.4.

1. The use of $\alpha = 1.5$ does result in a loss in efficiency compared to the other two designs. While the prediction is better in the design center, it suffers at the perimeter. In fact, one gains some insight on how much efficiency is lost with $\alpha = 1.5$ by observing that the value of max $v(\mathbf{x})$ at the design perimeter is roughly 65. The two horizontal lines are at p and $2p$, representing G-efficiencies of 100% and 50%. The G-efficiency for the $\alpha = 1.5$ design is roughly $(21 / 65) \times 100 \cong 32\%$.

2. While $\alpha = 2.0$ is the rotatable value, the use of $\alpha = \sqrt{5}$ and the resulting loss of rotatability does not produce a severe instability in $v(\mathbf{x})$. The same appeared to be true for the $\alpha = \sqrt{3}$ CCD in Figure 8.3. Again, this VDG reflects the fact that for the CCD, allowing the value of α to be pushed to the edge of the design perimeter, namely to $\sqrt{k}$, is

preferable, even if rotatability is compromised. Clearly the loss in rotatability is a trivial matter here and for the CCD in Figure 8.3. The max $v(\mathbf{x})$ plot for design S or design R reflects a G-efficiency of roughly 85%, which is consistent with the results in Table 8.3.

One pragmatic point needs to be made here regarding design S in Figure 8.4. For $k = 5$ and other large values, the "best" CCD requires $\alpha = \sqrt{k}$. This requires the levels set by the investigator to be $-2.236, -1, 0, 1, 2.236$. This design may, in fact, produce practical problems because the axial level is much larger than the factorial level. Indeed, the evenly spaced levels of design R may be more reasonable. For $k = 2, 3,$ and 4, this is not a serious problem.

Example 8.5 In previous sections we mentioned the importance of the class of hybrid designs in cases where the researcher requires the use of a small design size. The efficiencies of the hybrid design in Table 8.3 are quite good for saturated or near-saturated designs, though this may not be apparent because we showed no comparisons with other designs of small sample size. In Figure 8.5 we show overlayed VDGs of a hybrid 416B with four center runs and a $k = 4$ SCD with three center runs. The samll composite contains a resolution III 2^{4-1} with axial value $\alpha = 2.0$.

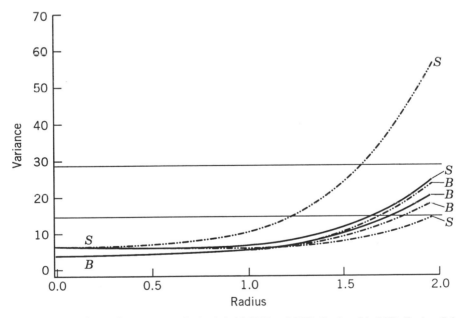

Figure 8.5 Variance dispersion graphs for hybrid 416B and SCD. Design S is SCD. Design B is hybrid 416B. $N = 19$ for the SCD and $N = 20$ for 416B.

From the p and $2p$ lines it is clear that the 416B is considerably more efficient. The G-efficiency of the 416B is roughly 60%, while the G-efficiency for the SCD is only slightly above 25%. More importantly, the max $v(\mathbf{x})$ for the SCD becomes large very "quickly." The discrepancy between the designs is not merely at the design perimeter.

Example 8.6   Figure 8.6 gives overlayed VDGs for the hybrid 310 with a center run added and the hybrid 311B. One should note from the efficiency table in Table 8.3 that both the D- and G-efficiency for the 311B are quite good compared to the 310. Indeed, from the figure it is clear that the 311B is nearly rotatable with high G-efficiency. However, what efficiency figures do not reflect is that prediction variance is much smaller for the 310 near the design center. Incidentally, the picture suggests that an additional center run for the 311B would substantially reduce $v(\mathbf{x})$ in the center. This, in fact, is the case. An additional run in the center reduces $v(\mathbf{x})$ to approximately the value achieved by the 310. Interestingly, this increases $v(\mathbf{x})$ at the perimeter slightly, *thereby reducing the G-efficiency*. Thus, we have often seen situations of a design (311B + one center run) that has a smaller G (and D in this case) efficiency than does a clearly inferior design (311B). Figure 8.7 shows a comparison of the same two designs as in Figure 8.6 with the additional center run given to the 311B.

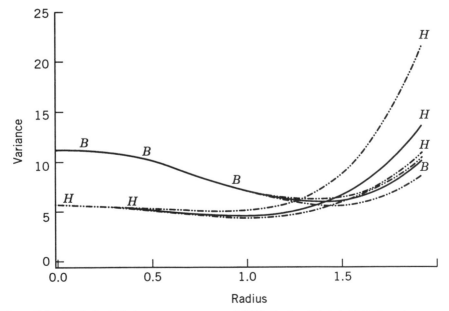

Figure 8.6   VDGs for 310 plus one center run and 311B. Design B is hybrid 311B and design H is hybrid 310.

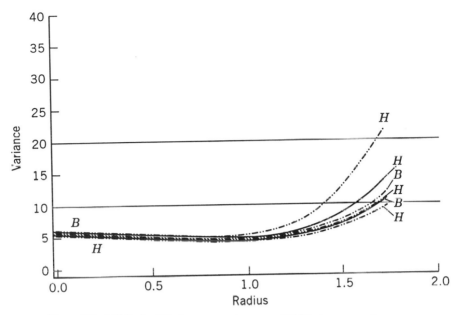

Figure 8.7   VDGs for 310 plus one center run and 311B plus one center run.

VDG for Cuboidal Designs

All of the preceding discussion and illustration deals with variance dispersion graphs for spherical regions. In this case it is natural to observe values of $v(\mathbf{x})$ averaging over the surfaces of spheres. Then the minimum and maximum of $v(\mathbf{x})$ is taken over the surfaces of spheres. However, it is not natural to deal with surfaces of spheres when the design is cuboidal, and hence the natural region of interest is a cube. As a result, the variance dispersion graph contains the following components:

1. A plot of the cuboidal variance V^{r*} against r, the half-size of the edge of a cube: V^{r*} is defined as

$$V^{r*} = \frac{N\psi^*}{\sigma^2} \int_{U_r^*} \left[\frac{\text{Var } \hat{y}(\mathbf{x})}{\sigma^2} \right] d\mathbf{x}$$

    where U_r^* is the surface area of the cube with half-size of edge r and $\int_{U_r^*}$ implies integration over the surface area of the same cube. In our illustrations r is called "radius." The term $\psi^* = (\int_{v_r^*} d\mathbf{x})^{-1}$

2. $\max_{\mathbf{x} \in U_r^*}[v(\mathbf{x})]$
3. $\min_{\mathbf{x} \in U_r^*}[v(\mathbf{x})]$

As a result, the average and range of $v(\mathbf{x})$ is depicted over the surface of cubes that are nested inside the cube defined by the design region. Figure 8.8

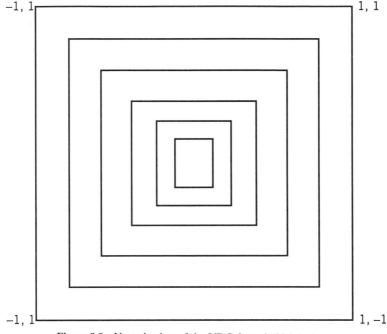

Figure 8.8   Nested cubes of the VDG for cuboidal designs.

gives the reader an impression of what is displayed by the VDG. Averages, maxima, and minima are computed for half-size of edges from 0 to 1.0. These values are plotted in a two-dimensional graph. Spheres nested inside the design cube can be used as the media for the graphs. See Myers et al. (1992b). However, cubes nested inside the design cube are more natural, just as rotatability [i.e., constant $v(\mathbf{x})$ on spheres] is a more important criterion for spherical regions of interest than for cuboidal regions.

Comparison of 3^3 Factorial with Face Center Cube

Although the CCD involves considerably fewer design points than the three-level factorial design, for $k = 3$, they are naturally rival designs when the design region is a cube. Of course, the scaled prediction variance will involve a sample size weighting of $N = 27$ for the factorial design and $N = 15$ for the $\alpha = 1.0$ CCD—that is, the face center cube. Figure 8.9 shows the VDGs for the two designs. Notice that the VDGs reflect that the designs are quite similar in performance, although the face center cube is more efficient. Certainly the face center cube is an important design for cuboidal regions. However, the VDG suggests that for the case of three design variables, if cost considerations allow a design size of $N = 27$, the full 3^3 factorial is a fine alternative.

Variance dispersion graphs for cuboidal regions are no less important than those for spherical regions. Both are useful tools for comparing competing

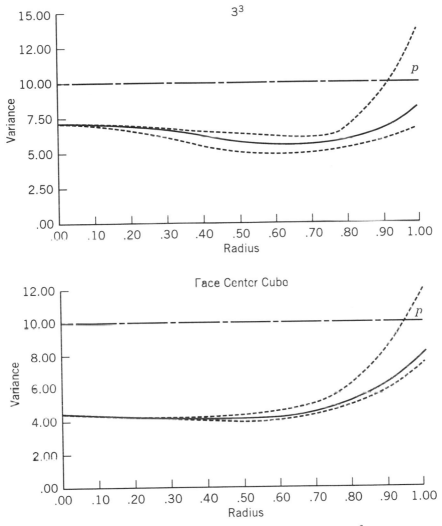

Figure 8.9   Variance dispersion graphs for face center cube and 3^3 factorial.

designs. The VDG will surface again later in this chapter as we focus on computer-generated designs. Information regarding software for construction of VDG can be found in Vining (1993) and Rozum and Myers (1991).

8.3   COMPUTER-GENERATED DESIGN IN RSM

Computer-generated design of experiments was the natural extension of design optimality as the computer era in statistics unfolded in the late 1960s and early 1970s. Today many computer packages generate designs for the user. While many criteria are available, the criterion that is used most often among all others is D-optimality or, rather, D-efficiency. Another variance-oriented criterion often used is based on A-efficiency. In addition, some computer packages offer a type of criterion that involves the use of $v(\mathbf{x})$, the scaled prediction variance. There were many important works that provided the fundamental theory that led to the current state of technology in computer-generated design. The work of Dykstra (1966), Covey-Crump and Silvey (1970), and Hebble and Mitchell (1972) represent but a few important contributions.

The computer is used to aid the researcher in designing RSM experiments in several ways. Some of these are as follows:

1. *To design an Experiment from Scratch.* The user inputs a model, required sample size, set of candidate points, ranges on the design variables, and other possible constraints (e.g., blocking information).

2. *To augment or Repair a Current Experimental Design.* The input information is often the same as in (1).

3. *To replace Points in a Current Experimental Design.* Often input information is as in (1).

4. *To construct the "Ideal" Standard Experimental Design.* This includes the CCD or the BBD.

One should certainly notice the reference to a list of candidates points. The implication is that the computer picks from this list for design augmentation or constructing a design from scratch. If the candidate list is not supplied, often the package will select as a default a grid of points of its choice. Thus, the design chosen certainly is a discrete design, and the goal is to select the design with the highest efficiency, say D-efficiency.

8.3.1 Important Relationship Between Prediction Variance and Design Augmentation for D-Optimality

An important relationship exists between the coordinates of points that augment an existing design via D-optimality and the concept of prediction variance. In fact, it is this relationship that is the foundation of computer-aided design of experiments. The result has to do with the selection of the $(r + 1)$th point in an existing design containing r points. The coordinates to be chosen will be those such that $|\mathbf{M}|$ is maximized. As a result, the $(r + 1)$th point chosen is that which produces a *conditional D-optimality*—that is, that which gives D-optimality given the initial r-point design. The result is as follows:

Given an r-point design with design and model characterized by $\mathbf{X}_r$, the point $\mathbf{x}_{r+1}$ which maximizes $|\mathbf{X}'_{r+1}\mathbf{X}_{r+1}|$ is that for which $\mathbf{x}_{r+1}^{(m)'}(\mathbf{X}'_r\mathbf{X}_r)^{-1}\mathbf{x}_{r+1}^{(m)}$ is maximized. Hence $\mathbf{X}_{r+1}$ is the $\mathbf{X}$ matrix containing the $(r + 1)$th point.

Details regarding the development of this result are found in Appendix 10.

The above result is not only interesting but also is very useful. It essentially says that for design augmentation, the *next point* that is most appropriate in a *D*-optimality sense is that for which the prediction is worse when one uses the current design for prediction. In other words, $\mathbf{x}_{r+1}$ is a type of "soft spot," and putting this spot in the design produces the design which is conditionally *D*-optimal.

Nearly all of the software algorithms make use of this type of result. It is obvious how it is used for design augmentation. In this regard, one must keep in mind that a candidate list of points is used; as a result, the augmentation is chosen, sequentially one at a time, from points in the candidate list. When a design is chosen "from scratch," the same result applies. Generally an algorithm starts with a set of points that allows parameter estimation; then a procedure involving deletion and augmentation prevails, with prediction variance playing a major role. It should be strongly emphasized that the goal of the computer-generated design algorithms is to produce a design that gives approximately the highest *D*-efficiency. Different packages may, indeed, produce different designs with the same input information. The following is a trivial example that allows the reader to observe the use of the result.

Example 8.7 Consider an initial design containing a 2^2 factorial:

x_1	x_2
-1	-1
-1	1
1	-1
1	1

The model is a first-order model in two design variables. Now, suppose the candidate list of points are those of a 3^2 factorial design. The next point to be chosen is that for which $\mathbf{x}^{(m)\prime}(\mathbf{X}'\mathbf{X})^{-1}\mathbf{x}^{(m)}$ is largest. Here, of course, $\mathbf{x}^{(m)\prime} = [1, x_1, x_2]$ and $(\mathbf{X}'\mathbf{X})^{-1} = \frac{1}{4}\mathbf{I}_3$, reflecting the fact that the initial design is first-order orthogonal. As a result, the expression Var $\hat{y}(\mathbf{x})/ \sigma^2 = \mathbf{x}^{(m)\prime}(\mathbf{X}'\mathbf{X})^{-1}\mathbf{x}^{(m)}$ is given by

$$[1, x_1, x_2]\begin{bmatrix} \frac{1}{4} & 0 & 0 \\ 0 & \frac{1}{4} & 0 \\ 0 & 0 & \frac{1}{4} \end{bmatrix}\begin{bmatrix} 1 \\ x_1 \\ x_2 \end{bmatrix}$$

$$= \tfrac{1}{4}\left[1 + x_1^2 + x_2^2\right]$$

As a result, the values of Var $\hat{y}(\mathbf{x})/\sigma^2$ for the candidate list are

x_1	x_2	Var $\hat{y}(\mathbf{x})/\sigma^2$
-1	-1	$\frac{3}{4}$
-1	1	$\frac{3}{4}$
1	-1	$\frac{3}{4}$
1	1	$\frac{3}{4}$
0	-1	$\frac{1}{2}$
0	1	$\frac{1}{2}$
-1	0	$\frac{1}{2}$
1	0	$\frac{1}{2}$
0	0	$\frac{1}{4}$

As a result, the best augmentation of the 2^2 factorial is another factorial point, and it makes no difference which one of the four is added. An interesting point here is that if the center run is added, the design remains orthogonal; but a five-run nonorthogonal design is preferred (in the D-optimal sense).

Example 8.8 Consider a situation in which one wishes to fit the model

$$y = \beta_0 + \beta_1 x_1 + \beta_2 x_2 + \beta_3 x_3 + \beta_{12} x_1 x_2 + \varepsilon$$

The initial design is the irregular fraction:

x_1	x_2	x_3
-1	-1	-1
1	1	-1
1	-1	1
-1	1	1
1	1	1

This is a five-point design and all parameters are estimable. Suppose the list of candidate points includes the complete 2^3 design. Computation of Var $\hat{y}(\mathbf{x})/\sigma^2$ is given below for the candidate list:

x_1	x_2	x_3	Var $\hat{y}(\mathbf{x})/\sigma^2$
-1	-1	-1	1.0
-1	-1	1	3.0
-1	1	-1	3.0
-1	1	1	1.0
1	-1	-1	3.0
1	-1	1	1.0
1	1	-1	1.0
1	1	1	1.0

As a result, the best augmentation of the initial design involves the use of any one of the points $(-1, -1, 1)$, $(-1, 1, -1)$, and $(1, -1, -1)$.

8.3.2 Illustrations Involving Computer-Generated Design

In this section we focus on a few illustrations of computer-generated designs with emphasis on the use of SAS and Design-Expert. This does not mean to imply that these are the only or even the best computer packages. In all of these cases the maximization of D-efficiency is used as the criterion, though as the reader will observe, in many cases the efficiency figures for other criteria are listed. Following the illustrations, the appropriateness, advantages, and disadvantages of computer-generated design will be addressed. Our first example is a fairly simple one in which SAS was used to generate a design. Based on our discussions in this and previous chapters, the reader very likely could have correctly conjectured the most appropriate design.

Example 8.9 Suppose there are $k = 3$ design variables and an intelligent "guess" gives the appropriate model

$$y = \beta_0 + \beta_1 x_1 + \beta_2 x_2 + \beta_3 x_3 + \beta_{11} x_1^2 + \varepsilon$$

The levels being considered by the researcher are (coded)

x_1: 5 levels,   1,   0.5, 0, 0.5, 1
x_2: 2 levels,   1, 1
x_3: 2 levels, -1, 1

The design is to use $N - 12$ design points.

The SAS-QC procedures PROC PLAN and PROC OPTEX were used. Table 8.4 shows a portion of the output. The terminology "Examining Design Number 1" indicates that SAS offers several designs. The one included here is the first listed. Note that the design in the output is a $3 \times 2 \times 2$ factorial. As the reader might expect, x_1 is held at three rather than five levels. Also, notice the D- and A-efficiencies. The design chosen is the one with the highest D-efficiency. It turns out that the G-efficiency of the design is 100%. In fact, the maximum value of $N \operatorname{Var} \hat{y}(x)/\sigma^2$ at any design location is five, the number of model parameters. This becomes clear when one notices that the "maximum prediction variance over candidates" is 0.4167. The candidate list here is the full $5 \times 2 \times 2$ factorial, and of course the final design is contained. The maximum prediction variance indicated in the output is not scaled by sample size. As a result, for this design we have

$$\max \frac{N \operatorname{Var} \hat{y}(\mathbf{x})}{\sigma^2} = \max v(\mathbf{x})$$

$$= 12(0.4167)$$

$$= 5.0$$

Table 8.4   SASQC Output for $N = 12$ Run Design of Example 8.9

Examining Design Number 1
Log determinant of the information matrix = 1.0515E + 01
Maximum prediction variance over candidates = 0.4167
Average prediction variance over candidates = 0.3885
Average variance of coefficients = 0.1833
D-Efficiency = 68.2558
A-Efficiency = 45.4545

Point Number	x_1	x_2	x_3
1	-1.000000	-1.000000	-1.000000
2	-1.000000	-1.000000	1.000000
3	-1.000000	1.000000	-1.000000
4	-1.000000	1.000000	1.000000
5	0	-1.000000	-1.000000
6	0	-1.000000	1.000000
7	0	1.000000	-1.000000
8	0	1.000000	1.000000
9	1.000000	-1.000000	-1.000000
10	1.000000	-1.000000	1.000000
11	1.000000	1.000000	-1.000000
12	1.000000	1.000000	1.000000

As a result, the design is G-optimal. Note also that the "average prediction variance over candidates" is given. This is an attempt at addressing the Q-optimality criterion. However, unlike the case of G-optimality, one has nothing with which to compare this value. One should keep in mind that the user should not necessarily be alarmed with what appears to be relatively low D- and A-efficiencies. Recall that discrete designs with fixed sample sizes and other constraints rarely achieve 100% efficiency; in fact, they often do not come close.

Example 8.10   Consider the same scenario as in the previous example except that the user can afford to use $N = 15$ runs. The design produced by SAS-QC is given in Table 8.5. Note again that x_1 contains only three levels though five levels were input. Note that the D- and A-efficiencies are lower. The maximum value of $v(\mathbf{x})$ among the candidates is

$$(0.4118)15 = 6.18$$

As a result, the G-efficiency is $(5/6.18)100 \cong 80\%$. One should keep in mind

Table 8.5 SASQC Output for $N = 15$ Run Design of Example 8.10

Examining Design Number 1
Log determinant of the information matrix = 1.1556E + 01
Maximum prediction variance over candidates = 0.4118
Average prediction variance over candidates = 0.3284
Average variance of coefficients = 0.1645
D-Efficiency = 67.2495
A-Efficiency = 40.5365

Point Number	x_1	x_2	x_3
1	-1.000000	-1.000000	-1.000000
2	-1.000000	-1.000000	1.000000
3	-1.000000	-1.000000	1.000000
4	-1.000000	1.000000	-1.000000
5	-1.000000	1.000000	-1.000000
6	-1.000000	1.000000	1.000000
7	0	-1.000000	1.000000
8	0	-1.000000	1.000000
9	0	1.000000	-1.000000
10	0	1.000000	-1.000000
11	1.000000	-1.000000	-1.000000
12	1.000000	-1.000000	-1.000000
13	1.000000	1.000000	1.000000
14	1.000000	1.000000	1.000000
15	1.000000	1.000000	1.000000

that despite the lower efficiencies, the 15-run design is superior (but not on a per observation basis) to the 12-run design. Note the smaller value for "average variance of coefficients." This is one value that is not scaled by sample size.

Example 8.11   In this example we provide another illustration of the use of computer-generated designs. This example serves as an illustration of some of the *disadvantages* of computer-generated design and design optimality in general.

Suppose we have four design variables and the input involves three levels on each coded to -1, 0, $+1$. Thus, the candidate list will be the $3^4 = 81$ design points. Suppose the practitioner has difficulty choosing a model for purpose of design construction but decides on

$$E(y) = \beta_0 + \beta_1 x_1 + \beta_2 x_2 + \beta_3 x_3 + \beta_4 x_4$$

$$+ \beta_{12} x_1 x_2 + \beta_{23} x_2 x_3 + \beta_{11} x_1^2 + \beta_{44} x_4^2$$

The practitioner can affort $N = 25$ runs. The D-optimal design produced by SAS is

Observation	x_1	x_2	x_3	x_4
1	-1	-1	-1	-1
2	-1	-1	-1	0
3	-1	-1	-1	0
4	-1	-1	1	-1
5	-1	-1	1	1
6	-1	1	-1	-1
7	-1	1	-1	0
8	-1	1	1	1
9	-1	1	1	1
10	0	-1	-1	1
11	0	-1	1	0
12	0	-1	1	1
13	0	1	-1	-1
14	0	1	-1	0
15	0	1	1	-1
16	0	1	1	0
17	1	-1	-1	-1
18	1	-1	-1	0
19	1	-1	-1	1
20	1	-1	1	-1
21	1	-1	1	0
22	1	1	-1	1
23	1	1	-1	1
24	1	1	1	-1
25	1	1	1	0

The D-efficiency reported by SAS is 62.35% and the A-efficiency is 40.75%. The G-efficiency is 94.13%.

It is of interest to compare this design to a more standard design. In fact, a CCD with $\alpha = 1.0$ and one center run satisfies the $N = 25$ requirements. Of course, the CCD is totally symmetric and requires three levels of each variable and estimates coefficients for all six two-factor interactions equally well. In addition, of course, all pure quadratic coefficients are estimated equally well. One may use SAS-QC to evaluate efficiencies for any design/model combination. As one would expect, the CCD is not as efficient for the incomplete second-order model listed above. In fact, the efficiencies are:

$$D_{\text{eff}} = 52.13\%$$

$$A_{\text{eff}} = 31.59\%$$

$$G_{\text{eff}} = 94.09\%$$

However, how often is the practitioner certain of the model prior to designing an experiment, particularly in an RSM scenario when most (or all) factors are continuous? Suppose we consider a situation in which, after the data are collected, the model that is fit to the data is given by

$$E(y) = \beta_0 + \beta_1 x_1 + \beta_2 x_2 + \beta_3 x_3 + \beta_4 x_4 + \beta_{12} x_1 x_2 + \beta_{34} x_3 x_4 \quad (8.23)$$

Of course, the design constructed by SASQC is no longer D-optimal. In fact, a highly efficient two-level design could be constructed from the computer. However, one must live with the design in Equation (8.22). Now, the CCD is also not designed for the model of equation (8.23), particularly because there are no quadratic terms. The efficiencies are:

	D	A	G
Computer-generated	75.88	67.86	82.07
CCD	72.96	72.31	85.04

As a result, the CCD is just as efficient as, (if not more so than), the computer-generated design. Now, suppose the actual model, unknown to the practitioner, involves a quadratic term in either x_2 or x_3. Here, the computer-generated design of Equation (8.22) is a singular design; that is, the coefficients of x_2^2 and x_3^2 are not estimable. The point here is that the CCD has a high D-efficiency and G-efficiency for the full second-order model and is often reasonably robust to model misspecification when compared to a computer-generated design. One cannot always be certain that a computer-generated D-optimal design allows estimation of all coefficients if the model that is eventually fit differs from that used in the input information.

Example 8.12: Use of Qualitative Variables   Computer-generated designs are a powerful tool when one encounters qualitative variables that must be modeled with the usual continuous quantitative design variables. Here, of course, the user must be concerned about interaction that exists between the continuous and qualitative variables. Later in this text we deal with qualitative variables in more detail.

Consider a situation in which there are two standard continuous design variables with three levels of a qualitative variable; call these levels A, B, and C. The model assumed is second order in x_1 and x_2, and the qualitative variable interacts with x_1 and x_2. The required design size is $N = 15$. Formally, the model is written

$$y = \beta_0 + \beta_1 x_1 + \beta_2 x_2 + \beta_{12} x_1 x_2 + \beta_{11} x_1^2 + \beta_{22} x_2^2 + \gamma_1 z_1 + \gamma_2 z_2$$

$$+ \, \delta_{11} z_1 x_1 + \delta_{21} z_2 x_1 + \delta_{12} z_1 x_2 + \delta_{22} z_2 x_2 + \varepsilon$$

Here

$$z_1 = 1 \quad \text{if } A \text{ is the discrete level}$$

$$\quad = 0 \quad \text{elsewhere}$$

$$z_2 = 1 \quad \text{if } B \text{ is the discrete level}$$

$$\quad = 0 \quad \text{elsewhere}$$

Notice that indicator variables are used to account for the qualitative factor. The $N = 15$ computer generated design using D-optimality is given by

$$\mathbf{D} = \begin{bmatrix}
x_1 & x_2 & x_3 \\
-1 & -1 & C \\
-1 & -1 & B \\
-1 & -1 & A \\
-1 & 1 & C \\
-1 & 1 & B \\
-1 & 1 & A \\
1 & -1 & C \\
1 & -1 & B \\
1 & -1 & A \\
1 & 1 & C \\
1 & 1 & B \\
1 & 1 & A \\
0 & 0 & A \\
0 & 1 & B \\
-1 & 0 & B
\end{bmatrix}$$

Note the "crossing" of the 2^2 factorial points with the three levels A, B, and C in order to capture the interactions $x_1 z_1$, $x_1 z_2$, $x_2 z_1$, $x_2 z_2$, and $x_1 x_2$. The "center" and two axial points in x_1 and x_2 allow further information for estimation of x_1^2 and x_2^2 terms.

Use of D-efficiency under constraints on lack of fit and pure error degrees of freedom via computer packages is a continuing attempt to make the "automated design" concept practical. Quite often the constraints revolve around total sample size, levels, and so on. However, there are algorithms that generate a most D-efficient design where the constraints are built around pure error and lack-of-fit degrees of freedom. For example, suppose that in a $k = 3$ situation it is of interest to design an experiment with two lack-of-fit degrees of freedom and two pure error degrees of freedom. For a second-order model this corresponds to 12 distinct design points and 15 total runs. The resulting design, produced by Design-Expert, is given in Table 8.6. Note from Table 8.6 that the design allows a single center run and has replication at two distinct factorial points. While we are accustomed to seeing

Table 8.6   Output from Design-Expert for D-Optimal Design with k = 3, Two Lack-of-Fit Degrees of Freedom, and Two Pure Error Degrees of Freedom

A: x_1	B: x_2	C: x_3
-1.000	-1.000	-1.000
1.000	-1.000	-1.000
1.000	-1.000	-1.000
0.000	0.000	-1.000
-1.000	1.000	-1.000
-1.000	1.000	-1.000
1.000	1.000	-1.000
-1.000	0.000	0.000
0.000	0.000	0.000
0.000	1.000	0.000
-1.000	-1.000	1.000
-1.000	-1.000	1.000
-1.000	1.000	1.000
1.000	1.000	1.000
0.500	-0.500	-0.500

designs with center runs replicated, there are certainly advantages to having replication at *different points* so some checking on the homogeneous variance assumption can be made.

8.4   SOME FINAL COMMENTS CONCERNING DESIGN OPTIMALITY AND COMPUTER-GENERATED DESIGN

As we indicated earlier in this chapter, the current status of computer generated design is a result of the theoretical work of Keifer and co-workers and a decade of work by many who essentially attempted to make the concept usable by practitioners. There is no question that computer-aided design of experiments can be of value. However, in RSM work there are several reasons to proceed with caution. Those who successfully use design computer packages to aid in constructing RSM designs are those who do not use it as a "black box." Computer-aided designs can be useful if the user allows it to complement the available arsenal of standard RSM designs. The user who is likely to fail with computer-aided designs is the one who is not armed with important knowledge about design optimality and standard RSM designs.

The development of optimal design theory is couched in the use of a continuous design measure which determines the proportion of runs that should be made at a number of points in a predetermined design space. In practice, of course, the concept of continuous designs is impractical. However, the idea is to determine discrete designs that approximate the optimal continuous designs. The idea of candidate lists and sequential formation of the design according to the important augmentation relationship of Section 8.3.1 provide computational ease and make computer-aided design practical. However, perhaps because of this useful relationship, D-optimality or, rather, the maximization of D-efficiency has become the criterion that is used more than any other. One must keep in mind that the D-criterion is necessarily quite narrow, based on crucial and often unrealistic assumptions, and does not address many of the 10 important goals of RSM designs discussed in Chapter 7. Many of these 10 goals address the notion of *design robustness*—that is, robustness to model misspecification, robustness to errors in control, and so on. In addition, they address the distribution of Var $\hat{y}(\mathbf{x})/\sigma^2$, which is very important in RSM studies. The idea of maximization of D-efficiency does not directly deal with the prediction variance, and, indeed, often the computer-generated design based on the D-criterion does not result in an attractive distribution of Var $\hat{y}(\mathbf{x})/\sigma^2$.

The following are, formally, some items that should breed caution among practitioners in the use of computer-generated D-efficient designs:

1. The user must indicate a model. Model editing is part of any modeling strategy. Often the model is not known. If the fitted model is not that which is input to the computer, the computer-generated design is *not* that which maximizes D-efficiency, and, indeed, it may not be a good design at all.

2. Designs generated from the D-criterion do not address prediction variance.

3. For second-order models, the D-criterion often does not allow any (or many) center runs. This often leaves large values of Var $\hat{y}(\mathbf{x})/\sigma^2$ in the design center.

In RSM analyses, the user should search for an appropriate standard design discussed in Chapter 7 and 8. Nevertheless, there will be circumstances under which a computer-generated design will be helpful. Constraints regarding sample sizes, the use of qualitative variables, and cases where the user insists on unusual combination of ranges are situations in which computer-aided design and the resulting D-criterion may be very helpful. However, the user should be cautious regarding the input model and should be aware of the pitfalls we have discussed here.

EXERCISES

8.1 There is considerable interest in the use of saturated or near saturated second order designs. One interesting comparison involves the hexagon

($k = 2$) and the small composite design. Both involve six design points (plus center runs). In this exercise, we make comparisons that take into account variances of coefficients and prediction variance. However, before any two designs can be compared, they must be adjusted so they have the same scale. Consider the hexagon of Equation (7.22) and the $k = 2$ small composite design listed in Section 8.1. For proper comparisons we have

$$\mathbf{D}_{SCD} = \begin{array}{cc} x_1 & x_2 \\ \left[\begin{array}{cc} -1 & -1 \\ 1 & 1 \\ -\sqrt{2} & 0 \\ \sqrt{2} & 0 \\ 0 & -\sqrt{2} \\ 0 & \sqrt{2} \\ 0 & 0 \\ 0 & 0 \end{array}\right] \end{array}$$

$$\mathbf{D}_{HEX} = \begin{array}{cc} x_1 & x_2 \\ \left[\begin{array}{cc} \sqrt{2} & 0 \\ \sqrt{2}/2 & \sqrt{3}/2 \\ -\sqrt{2}/2 & \sqrt{3}/2 \\ -\sqrt{2} & 0 \\ -\sqrt{2}/2 & -\sqrt{3}/2 \\ \sqrt{2}/2 & -\sqrt{3}/2 \\ 0 & 0 \\ 0 & 0 \end{array}\right] \end{array}$$

For this scaling, both designs are on spheres of radius $\sqrt{2}$ (apart from center runs).

(a) Compute $N \, \mathrm{Var} \, b_0/\sigma^2$, $N \, \mathrm{Var} \, b_{ii}/\sigma^2$, $N \, \mathrm{Var} \, b_i/\sigma^2$, $N \, \mathrm{Var} \, b_{ij}/\sigma^2$ for both designs. Comment.

(b) Compute $N \, \mathrm{Var} \, \hat{y}(\mathbf{x})/\sigma^2$ at the design center and at all design points for both designs. Comment.

(c) From part (b), is there an illustration of the rotatability of the hexagon?

(d) Is the small composite design rotatable? Explain!

8.2 Consider a comparison between the $k = 4$ SCD of Equation (8.5) with the $k = 4$ CCD ($N = 24 + n_c$). Use $n_c = 4$ center runs for each design and use $\alpha = 2$ for both designs.

(a) Do the designs need to be rescaled? Explain!

(b) Make the same type of comparison as in Exercise 8.1.

8.3 Perhaps you noticed some evidence in Exercise 8.1 that the $k = 2$ SCD was not rotatable. Actually the conditions required for rotatability cannot be met for any SCD that involves a resolution III fraction of a 2^k. Prove this result. *Hint*: Look at odd moments.

8.4 Consider Table 8.2 that deals with the SCD. Construct a small composite design for $k = 5$, 6, and 7. Is the $k = 5$ SCD listed in Table 8.2 rotatable?

8.5 The SCD that involves the PBD represents a nice alternative to the CCD for $k \geq 4$. It produces a nice design size that is quite economical when compared to what is usually recommended. Because the design size is considerably above saturation, the design enjoys efficiency that is higher than the *smaller* SCD listed in Table 8.2.

   (a) Consider the case of $k = 6$. Construct a SCD by using a 24-run PBD combined with axial points at $\alpha = \sqrt{6}$ plus $n_c = 4$ center runs. Compare, in terms of the scaled variances of regression coefficients, with the SCD in Table 8.2 with $n_c = 2$.

   (b) Compare the larger SCD in (a) with a CCD containing
       (i) 2^{6-1}_{VI}
       (ii) axial points at $\alpha = \sqrt{6}$
       (iii) $n_c = 4$
   Be sure that the design from Table 8.2 contains a resolution III fraction and $\alpha = \sqrt{6}$.

8.6 The Koshal design is an interesting experimental plan that was not derived with statistical analysis in mind. Analysts are often surprised with the applicability and the efficiency of the design, particularly when information regarding interaction is either known or considered to be unimportant. The Koshal design is a natural competitor to the saturated or near-saturated SCD. Compare the $k = 3$ Koshal design with the $k = 3$ ($N = 10 + n_c$) SCD. Again, use scaled variances of regression coefficients. In addition, use N Var $\hat{y}(\mathbf{x}) / \sigma^2$ at the following locations in the *scaled design units*:

    $(0, 0, 0)$
    $(1, 0, 0)$
    $(-1, 0, 0)$
    $(1.5, 0, 0)$
    $(-1.5, 0, 0)$
    $(1, 1, 1)$
    $(-1, -1, -1)$

The design scaling should be as follows:

$$
D_{SCD} = \begin{bmatrix}
x_1 & x_2 & x_3 \\
-1 & -1 & -1 \\
1 & 1 & -1 \\
1 & -1 & 1 \\
-1 & 1 & 1 \\
-\sqrt{3} & 0 & 0 \\
\sqrt{3} & 0 & 0 \\
0 & -\sqrt{3} & 0 \\
0 & \sqrt{3} & 0 \\
0 & 0 & -\sqrt{3} \\
0 & 0 & \sqrt{3} \\
0 & 0 & 0 \\
0 & 0 & 0
\end{bmatrix}
$$

$$
D_{Kosh} = \begin{bmatrix}
x_1 & x_2 & x_3 \\
\sqrt{3/2} & 0 & 0 \\
0 & \sqrt{3/2} & 0 \\
0 & 0 & \sqrt{3/2} \\
-\sqrt{3/2} & 0 & 0 \\
0 & -\sqrt{3/2} & 0 \\
0 & 0 & -\sqrt{3/2} \\
\sqrt{3/2} & \sqrt{3/2} & 0 \\
\sqrt{3/2} & 0 & \sqrt{3/2} \\
0 & \sqrt{3/2} & \sqrt{3/2}
\end{bmatrix}
$$

Comment on your results.

8.7 Compare, in terms of scaled regression coefficients, the $k = 3$ 311B (use $n_c = 3$) with the CCD and SCD values in Table 8.1. The 311B must, of course, be scaled properly. Thus, divide the 311B by $\sqrt{2}$. Comment on your results.

8.8 Consider the following first order design

$$
D = \begin{bmatrix}
x_1 & x_2 & x_3 \\
-1 & -1 & -1 \\
1 & 1 & -1 \\
1 & 1 & -1 \\
1 & -1 & 1 \\
-1 & -1 & 1 \\
-1 & 1 & -1 \\
1 & 1 & 1 \\
-1 & -1 & 1
\end{bmatrix}
$$

(a) Is the design first-order orthogonal? Explain?

(b) Give the D-efficiency of this design.

(c) Give the Q-efficiency of the design.

(d) Give the A-efficiency of the design. *Hint*: The A-efficiency of a design is given by

$$\frac{\underset{\zeta}{\text{Min}} \ \text{tr}[\mathbf{M}(\zeta)]^{-1}}{N \ \text{tr}(\mathbf{X'X})^{-1}}$$

Now, of course, for a first-order model with range $[-1, +1]$ on each design variable we have

$$\underset{\zeta}{\text{Min}} \ \text{tr}[\mathbf{M}(\zeta)]^{-1} = p$$

8.9 Consider a model containing first-order and two-factor interaction terms in three design variables—that is,

$$\hat{y} = b_0 + b_1 x_1 + b_2 x_2 + b_3 x_3 + b_{12} x_1 x_2 + b_{13} x_1 x_3 + b_{23} x_2 x_3$$

In Section 8.2.1 we demonstrated that a resolution III two-level design is D, A, G, and Q-optimal for a first-order model. The levels must be at ± 1 extremes.

(a) Show that all variances of coefficients are minimized with a resolution IV two-level design with levels at ± 1 extremes.

(b) Using the result in (a), show that the members of the same resolution IV class are Q-optimal.

(c) Show that these designs are also G-optimal. The region R is, of course, the unit cube. Make use of the result [Equation (8.12)]

$$\underset{\mathbf{x} \in R}{\text{Max}} \ [v(\mathbf{x})] \geq p$$

and show that, for the class of designs in question,

$$\underset{\mathbf{x} \in R}{\text{Max}} \ [v(\mathbf{x})] = p$$

where, of course, $p = 7$.

8.10 The class of PBD is often used to fit first-order models. However, there are instances in which models containing interactions can be accommodated by a PBD. Consider a $k = 3$ situation with the model

$$\hat{y} = b_0 + b_1 x_1 + b_2 x_2 + b_3 x_3 + b_{12} x_1 x_2 + b_{13} x_1 x_3$$

The practitioner is convinced that this model well describes the system.

(a) Construct a $N = 12$, $k = 3$ PBD.

(b) Compute the D-efficiency of the PBD.

(c) Computer the A-efficiency.

8.11 Consider the following Koshal design for $k = 3$:

$$
D = \begin{array}{c} \begin{array}{ccc} x_1 & x_2 & x_3 \end{array} \\ \left[\begin{array}{ccc} 0 & 0 & 0 \\ 1 & 0 & 0 \\ 0 & 1 & 0 \\ 0 & 0 & 1 \\ 1 & 1 & 0 \\ 1 & 0 & 1 \\ 0 & 1 & 1 \end{array} \right] \end{array}
$$

Assume that the region of design operability is $[-1, +1]$ on each design variable. Consider the model

$$\hat{y} = b_0 + b_1 x_1 + b_2 x_2 + b_3 x_3 + b_{12} x_1 x_2 + b_{13} x_1 x_3 + b_{23} x_2 x_3$$

(a) Compute the D-efficiency for this design for the above model.

(b) Compute the A-efficiency.

8.12 Often a standard second-order design is planned because the practitioner expects that the fitted model will, indeed, be second order. However, when the analysis is conducted, it is determined that the model is considerably less than order 2. As an example, suppose a $k = 2$ CCD with $\alpha = 1.0$ and two center runs is used and all second-order effects (quadratic and interaction) are very insignificant. The design is used eventually for a first-order model.

(a) What is the D-efficiency for a $k = 2$ CCD with $\alpha = 1.0$ and two center runs for a first-order model? Comment.

(b) How does the D-efficiency change if there are no center runs?

(c) Is the difference between the results in (a) and (b) expected?

(d) Give the A-efficiency for both designs.

8.13 Please refer to Exercise 8.12. Give the D-efficiency of a $k = 3$ BBD for the model

$$\hat{y} = b_0 + b_1 x_1 + b_2 x_2 + b_3 x_3 + b_{12} x_1 x_2 + b_{13} x_1 x_3 + b_{23} x_2 x_3$$

8.14 Consider the following experimental design:

$$\mathbf{D} = \begin{bmatrix} x_1 & x_2 & x_3 \\ -1 & -1 & -1 \\ 1 & -1 & -1 \\ -1 & 1 & -1 \\ -1 & -1 & 1 \\ 1 & 1 & -1 \\ 1 & -1 & 1 \\ -1 & 1 & 1 \\ 0 & 0 & 0 \\ 0 & 0 & 0 \end{bmatrix}$$

A second stage is used as it becomes obvious that the model is quadratic.

Use a computer-generated design package to augment the above design with six more design points in order to accommodate a complete second-order model.

8.15 Consider a first-order orthogonal design with ± 1 levels. Show that the addition of center runs *must* lower the D-efficiency. Show that the same is true for A-efficiency and for G-efficiency.

8.16 Consider a $k = 4$ scenario. The practitioner plans to use 20 design points in order to accommodate a complete second-order model. However, it is not clear what the true model is prior to conducting the experiment. As a result, the design selected should be one that is fairly "robust"—that is, one that will be reasonably efficient even if the model on which the design is based is not correct.

(a) Using the 3^4 factorial as the candidate list, construct the D-optimal design using a second-order model as the basis.

(b) Suppose the edited model contains all linear terms and the interactions $x_1 x_2$ and $x_1 x_3$. Give the D-efficiency of the design for this model.

(c) Suppose, rather than use the computer-generated design, the practioner uses a SCD containing
  (i)   A PBD with 12 design points
  (ii) Eight axial points with $\alpha = 1.0$.
  Give the D-efficiency of this design for a second-order model.

(d) Give the D-efficiency of this design for the reduced model given in (b).

(e) Comment on the above.

8.17 Consider Table 8.3. Notice that the CCD has high D-efficiencies but low G-efficiencies when the design has small numbers of center runs. Then when the number of center runs increase, the G-efficiency increases substantially. Explain why this is happening. Look at the variance dispersion graphs.

8.18 Consider a second-order model with four design variables and levels $-2, -1, 0, 1, 2$ for each design variable. However, due to experimental constraints the candidate list should include:

    (i) A 3^4 factorial with levels $-1, 0, 1$
    (ii) Eight axial points at $\alpha = 2$
    (iii) A center point

Suppose 30 runs are to be used.

    (a) Using a computer-generated design package, give the D-optimal design.

    (b) Compare your design in (a) with a CCD with $\alpha = 2$ and $n_c = 6$. Use $N \operatorname{Var} \hat{y}(\mathbf{x})$ at the center of the design for your comparison.

    (c) Compare again using the variances of the predicted values averaged over all of the candidate points.

8.19 Consider again the two designs compared in Exercise 8.18. Suppose that the "true" model involves all linear main effects and two-factor interactions only. Compare the D-efficiencies and the A-efficiencies.

8.20 Consider a hybrid 311B with three center runs. As an alternative, consider a BBD for $k = 3$ with $n_0 = 3$ center runs. Answer the following questions:

    (a) Which has the highest D-efficiency for a second-order model?

    (b) Which has the highest D-efficiency if, in fact, the true model is first order? First order with two-factor interactions?

8.21 Show that for the case of a first-order model in k variables and spherical region of interest, a resolution III design with levels at ± 1 extremes is a Q-optimal design. *Hint*: For this region, $(1/K)\int_R x_i \, d\mathbf{x}$ and $(1/K)\int_R x_i x_j \, d\mathbf{x}$ are both zero.

CHAPTER 9

Miscellaneous Response Surface Topics

In this chapter we deal with a few special response surface topics. The title of this chapter should not imply that the topics are unimportant or only slightly relevant. We begin with a discussion of the impact of bias in the fitting of response surface models.

9.1 IMPACT OF MODEL BIAS ON THE FITTED MODEL AND DESIGN

In Chapter 7 we discussed the role of model bias on the test of lack of fit and the estimate of the error variance, σ^2. We learned that if the fitted polynomial approximation is given by

$$\hat{\mathbf{y}} = \mathbf{X}\mathbf{b} \tag{9.1}$$

and the true model of $E(\mathbf{y})$ is better approximated by

$$E(\mathbf{y}) = \mathbf{X}\boldsymbol{\beta} + \mathbf{Z}\boldsymbol{\gamma} \tag{9.2}$$

then the estimator $\mathbf{b}$ is biased and the estimator of σ^2, the error mean square in the analysis is biased upward. In fact, it is the bias in the mean squared error that one is detecting in a standard lack-of-fit test.

The emphasis here is that there are two types of model errors. In the postulation model

$$\mathbf{y} = \mathbf{X}\boldsymbol{\beta} + \boldsymbol{\varepsilon}$$

$\mathbf{X}\boldsymbol{\beta}$ is merely an approximation; errors associated with the model can be put

402

into two categories:

1. Random errors—that is, elements in ε (produces variance σ^2)
2. Systematic errors—that is, bias errors of the type $\mathbf{Xb} - E(\mathbf{y})$

Apart from the simple lack-of-fit test, the entire emphasis in design and analysis has been placed on variability produced by random variation. In particular, all experimental design criteria in Chapter 8 deal with variance oriented criteria; that is, D, G, A, and Q (or I.V.) optimality all deal with criteria that relate to $(\mathbf{X'X})$ or the moment matrix $(\mathbf{X'X}/N)$. In what follows we will link the Q-criterion with another involving model bias.

What Effect Does Design Have on Variance and Bias?
The most effective way to demonstrate the effect of design on variance and bias is to illustrate with a simple $k = 1$ scenario. Suppose the **true model** $f(x)$ is given by

$$f(x) = \beta_0 + \beta_1 x + \beta_{11} x^2 = 1 + x + x^2 \tag{9.3}$$

and it is approximated by a straight line with the formal fitted model

$$\hat{y}(x) = b_0 + b_1 x$$

Suppose for sake of simplicity that we assume that the region of interest and region of operability are the same, namely, the interval $[-1, +1]$. In addition, suppose the design contains a single run at each of the endpoints, namely, -1 and $+1$. As a result, the variance transmitted to the predicted value, namely $v(x) = N \operatorname{Var} \hat{y}$, is given by (assume $\sigma = 1$)

$$v(x) = 2\left[\operatorname{Var} b_0 + x^2 \operatorname{Var} b_1\right]$$

$$= 2\left[\frac{1}{2} + x^2 \operatorname{Var} b_1\right]$$

$$= 2\left[\frac{1}{2} + \frac{x^2}{2}\right] \tag{9.4}$$

Here, of course, $\operatorname{Var} b_1 = \sigma^2 / \sum_{i=1}^{2}(x_i)^2 = \frac{1}{2}$. As a result,

$$v(x) = 1 + x^2 \tag{9.5}$$

The measure of bias that is consistent in units with $v(x)$ is given by

$$d(x) = N[E\hat{y}(x) - f(x)]^2 \tag{9.6}$$

The term $E[\hat{y}(x)] = E(b_0) + xE(b_1)$. Recall from Chapter 7 that if one fits an incomplete model $\hat{y} = \mathbf{Xb}$, when in fact $E(\mathbf{y}) = \mathbf{X\beta} + \mathbf{Z\gamma}$,

$$E(\mathbf{b}) = \mathbf{\beta} + (\mathbf{X'X})^{-1}\mathbf{X'Z\gamma}$$

Indeed, in our case

$$\mathbf{\beta} = \begin{bmatrix} 1 \\ 1 \end{bmatrix}$$

$$(\mathbf{X'X})^{-1} = \begin{bmatrix} \frac{1}{2} & 0 \\ 0 & \frac{1}{2} \end{bmatrix}$$

$$\mathbf{X'Z} = \begin{bmatrix} 1 & 1 \\ 1 & -1 \end{bmatrix} \begin{bmatrix} (1)^2 \\ (-1)^2 \end{bmatrix} = \begin{bmatrix} 2 \\ 0 \end{bmatrix}$$

Thus, allowing that $\gamma = \beta_{11} = 1$, we obtain

$$E\begin{bmatrix} b_0 \\ b_1 \end{bmatrix} = \begin{bmatrix} \beta_0 + 1 \\ \beta_1 \end{bmatrix}$$

As a result, in this case the bias is transmitted only to the intercept estimate. Thus $E[\hat{y}(x)] = \beta_0 + 1 + \beta_1 x = 2 + x$. Thus from Equation (9.6) we obtain

$$d(x) = 2\left[\beta_0 + 1 + \beta_1 x - \left(\beta_0 + \beta_1 x + x^2 \right) \right]^2$$

$$= 2\left[1 - x^2 \right]^2 \tag{9.7}$$

The results in Equations (9.5) and (9.7) represent the two components that contribute to "error" in predicted values $\hat{y}(x)$ in the region $[-1, +1]$. Now, consider Figure 9.1. The actual function $f(x)$ is plotted along with $E[\hat{y}(x)] = 2 + x$. The differences, or vertical deviations, represents "model bias" values that are incurred with this endpoint design, namely, the two point design at -1 and $+1$.

Now, suppose we consider an alternative design, namely, a two-point design at the points -0.6 and $+0.6$. Again, in the fitted model $\hat{y}(x) = b_0 + b_1 x$, b_1 is found to be unbiased for β_1 but b_0 is biased. Following the same

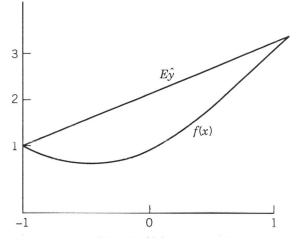

Figure 9.1 True function $f(x)$ and $E\hat{y}(x)$ for design with points at -1, $+1$.

route as before, we have

$$
E\begin{bmatrix} b_0 \\ b_1 \end{bmatrix} = \begin{bmatrix} \beta_0 \\ \beta_1 \end{bmatrix} + \begin{bmatrix} \dfrac{1}{2} & 0 \\ 0 & \dfrac{0.5}{(0.6)^2} \end{bmatrix} \begin{bmatrix} 1 & 1 \\ -0.6 & 0.6 \end{bmatrix} \begin{bmatrix} (0.6)^2 \\ (0.6)^2 \end{bmatrix} \beta_{11}
$$

$$
E\begin{pmatrix} b_0 \\ b_1 \end{pmatrix} = \begin{bmatrix} \beta_0 + 0.36 \\ \beta_1 \end{bmatrix}
$$

Note that the bias in the intercept is not as great as in the case of the previous or endpoint design. Thus the squared bias $d(x)$ is given by

$$
d(x) = 2\left[E(b_0 + b_1 x) - (\beta_0 + \beta_1 x + \beta_{11} x^2) \right]^2
$$

$$
= 2\left[\beta_0 + 0.36 + \beta_1 x - \beta_0 - \beta_1 x - \beta_{11} x^2 \right]^2
$$

$$
= 2\left[0.36 - x^2 \right]^2
$$

The expected value of the fitted approximation, namely $E[\hat{y}(x)]$, is given by

$$
E[\hat{y}(x)] = 1.36 + x
$$

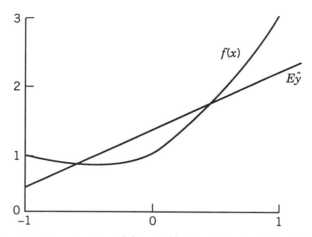

Figure 9.2   True function $f(x)$ and $E\hat{y}(x)$ for design at points $-0.6, 0.6$.

Figure 9.2 displays $f(x)$ and $E[\hat{y}(x)]$ in the case of the second two-point design.

Figures 9.1 and 9.2 reveal how the structure of the bias values $E[\hat{y}(x)] - f(x)$ are affected by the two designs in different ways. In the case discussed here the endpoint design results in rather small bias errors only near the endpoints where the design points are. Near the design center very large bias errors occur because there are no data there. As a result, it would appear reasonable that if design points are taken closer to the center, as in the case of the second design, the bias errors should be smaller. In the case of the second design, very small errors occur near the design points and moderate errors are revealed near the design center and at the design perimeter. Clearly, designs with points that are pushed closer to the center of the design afford considerably more protection against bias errors than does the endpoint design, though of course the latter is a minimum variance design. This suggests that when one "fears" model inadequacy perhaps there is a need to consider some type of compromise design because a strategy involves protection against bias errors works counter to a strategy designed to minimize variance.

9.2   A DESIGN CRITERION INVOLVING BIAS AND VARIANCE

The scientist or engineer who conducts a response surface methodology (RSM) study is not always aware when the practitioner should fit a second-order model. However, very often only a moderate amount of model inadequacy (in magnitude less than is detectable by a lack-of-fit test) will provide errors that are as serious as variation due to random errors.

It is natural to consider combining bias and variance through the mean squared error criterion

$$E[\hat{y}(\mathbf{x}) - f(\mathbf{x})]^2 \tag{9.8}$$

Here we use $\mathbf{x}$ as a vector valued to account for multiple design variables. Box and Draper (1959, 1963) introduced an average mean squared error criterion in which the mean squared error in Equation (9.8) is averaged over a region of interest in the design variables. They also showed some compelling evidence that bias due to model misspecification should not be ignored when one chooses an experimental design for response surface exploration. Indeed, there are situations when model bias is the dominant component in the mean squared error formulation. As we saw in the development in Section 9.1, it often occurs that the minimum variance approach applies a premium to pushing points to the edge of the region of interest (much like in a D-optimal approach), whereas a minimum bias approach calls for more placement of points that are an intermediate distance from the design center. The details depend a great deal on (a) the model that one is fitting and (b) the model that one chooses to protect against. The latter model usually involves model terms that are of an order higher than that of the fitted model.

Generalization of the Average Mean Squared Error

Consider the mean squared error of Equation (9.8). This "average squared error loss criterion" is written so as to be a function of $\mathbf{x}$, the location at which one is predicting the response with $\hat{y}(\mathbf{x})$. Now, the mean squared error in Equation (9.8) can be partitioned as

$$E[\hat{y}(\mathbf{x}) - f(\mathbf{x})]^2 = E\{[\hat{y}(\mathbf{x}) - E\hat{y}(\mathbf{x})] + [E\hat{y}(\mathbf{x}) - f(\mathbf{x})]\}^2$$

$$= E[\hat{y}(\mathbf{x}) - E\hat{y}(\mathbf{x})]^2 + [E\hat{y}(\mathbf{x}) - f(\mathbf{x})]^2 \tag{9.9}$$

The cross-product term in the above squaring operation is zero. The two terms on the right-hand side of Equation (9.9) are, respectively, the variance and the squared bias of $\hat{y}(\mathbf{x})$.

It should be clear from the foregoing and other developments in this text that there is a tradeoff in dealing with variance and bias. It should also be apparent that Equation (9.9) does not afford the user a single number criterion because the mean squared error depends on $\mathbf{x}$, the point at which one predicts. However, Box and Draper suggested the use of an **average mean squared error** (AMSE), where the quantity $E[\hat{y}(\mathbf{x}) - f(\mathbf{x})]^2$ in Equation (9.9) is averaged over $\mathbf{x}$ in a preselected region of interest. In other words, a quantity

$$\text{AMSE} = \frac{NK}{\sigma^2} \int_R E[\hat{y}(\mathbf{x}) - f(\mathbf{x})]^2 \, d\mathbf{x} \tag{9.10}$$

was offered as a single number criterion. The integral $\int_R$ along with the multiplication by $K = 1/\int_R d\mathbf{x}$, the inverse of the volume of the region, implies that the quantity AMSE is an average of the MSE over R. The quantity N/σ^2 is merely a scale factor in the same spirit as that provided by $N \text{Var } \hat{y}(\mathbf{x})/\sigma^2$ which has been used throughout the text. The AMSE can be subdivided as follows:

$$\text{AMSE} = \frac{NK}{\sigma^2} \int_R E[\hat{y}(\mathbf{x}) - E\hat{y}(\mathbf{x})]^2 \, d\mathbf{x} + \frac{NK}{\sigma^2} \int_R [E\hat{y}(\mathbf{x}) - f(\mathbf{x})]^2 \, d\mathbf{x}$$

$$= \text{APV} + \text{ASB} \tag{9.11}$$

where APV stands for **average prediction variance** and ASB stands for **average squared bias**. Here, of course, the averaging is done over the region of interest R. In what follows we focus heavily on designs that minimize APV and ASB, with strong emphasis on the latter. Note that minimization of APV is equivalent to the Q-optimality criterion discussed in Chapter 8.

The partition of mean squared error into variance and squared bias should bring to mind the material in Chapter 7 dealing with lack of fit. The structure of bias stems from the assumption that the model

$$\hat{\mathbf{y}} = \mathbf{X}_1 \mathbf{b}_1$$

is fit and the "true" model is a polynomial of a higher degree. Here we will alter the somewhat simplified notation of Chapter 7 and introduce notation that will be more convenient in the general setting here. The matrix $\mathbf{X}_1$ contains model terms of a low-order approximation of the true structure, and the matrix $\mathbf{X}_2$ contains additional higher-order terms that one wishes to protect aganst in designing the experiment. Thus

$$\mathbf{f} = E(\mathbf{y}) = \mathbf{X}_1 \boldsymbol{\beta}_1 + \mathbf{X}_2 \boldsymbol{\beta}_2$$

Here, of course, $\mathbf{y}$ is the $N \times 1$ vector of observations. From material in Chapter 7, we know that the estimator $\mathbf{b}_1$ is biased, namely,

$$E(\mathbf{b}_1) = \boldsymbol{\beta}_1 + \mathbf{A}\boldsymbol{\beta}_2 \tag{9.12}$$

where, in our new notation,

$$\mathbf{A} = (\mathbf{X}_1'\mathbf{X}_1)^{-1}\mathbf{X}_1'\mathbf{X}_2$$

As a result, the bias portion of AMSE, namely ASB, can be written

$$\text{ASB} = \frac{NK}{\sigma^2} \int_R \left[\mathbf{x}_1' E(\mathbf{b}_1) - \left[(\mathbf{x}_1'\boldsymbol{\beta}_1 + \mathbf{x}_2'\boldsymbol{\beta}_2) \right] \right]^2 d\mathbf{x} \qquad (9.13)$$

where $\mathbf{x}_1$ contains p_1 model terms that are fitted and $\mathbf{x}_2$ contains the p_2 terms that are ignored. By invoking Equation (9.12), Equation (9.13) reduces to

$$\text{ASB} = \frac{NK}{\sigma^2} \int_R \left[(\mathbf{x}_1'\mathbf{A} - \mathbf{x}_2')\boldsymbol{\beta}_2 \right]^2 d\mathbf{x}$$

$$= \frac{NK}{\sigma^2} \int_R \boldsymbol{\beta}_2'[\mathbf{A}\mathbf{x}_1 - \mathbf{x}_2][\mathbf{x}_1'\mathbf{A} - \mathbf{x}_2']\boldsymbol{\beta}_2$$

$$= \frac{NK}{\sigma^2} \int_R \left[\boldsymbol{\beta}_2'\mathbf{A}\mathbf{x}_1\mathbf{x}_1'\mathbf{A}\boldsymbol{\beta}_2 - 2\boldsymbol{\beta}_2'\mathbf{x}_2\mathbf{x}_1'\mathbf{A}\boldsymbol{\beta}_2 + \boldsymbol{\beta}_2'\mathbf{x}_2\mathbf{x}_2'\boldsymbol{\beta}_2 \right] d\mathbf{x} \quad (9.14)$$

For the case of APV we have

$$\text{APV} = \frac{NK}{\sigma^2} \int_R \mathbf{x}_1'(\mathbf{X}'\mathbf{X})^{-1}\mathbf{x}_1 \, d\mathbf{x}$$

9.2.1 The Case of a First-Order Fitter Model and Cuboidal Region

For the situation where the fitted model is first order, we have

$$\hat{y}(\mathbf{x}) = b_0 + \sum_{i=1}^{k} b_i x_i$$

and hence $\mathbf{x}_1' = [1, x_1, \ldots, x_k]$. If we further assume that the researcher is interested in "protecting" against a true model that is of second order, then

$$\mathbf{X}_1 = \begin{bmatrix} 1 & x_{11} & \cdots & x_{1k} \\ 1 & x_{21} & \cdots & x_{2k} \\ \vdots & \vdots & \vdots & \vdots \\ 1 & x_{N,1} & \cdots & x_{N,k} \end{bmatrix}$$

and

$$\mathbf{X}_2 = \begin{bmatrix} x_{11}^2 & \cdots & x_{1k}^2 & x_{11}x_{12} & \cdots & x_{1,k-1} & x_{1,k} \\ \vdots & & \vdots & \vdots & & \vdots & \vdots \\ \vdots & & \vdots & \vdots & & \vdots & \vdots \\ x_{N,1}^2 & \cdots & x_{N,k}^2 & x_{N1}x_{N2} & \cdots & x_{N,k-1} & x_{N,k} \end{bmatrix} \qquad (9.15)$$

Let us assume a cuboidal region of interest and region of operability on the variables—that is, $-1 \le x_i \le 1$ for $1 = 1, 2, \ldots, k$. If we consider variance alone, we can borrow from Chapter 8 where we learned that the Q-optimal design is an orthogonal design in which all levels are set at the ± 1 extremes. However, for the case of bias, one would expect from the discussion in Section 9.1 that levels should be placed in from the ± 1 extremes.

Suppose we consider the minimization of ASB in Equation (9.14) by choice of design. Using the fact that $\mathbf{A} = (\mathbf{X}_1'\mathbf{X}_1)^{-1}\mathbf{X}_1'\mathbf{X}_2 = \mathbf{M}_{11}^{-1}\mathbf{M}_{12}$, where $\mathbf{M}_{11} = \mathbf{X}_1'\mathbf{X}_1/N$ and $\mathbf{M}_{12} = \mathbf{X}_1'\mathbf{X}_2/N$ are *moment matrices*. We have

$$\text{ASB} = \frac{NK}{\sigma^2} \int_R \boldsymbol{\beta}_2' \left[\mathbf{M}_{12}'\mathbf{M}_{11}^{-1}\mathbf{x}_1\mathbf{x}_1'\mathbf{M}_{11}^{-1}\mathbf{M}_{12} - 2\mathbf{x}_2\mathbf{x}_1'\mathbf{M}_{11}^{-1}\mathbf{M}_{12} + \mathbf{x}_2\mathbf{x}_2' \right] \boldsymbol{\beta}_2 \, d\mathbf{x}$$

At this point it is convenient to introduce the concept of **region moments**. Letting $\boldsymbol{\alpha}_2 = \sqrt{N}\boldsymbol{\beta}_2/\sigma$, ASB can be written as

$$\text{ASB} = \boldsymbol{\alpha}_2' \left[\mathbf{M}_{12}'\mathbf{M}_{11}^{-1}\boldsymbol{\mu}_{11}\mathbf{M}_{11}^{-1}\mathbf{M}_{12} - 2\boldsymbol{\mu}_{12}'\mathbf{M}_{11}^{-1}\mathbf{M}_{12} + \boldsymbol{\mu}_{22} \right] \boldsymbol{\alpha}_2 \quad (9.16)$$

where

$$\boldsymbol{\mu}_{11} = K\int_R \mathbf{x}_1\mathbf{x}_1' \, d\mathbf{x}, \quad \boldsymbol{\mu}_{22} = K\int_R \mathbf{x}_2\mathbf{x}_2' \, d\mathbf{x}, \quad \text{and} \quad \boldsymbol{\mu}_{12} = K\int_R \mathbf{x}_1\mathbf{x}_2'.$$

The quantities $\boldsymbol{\mu}_{11}$ and $\boldsymbol{\mu}_{22}$ are region moment matrices. The matrix $\boldsymbol{\mu}_{11}$ contains first and second moments. In fact,

$$\boldsymbol{\mu}_{11} = K\int_R \mathbf{x}_1\mathbf{x}_1' = K\int_R \begin{bmatrix} 1 \\ x_1 \\ x_2 \\ \vdots \\ x_k \end{bmatrix} [1, x_1, x_2, \ldots, x_k] \, d\mathbf{x}$$

Because $\int_R$ implies $\int_{-1}^{1}\int_{-1}^{1} \cdots \int_{-1}^{1}$, this first- and second-order region moment matrix μ_{11} is given by

$$\boldsymbol{\mu}_{11} = \begin{bmatrix} 1 & 0 & \cdots & 0 \\ \hline 0 & & & \\ 0 & & & \\ \vdots & & 1/3I_k & \\ 0 & & & \end{bmatrix}$$

The matrix $\boldsymbol{\mu}_{22}$, in a similar fashion, displays region moments through order four, while $\boldsymbol{\mu}_{12}$ contains region moments through order three.

An easy prescription for the form of the design that minimizes ASB is obtained by writing the average squared bias in Equation (9.15) as

$$ASB = \alpha_2' \big[\big(\mu_{22} - \mu_{12}' \mu_{11}^{-1} \mu_{12} \big)$$

$$+ \big(M_{11}^{-1} M_{12} - \mu_{11}^{-1} \mu_{12} \big)' \mu_{11} \big(M_{11}^{-1} M_{12} - \mu_{11}^{-1} \mu_{12} \big) \big] \alpha_2 \quad (9.17)$$

In the second term in Equation (9.17) the matrix μ_{11} is positive definite. See Box and Draper (1987) and Myers (1976). As a result, a sufficient condition for a minimum bias design (i.e., a design that minimizes ASB) is to let

$$M_{11} = \mu_{11} \quad \text{and} \quad M_{12} = \mu_{12} \quad (9.18)$$

As a result, the minimum bias design is one in which $ASB = \alpha_2'[\mu_{22} - \mu_{12}'\mu_{11}^{-1}\mu_{12}]\alpha_2$. Please note that selection of an appropriate design reduces to the selection of design moment matrice M_{11} and M_{12}. In the case where a first-order model is being fit and one seeks to protect against a second-order model, M_{11} contains moments through order two and M_{12} contains moments through order three. However, please note that Equation (9.17) allows for the minimum ASB in general—that is, if one fits a model with the form $\hat{y} = X_1 b_1$ and, indeed, protection is sought against a model $y = X_1 \beta_1 + X_2 \beta_2$.

Single Design Variable—Minimum Bias Design

It is useful and constructive to make one of the general expressions in Equation (9.18) for the special case of a single design variable. Here, of course, the "design" becomes the selection of the spacing of points in the interval $[-1, +1]$ in the design variable. As a result, we assume a fitted first-order model

$$\hat{y} = b_0 + b_1 x$$

when the true structure is best approximated by

$$E(y) = \beta_0 + \beta_1 x + \beta_{11} x^2$$

We have an experiment that results in

$$X_1 = \begin{bmatrix} 1 & x_1 \\ 1 & x_2 \\ \vdots & \vdots \\ 1 & x_N \end{bmatrix}$$

with

$$\mathbf{X}_2 = \begin{bmatrix} x_1^2 \\ x_2^2 \\ \vdots \\ x_N^2 \end{bmatrix}.$$

The relevant moment matrices are

$$\mathbf{M}_{11} = \frac{\mathbf{X}_1' \mathbf{X}_1}{N} = \begin{bmatrix} 1 & [1] \\ [1] & [11] \end{bmatrix}$$

and

$$\mathbf{M}_{12} = \begin{bmatrix} [11] \\ [111] \end{bmatrix}$$

Here, of course, $[11] = (1/N)\Sigma_u x_u^2$; $[1] = (1/N)\Sigma_u x_u$; and $[111] = (1/N)\Sigma_u x_u^3$. Now, the region moment matrices $\boldsymbol{\mu}_{11}$ and $\boldsymbol{\mu}_{22}$ are of the same dimension and structure as their design moment counterparts $\mathbf{M}_{11}$ and $\mathbf{M}_{12}$. In fact,

$$\boldsymbol{\mu}_{11} = \frac{1}{2}\int_{-1}^{1} \begin{bmatrix} 1 \\ x \end{bmatrix} \begin{bmatrix} 1 & x \end{bmatrix} dx = \begin{bmatrix} 1 & 0 \\ 0 & \frac{1}{3} \end{bmatrix}$$

and

$$\boldsymbol{\mu}_{12} = \frac{1}{2}\int_{-1}^{1} \begin{bmatrix} 1 \\ x \end{bmatrix} \begin{bmatrix} x^2 \end{bmatrix} dx = \begin{bmatrix} \frac{1}{3} \\ 0 \end{bmatrix}$$

For the interval $[-1, +1]$ which represents the region of interest in x, the *odd region moments are all zero*. The quantity $\frac{1}{3}$ in $\boldsymbol{\mu}_{11}$ and $\boldsymbol{\mu}_{12}$ is the value of the second pure region moment $\frac{1}{2}\int_{-1}^{1} x^2 \, dx$. As a result, a very simple application of the results in Equation (9.18) suggests:

A minimum bias (minimum ASB) design for a fitted first-order model with protection against a model of second order can be achieved by having $[1] = 0$, $[11] = \frac{1}{3}$ and $[111] = 0$.

The above result is not at all counterintuitive and, in fact, is certainly the kind of result one expects from the discussion in Section 9.1. Clearly the *minimum variance design* (i.e., the design that minimizes APV for this case) will have $[1] = 0$ and $[11] = 1.0$ (all points pushed to the ± 1 levels). The minimum bias design requires levels to be placed in from the ± 1 extremes.

For the case of the single design variable, there are many designs that meet the moment conditions that result in minimum bias—that is, minimum average squared bias. The requirement of zero first and third moments requires only an equal sample size design with levels spaced symmetrically around the center of the design interval. It is interesting, then, that the design that best protects against the quadratic model can take so many forms. For example, a *two-level design* with, say, N observations (N even) requires

$$N/2 \text{ observations at } x = \sqrt{\tfrac{1}{3}} = 0.58,$$

$$N/2 \text{ observations at } -\sqrt{\tfrac{1}{3}} = -0.58$$

Of course there are obvious advantages in using more than two levels. The extra degrees of freedom for lack of fit are certainly useful. Thus, for three levels and $N/3$ runs at each level, the design is given by

$$\frac{N}{3} \text{ at } x = a; \qquad \frac{N}{3} \text{ at } x = 0, \qquad \frac{N}{3} \text{ at } x = -a$$

where, to allow $[11] = \tfrac{1}{3}$, we have

$$\tfrac{2}{3}a^2 = \tfrac{1}{3}$$

and thus

$$a = \sqrt{\tfrac{1}{2}} = 0.707$$

The reader should be able to understand by now the uncommon flexibility in constructing the minimum bias design. One can have various numbers of center runs and divide the remaining runs between two levels symmetric around $x = 0$. For example, suppose 15 runs are allowed. Suppose, also, that three runs are to be used in the center. The remaining 12 runs are then evenly divided between $-a$ and a where

$$\left(\tfrac{4}{5}\right)a^2 = \tfrac{1}{3}$$

and thus

$$a = \sqrt{\tfrac{5}{12}}$$

$$\cong 0.644$$

Thus, it is important to be sure that all three moment conditions hold. The minimum bias design need not have equally spaced levels. For example, consider the design with levels $-a_2$, $-a_1$, a_1, and a_2 (no center runs) with $N/4$ runs at each level. That is, the design is as below:

$$\overline{\quad -a_2 \qquad -a_1 \qquad 0 \qquad a_1 \qquad a_2 \quad}$$

The values a_1 and a_2 are such that

$$\frac{a_1^2 + a_2^2}{2} = \tfrac{1}{3}$$

There are many choices for a_1 and a_2 and thus many four-level designs. One example is $a_1 = \sqrt{\tfrac{1}{6}}$ and $a_2 = \sqrt{\tfrac{1}{2}}$, and another is $a_1 = \tfrac{1}{2}$ and $a_2 = \sqrt{\tfrac{5}{12}}$.

Multiple Design Variables—Minimum Bias Design

The use of Equation (9.18) is quite general and certainly applies in the case of the fitted model

$$\hat{y} = b_0 + b_1 x_1 + b_2 x_2 + \cdots + b_k x_k$$

Suppose we wish to protect against a full second-order model characterized by $\mathbf{X}_2$ in Equation (9.15). Suppose the region of interest is a cuboidal region with $-1 \le x_i \le 1$ for $i = 1, 2, \ldots, k$. We know that the pertinent region moments are

$$\boldsymbol{\mu}_{11} = \begin{bmatrix} 1 & \cdots & 0 & 0 \\ \vdots & & & \\ & & \tfrac{1}{3} I_k & \\ 0 & & & \end{bmatrix}$$

$$\boldsymbol{\mu}_{12} = 0$$

As a result, if we again accomplish minimum ASB by equating design moments to region moments through order three, we require $M_{11} = \mu_{11}$ and $M_{12} = \mu_{12}$. This results in setting all odd moments through order three equal to zero. It should be noted that

$$\mu_{12} = \frac{1}{K} \int_R \begin{bmatrix} 1 \\ x_1 \\ x_2 \\ \vdots \\ x_k \end{bmatrix} \begin{bmatrix} x_1^2 & x_2^2 & \cdots & x_k^2 & x_1 x_2, x_1 x_3, \ldots, x_k x_{k-1} \end{bmatrix} d\mathbf{x}$$

and thus $\boldsymbol{\mu}_{12}$ is a $k + 1$ by $k + k(k - 1)/2$ matrix containing second and third moments. The moment matrix $\mathbf{M}_{12}$ is a matrix with the same dimension and structure. As a result, the minimum ASB can be obtained by setting

$$[i] = 0, \qquad i = 1, 2, \ldots, k$$

$$[ii] = \tfrac{1}{3}, \qquad i = 1, 2, \ldots, k$$

$$[iii] = 0, \qquad i = 1, 2, \ldots, k$$

$$[iij] - 0, \qquad i \neq j$$

$$[ij] - 0, \qquad i \neq j$$

$$[ijk] = 0, \qquad i \neq j \neq k$$

The two-level full factorial design in k variables contains the type of symmetries that result in the above odd moments being zero. In order to achieve a design with each second pure moment equal to $\tfrac{1}{3}$, the level of the factorial points must be $\perp g$, where $g - \sqrt{\tfrac{1}{3}} - 0.58$.

What About Fractional Factorials and Center Runs?

It is clear that complete 2^k factorials with levels "brought in" considerably from the ± 1 extreme in the case of a cuboidal region satisfy the moment conditions given in Equation (9.19) for a minimum bias design. But, will fractional factorials suffice? For the case of odd design moments, the regular 2^{k-p} fractional factorial of resolution $\geq$ III, or even the Plackett–Burman designs, produce moments $[i]$, $[iii]$, $[ij]$ and $[iij]$ which are zero. However, the resolution III design *does not result in* $[ijk] = 0$. This should be obvious because all x_i will not be orthogonal to all $x_j x_k$ $(i \neq j \neq k)$ for the case of resolution III. However, the following statement can be made:

A minimum bias design in a cuboidal region for a first-order model with protection against a model of second order is achieved with a 2^k factorial design or a 2^{k-p} fraction with resolution $\geq$ IV. If the extremes of the cuboidal region are scaled to ± 1, the design levels will be at ± 0.58.

Center runs are very useful when one desires to afford protection against curvature. The conditions in Equation (9.19) on the odd moments are not altered by the use of center runs, and the condition on the second pure moment is achieved with the intuitively appealing result that the design levels can be moved closer to the ± 1 extremes. This will likely be an attractive alternative to the researcher. For example, suppose one uses a $\tfrac{1}{2}$ of a 2^4 factorial with $n_0 = 4$ center runs. The extremes for the region of interest are scaled to ± 1, and a minimum bias design is given by

$$
\mathbf{D} = \begin{bmatrix}
x_1 & x_2 & x_3 & x_4 \\
-g & -g & -g & -g \\
g & g & -g & -g \\
g & -g & g & -g \\
g & -g & -g & g \\
-g & g & g & -g \\
-g & g & -g & g \\
-g & -g & g & g \\
g & g & g & g \\
0 & 0 & 0 & 0 \\
0 & 0 & 0 & 0 \\
0 & 0 & 0 & 0 \\
0 & 0 & 0 & 0
\end{bmatrix}
$$

with g determined so that $[ii] = \frac{1}{3}$. In fact,

$$
g = \sqrt{\left(\tfrac{12}{8}\right)\left(\tfrac{1}{3}\right)}
$$

$$
= 0.707
$$

As a second example consider a $\frac{1}{2}$ of a 2^5 factoral with $n_c = 8$ center runs. Using $\pm g$ notation to indicate design levels, a minimum bias design is produced for a cuboidal region of interest scaled to ± 1 by using a resolution IV or V fraction with

$$
g = \sqrt{\left(\tfrac{24}{16}\right)\left(\tfrac{1}{3}\right)}
$$

$$
= 0.707
$$

It should be clear to the reader that the selection of the region of interest is important. Minimizing the bias due to an underspecification of the model requires that points be placed in from the extremes of that region, but the criterion (i.e., "squared bias") is integrated or "averaged" over the entire region. While the cuboidal region of interest is certainly quite common in practice, there are still instances in which a spherical region is the operative region of interest. In the next section we discuss the protection against bias due to the presence of a second-order model when the region of interest is spherical.

9.2.2 Minimum Bias Designs for a Spherical Region of Interest

Consider again a fitted first-order model with protection against a second-order model, but the squared bias in prediction is average over a *spherical region*. That is, there are ranges in the variables that aid in structuring the region of interest, but locations that represent simultaneous extremes in the variables are not included. As we discussed at length in Chapter 7, this condition suggests a region not unlike that of a sphere. For simplicity in what follows, we will assume that the region of interest is the unit sphere—that is, the region

$$\sum_{i=1}^{k} x_i^2 \leq 1 \tag{9.20}$$

Thus the variables are centered so that the center of the region is $(0, 0, \ldots, 0)$ and the radius of the sphere is 1.0.

As in the case of the cuboidal region, the minimum bias design for the spherical region can be obtained by achieving moment conditions that are governed by Equation (9.18). Again this involves equating design moments to region moments through order three. For the region given by Equation (9.20), all odd region moments are zero and the second pure region moment is given by $K\int_R x_i^2 \, d\mathbf{x} = 1/(k+2)$. One quickly can notice that for the special case in which $k = 1$, the value is $\frac{1}{3}$. The minimum bias design for the spherical region of Equation (9.20) is achieved by a design with moment conditions:

$$[i] = [ij] = [iii] = [iij] = [ijk] = 0 \quad (i \neq j \neq k)$$
$$[ii] = 1/(k+2) \quad (i = 1, 2, \ldots, k) \tag{9.21}$$

Again it becomes clear that the points the produce the minimum bias design are set in from the perimeter of the sphere. In addition, the symmetry of the region, like the cube, requires a design symmetry that forces odd design moments through order three to be zero. Once again the two-level factorial meets all design requirements, with the $[ii] = 1/(k+2)$ produced by the choice of the specific factorial levels $\pm g$.

Example 9.1 It is important for the reader to understand the impact of the scaling specified by the region $\sum_{i=1}^{k} x_i^2 = 1.0$. This implies that the factorial points that are placed at the perimeter of the region reside at $\pm 1/\sqrt{k}$. The points ± 1 are outside the region of interest or operability in the metric invoked by the region. Suppose then that $k = 4$ and a 2^4 factorial with $n_c = 5$ center runs is employed. The zero odd moments satisfy the requirements of a minimum bias design according to Equation (9.21). The second pure moment $[ii]$ should be

$$[ii] = \frac{1}{k+2} = \frac{1}{6}$$

As a result, with $n_e = 5$ center runs, the 2^4 factorial contains the design levels $\pm g$ at

$$g = \sqrt{\left(\tfrac{1}{6}\right)\tfrac{21}{16}}$$

$$= \sqrt{0.21}$$

$$\cong 0.46$$

As a result, the minimum bias design requires factorial points that are placed at 0.46, whereas the factorial points at the perimeter of the design region are scaled to $1/\sqrt{4} = 0.5$.

9.2.3 Simultaneous Consideration of Bias and Variance

Early in Section 9.2 we introduced the notion of average mean squared error (AMSE) and the partition in Equation (9.11) into APV and ASB. From that point we have focused strictly on bias and thus the minimization of ASB. However, at this point let us consider the use of the entire AMSE. From Equations (9.11) and (9.16), we have

$$\text{AMSE} = \alpha'_2 \big[\big(\mu_{22} - \mu'_{12}\mu_{11}^{-1}\mu_{12} \big)$$

$$+ \big(M_{11}^{-1}M_{12} - \mu_{11}^{-1}\mu_{12} \big)' \mu_{11} \big(M_{11}^{-1}M_{12} - \mu_{11}^{-1}\mu_{12} \big) \big] \alpha_2$$

$$+ \frac{NK}{\sigma^2} \int_R \text{Var}\, \hat{y}(\mathbf{x})\, d\mathbf{x} \qquad (9.22)$$

As before we are assuming, in this general formulation, that we are fitting a model $\hat{y} = \mathbf{x}'_1 \mathbf{b}_1$ and the "true" model is given by $E(y) = \mathbf{x}'_1 \boldsymbol{\beta}_1 + \mathbf{x}'_2 \boldsymbol{\beta}_2$. Of course our emphasis will be focused on $\mathbf{x}'_1 \boldsymbol{\beta}_1$ representing a first-order model while $\mathbf{x}'_2 \boldsymbol{\beta}_2$ contains second-order terms. The average prediction variance can be written

$$\frac{NK}{\sigma^2} \int_R \text{Var}\, \hat{y}(\mathbf{x})\, d\mathbf{x} = K \int_R \mathbf{x}'_1 M_{11}^{-1} \mathbf{x}_1\, d\mathbf{x} = \text{tr}\big[M_{11}^{-1}\mu_{11} \big] \qquad (9.23)$$

Clearly a desirable design criterion is to select the design that minimizes AMSE. AMSE is a function of design moments, region moments, and α_2, a standardized vector of parameters. It was clear in the previous section that the bias component can be minimized in spite of α_2. *However, AMSE cannot be minimized with respect to design moments without knowledge of α_2.* As a result, the practical use of AMSE is difficult at best. However, it is of interest to determine what type of role bias plays in the AMSE and, in particular, how effective the minimum bias design is in controlling AMSE. The following section deals with these matters and discusses the use of compromise designs

—that is, designs that are not minimum AMSE but are approximately minimum AMSE.

9.2.4 How Important Is Bias?

Let us consider the relative performance of te minimum variance and minimum bias designs for $k = 1$ for various magnitudes of bias. In other words, let us endeavor to answer the question regarding which design is able to more closely emulate the performance of the optimal design, i.e., that which minimizes AMSE? In Table 9.1, various degrees of bias are listed in the form of $\sqrt{N}\beta_{11}/\sigma$, the standardized regression coefficient of the quadratic term. Here, of course, we are assuming a fitted model

$$\hat{y} = b_0 + b_1 x$$

and a "true" model

$$E(y) = \beta_0 + \beta_1 x + \beta_{11} x^2$$

with R scaled to $[-1, +1]$. For each value of $\sqrt{N}\beta_{11}/\sigma$, the optimal (minimization of AMSE) was computed and thus the minimum value of AMSE was also computed. In addition, the AMSE for the minimum bias ($[11] = \frac{1}{3}$) design and the AMSE for the minimum variance design ($[11] = 1.0$) were computed. These items are listed as well as $\sqrt{N}\beta_{11}/\sigma$. The degree of bias is quantified by the ratio APV/ASB—that is, the ratio of the variance portion of the AMSE to the bias contribution in the case of the optimal design.

Table 9.1 Values of Optimal [11] and AMSE for Optimal Design, Minimum Variance Design, and Minimum Bias Design

$\sqrt{N}\beta_{11}/\sigma$	APV/ASB	[11] (Optimum)	AMSE (Optimum)	AMSE ([11] = 1.0)	AMSE ([11] = $\frac{1}{3}$)
0		1.0	1.333	1.333	2.0
0.50	10	0.999	1.466	1.466	2.022
1.0	7	0.687	1.699	1.867	2.088
1.5	5	0.565	1.910	2.533	2.205
2.0	3.7	0.500	2.133	3.466	2.352
2.5	3	0.462	2.382	4.672	2.551
3.0	2	0.432	2.666	6.135	2.805
4.0	1.5	0.399	3.333	9.872	3.422
4.5	1.0	0.388	3.718	11.521	3.798
5.0	0.9	0.379	4.153	14.677	4.222
7.0	0.4	0.359	6.316	27.465	6.355

Table 9.2   AMSE Values for the Optimal Design and Compromise Design of [11] = 0.4

$\sqrt{N}\beta_{11}/\sigma$	AMSE (Optimal)	AMSE ([11] = 0.40)
0	1.33	1.833
0.5	1.467	1.86
1.0	1.699	1.927
1.5	1.910	2.043
2.0	2.133	2.206
2.5	2.382	2.416
3.0	2.666	2.67
4.0	3.333	3.327
5.0	4.153	4.167
7.0	6.316	6.406

An observation of Table 9.1 makes it clear that bias is, indeed, an important consideration. Even when the bias makes a relatively small contribution—for example, when APV/ASB = 5 or APV/ASB = 3.7—the minimum bias design results in a smaller AMSE than that which results from the minimum variance design. In addition, it is clear that the minimum bias approach is considerably more robust than the minimum variance design. The minimum bias design is never far from optimal (in terms of mean squared error), whereas the minimum variance design performs quite poorly when the bias present is such that APV/ASB < 5.

Table 9.1 suggests that perhaps a compromise design for which [11] is slightly larger than the minimum bias value of $\frac{1}{3}$ might be appropriate because it should be extremely robust—that is, very close in AMSE to that of the optimal design. A design with a second moment [11] of 0.40–0.45 is extremely robust. An impression can be observed from Table 9.2 which shows the performance of the design with [11] = 0.40. Obviously, the use of [11] = 0.45 or 0.5 would produce an even better performance for very small values of the bias parameter $\sqrt{N}\beta_{11}/\sigma$.

Tables similar to Tables 9.1 and 9.2 for $k > 1$ design variables can be constructed. The results illustrate essentially the same lesson; namely, if there is any suspicion at all that there is curvature in the system, the minimum variance design is very likely to be counterproductive. In fact, a reasonable rule of thumb for accounting for possible model bias is to use a design with zero moments through order three and second pure moments [ii] that are 20–25% higher than that suggested by the minimum bias design. This design prescription serves as a reasonable compromise and will provide a design that will not be far away in performance from the optimal (minimum mean squared error) design over a wide range of bias in the system.

The above design moment recommendations can sometimes be a bit difficult to implement in practice. The problem that arises often stems from uncertainty regarding what is truly the region of interest. Even in the case of what might be perceived to be a cuboidal region, the exact locations in the design variables that define the corners of the region are not always known. Certainly in preliminary studies where variable screening is involved, choices of ranges of variables often require much thought. As a result, the recommendations made here should be confined to situations in which the researcher is not engaged in variable screening but is reasonably certain regarding the relative importance of design variables and the region in which prediction or predicted values are sought. Thus, of course, protection is achieved by designing an experiment that is resolution $\geq$ IV, and the design levels are brought in from the perimeter of the region in which one is required to predict response.

9.3 RSM IN THE PRESENCE OF QUALITATIVE VARIABLES

Quite often in practice, response surface studies need to be conducted even though at least one factor is qualitative in nature. Examples are numerous. In an extraction experiment an engineer may wish to model the amount of extraction against temperature and time. The eventual goal may be to determine temperature and time values that maximize extraction. However, a third factor, type of extraction solvent, is natural to consider. One may view this qualitative factor as having two levels because ethanol and toluene are candidates for "solvent type." In a corrosion fatigue study, shear stress and humidity are natural factors to study in a model with magnitude of corrosion serving as the response. However, the scientist or engineer may wish to study the qualitative factor "type of coating" and use uncoated and chromatic-type coating as the two levels. As in the case of the first example, values of the quantitative variables need to be determined for which corrosion is minimized. However, one must certainly consider the role of coating in the modeling and the optimization process. In addition, the role of coating or, for that matter, the role of any qualitative variable, must be taken into account in the experimental design that is chosen.

In an RSM study involving qualitative variables, it stands to reason that the goal of the experiment for the qualitative variable (or variables) is no different from that of the quantitative variables. That is, they must be included in the model and designs, and they must be involved in any response prediction or optimization using the model. The reader must bear in mind that *qualitative variables in an RSM are not like blocks*. They are not nuisance factors. Indeed, one must account for possible interaction between qualitative factors and standard RSM model terms involving quantitative factors. As we shall see in what follows, the nature of the interaction between qualitative and quantitative variables determines the complexity of both the design and analysis. We begin by considering the case where the standard quantitative design variables are modeled with a first-order model or a model containing first-order-plus-interaction terms.

9.3.1 Models First Order in the Quantitative Design Variables (Two-Level Design)

Qualitative factors are nicely modeled in a regression or RSM setting by using **dummy**, or **indicator** *variables*. This concept was discussed in Chapter 2. Here the presence of a level of a qualitative factor is accompanied by a value of 1.0, and absence is accompanied by a value of 0. This is much like the models we employed in Chapter 7 with blocking in RSM. For example, for a first-order model in two quantitative design variables x_1, x_2 and one qualitative variable, say z, where all three variables contain two levels in a 2^3 structure, we have

$$y_i = \beta_0 + \beta_1 x_{i1} + \beta_2 x_{i2} + \gamma_1 z_i + \varepsilon_i \qquad (i = 1, 2, \ldots, N) \qquad (9.24)$$

where $z_i = 1$ if the ith design point contains the first level (choice is arbitrary) and $z_i = 0$ if the design point contains the second level. Thus, the **X** matrix is given by

$$
\mathbf{X} = \begin{matrix}
& x_1 & x_2 & z \\
\begin{bmatrix}
1 & -1 & -1 & 0 \\
1 & -1 & 1 & 0 \\
1 & 1 & -1 & 0 \\
1 & 1 & 1 & 0 \\
1 & -1 & -1 & 1 \\
1 & -1 & 1 & 1 \\
1 & 1 & -1 & 1 \\
1 & 1 & 1 & 1
\end{bmatrix}
\end{matrix} \qquad (9.25)
$$

An alternative coding for the qualitative variable in this case is to *center* the indicator variable z and scale so that the model becomes

$$y_i = \beta_0^* + \beta_1 x_{i1} + \beta_2 x_{i2} + \gamma_1^* \frac{(z_i - \bar{z})}{\frac{1}{2}} + \varepsilon_i$$

This produces the -1, $+1$ values for the two levels of the qualitative variable, and the **X** matrix is the familiar

$$
\mathbf{X} = \begin{matrix}
& x_1 & x_2 & z^* \\
\begin{bmatrix}
1 & -1 & -1 & -1 \\
1 & -1 & 1 & -1 \\
1 & 1 & -1 & -1 \\
1 & 1 & 1 & -1 \\
1 & -1 & -1 & 1 \\
1 & -1 & 1 & 1 \\
1 & 1 & -1 & 1 \\
1 & 1 & 1 & 1
\end{bmatrix}
\end{matrix} \qquad (9.26)
$$

In terms of ease of analysis and interpretation, there really is no real preference between the $(0, 1)$ and the $(-1, +1)$ coding. However, those who are accustomed to studying the virtues of the two-level factorial structure usually prefer the $(-1, +1)$ coding. Obviously, a two level factorial or Resolution $\geq$ III fraction is a variance optimal design as before. The need to have a diagonal $(\mathbf{X'X})$ matrix remains a steadfast requirement in order to achieve 100% efficiency. One may notice that the $\mathbf{X'X}$ created by the design implied by Equation (9.25) is not diagonal. However, centering z produces "orthogonality with the intercept," which results in a diagonal $\mathbf{X'X}$.

In the case of more than one qualitative factor, the two-level factorial still is quite useful. For example, a $\frac{1}{2}$ fraction of a 2^4 factorial with two qualitative factors results in

$$\mathbf{X}_{(0,1)} = \begin{array}{cccc} x_1 & x_2 & z_1 & z_2 \\ \begin{bmatrix} 1 & -1 & -1 & 0 & 0 \\ 1 & 1 & 1 & 0 & 0 \\ 1 & 1 & -1 & 1 & 0 \\ 1 & 1 & -1 & 0 & 1 \\ 1 & -1 & 1 & 1 & 0 \\ 1 & -1 & 1 & 0 & 1 \\ 1 & -1 & -1 & 1 & 1 \\ 1 & 1 & 1 & 1 & 1 \end{bmatrix} \end{array},$$

$$\mathbf{X}_{(-1,+1)} = \begin{array}{cccc} x_1 & x_2 & z_1^* & z_2^* \\ \begin{bmatrix} 1 & -1 & -1 & -1 & -1 \\ 1 & 1 & 1 & -1 & -1 \\ 1 & 1 & -1 & 1 & -1 \\ 1 & 1 & -1 & -1 & 1 \\ 1 & -1 & 1 & 1 & -1 \\ 1 & -1 & 1 & -1 & 1 \\ 1 & -1 & -1 & 1 & 1 \\ 1 & 1 & 1 & 1 & 1 \end{bmatrix} \end{array}$$

The interpretation of regression coefficients in the first-order model is extremely important when the model contains qualitative variables. In the case of a two-level design, interpretation is quite simple. The following example is an illustration.

Example 9.2 Consider an extraction process in which the factors are time, temperature, and type of solvent. Solvents A and B are used, and two levels of time and temperature are involved in the experiments. The fitted first-order model involves the levels -1 and $+1$ for the low and high levels, respectively, of the quantitative variables. Solvent A is coded with -1 and solvent B is coded with $+1$. The fitted first-order model is given by

$$\hat{y} = 17.5 + 4.7x_1 + 10x_2 - 3z^* \tag{9.27}$$

where $\hat{y}$ is estimated amount of extraction, and z^* is the coded solvent level. The quantities x_1 and x_2 are time and temperature, respectively. For maximization of response, solvent A should be used because $z^* = -1$ maximizes $\hat{y}$. The use of constrained optimization of $\hat{y}$ with respect to x_1 and x_2 or steepest ascent can certainly proceed accordingly. Now, let us suppose that the fitted function in Equation (9.27) is computed from a 2^3 factorial or a 2_{IV}^{3-1} fraction. If the $(0, 1)$ coding had been used, the result would have been

$$\hat{y} = \left[17.5 + 6\left(\tfrac{1}{2}\right)\right] + 4.7x_1 + 10x_2 - 6z$$

$$= 20.5 + 4.7x_1 + 10x_2 - 6z \tag{9.28}$$

The change in intercept is accounted for by the difference in coding between z and z^*. Note that predicted values are equivalent for models (9.27) and (9.28). In addition, the conclusion drawn regarding the relative roles of the solvent is the same.

Notice that it is convenient to view the model with multiple levels of a qualitative factor as multiple RMS models with a different intercept for each level of the qualitative factor. For example, for the model of Equations (9.27) and (9.28), we have a first-order response function with an intercept of 20.5 for solvent A and 14.5 for solvent B. However, it must be emphasized that this way of viewing the role of qualitative factors only applies if there is no interaction between the qualitative and quantitative factors.

9.3.2 First-Order Models with More Than Two Levels of the Qualitative Factors

Quite often the analyst is faced with more than one qualitative variable and more than two levels of at least one of the qualitative factors. The $(0, 1)$ indicator works quite nicely. Again, of course, let us emphasize that we are assuming here that *no interactions exists among qualitative and quantitative factors*. Lack of interaction implies, of course, that the coefficients of quantitative factors do not vary across levels of the qualitative factors, and thus changing the levels of the qualitative factors only changes the intercept.

Suppose we have two qualitative factors, both at three levels. Suppose a single quantitative factor x is present at two levels and a $3^2 \times 2$ factorial experiment is used. The coding is given by

$$
\begin{array}{ccccc|c}
x & z_1 & z_2 & w_1 & w_2 & \text{Levels of qualitative factors} \\
-1 & 1 & 0 & 0 & 0 & (1,3) \\
-1 & 1 & 0 & 0 & 1 & (1,2) \\
-1 & 1 & 0 & 1 & 0 & (1,1) \\
-1 & 0 & 1 & 0 & 0 & (2,3) \\
-1 & 0 & 1 & 0 & 1 & (2,2) \\
-1 & 0 & 1 & 1 & 0 & (2,1) \\
-1 & 0 & 0 & 0 & 0 & (3,3) \\
-1 & 0 & 0 & 0 & 1 & (3,2) \\
-1 & 0 & 0 & 1 & 0 & (3,1)
\end{array}
$$

The balance of the design would involve $x = 1$ with a repeat of the values of the w's and z's. The z_1, z_2 coding represents the first qualitative factors at three levels. The w_1, w_2 coding represents the second qualitative factor. For the first qualitative factor, a 1 in the z_1 position implies the presence of the first level, and a 1 in the z_2 position implies the presence of the second level. A $(0,0)$ combination for (z_1, z_2) implies the presence of the third level. As a result, one can view this coding as allowing, quite arbitrarily, a zero contribution to the model for the third level, and the coefficients of z_1 and z_2, respectively, as the contribution from levels 1 and 2.

One can generalize the coding above to any number of levels. For l levels of a qualitative variable, there will be $l - 1$ indicator variables, with $(0, 0, \ldots, 0)$ arbitrarily representing the lth level. For example, for four levels, we have

z_1	z_1	z_3	
1	0	0	1st level
0	1	0	2nd level
0	0	1	3rd level
0	0	0	4th level

9.3.3 Models with Interaction Among Qualitative and Quantitative Variables

As we indicated earlier, no interaction involving the qualitative variables merely implies that the qualitative variables alter only the model intercept. However, the presence of interaction implies that certain coefficients change as one changes levels of the qualitative variables. For example, consider again the case of the extraction process with time, temperature, and solvent (A or B) as the factors. Suppose one wishes to include all two-factor interactions in the model. Following the notation of Equation (9.27), we have

$$\hat{y} = b_0 + b_1 x_1 + b_2 x_2 + b_{12} x_1 x_2 + cz^* + d_1 x_1 z^* + d_2 x_2 z^* \quad (9.29)$$

Here we are using the $(+1, -1)$ coding on the solvent, with solvent A coded as -1. This important model suggests that for solvent A the model is written

$$\hat{y} = (b_0 - c) + (b_1 - d_1) x_1 + (b_1 - d_2) x_2 + b_{12} x_1 x_2 \qquad \text{(Solvent A)}$$

whereas for solvent B

$$\hat{y} = (b_0 + c) + (b_1 + d_1) x_1 + (b_2 + d_2) x_2 + b_{12} x_1 x_2 \qquad \text{(Solvent B)}$$

Thus, the intercept varies due to the main effect term in the qualitative factor and the linear terms vary due to the interaction between qualitative factor and the linear terms. The use of interactions involving qualitative variables should result from suspicions that certain model terms will vary depending on the level of the qualitative factor. For example, the model in Equation (9.29) should be fit when the scientist expects that the linear model terms in time and temperature will be different for the two solvents.

When the portion of the model involving the quantitative variables contains first-order terms or first-order-plus-interaction terms, the two-level designs involving the quantitative factors are certainly useful. When sample size constraints do not allow the use of a standard design, computer-generated design packages are available. However, some special designs deserve attention.

9.3.4 Design Considerations: First-Order Models with and without Interaction

As we indicated in the previous section, there are many instances in which standard designs are not suitable when both qualitative and quantitative factors are present. This is particularly a problem when one is considering the use of second-order terms in the quantitative variables. The choice of design should be very much a function of the structure of the interaction between the qualitative factors and the response surface model terms in the quantitative factors. Draper and John (1988) shed light on this problem. They emphasize the following points:

1. The design must allow for estimation of quantitative factor model terms.
2. The design should allow for the effect that changes in the qualitative variables have on the model terms—that is, the interaction between qualitative variables and response surface model terms.
3. For any particular fitted model the design should allow for significance testing to determine if a simple model is adequate.

Property 3 is particularly important because it is often difficult to determine, prior to taking data, the structure of the interaction between qualitative and quantitative variables. As a result, the design should allow for various possibilities. In other words, the design should be sufficiently robust to allow for the fitting of more than one model type.

Example 9.3 Suppose there are two quantitative variables and one qualitative variable, z, at two levels. Suppose the experimenter postulates the model

$$y = \beta_0 + \gamma_1 z + \beta_1 x_1 + \beta_2 x_2 + \varepsilon \qquad (9.30)$$

In other words, it is felt that only a change in intercept is required as the level of the qualitative variable changes. Clearly a 2^3 factorial would allow for estimation of the model in Equation (9.30) as well as the model that requires that the linear coefficients of x_1 and x_2 change at the different levels of z. In addition, a 2^3 factorial with, say, two center runs allows for testing to determine if quadratic terms are needed. But, if one cannot afford the full 2^3 (or 2^3 plus center runs), what smaller designs may be available? Figure 9.3 shows five different five-point designs that allow for estimation of the model in Equation (9.30) but afford an extra degree of freedom.

In each of the five designs there is a structure containing a 2^2 factorial and center-type design, and the five design points are assigned to the two levels of

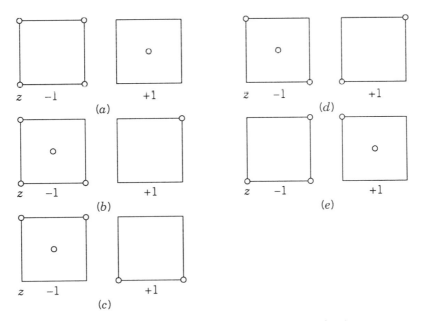

Figure 9.3 Five designs for estimation of Equation (9.30).

z in various ways. Equation (9.30) required data at both levels of z, as well as the estimability of β_1 and β_2 (first-order model) when these design points are collapsed across the two levels of z. All five designs possess these two properties.

Now, when one compares these five designs it becomes apparent that all five allow estimation of $\beta_{12}x_1x_2$. However, designs (c) and (e) allow $\delta_1 x_1 z$ or $\delta_2 x_2 z$. Designs (a), (b), and (d) allow no interaction between linear quantitative terms and z. As a result, one can argue that designs (c) and (e) are most sensitive to departures from the simple model of Equation (9.30).

It is of interest to focus on design (e) further. Consider the $\mathbf{X}$ matrix for the fitting of Equation (9.30), augmented by additional terms. It should be made clear that only one of the extra terms can be fit.

$$
\mathbf{X}_e =
\begin{bmatrix}
 & z & x_1 & x_2 & & x_1x_2 & x_1z & x_2z \\
1 & -1 & -1 & -1 & : & 1 & 1 & 1 \\
1 & -1 & 1 & -1 & : & -1 & -1 & 1 \\
1 & -1 & 1 & 1 & : & 1 & -1 & -1 \\
1 & 1 & 0 & 0 & : & 0 & 0 & 0 \\
1 & 1 & -1 & 1 & : & -1 & -1 & 1
\end{bmatrix}
$$

Any (but only one) of the terms x_1x_2, x_1z, and x_2z can be estimated. Now consider design (d) in Figure 9.3. Here we also augment the $\mathbf{X}$ matrix with x_1x_2, x_1z and x_2z.

$$
\mathbf{X}_d =
\begin{bmatrix}
 & x_1 & x_2 & z & & x_1x_2 & x_1z & x_2z \\
1 & -1 & 1 & -1 & : & -1 & 1 & -1 \\
1 & 0 & 0 & -1 & : & 0 & 0 & 0 \\
1 & 1 & -1 & -1 & : & -1 & -1 & 1 \\
1 & -1 & -1 & 1 & : & 1 & -1 & -1 \\
1 & 1 & 1 & 1 & : & 1 & 1 & 1
\end{bmatrix}
$$

Notice that x_1x_2 is orthogonal to x_1 and x_2 but highly correlated with z. Note that $x_1z = x_2$ and $x_2z = x_1$, so neither interaction with the qualitative variable can be estimated. In case of design (c)

$$
\mathbf{X}_c =
\begin{bmatrix}
 & x_1 & x_2 & z & & x_1x_2 & x_1z & x_2z \\
1 & -1 & 1 & -1 & : & -1 & 1 & -1 \\
1 & 1 & 1 & -1 & : & 1 & -1 & -1 \\
1 & 0 & 0 & -1 & : & 0 & 0 & 0 \\
1 & -1 & -1 & 1 & : & 1 & -1 & -1 \\
1 & 1 & -1 & 1 & : & -1 & 1 & -1
\end{bmatrix}
$$

Note that the weakness of design (c) is that x_2 and z are highly correlated. This implies that x_2z is highly correlated wth the constant. This results in poor estimation of x_2z.

The purpose of studying designs (a) through (e) in Figure 9.3 is to show that designs that are smaller than the full factorial (i.e., the 2^3) can be used to fit a model containing the linear effects on x_1 and x_2 as well as the term γz which produces a shift in the intercept. In addition, two of these designs allow an extra degree of freedom for lack of fit which can produce information about interaction, including the very important information on interaction between quantitative variables and the qualitative variable z. Of course, for the five-point designs discussed here, replication would be necessary in order to obtain a test for lack of fit.

For the case of more than two quantitative variables and qualitative variables that are at two levels, the two-level factorials and resolution $\geq V$ fractions are, of course, optimal. We indicate resolution V because this gives assurance that all two-factor interactions involving the qualitative and quantitative factors are estimable. However, there are cases in which a resolution $< V$ fraction can be used and still result in estimability of all two-factor interactions involving qualitative and quantitative variables. The trick is to use designs which are of so-called "mixed resolution." That is, allow a smaller resolution that applies in the case of quantitative variables than that which applies in dealing with interactions involving quantative and qualitative variables. The notion of mixed resolution will surface again in Chapter 10. The following example illustrates the point.

Example 9.4   Consider an example in which five quantitative variables are present at two levels and a single qualitative variable is present at two levels. The researcher feels confident that a first-order model is reasonable for the quantitative variables. However, it is not clear whether or not the linear coefficients of x_1, x_2, x_3, x_4, and x_5 vary from one level to another of the qualitative variable. Thus, the model to be fit is

$$y = \beta_0 + \sum_{i=1}^{5} \beta_i x_i + \gamma z + \sum_{i=1}^{5} \delta_i x_i z + \varepsilon \qquad (9.31)$$

This involves 12 model terms. Let us suppose that a full 2^6 or 2^{6-1} design cannot be used. A 2^{6-2} design (16 design points) with resolution V cannot be constructed. However, the model of Equation (9.31) can be fit with a 2^{6-2} factorial with defining relations

$$x_1 x_2 x_3 = I$$

$$x_1 x_3 x_4 x_5 z = I$$

resulting in a third defining interaction $x_2 x_4 x_5 z$. The $\frac{1}{4}$ fraction with these defining relations will allow the fitting of Equation (9.31) and still leave four degrees of freedom for lack of fit. The resolution on this design is III but it allow estimation of the $x_j z$ ($j = 1, 2, \ldots, 5$). As a result, all of these interactions are orthogonal with other model terms.

   While we have illustrated our point with special cases here, we should reemphasize that computer-generated designs can be very useful in design construction, particularly when more than two levels are needed in the qualitative variables. In the next section we move on with design considerations but deal with cases where the model is second order in the quantitative variables.

9.3.5 Design Considerations: Second-Order Models in the Quantitative Variables

The choice of an appropriate design can be considerably more difficult in the case of a second-order model when qualitative variables are present. Model uncertainty regarding the appropriate interactions with the qualitative variables often presents problems. Knowledge regarding terms such as $x_i^2 z_j$, $x_i x_l z_j$, and so on, is hard to come by. A safe approach is to use a standard second-order design for all combinations of the qualitative variables. This can be used to fit the model

$$y = \beta_0 + \sum_i \beta_i x_i + \sum_{i \leq j} \beta_{ij} x_i x_j + \sum_i \gamma_i z_i$$

$$+ \sum_i \sum_j \delta_{ij} x_i z_j + \sum_l \sum_{i<j} \sum \rho_{ijl} x_i x_j z_l + \sum_l \sum_i \eta_i x_i^2 z_l + \varepsilon \quad (9.32)$$

The model of Equation (9.32) contains all first- and second-order terms as well as interactions between all qualitative variables and all first- and second-order terms in the quantitative variables. Of course, in many situations the design that is required for the model of Equation (9.32) is much too costly. In what follows we illustrate the availability of small designs for fitting subsets of the "complete" model in Equation (9.32). Our illustration exploits the central composite design.

   Example 9.5   Let us suppose that there are two quantitative design variables and, again, one qualitative variable at two levels, coded at -1 and $+1$. The experimenter feels confident that the "equal slopes" assumption is not valid and thus the fitted model should at least contain

$$y = \beta_0 + \beta_1 x_1 + \beta_2 x_2 + \gamma z + \delta_1 x_1 z + \delta_2 x_2 z + \varepsilon \quad (9.33)$$

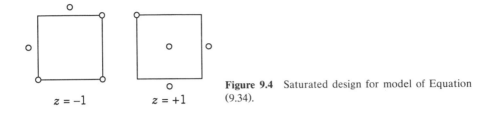

$z = -1$ $z = +1$

Figure 9.4   Saturated design for model of Equation (9.34).

However, there is some concern that there is curvature in the system so the design should be allowed to accommodate

$$y = \beta_0 + \beta_1 x_1 + \beta_2 x_2 + \gamma z + \delta_1 x_1 z + \delta_2 x_2 z$$
$$+ \beta_{12} x_1 x_2 + \beta_{11} x_1^2 + \beta_{22} x_2^2 + \varepsilon u \qquad (9.34)$$

The model of Equation (9.34) is a reasonable model in many circumstances. It allows for changes in the structure of the response surface across the levels of z but assumes that the nature of any existing curvature doesn't change. This model requires a minimum of nine design points. Draper and John (1988) point out that if design (e) in Figure 9.3 is augmented with four axial points, the model of Equation (9.34) can be fit. The resulting design is given in Figure 9.4. Notice that a first-order model can be fit at each level of z. This, of course, allows for interaction of z with x_1 and x_2. Now, the design in Figure 9.4 is saturated for the important model of Equation (9.34). This, of course, affords no lack-of-fit information on terms such as $x_1^2 z$, $x_2^2 z$, and $x_1 x_2 z$. However, a further augmentation of the design in Figure 9.4 will allow the use of the model in Equation (9.32), which allows interaction of z with all terms in the second-order model. When two central composite designs are too costly, Figure 9.5 represents a nice economical design that allows two additional degrees of freedom for lack of fit. Note that seven points are allocated to each level of z. Also, a second-order model can be fit to each level of z.

Example 9.6   Draper and John (1988) show an example in three quantitative variables and a single qualitative variables at two levels. The experimenter wishes to fit a model that is second order in the three quantitative

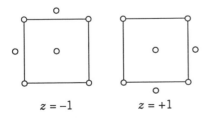

$z = -1$ $z = +1$

Figure 9.5   Augmentation of design in Figure 9.3 for model in equation (9.32).

variables and reflects interaction between z and the linear terms. In other words, the model is like that in Equation (9.34) except that three quantitative variables are involved. However, some information is sought on the $x_i^2 z$ and $x_i x_j z$ $(i \neq j)$ terms. Thus, the model that is fit is given by

$$
y = \beta_0 + \sum_{j=1}^{3} \beta_j x_j + \sum_{j=1}^{3} \beta_{jj} x_j^2 + \sum_{i<j} \beta_{ij} x_j x_i + \gamma z + \sum_{j=1}^{3} \delta_j x_j z + \varepsilon
$$

$$(9.35)$$

The design requires at least 14 design points. Again, to estimate the xz terms a first-order model must be available for each level of z. A second-order model must be available in the total design—that is, across both levels of z. We shall display the design matrices for three candidate designs. Consider the first, design 1, with design matrix given by

$$
\mathbf{D}_1 =
\begin{array}{c}
\begin{array}{cccc}
x_1 & x_2 & x_3 & z
\end{array} \\
\left[
\begin{array}{cccc}
-1 & -1 & -1 & -1 \\
1 & -1 & -1 & 1 \\
-1 & 1 & -1 & 1 \\
-1 & -1 & 1 & 1 \\
1 & 1 & -1 & -1 \\
1 & -1 & 1 & -1 \\
-1 & 1 & 1 & -1 \\
1 & 1 & 1 & 1 \\
-\alpha & 0 & 0 & -1 \\
\alpha & 0 & 0 & 1 \\
0 & -\alpha & 0 & -1 \\
0 & \alpha & 0 & 1 \\
0 & 0 & -\alpha & 1 \\
0 & 0 & \alpha & -1 \\
0 & 0 & 0 & 1 \\
0 & 0 & 0 & -1
\end{array}
\right]
\end{array}
$$

Note that a resolution III fractional factorial appears at each level of z. Then, of course, the full central composite design, taken across levels of z, allows a second-order model to be fit. As a result, the model of Equation (9.35) can be fit.

For the second design, the distribution of factorial points is identical to that of design (1). However, the axial points are distributed among the levels of z differently than that of design (1). For the second and third design, we will use the a, b, c, ab, and so on, notation for the factorial design points with A implying x_1, B implying x_2, and C implying x_3.

$$
\mathbf{D}_2 =
\begin{array}{c}
 \\
 \\
 \\
 \\
 \\
 \\
 \\
 \\
 \\
 \\
 \\
 \\
 \\
 \\
 \\
 \\
\end{array}
\begin{bmatrix}
(1) & -1 \\
ab & -1 \\
ac & -1 \\
bc & -1 \\
a & 1 \\
b & 1 \\
c & 1 \\
abc & 1 \\
-\alpha00 & -1 \\
\alpha00 & -1 \\
0-\alpha0 & -1 \\
0\alpha0 & 1 \\
00-\alpha & 1 \\
00\alpha & 1 \\
000 & -1 \\
000 & 1 \\
\end{bmatrix}
\begin{array}{c} z \end{array}
$$

$$
\mathbf{D}_3 =
\begin{bmatrix}
(1) & -1 \\
a & -1 \\
b & -1 \\
c & -1 \\
ab & -1 \\
ac & -1 \\
bc & -1 \\
abc & -1 \\
-\alpha00 & 1 \\
\alpha00 & 1 \\
0-\alpha0 & 1 \\
0\alpha0 & 1 \\
00-\alpha & 1 \\
00\alpha & 1 \\
000 & -1 \\
000 & 1 \\
\end{bmatrix}
\begin{array}{c} z \end{array}
$$

Note that the distribution of points among the levels of z is quite different for design $\mathbf{D}_3$. All three designs allow the estimation of model (9.35). In addition, all three allow lack-of-fit information for interactions among the qualitative variable and second-order terms. Specifically, $\mathbf{D}_3$ provides information on $x_i^2 z$ terms, whereas $\mathbf{D}_1$ and $\mathbf{D}_2$ allow information on $x_i x_j z$ and $x_i^2 z$ terms. Of course, the test for lack of fit cannot be made without replication variance.

9.3.6 Use of Computer Generated Design

In the foregoing we emphasize specific two and three variable designs that make use of the central composite structure and nicely allow qualitative variables at two levels to be used in a response surface study. However, quite often the analyst must resort to computer-generated design to accommodate situations involving more than three quantitative variables and more than two levels of the qualitative variables. The following two examples illustrate the use of SAS-QC for construction of response surface designs involving both qualitative and quantitative variables.

   Example 9.7   Consider a situation in which there are four quantitative variables and one qualitative variable at two levels. It is felt as if a second-order model in the quantitative variables is appropriate. In order to account for possible changes in the structure of the response surface when changes are made in the qualitative variable, the analyst decides to model the terms $x_i \cdot z$ for $i = 1, 2, 3, 4$, as well as the main effect in z. As a result, the model should contain the 20 terms

$$y = \beta_0 + \sum_{i=1}^{4} \beta_i x_i + \sum_{j=1}^{4} \beta_{jj} x_j^2 + \sum_{j<k} \sum \beta_{jk} x_j x_k$$

$$+ \gamma z + \sum_{J=1}^{4} \delta_j x_j z + \varepsilon \tag{9.36}$$

The candidate list of design points is a grid involving a full central composite design with levels -2, -1, 0, 1, 2 in the x's crossed with -1, $+1$, the two levels of the qualitative variable. This grid contains 50 candidate points. The analyst can afford only 24 total runs. The following is the design given using G-efficiency:

The SAS System

Observation	x_1	x_2	x_3	x_4	z
1	-2	0	0	0	-1
2	-1	-1	-1	-1	-1
3	-1	-1	-1	-1	1
4	-1	-1	1	-1	1
5	-1	-1	1	1	1
6	-1	1	-1	-1	-1

The SAS System

Observation	x_1	x_2	x_3	x_4	z
7	-1	1	-1	1	1
8	-1	1	1	-1	1
9	-1	1	1	1	-1
10	0	-2	0	0	1
11	0	0	-2	0	-1
12	0	0	0	0	1
13	0	0	0	2	-1
14	0	0	2	0	1
15	0	2	0	0	1
16	1	-1	-1	-1	-1
17	1	-1	1	1	1
18	1	-1	1	1	-1
19	1	-1	1	1	1
20	1	1	-1	-1	1
21	1	1	-1	1	1
22	1	1	1	-1	-1
23	1	1	1	-1	1
24	2	0	0	0	-1

The following is the design generated using the A-criterion:

The SAS System

Observation	x_1	x_2	x_3	x_4	z
1	-1	-1	-1	-1	-1
2	-1	-1	-1	1	1
3	-1	-1	1	-1	1
4	-1	-1	1	1	-1
5	-1	1	-1	-1	1
6	-1	1	-1	1	-1
7	-1	1	1	-1	-1
8	-1	1	1	1	1
9	0	-2	0	0	-1
10	0	0	-2	0	-1
11	0	0	0	-2	-1
12	0	0	0	0	-1
13	0	0	0	0	1
14	0	0	0	2	-1
15	0	0	2	0	-1
16	0	2	0	0	-1
17	1	-1	-1	-1	1
18	1	-1	1	-1	-1
19	1	-1	1	1	1
20	1	1	-1	-1	-1

The SAS System

Observation	x_1	x_2	x_3	x_4	z
21	1	1	−1	1	1
22	1	1	1	−1	1
23	1	1	1	1	−1
24	2	0	0	0	−1

The following is the design constructed with the use of the D-criterion:

Observation	x_1	x_2	x_3	x_4	z
1	−2	0	0	0	−1
2	−1	−1	−1	−1	1
3	−1	−1	−1	1	1
4	−1	−1	1	−1	1
5	−1	−1	1	1	1
6	−1	1	−1	−1	1
7	−1	1	−1	1	1
8	−1	1	1	−1	1
9	−1	1	1	1	1
10	0	−2	0	0	−1
11	0	0	−2	0	−1
12	0	0	0	−2	−1
13	0	0	0	0	1
14	0	0	0	2	−1
15	0	0	2	0	−1
16	0	2	0	0	−1
17	1	−1	−1	−1	1
18	1	−1	1	−1	1
19	1	−1	1	1	1
20	1	1	−1	−1	1
21	1	1	−1	1	1
22	1	1	1	−1	1
23	1	1	1	1	1
24	2	0	0	0	−1

Notice that the designs are similar. In fact, the G-criterion contains a 2^{5-1} fraction plus seven of the eight axial points with z at -1 four times and $+1$ three times. The final run is a center run in the x's with z at $+1$.

The A-criterion generated 15 of the 16 runs in the 2^{5-1} with seven axial points and two center runs in the x's with z split at -1 and $+1$. The D-criterion gives an interesting design. It contains 15 of the 16 points of the 2^4 factorial in the x's, but z remains at a constant level, namely $z = 1$. The design also contains all axial points in the x's while z is at -1. Then a center run is revealed with $z = +1$. It is noteworthy that this design is very much like the "blocking design" for two blocks that was discussed in Chapter 8 where the factorial portion is assigned to one block and the axial portion to a second block. In fact, with $\alpha = 2.0$ as the axial parameter the design nearly satisfies the orthogonal blocking definition.

Example 9.8 Consider a situation in which there are three quantitative variables x_1, x_2, and x_3, and two qualitative variables, z_1 and z_2, each at two levels. Again, assume that the practitioner wishes to fit a full second-order model in the three quantitative variables as well as the terms $z_1 x_j$, $z_1 x_j^2$, $z_2 x_j$, $z_2 x_j^2$, z_1, and z_2 for $j = 1, 2, 3$. There are 24 model terms to be fit, including β_0. The candidate list for SAS QC is a grid containing the 60 points defined by the central composite with levels $-\sqrt{3}$, -1, 0, 1, $\sqrt{3}$ (15 points) crossed with the 2^2 factorial with z_1 and z_2, each coded at ± 1. Let us assume that 30 runs can be used.

The reader should note that in this example we are not assuming a separate response surface for all four combinations of the qualitative variables. We are in fact assuming that the $x_i x_j$ ($i \neq j$) terms are the same at the four combinations. Indeed, a model assuming separate response surfaces would require 16 additional model terms containing $x_i x_j z_k$ terms ($i \neq j$), $z_1 z_2 x_i$ terms, $z_1 z_2 x_i^2$ terms, and $z_1 z_2 x_i x_j$ terms ($i \neq j$). Thus, in our model we assume stability of $x_i x_j$ across z_1 and z_2 and no interaction among z_1 and z_2. Again, the criteria gave similar designs. The A-criterion design is given by the following:

The SAS System

Observation	x_1	x_2	x_3	z_1	z_2
1	1.732	0.000	0.000	-1	1
2	-1.732	0.000	0.000	1	-1
3	-1.000	-1.000	-1.000	-1	-1
4	-1.000	-1.000	-1.000	1	1
5	-1.000	-1.000	1.000	-1	-1
6	-1.000	1.000	-1.000	-1	1
7	-1.000	1.000	-1.000	1	-1
8	-1.000	1.000	1.000	-1	-1
9	-1.000	1.000	1.000	1	1
10	0.000	-1.732	0.000	-1	1
11	0.000	-1.732	0.000	1	-1
12	0.000	0.000	-1.732	1	1
13	0.000	0.000	0.000	-1	-1
14	0.000	0.000	0.000	-1	1
15	0.000	0.000	0.000	1	-1
16	0.000	0.000	0.000	1	1
17	0.000	0.000	1.732	-1	-1
18	0.000	0.000	1.732	-1	1

The SAS System

Observation	x_1	x_2	x_3	z_1	z_2
19	0.000	0.000	1.732	1	−1
20	0.000	1.732	0.000	−1	−1
21	0.000	1.732	0.000	1	1
22	1.000	−1.000	−1.000	−1	1
23	1.000	−1.000	−1.000	1	−1
24	1.000	−1.000	1.000	−1	−1
25	1.000	−1.000	1.000	1	1
26	1.000	1.000	−1.000	−1	−1
27	1.000	1.000	1.000	−1	1
28	1.000	1.000	1.000	1	−1
29	1.732	0.000	0.000	−1	−1
30	1.732	0.000	0.000	1	1

The *G*-criterion gave the following design:

The SAS System

Observation	x_1	x_2	x_3	z_1	z_2
1	−1.732	0.000	0.000	−1	1
2	−1.732	0.000	0.000	1	−1
3	−1.732	0.000	0.000	1	1
4	−1.000	−1.000	−1.000	−1	1
5	−1.000	−1.000	1.000	−1	−1
6	−1.000	1.000	−1.000	−1	−1
7	−1.000	1.000	−1.000	1	1
8	−1.000	1.000	1.000	1	−1
9	0.000	−1.732	0.000	1	−1
10	0.000	−1.732	0.000	1	1
11	0.000	0.000	−1.732	1	−1
12	0.000	0.000	0.000	−1	−1
13	0.000	0.000	0.000	−1	1
14	0.000	0.000	0.000	1	−1
15	0.000	0.000	0.000	1	1
16	0.000	0.000	1.732	−1	−1
17	0.000	0.000	1.732	−1	1
18	0.000	0.000	1.732	1	1
19	0.000	1.732	0.000	−1	−1
20	0.000	1.732	0.000	−1	1
21	1.000	−1.000	−1.000	−1	−1
22	1.000	−1.000	−1.000	1	1
23	1.000	−1.000	1.000	−1	−1
24	1.000	−1.000	1.000	−1	1

The SAS System

Observation	x_1	x_2	x_3	z_1	z_2
25	1.000	1.000	−1.000	−1	1
26	1.000	1.000	−1.000	1	−1
27	1.000	1.000	1.000	−1	−1
28	1.000	1.000	1.000	1	1
29	1.732	0.000	0.000	1	−1
30	1.732	0.000	0.000	1	1

The D-criterion gave the design below:

The SAS System

Observation	x_1	x_2	x_3	z_1	z_2
1	−1.732	0.000	0.000	−1	1
2	−1.732	0.000	0.000	1	−1
3	−1.732	0.000	0.000	1	1
4	−1.000	−1.000	−1.000	−1	1
5	−1.000	−1.000	1.000	−1	−1
6	−1.000	1.000	1.000	1	1
7	−1.000	1.000	1.000	−1	1
8	0.000	−1.732	0.000	−1	1
9	0.000	−1.732	0.000	1	−1
10	0.000	−1.732	0.000	1	1
11	0.000	0.000	−1.732	−1	1
12	0.000	0.000	−1.732	1	−1
13	0.000	0.000	−1.732	1	1
14	0.000	0.000	0.000	−1	−1
15	0.000	0.000	0.000	−1	1
16	0.000	0.000	0.000	1	−1
17	0.000	0.000	0.000	1	1
18	0.000	0.000	1.732	−1	1
19	0.000	0.000	1.732	1	−1
20	0.000	0.000	1.732	1	1
21	0.000	1.732	0.000	−1	1
22	0.000	1.732	0.000	1	−1
23	0.000	1.732	0.000	1	1
24	1.000	−1.000	−1.000	−1	−1
25	1.000	−1.000	1.000	−1	−1
26	1.000	1.000	−1.000	−1	−1
27	1.000	1.000	1.000	−1	−1
28	1.732	0.000	0.000	−1	1
29	1.732	0.000	0.000	1	−1
30	1.732	0.000	0.000	1	1

Example 9.9 The factors that influence the surface finish of a metal part are the feed rate (inches per minute) and the depth of cut in inches. An experiment is run in which feed rate (x_1) and depth of cut (x_2) are varied in a 3×4 factorial structure. However, two types of materials were used in the experiment. Two observations were taken at each of the 24 combinations. The coding used for x_1 and x_2 is given by

$$x_1 = \frac{\text{Feed rate} - 0.25}{0.05}$$

$$x_2 = \frac{\text{Depth} - 0.25}{0.05}$$

The data along with the natural and coded levels are given by

		\multicolumn{4}{c}{Depth of Cut}	
Feed Rate		0.15 (−2)	0.20 (−1)	0.30 (+1)	0.35 (+2)	
0.20	(−1)	74	79	89	102	
		78	82	94	98	
0.25	(0)	98	97	98	105	Material 1
		91	93	105	102	($z = -1$)
0.30	(+1)	114	115	122	133	
		108	111	117	138	

		\multicolumn{4}{c}{Depth of Cut}	
Feed Rate		0.15 (−2)	0.20 (−1)	0.30 (+1)	0.35 (+2)	
0.20	(−1)	63	73	77	101	
		68	74	79	103	
0.25	(0)	74	85	83	105	Material 2
		77	81	87	104	($z = +1$)
0.30	(+1)	100	105	111	118	
		97	108	107	122	

The goal of the experiment is to build response surfaces for each material and to determine the combination of x_1 and x_2 which maximizes the response for each material. The model initially fit to the data is given by

$$\hat{y} = b_0 + b_1 x_1 + b_2 x_2 + b_{11} x_1^2 + b_{22} x_2^2 + b_{12} x_1 x_2 + \hat{\gamma} z$$

$$+ \hat{\delta}_1 x_1 z + \hat{\delta}_2 x_2 z + \hat{\rho}_1 x_1^2 z + \hat{\rho}_2 x_2^2 z \qquad (9.37)$$

Table 9.3 PROC REG for Model of Equation (9.37) for Example 9.9

The SAS System
Model: MODEL 1
Dependent variable: y

Analysis of Variance

Source	Degrees of Freedom	Sum of Squares	Mean Square	F-Value	Probability $> F$
Model	10	13132.12500	1313.21250	58.375	0.0001
Error	37	832.35417	22.49606		
Corrected total	47	13964.47917			

Root mean squared error:		4.74300	R square:	0.9404	
Dependent mean:		96.89583	Adjusted R-square:	0.9243	
CV:		4.89495			

Parameter Estimates

Variable	Degrees of Freedom	Parameter Estimate	Standard Error	t for H_0: Parameter $= 0$	Probability $> |t|$
INTERCEP	1	89.638889	1.64555732	54.473	0.0001
FR	1	15.500000	0.83845204	18.486	0.0001
DC	1	5.316667	0.43297477	12.279	0.0001
FRDC	1	−0.937500	0.53028363	−1.768	0.0853
FRFR	1	5.937500	1.45224154	4.089	0.0002
DCDC	1	1.319444	0.45639548	2.891	0.0064
M	1	−6.944444	1.64555732	−4.220	0.0002
MFR	1	−1.125000	0.83845204	−1.342	0.1879
MDC	1	0.783333	0.43297477	1.809	0.0786
MFRFR	1	1.187500	1.45224154	0.818	0.4188
MDCDC	1	0.402778	0.45639548	0.883	0.3832

The response surface above allows for the development of separate response surfaces for each of the two as distinct linear and pure quadratic terms. Table 9.3 displays a SAS PROC REG showing regression coefficients, standard errors, t-statistics, and P-values. Strictly speaking, the factorial arrangement allowed an additional term describing the interaction $x_1 x_2 z$. However, this term was not used.

The descriptions of the model terms are FR (feed rate), DC (depth of cut), and M (materials). Based on the P-values, the terms $x_1^2 z$ and $x_2^2 z$ were eliminated from the model. The final model appears in the PROC REG in Table 9.4. From the response surface model shown in Table 9.4 we can write the following response surface models for the two materials. Here we set $z = -1$ and $+1$ in the model listed in Table 9.4.

Material 1 ($z = -1$):

$$\hat{y} = 94.78 + 16.63\text{FR} + 4.53\text{DC} - 0.94\text{FR} * \text{DC} + 5.94\text{FR}^2 + 1.32\text{DC}^2$$

Table 9.4   PROC REG for Reduced Model for Surface Finish Study of Example 9.9

The SAS System
Model: MODEL 1
Dependent variable: y

Analysis of Variance

Source	Degrees of Freedom	Sum of Squares	Mean Square	F-Value	Probability $> F$
Model	8	13099.56250	1637.44531	73.834	0.0001
Error	39	864.91667	22.17735		
Corrected total	47	13964.47917			

Root mean squared error:	4.70928	R-square: 0.9381	
Dependent mean:	96.89583	Adjusted R-square: 0.9254	
CV:	4.86015		

Parameter Estimates

Variable	Degrees of Freedom	Parameter Estimate	Standard Error	t for H_0: Parameter = 0	Probability $> |t|$
INTERCEP	1	89.638889	1.63385920	54.863	0.0001
FR	1	15.500000	0.83249156	18.619	0.0001
DC	1	5.316667	0.42989679	12.367	0.0001
FRDC	1	−0.937500	0.52651389	−1.781	0.0828
FRFR	1	5.937500	1.44191768	4.118	0.0002
DCDC	1	1.319444	0.45315101	2.912	0.0059
M	1	−5.145833	0.67972651	−7.570	0.0001
MFR	1	−1.125000	0.83249156	−1.351	0.1844
MDC	1	0.783333	0.42989679	1.822	0.0761

Material 2 ($z = +1$):

$$\hat{y} = 84.49 + 14.38\text{FR} + 6.10\text{DC} - 0.94\text{FR} * \text{DC} + 5.94\text{FR}^2 + 1.32\text{DC}^2$$

Note that the two response surfaces have distinct intercepts and linear terms. Figures 9.6 and 9.7 show response contours for the two materials. The two response surfaces have very similar characteristics. For all values of feed rate and depth of cut, material 1 has a higher surface finish than material 2. As the data suggest, high surface finish is predicted with large values of feed rate and depth of cut.

9.4   FURTHER COMMENTS ABOUT QUALITATIVE VARIABLES

We have warned that when the problem requires the use of qualitative and quantitative variables in the same problem, the choice of a model can be quite different. It is often not clear what structure should be assumed

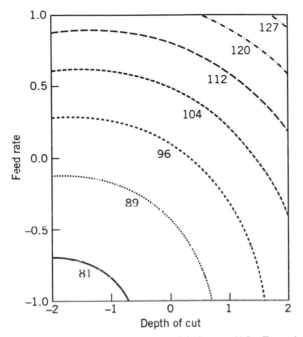

Figure 9.6 Response contours for material 1 ($z = -1$) for Example 9.9.

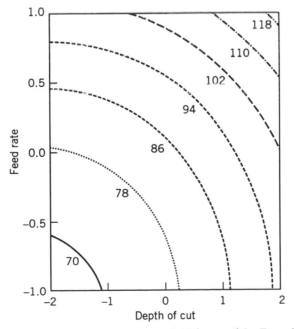

Figure 9.7 Response contours for material 2 ($z = +1$) for Example 9.9.

regarding the interaction between the quantitative and qualitative variables. In the case of a second-order model in the quantitative variables, it is not at all uncommon for one to encounter completely separate response surfaces for each level of the qualitative variable (or variables). As a result, when computer-generated design is used, one should be "liberal" with the choice of model in order to allow for such terms as $x_i x_j z$ and $x_i^2 z$.

9.5 ERRORS IN CONTROL OF DESIGN LEVELS

In practical situations one plans to design experiments according to some master plan that entails the choice of a combination of design levels and a random assignment of these design levels to experimental units. However, it is often **difficult to control design levels**. This is true even in laboratory situations. This suggests a need to determine how one can assess the impact of **errors in control** of design levels. For an excellent account see Box (1963).

The approach to assessment of errors in design levels in an RSM setting depends on the structure of the error. In some (perhaps most) instances there is a chosen level according to the design plan. This "aimed at" level is reported and used in the analysis even though it is known that it is subject to error. The actual level that is experienced by the process is not known. We call this the "fixed design level" case.

Fixed Design Level Case
One must keep in mind that the actual levels are not observed in this case. The deviation between the actual and planned level is a random variable. That is, for the uth run of the jth design variable we have

$$w_{ju} = x_{ju} + \theta_{ju} \qquad (u = 1, 2, \ldots, N; j = 1, 2, \ldots, k) \qquad (9.38)$$

where w_{ju} is the actual level, x_{ju} is the fixed and planned level, and θ_{ju} is a random "error in control." It may be assumed that

$$E(\theta_{ju}) = 0$$

$$\mathrm{Var}(\theta_{ju}) = \sigma_j^2 \qquad (u = 1, 2, \ldots, N; j = 1, 2, \ldots, k)$$

For the case of multiple design levels, the variances $\sigma_1^2, \sigma_2^2, \ldots, \sigma_k^2$ are the variances of errors in control.

First-Order Model
Consider the fixed level case when one fits a first-order model. The model, in terms of the true, yet unknown, levels are given by

$$y = \beta_0 + \beta_1 w_1 + \beta_2 w_2 + \cdots + \beta_k w_k + \varepsilon \tag{9.39}$$

Because the analysis will be done on the planned levels it is instructive to write the model in Equation (9.39) as

$$y = \beta_0 + \beta_1(x_1 + \theta_1) + \beta_2(x_2 + \theta_2) + \cdots + \beta_k(x_k + \theta_k) + \varepsilon$$

$$= \beta_0 + \sum_{j=1}^{k} \beta_j x_j + \left\{ \varepsilon + \sum_{j=1}^{k} \beta_j \theta_j \right\}$$

$$= \beta_0 + \sum_{j=1}^{k} \beta_j x_j + \varepsilon^* \tag{9.40}$$

Of course the θ_j are random variables as described by Equation (9.38), but the x_j are fixed (predetermined by the design choice). As a result, Equation (9.40) represents an almost standard RSM model with model error

$$\varepsilon^* = \varepsilon + \sum_{j=1}^{k} \beta_j \theta_j \tag{9.41}$$

It becomes obvious from Equation (9.41) that the errors associated with controlling design variables are transmitted to the model error and the result is an inflated error variance. As a result, if the model in Equation (9.40) is used and least squares is used to estimate $\beta_0, \beta_1, \beta_2, \ldots, \beta_k$ with ordinary least squares we have the following:

1. The regression coefficients are unbiased.
2. The variance–covariance matrix of $\mathbf{b} = (\mathbf{X'X})^{-1}\mathbf{X'y}$ is given by $(\mathbf{X'X})^{-1}\sigma^{*2}$, where $\sigma^{*2} = \sigma^2 + \sum_{j=1}^{k} \beta_j^2 \sigma_j^2$ assuming that the θ_j are independent. This results in inflated error variances. Depending on the magnitude of the σ_j^2, the inflation may be severe.
3. Standard RSM (e.g., variable screening) and steepest ascent are valid, although they may be rendered considerably less effective if the σ_j^2 are large. Of course, if the σ_j^2 are extremely small compared to the model variance σ^2, these errors will not result in major difficulties.

It should be emphasized that the conclusions drawn in 1 and 2 above do depend on the assumptions made regarding the θ_j. Of particular interest is the assumption that the θ_j are unbiased—that is, $E(\theta_j) = 0$. This assumption is certainly a reasonable one in most cases. However, if it does not hold, the least squares estimator of β_0 becomes biased.

Second-Order Model

In the case of the second-order model, errors in control result in more difficulties than one experiences in the first-order case if the error variances are not negligible. Consider again Equation (9.38) and the second-order model given by

$$
\begin{aligned}
y &= \beta_0 + \sum_{j=1}^{k} \beta_j w_j + \sum_{j=1}^{k} \beta_{jj} w_j^2 + \sum\sum_{i<j} \beta_{ij} w_i w_j + \varepsilon \\
&= \beta_0 + \sum_{j=1}^{k} \beta_j (x_j + \theta_j) + \sum_{j=1}^{k} \beta_{jj} (x_j + \theta_j)^2 \\
&\quad + \sum\sum_{i<j} \beta_{ij} (x_i + \varepsilon_i)(x_j + \theta_j) + \varepsilon \\
&= \beta_0 + \sum_{j=1}^{k} \beta_j x_j + \sum_{j=1}^{k} \beta_{jj} x_j^2 + \sum\sum_{i<j} \beta_{ij} x_i x_j \\
&\quad + \left\{ \varepsilon + \sum_{j=1}^{k} \beta_j \theta_j + \sum_{j=1}^{k} \beta_{jj} \theta_j (2x_j + \theta_j) \right. \\
&\qquad\qquad \left. + \sum\sum_{i<j} \beta_{ij} \left[\theta_i x_j + x_i \theta_j + \theta_i \theta_j \right] \right\} \\
&= \beta_0 + \sum_{j=1}^{k} \beta_j x_j + \sum_{j=1}^{k} \beta_{jj} x_j^2 + \sum\sum_{i<j} \beta_{ij} x_i x_j + \varepsilon^* \qquad (9.42)
\end{aligned}
$$

The model in Equation (9.42) is best analyzed by considering the expectation and variance of ε^*. Once again we have a standard RSM model with error ε^*. However, in this case the properties of ε^* are more complicated. The expected value of ε^* is given by

$$
E(\varepsilon^*) = \sum_{j=1}^{k} \beta_{jj} \sigma_j^2
$$

Because $E(\varepsilon^*) \neq 0$ and the "error bias" $\sum_{j=1}^{k} \beta_{jj} \sigma_j^2$ is independent of the x_j, the least squares estimator of β_0 is biased by $\sum_{j=1}^{k} \beta_{jj} \sigma_j^2$.

In order to better illustrate impact in the case of the second-order model we rewrite ε^* in Equation (9.42) as

$$\varepsilon^* = \varepsilon + \sum\sum_{i<j} \theta_i(\beta_{ij}x_j) + \sum\sum_{i<j} \theta_j \beta_{ij}x_i + \sum\sum_{i<j} \beta_{ij}\theta_i\theta_j$$

$$+ \sum_{j-1}^{k} \theta_j \beta_j + 2\sum_{j=1}^{k} \beta_{jj}\theta_j x_j + \sum_{j=1}^{k} \beta_{jj}\theta_j^2 \qquad (9.43)$$

If one considers the variance of ε^* in Equation (9.43), it becomes apparent that:

1. The error variance is inflated by errors in control transmitted to ε^* just as in the first-order case.
2. The error variance will no longer be homogeneous. Rather, $\text{Var}(\varepsilon^*)$ depends on design levels.

While we will not bother to display $\sigma^{*2} = \text{Var}\,\varepsilon^*$, it should be apparent that terms such as $\sum_{j=1}^{k} \beta_j^2\sigma_j^2, \sum_{j=1}^{k} \beta_{jj}^2[E(\theta_j^4) - \sigma_j^4]$, and $\sum_{i<j} \beta_{ij}^2\sigma_i^2\sigma_j^2$ inflate the error variance whereas terms such as $4\sum_{j=1}^{k} \beta_{jj}^2 x_j^2\sigma_j^2, \sum\sum_{i<j} \beta_{ij}^2 x_j^2\sigma_i^2$, and $\sum\sum_{i<j} \beta_{ij}^2 x_i^2\sigma_j^2$ render the error variance nonhomogeneous.

Further Comments

The degree of difficulty caused by errors in control obviously depends on the model complexity and on the size of the error variances $\sigma_1^2, \sigma_2^2, \ldots, \sigma_k^2$. If the latter are dominated by the model error variance σ^2, then the impact of errors in control is negligible.

9.6   RESTRICTIONS IN RANDOMIZATION IN RSM

In Chapters 3, 4, 7, and 8, we discussed response surface designs. Designs for both first- and second-order models were presented and illustrated. Apart from the material on blocking designs in Chapter 3 and 7, it is assumed that the experimental runs listed in the design plan are accomplished in a **completely randomized** fashion. That is, design points are assigned randomly to experimental units. In the blocking designs where different design points are assigned to different blocks, the **restriction in randomization** requires that the design point assigned to a particular block be allotted randomly to experimental units *within that block*. An example is the now very familiar 2^3 factorial plan in 2 blocks, namely,

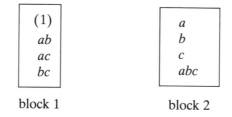

block 1   block 2

Here, of course, if the practitioner regards the experimental unit to be *time*, the design points (1), *ab*, *ac*, and *bc* in block 1 are run in random order. The same is done for block 2. Strictly speaking, this design is called a **randomized incomplete block design,** the incomplete being derived from the fact that not all design points in the experimental plan are assigned to the same block. The randomized incomplete block design as we have described it here is quite practical in many situations because one cannot often afford to use large "block sizes" in RSM situations. The same type of design and randomization scheme is employed for the second-order blocking designs discussed and developed in Chapter 7.

The notion of restrictions in randomization are quite natural for blocking designs in RSM and other fields where design of experiments is an important tool. The entities that we call "blocks" appear as necessary components of the total plan, and an assignment to blocks that produces a kind of balance (orthogonal blocking) is sought. The analysis is quite straightforward, and the design is structured so that only unimportant effects are compromised. However, the use of the standard blocking designs is not the only source of restricted randomization. At times the design that is chosen for its efficiency in a completely randomized setting must be used (due to practical considerations), in a "restrictions in randomization" setting.

9.6.1 The Dilemma of "Difficult to Change" Factors

Quite often in RSM a completely randomized design (CRD) is very difficult to conduct. This is particularly true where robust parameter design (see Chapter 10) is involved. A completely randomized design can be quite impractical in cases where there are factors whose levels are difficult to change or difficult to control. At times, process temperature fits this description. In a large-scale experiment in which temperature is controlled in an oven, it is convenient and less costly to execute all experimental runs with low-temperature followed by the high-temperature runs (say in a two-level design case). This would eliminate the act of moving temperature back and forth in the oven according to the assigned randomization scheme. The experimenter would be reluctant to carry out a CRD in this case. In any application there are always factors that are difficult to change.

9.6.2 Split Plot Structures

There are occasions when the difficult to change nature of one or more factors necessitates the use of a **split plot** design in RSM. For discussion of the nature of a split plot design, see, for example, Yates (1935, 1937), Cox (1958), and Lentner and Bishop (1993) and Montgomery (1991b). For a discussion of the split plot in RSM situations, see Box and Jones (1992).

A split plot design traditionally has a factorial structure with randomization restrictions. The design has two different experimental units called **whole plots** and **subplots** and hence two separate randomization procedures. The former is the larger experimental unit of the two. The CRD has one type of experimental unit and, of course, complete (unrestricted) randomization. The difficult-to-change factor normally has its levels assigned randomly to the whole plots used in the experiment. The whole plots are divided into smaller subplots. The factor (or factors) not difficult to change have levels assigned randomly to the subplot experimental units. As a result, each whole plot receives only one level of the difficult to control factor (the whole plot factor) but all levels of the not-difficult-to-control factor (the subplot factor). Consider, for example, a simple 2×3 factorial structure in which two levels of temperature are crossed with three concentration levels of a catalyst. Assume that all runs at high temperature are executed first and time is the experimental unit. One might regard the design to be as follows:

Replication 1

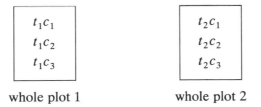

whole plot 1   whole plot 2

The levels t_1 and t_2 are, respectively, low and high temperature, whereas c_1, c_2, and c_3 are concentration levels. One may view the whole plots as incomplete blocks where temperature, the whole plot factor, is confounded with blocks. The order of the concentration within whole plots is random.

The above factorial structure involves restricted randomization and thus is not a CRD. If the design were replicated (i.e., executed say r times), then the setup is a split plot design and the ANOVA degrees of freedom are as follows:

Source	Degrees of Freedom
Replications	$r - 1$
Temperature	1
Temperature × replication	$r - 1$ (whole plot error)
Concentration	2
Concentration × temperature	2
Concentration × replications	$2(r - 1)$ ⎫ (subplot error)
Concentration × temperature × replication	$2(r - 1)$ ⎭
Total	$6r - 1$

The two errors are a result of the two types of randomization. There are a few items worth highlighting here before we project this into an RSM model building setting:

1. The degrees of freedom for testing subplot effects (concentration, concentration × temperature) are larger than those for testing the whole plot factor (temperature).
2. The assignment of subplot levels (concentration) can be viewed as a secondary replicated randomized complete block design.

From item 1 it is clear that any tests on subplot effects will be more efficient due to the larger number of error degrees of freedom.

Obviously, if the experiment is designed as discussed here and yet analyzed as if it were a CRD, the conclusions may be quite different. The same factorial structure with a *completely randomized* execution would yield the following ANOVA degrees of freedom:

Source	Degrees of Freedom
Temperature	1
Concentration	2
Temperature × concentration	2
Error	$6(r - 1)$
Total	$6r - 1$

Here there are $6(r - 1)$ degrees of freedom for testing all three effects. As a result, it is quite misleading to mistakenly analyze an experiment as if it were run as a CRD when in fact the design is executed as a split plot. In a completely randomized setting, an appropriate RSM model might be written as

$$y_u = \beta_0 + \beta_t t_u + \beta_c c_u + \beta_{cc} c_u^2 + \beta_{tc} t_u c_u + \varepsilon_u, \qquad u = 1, 2, \ldots, N = 6r$$

where c_u and t_u are levels of concentration and temperature, respectively. Here we have the usual ε_u as i.i.d. $N(0, \sigma)$. We have not chosen to include a term in $t_u c_u^2$ which along with $c_u t_u$ accounts for the two concentration $\times$ temperature degrees of freedom in the ANOVA breakdown above. The variance σ^2 is the single error variance whose estimate possesses $6(r - 1)$ degrees of freedom.

In the foregoing we have displayed the contrast in the different layout degrees of freedom between the CRD design execution and the split plot execution, the latter featuring restriction in randomization. However, it should be pointed out that in RSM the replication of a factorial experiment in a completely randomized fashion is not common in an industrial setting, though in this case with only two factors in a 2×3 structure, the use of $r = 2$ or 3 would not involve an inordinate number of design runs.

Model for Split Plot Error Structure in an RSM Setting

We have illustrated, through a look at ANOVA degree of freedom tables, that the restriction in randomization requires two different error terms whereas the CRD requires but one. With the 3×2 factorial illustration we have reviewed the pertinent RSM model and noted that the "single error" in the CRD is derived from the standard error variance σ^2 when homogeneous variance assumptions are made. It is instructive to consider the RSM model when difficult to change factors force the researcher into a randomization scheme like that of a split plot. From the model one can sense the need for the two error variances that necessitate two error terms in the ANOVA.

Consider an RSM situation in which there are hard-to-control or hard-to-change variables $z_1, z_2, \ldots, z_z$ (these may be noise variables as described in Chapter 10) and other, more standard RSM variables $x_1, x_2, \ldots, x_x$. *Let us also assume that we have a true split plot randomization structure.* This requires that all combinations of the whole plot factor be crossed with the same combinations of the subplot factors. The obvious example is a complete factorial. Suppose there are a unique combination of the z's serving as whole plot factor combinations and b unique combinations of the x's serving as subplot factor combinations. Call the whole plot combinations $\mathbf{z}_1, \mathbf{z}_2, \ldots, \mathbf{z}_a$; they are randomly assigned to the available whole plots (say time periods) while the x-combinations are called $\mathbf{x}_1, \mathbf{x}_2, \ldots, \mathbf{x}_b$, and they are assigned to the b available subplots within each whole plot. Consider an example, say, with four factors z, x_1, x_2, and x_3 in a 2^4 factorial structure with z being difficult to change. For notational ease let z, x_1, x_2, and x_3 be denoted by A, B, C, and D. The split plot structure is then

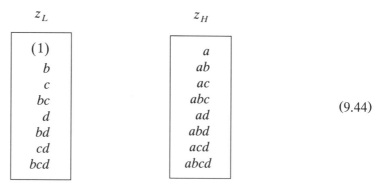

$$
\begin{array}{cc}
z_L & z_H \\
\hline
(1) & a \\
b & ab \\
c & ac \\
bc & abc \\
d & ad \\
bd & abd \\
cd & acd \\
bcd & abcd \\
\hline
\end{array}
\qquad (9.44)
$$

where H and L denote low and high, respectively, for z (factor A). There are two whole plots and eight subplots observations within each whole plot.

In general, then, the design matrix becomes

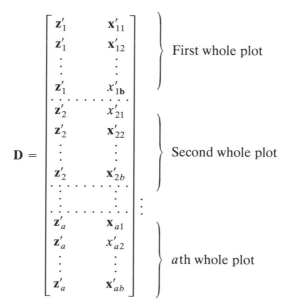

$$
\mathbf{D} = \begin{bmatrix}
\mathbf{z}'_1 & \mathbf{x}'_{11} \\
\mathbf{z}'_1 & \mathbf{x}'_{12} \\
\vdots & \vdots \\
\mathbf{z}'_1 & \mathbf{x}'_{1b} \\
\hdashline
\mathbf{z}'_2 & \mathbf{x}'_{21} \\
\mathbf{z}'_2 & \mathbf{x}'_{22} \\
\vdots & \vdots \\
\mathbf{z}'_2 & \mathbf{x}'_{2b} \\
\hdashline
\vdots & \vdots \\
\hdashline
\mathbf{z}'_a & \mathbf{x}'_{a1} \\
\mathbf{z}'_a & \mathbf{x}'_{a2} \\
\vdots & \vdots \\
\mathbf{z}'_a & \mathbf{x}'_{ab}
\end{bmatrix}
\begin{array}{l}
\left.\rule{0pt}{3.5em}\right\} \text{First whole plot} \\
\left.\rule{0pt}{3.5em}\right\} \text{Second whole plot} \\
\left.\rule{0pt}{3.5em}\right\} a\text{th whole plot}
\end{array}
$$

One must keep in mind that for a given whole plot, say the ith whole plot, the subplot factor levels $x'_{i1}, x'_{i2} \ldots x'_{ib}$ represent the aforementioned combinations $x_1, x_2, \ldots, x_b$. The first subscript denotes the whole plot in question. The RSM model then becomes

$$y_{ij} = \beta_0 + \gamma' z_i + \beta' x_{ij} + z'_i B_z z_i + x'_{ij} B_x x_{ij} + z'_i \Delta x_{ij} + \delta_i + \varepsilon_{ij}$$

$$i = 1, 2, \ldots, a; \; j = 1, 2, \ldots, b; \; \delta_i \sim \text{i.i.d. } N(0, \sigma_\delta^2); \; \varepsilon_{ij} \sim \text{i.i.d. } N(0, \sigma^2)$$

$$(9.45)$$

and $\text{Cov}(\varepsilon_i, \delta_{ij}) = 0$ for all i and j. Here we have written the model in the general second-order model framework. Obviously, for many scenarios (as in the illustration presented by display equation (9.44)) the quadratic terms may .not be present. The Δ matrix contains interaction coefficients between the x's and z's. One should focus on the errors δ_i and ε_{ij}. The error δ_i is the **whole plot error** ε_{ij} **is the subplot error.** Two different types of experimental units necessitate two separate errors and, of course, two error variances.

In the next subsection, we discuss the ramifications of testing and estimation of RSM settings in which a split plot structure exists in the design.

9.6.3 RSM Estimation and Testing Under a Split Plot Structure

The non-RSM classical split plot structure with replication is illustrated in the temperature-concentration scenario in 9.6.2. In this case, one is primarily interested in testing whole plot and subplot effects. The two error mean squares are used as separate denominators for testing whole plot effects and subplot effects. The validity of these tests depend on the existence of replication. However, in an RSM setting, one is interested in more than testing main effects and interactions. Variable screening, model development, and estimation of optimum conditions are of primary importance. In addition, the existence of resources that allow replication of entire factorial experiments is rare in industrial settings. If replication is a practical option, then variable screening can be accomplished by making use of the two error terms. Estimation of regression coefficients is impacted by the existence of the two variance components σ^2 and σ_δ^2 whether one replicates the experiment or not.

To illustrate the problems associated with model estimation under the split plot structure, consider the model of Equation (9.45) in matrix form—that is,

$$y \sim N(\mu(x, \beta), V)$$

where $\mu(x, \beta)$ signifies $E(y)$ from the model in Equation (9.45) and $V = \text{Var}(y)$

is the variance covariance matrix of **y**. It should be clear from (9.45) that any two observations in the same whole plot are not independent but instead have covariance σ_δ^2 because they share the common error component. Any two observations not in the same whole plot are independent. The form of the V matrix is given by

$$
V = \begin{bmatrix}
\mathbf{T}_1 & 0 & \cdots & \cdots & 0 \\
0 & \mathbf{T}_2 & 0 & \cdots & 0 \\
0 & & \ddots & & \\
\vdots & \vdots & & \ddots & \\
\vdots & \vdots & & & \ddots \\
0 & 0 & & & \mathbf{T}_a
\end{bmatrix}
\tag{9.46}
$$

where $\mathbf{T}_i = \sigma^2 \mathbf{I}_{b \times b} + \sigma_\delta^2 \mathbf{1}_{b \times 1} * \mathbf{1}'_{1 \times b}$. As a result, the ordinary least squares estimation that we have emphasized throughout the text does not necessarily apply for estimation of coefficients. Rather, one should consider

$$
\mathbf{b} = \left(\mathbf{X}' \mathbf{V}^{-1} \mathbf{X} \right)^{-1} \mathbf{X}' \mathbf{V}^{-1} \mathbf{y}
\tag{9.47}
$$

where **X** is the entire model matrix associated with Equation (9.45) and **V** is the $ab \times ab$ variance covariance matrix listed in Equation (9.46). The estimators in Equation (9.47) are the **generalized least squares** estimators that are appropriate when one is under nonstandard conditions—that is, $\mathbf{V} \neq \sigma^2 \mathbf{I}$. See Montgomery and Peck (1992) or Myers (1990). Unfortunately, Equation (9.47) cannot be used directly because the analyst does not have knowledge of σ^2 and σ_δ^2 and thus cannot compute **V**. There are methods that allow estimation of σ^2 and σ_δ^2, but we will not discuss these here. See Davison et al. (1995) for a discussion.

There are design situations in which the estimator in Equation (9.47) is equivalent to the ordinary least squares estimator. See Davison et al. (1995). However, one still requires estimates of σ^2 and σ_δ^2 to obtain estimated standard errors for variable screening or other forms of model editing. One must keep in mind that if variable screening is done with two-level factorials and fractional factorials when a split plot structure is used in the design, the normal probability plotting must be executed differently than we outlined in Chapters 3 and 4. Whole plot effects (including interactions) should not be plotted on the same graphs as subplot effects (including subplot by whole plot interactions). The two different variance components and hence error terms result in different standard errors for the two different sets of effects. For details involved in RSM design and analysis in a split plot scenario, see Box and Jones (1992), Lucas and Ju (1992), and Davison et al. (1995).

EXERCISES

9.1 Consider the two simple designs discussed in Section 9.1. Namely, consider the following two-point designs for fitting a first-order model:
  (i) $N/2$ runs at $x = -1$; $N/2$ runs at $x = +1$
  (ii) $N/2$ runs at $x = -0.6$; $N/2$ runs at $x = +0.6$
  The region of interest $[-1, +1]$ in the design variable x.
  (a) Suppose the first-order model is correct. Compute

$$\left(N/2\sigma^2\right)\int_{-1}^{1} \text{Var } \hat{y}(\mathbf{x})\, dx$$

  for both designs.
  (b) Suppose the "true" model is

$$y = \beta_0 + \beta_1 x + \beta_2 x^2$$

  with $\sqrt{N}\beta_2/\sigma = 2.0$. Compute $(N/2\sigma^2)\int_{-1}^{1}\{\text{Var } \hat{y}(x) + [\text{Bias } \hat{y}(x)]^2\}\, dx$ for both designs.
  (c) Do the computations as in (b) above but let $\sqrt{N}\beta_2/\sigma = 4.0$.

9.2 Consider a fitted first-order model in four design variables:

$$\hat{y} = b_0 + b_1 x_1 + b_2 x_2 + b_3 x_3 + b_4 x_4$$

Suppose the region of interest in the natural variables is given by the following ranges:

ξ_1: $[100, 150]$
ξ_2: $[1, 2]$
ξ_3: $[1000, 2000]$
ξ_4: $[0, 1]$

The region is cuboidal. Suppose, in addition, that one is interested in protecting against a complete second-order model. A total of 10 runs are to be used.
  (a) Give a *minimum bias design in the natural variables*.
  (b) Suppose, instead of using the minimum variance design, the analyst achieves protection against model misspecfication by using a scaled second moment of 0.45. Again, two center runs are used in addition to the eight factorial points. *Give the appropriate design in the natural variables.*

9.3 Consider exercise 9.2. Suppose that a first-order model is fit but the practitioner is interested in protection against the model

$$E(y) = \beta_0 + \sum_{i=1}^{4} \beta_i x_i + \sum_{i<j} \sum \beta_{ij} x_i x_j$$

In other words, there is no fear of pure quadratic effects in this design region, but the protection is sought against *interaction terms only*. Consider the use of Equation (9.18) for the determination of a minimum bias design.

(a) What are the moment requirements for the minimum bias design in this case?

(b) Give a minimum bias design with 10 runs for this case (coded levels).

(c) Give the same design in the natural variables.

9.4 Refer to Exercise 9.3. With the suggested alternative model, is there any restriction on the second pure design moment for achieving a minimum bias design? Explain.

9.5 Consider the following general formulation: If the fitted model is first order, and one is interested in protection against a model that contains first-order terms plus two-factor interactions, the minimum bias design is one in which all odd moments through order three are zero. This means that the two-level design for minimum bias must be resolution $\geq$ IV. Prove that this statement is correct.

9.6 Prove the following statement: If the fitted model is first order in k variables and one is interested in protecting against a model containing linear terms and two-factor interactions, a *minimum mean squared error design* is a two-level resolution $\geq$ IV design with pure second moments (ii) as large as possible. As a result, the levels are to be placed at ± 1 extremes.

9.7 An experiment was conducted to study the effect of temperature and type of oven on the life of a particular component being tested. Four types of ovens and three temperature levels were used in the experiment. Twenty-four pieces were assigned randomly, two to each combination of treatments, and the following results were recorded:

Temperature (degrees)	Oven			
	O_1	O_2	O_3	O_4
500	227	214	225	260
	221	259	236	229
550	187	181	232	246
	208	179	198	273
600	174	198	178	206
	202	194	213	219

(a) Fit an appropriate response surface model using 1, 0, 1 coding for temperature. Be sure to determine if interaction exists among ovens and the response surface terms in temperature.

(b) Find the estimated optimum temperature—that is, that which maximizes oven life. Does it depend on oven type?

9.8 To estimate the degree of a suspension polyethylene, we extract polyethylene by using a solvent, and we compare the gel contents (gel proportion). This method is called the *gel proportion estimation method*. In the book *Design of Experiments for Quality Improvement* published by the Japanese Standards Association (1989), a study was conducted on the relationship between the amount of gel generated and three factors (solvent, extraction temperature, and extraction time) that influence amount of gel. The data are as follows:

Solvent	Temperature	Time		
		4	8	16
		94.0	93.8	91.1
Ethanol	120	94.0	94.2	90.5
		95.3	94.9	92.5
	80	95.1	95.3	92.4
		94.6	93.6	91.1
Toluene	120	94.5	94.1	91.0
		95.4	95.6	92.1
	80	95.4	96.0	92.1

(a) Build (and state) an appropriate response surface model in terms of coded temperature and time. Write a separate model for each solvent if necessary.

(b) Give estimate of conditions which maximize the amount of gel generated. Is there any indication of conditions for future experiments?

9.9 In an experiment conducted at the Department of Mechanical Engineering and analyzed by the Statistics Consulting Center at the Virginia Polytechnic Institute and State University, a sensor detects an electrical charge each time a turbine blade makes one rotation. The sensor then measures the amplitude of the electrical current. Five factors are: RPM (A), temperature (B), gap between blades (C), gap between blade and casing (D), and location of input (E). A $\frac{1}{2}$ fraction of a 2^6 factorial experiment is used, with defining contrasts being $ABCE$ and BCD. The data are as follows:

A	B	C	D	E	Response
-1	-1	-1	-1	-1	3.89
1	-1	-1	-1	1	10.46
-1	1	-1	-1	1	25.98
1	1	-1	-1	-1	39.88
-1	-1	1	-1	1	61.88
1	-1	1	-1	-1	3.22
-1	1	1	-1	-1	8.94
1	1	1	-1	1	20.29
-1	-1	-1	1	-1	32.07
1	-1	-1	1	1	50.76
-1	1	-1	1	1	2.80
1	1	-1	1	-1	8.15
-1	-1	1	1	1	16.80
1	-1	1	1	-1	25.47
-1	1	1	1	-1	44.44
1	1	1	1	1	2.45

(a) Build an appropriate response model, indicating different models for each of the two levels of location of input.

(b) Is the model improved by using a transformation on the response? Explain.

(c) What are recommended operating conditions or recommended conditions for future experiments?

9.10 Consider a situation in which one is interested in three levels of each of two quantitative design variables. These levels are -1, 0, 1. In addition, there are three levels of a single qualitative factor. It is felt that the design should accommodate a second-order model in the quantitative variables and interaction between the linear terms and the qualitative variable.

(a) Write out the model to be fitted.

(b) Using 18 experimental runs, construct, using a computer-generated design package, an appropriate design.

9.11 Consider the experimental designs shown in Figure 9.3. Suppose design (e) is used with four center runs under $+1$ of qualitative variable z, making a total of 8 runs. The model

$$\hat{y} - b_0 + c_1 z + b_1 x_1 + b_2 x_2$$

is fit, and lack of fit is significant. Four additional design points are sought in order to fit the model

$$\hat{y} = b_0 + c_1 z + b_1 x_1 + b_2 x_2 + b_{11} x_1^2 + b_{22} x_2^2$$
$$+ b_{12} x_1 x_2 + d_1 z x_1 + d_2 z x_2$$

What augmentation (use a computer generated design package) is D-optimal?

9.12 The following data was taken on two quantitative design variables and one two-level qualitative variable:

x_1	x_2	z	y
-1	1	-1	17.5
1	1	-1	22.1
0	0	-1	25.2
0	0	-1	26.5
0	0	$+1$	29.4
0	0	$+1$	31.3
-1	-1	$+1$	19.2
1	-1	$+1$	13.4

(a) Fit a model of the type

$$\hat{y} = b_0 + c_1 z + b_1 x_1 + b_2 x_2 + b_{12} x_1 x_2$$

(b) Edit the model and give recommendations on levels of x_1 and x_2 that maximize the response.

9.13 Consider the design in Figure 9.4. As indicated in the text, this design can be used for fitting the model in Equation (9.34) with model terms

x_1, x_2, x_1^2, x_2^2, and $x_1 x_2$, z and interactions $x_1 z$ and $x_2 z$. What is the D-efficiency of this design for the model of equation (9.34) for $k = 2$? What is the A-efficiency?

9.14 In Section 6.3.3, the method of ridge analysis is used to optimize a process that is designed to convert 1,2-propanediol to 2,5-dimethylpiperazine. The design is a central composite design with $k = 4$ and $\alpha = 1.4$. Suppose, in addition, that two different catalysts are being used. The design and response values are as follows:

x_1	x_2	x_3	x_4	z	y
-1	-1	-1	-1	-1	68.4
1	-1	-1	-1	1	28.5
-1	1	-1	-1	1	22.0
-1	-1	1	-1	1	15.7
-1	-1	-1	1	1	48.5
1	1	-1	-1	-1	28.5
1	-1	1	-1	-1	18.7
1	-1	-1	1	-1	65.8
-1	1	1	-1	-1	9.2
-1	1	-1	1	-1	32.9
-1	-1	1	1	-1	12.5
-1	1	1	1	1	49.8
1	-1	1	1	1	12.9
1	1	-1	1	1	30.9
1	1	1	-1	1	22.9
1	1	1	1	-1	28.2
-1.4	0	0	0	-1	32.1
1.4	0	0	0	-1	18.2
0	-1.4	0	0	-1	19.2
0	1.4	0	0	-1	50.5
0	0	-1.4	0	1	52.1
0	0	1.4	0	1	39.2
0	0	0	-1.4	1	30.7
0	0	0	1.4	1	40.7
0	0	0	0	1	39.5
0	0	0	0	1	38.2

(a) Compute an appropriate response surface model taking into account the two catalysts ($z = +1, -1$).

(b) Transfer the above to a separate response surface model for each catalyst.

9.15 Consider Exercise 3.9 in Chapter 3. Four factors were varied and their effects were measured on alloy cracking. One of the factors is "type of heat treatment method" (factor C) at two levels. This is a good example of a mix of qualitative and quantitative variables. Use the coding -1 and $+1$ for all factors, including factor C. Answer the following questions:

(a) Write out an appropriate model for relating the design variables to the response. Be sure the model is edited if necessary.

(b) Use your result in (a) to develop two response surface models, one for each heat treatment method.

9.16 Consider Exercise 4.12. A 2^{6-3} fractional factorial is used to fit a *first-order model*.

(a) Write out the first order model using -1, $+1$ coding on the qualitative variable, material type.

(b) Write out separate models for material type 1 and material type 2.

(c) What assumptions are you making in your developments in (b) above.

9.17 Consider the data in Exercise 4.14. Two qualitative variables appear.

(a) Fit a model containing all main effects and two-factor interactions.

(b) Edit the model to contain only significant terms.

(c) Use your model in (b) to write separate models for each combination of the qualitative variables.

CHAPTER 10

Response Surface Methods and Taguchi's Parameter Design

10.1 INTRODUCTION

In the 1980s, Genichi Taguchi [Taguchi and Wu (1980), Taguchi (1986, 1987)] introduced new ideas on quality improvement in the United States. He revealed an innovative parameter-design approach for reducing variation in products and processes. His methods and philosophy have generated considerable interest among quality engineers and statisticians. In fact, his methodology has gained usage at AT & T Bell Laboratories, Ford Motor Company, Xerox, and many other prominent giants of industry in America. Interest and curiosity has spread to the point where many statistical researchers have attempted to integrate the parameter design principles with well-established statistical techniques. This movement was generated by the result of much debate in the statistical community—not about the Taguchi philosophy, but, rather, about implementation and the technical nature of data analysis. As a result, the response surface approach has been suggested as a collection of tools that allow for the adoption of Taguchi principles while providing a more rigorous statistical approach to analysis.

10.2 WHAT IS PARAMETER DESIGN?

Parameter design or **robust parameter design** is essentially a principle that emphasizes proper choice of levels of controllable factors (*parameters* in Taguchi's terminology) in a process for the manufacturing of a product. The principle of choice of levels focuses to a great extent on variability around a . prechosen target for the process response. These controllable factors (controlled in any conducted experiments and in the design of the process or product) are called **control factors**. It is assumed that the majority of variability around target is caused by the presence of a second set of factors called noise factors or **noise variables**. *Noise factors are uncontrollable* in the design of the process. In fact, it is this lack of control that transmits variability to the process response. As a result, the term *robust parameter design* entails designing (not an experimental design) the process (select the levels of the controllable variables) so as to achieve **robustness** (insensitivity) to inevitable changes in the noise variables. One huge difference between Taguchi's view of process optimization and that which was operative in this country prior to Taguchi centers around the notion of *reducing process variability* and the concomitant principle that one should involve factors that cause process variability in the experimental design. These factors, the noise factors, are often functions of environmental conditions—for example, humidity, properties of raw materials, product aging, and so on. In certain applications they may involve factors that deal with the way the consumer handles or uses the product. Noise variables may be, and often are, controlled at the research or development level, but of course they cannot be controlled at the production level.

10.2.1 Examples of Noise Variables

It should not be inferred that Taguchi discovered the notion of noise variables. Prior to Taguchi, researchers certainly were aware that certain uncontrollable factors provide major sources of variability. However, Taguchi encouraged the formal use of noise variables in the experimental design and, as a result, allowed the subject of the analysis to involve process variability. The following is a list of candidates for noise variables in scenarios in which control variables are also listed.

Application	Control Variables	Noise Variables
Development of a cake mix	Amount of sugar, starch, and other ingredients	Oven temperature, baking time, amount of milk added
Development of a gasoline	Ingredients in the blend; other processing conditions	Type of driver, driving conditions, changes in engine type
Development of a tobacco product	Ingredients and concentrations; other processing conditions	Moisture conditions; storage conditions on tobacco
Development of a copy machine	Materials and other processing variables	Deviations from directions given to the consumer for use of machine
Large-scale chemical process	Processing conditions including nominal temperature	Deviations from nominal temperature; deviations from other processing conditions
Production of a box-filling machine for filling boxes of detergent	Surface area; geometry of the machine, (rectangular, circular)	Particle size of detergent

The above are only a few illustrations showing examples where the noise variables involved clearly cannot be controlled in the process, but can be controlled in an experimental situation—that is, in an experimental design effort. One important point should be made about the "large-scale chemical process" example. Many times the most important noise variables are tolerances in the control variables. In the chemical and some other industries, so-called controllable variables are not controlled at all. In fact, if there is a nominal temperature of, say, 250° F, during the large-scale production process, the temperature may vary between 235° F and 265° F, and the inability to control the process temperature produces unwanted variability in the product. As a result, nominal temperature (or nominal pressure) becomes a control variable and tolerance on temperature at levels $\pm 15°$ F is a noise variable.

10.2.2 Example of a Robust Product

It is sometimes easier to illustrate a concept with a contrived data set than with a more complex real-life data set. Consider a situation in which one is interested in establishing the geometry and surface area of a box-filling machine that achieves a rate of filling that is "closest" to 14 g/ sec of detergent. We will call surface area *factor A* (two levels) and geometry *factor B* (two levels). The single noise variable is the particle size of the detergent product. Three levels of the noise variable were used. The data are as follows:

Particle Size 1			Particle Size 2			Particle Size 3		
	B			*B*			*B*	
A	-1	$+1$	*A*	-1	$+1$	*A*	-1	$+1$
-1	13.7	13.7	-1	14.9	14.2	-1	17.4	14.4
$+1$	14.0	11.9	$+1$	16.0	11.8	$+1$	12.0	11.7

Clearly the product that is most insensitive to the noise factor, particle size, is given by $(+1, +1)$. The empirical standard deviation in fill rate is 0.1 g/sec. However, the mean filling rate across particle size is 11.8 and the target is 14.0. On the other hand, the product $(+1, -1)$ results in an average of 14.0, but the variability around the target is relatively large (stnd. dev. $= 2.0$) and thus the product is *not a robust product*. Consider the product $(-1, +1)$. The average filling rate is 14.1 and the variability around the target is quite small (standard deviation $= 0.36$). Thus, the optimum product among the four is $(-1, +1)$. It produces the best combination of bias error and variance error.

Strictly speaking, the "most robust" parameter design is that which is insensitive to changes in the noise variables. The implication of this definition is that the robust product minimizes variance transmitted by the noise variables. However, one cannot ignore the process mean. In this illustration one might consider the squared error loss criterion; that is, choose the product that minimizes

$$L = \sum (y - 14)^2$$

where the $\sum$ is taken over the levels of the particle size. In later sections where we deal with the response surface approach to robust parameter design, the process mean and variance will be considered simultaneously.

10.3 CROSSED ARRAY DESIGNS AND SIGNAL-TO-NOISE RATIOS

Taguchi's methodology for robust parameter design revolves around the use of orthogonal designs where an orthogonal array involving control variables is crossed with an orthogonal array for the noise variables. For example, in a $2^2 \times 2^2$, the 2^2 for the control variables is called the **inner array** and the 2^2 for the noise variables is called the **outer array**. The result is a 16 run design called the **crossed array**. Figure 10.1 represents a graphical representation.

The corners of the inner array represent $(-1, -1)$, $(-1, +1)$, $(+1, -1)$, and $(+1, +1)$ for the control variables. Each outer array is a 2^2 factorial in the noise variables. The dots in the outer arrays represent the locations of observations. Figure 10.2 shows another crossed array. There are 32 observations. The design is a $\frac{1}{2}$ fraction of 2^4 as the inner array which is crossed with a $\frac{1}{2}$ fraction of a 2^3 as the outer array.

Taguchi suggested that one may summarize the observations in each outer array with a summary statistic, one which provides information about the mean and variance, where, for example, in Figures 10.1 and 10.2, the summary statistic is computed across four observations. Thus, in Figure 10.2 there will be eight summary statistics. The summary statistics are called **signal-to-noise ratios** (SNRs), and the statistical analysis is done with the SNR

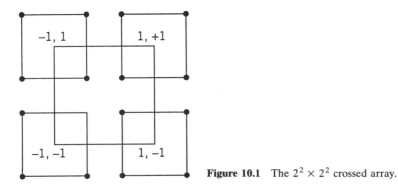

Figure 10.1 The $2^2 \times 2^2$ crossed array.

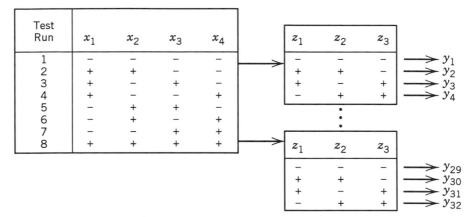

Figure 10.2   Another crossed array.

being the object of the analysis. There are many different SNR. However, there are four primary ones suggested by Taguchi. Specific SNR depend on the goal of the experiment. As in the case of response surface work, there are three specific goals:

1. *The smaller the better*. The experimenter wishes to minimize the response.
2. *The larger the better*. The experimenter wishes to maximize the respone.
3. *The target is best*. The experimenter wishes to achieve a particular target value.

The Smaller the Better

Taguchi treats the smaller the better case as if there is a target value of zero for the response. Thus, the quadratic loss function $E_z(y - 0)^2$ leads to the performance criterion

$$\text{SNR}_s = -10 \log \sum_{i=1}^{n} \left[\frac{y_i^2}{n} \right] \tag{10.1}$$

The rationale comes from the need to find **x**, the levels of the control variables that minimize the expected square error $E_z y^2$, where E_z implies expectation across the distribution of the noise variables. In the SNR calculation, $\sum_{i=1}^{n}$ implies summation over n response values at the outer array points. Thus, an SNR value will be present at each inner array design point. Because the $-10 \log$ transformation is used, one seeks to maximize SNR.

The Larger the Better

This case is treated in the same fashion as the smaller the better case, but y_i in Equation (10.1) is replaced by $1/y_i$. Thus, the SNR is motivated by the squared error criterion $E_z(1/y)^2$. As a result, the SNR is given by

$$\mathrm{SNR}_1 = -10 \log \sum_{i=1}^{n} \left[\frac{1}{y_i^2} \right] \bigg/ n \qquad (10.2)$$

Again, values of the control variables are sought that maximize SNR.

The Target Is Best

In this case, we are attempting to determine values of x that achieve a target value for the response, namely $y = t$. Deviations in either direction are undesirable. Two different SNR are to be considered. The appropriate SNR depends on the nature of the system. If the response mean and variance can be independently altered, Taguchi suggests that one or more *tuning factors* may be used to eliminate bias, that is, provide an adjustment that results in $E_z y = t$. These tuning factors are those that allow the analyst to alter the mean but leave the variance unchanged. As a result, the analysis is a two-step process: Select the tuning factors that renders $y - t$ and select levels of other control factors that maximize an SNR. Assuming that this can be accomplished, the obvious squared error loss function, $E_z(y - t)^2$, reduces to $\mathrm{Var}(y)$. Thus, the SNR used by Taguchi is given by

$$\mathrm{SNR}_{T_1} = -10 \log s^2 \qquad (10.3)$$

Here, of course, $s^2 = \sum_{i=1}^{n}(y_i - \bar{y})^2/(n-1)$. Thus, s^2 is the sample variance over the outer array design points. Again, values of x are chosen so that SNR is maximized.

A SNR ratio is suggested by Taguchi for cases in which the response standard deviation is related to the mean. It nicely accommodates situations in which the relationship is linear. In this case adjustment or tuning factors are sought which, again, eliminates bias but leaves the coefficient of variation σ_y/μ_y relatively unaffected. As a result, the natural SNR involves the sample coefficient of variation $s/\bar{y}$. In fact, Taguchi's SNR for this case is given by

$$\mathrm{SNR}_{T_2} = 10 \log \left(\frac{\bar{y}^2}{s^2} \right) \qquad (10.4)$$

Again, the SNR is to be maximized. So, again, the two-step procedure is used. Values of the tuning factor (or factors) are chosen which render $y = t$. These tuning factors are those that have an impact on the mean response but little or no impact on SNR.

What Is the Utility of the SNR?

The purpose of the SNR in Taguchi's approach to robust parameter design is to provide an easy-to-use performance criterion that takes process mean and variance into account. The terminology itself is rather interesting. Of the four major SNR formulations, only the one in Equation (10.4) involves a true signal to noise ratio. There is a wealth of literature in which SNRs are criticized. See, for example, Box (1988) and Welch et al. (1990). A panel discussion, edited by Nair et al. (1992), points out opinions by several researchers regarding the use of SNR. We would like to focus on one interesting characteristic of the SNRs here. While the intention is to use them as performance characteristics that take both response mean and variance into account, many authors have suggested the use of separate models for the process mean and variance as a way of achieving better understanding of the process itself. More details will be supplied on this approach in a later section.

The use of SNR does not guarantee that the subject matter scientist will secure valuable information regarding the roles of the process mean and process variance. This "uncoupling" of those control factors that affect mean (location effects) and control factors that affect variance (dispersion effects) is very important in the understanding of the process. For example, the SNR in Equation (10.4) can be written

$$\text{SNR}_{T_2} = 10 \log \bar{y}^2 + 10 \log s^2$$

So, of course, the maximization of this SNR does not isolate what control factors are location effects and what are dispersion effects. As another example, consider SNR$_s$ in Equation (10.1):

$$\text{SNR}_s = -10 \log\left(\sum y_i^2 / n \right)$$

While $\sum y_i^2 / n$ does deal with variability around the target of zero, it is clear that an analysis using this SNR cannot separate the location effects from dispersion effects. Indeed,

$$\frac{\sum y_i^2}{n} = \bar{y}^2 + \left(1 - \frac{1}{n}\right) s^2$$

so again, the mean and variance contributions in (SNR)$_s$ are confounded. In order to gain some insight into how different data structures and hence

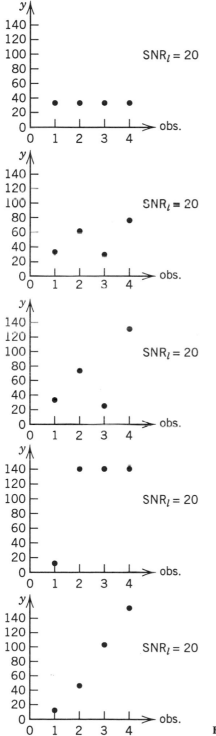

Figure 10.3 Five data sets with equivalent SNR$_l$.

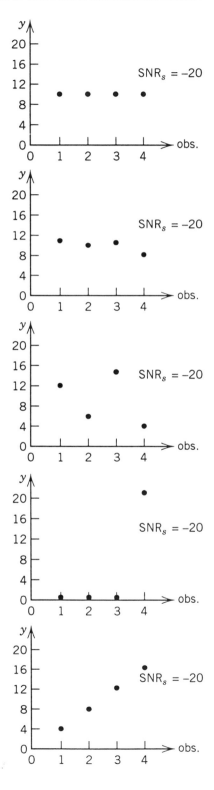

Figure 10.4   Five data sets with equivalent SNR$_s$

differing performances among processes can lead to the same SNR, consider Figure 10.3. These represent four different data sets that produce the same value of SNR_l, namely $SNR_l = 20$ [See Box (1988)]. Note how diverse the mean and variance properties are for these five cases. It would seem that the quite different mean and variance structures represent conditions that the scientist or engineer would like to detect. A similar set of data sets can be constructed for SNR_s. Consider Figure 10.4. All data sets here give an SNR_s of -20. Again, the data structures, including mean and variance, are quite different. Once again, it would seem as if the researcher would hardly call the five scenarios in Figure 10.4 equivalent performance, and yet, in terms of SNR_s they are equivalent.

As we indicated earlier in this section, there are other sources of criticism of the SNR put forth by Taguchi for analysis in a problem involving robust parameter design. However, because our development will eventually lead to the use of separate response surfaces for the mean and variance, we feel that we should let the focus rest with how SNRs can lead the scientist or engineer to be baffled about the separate structure of process mean and variance.

10.4 THE TAGUCHI APPROACH TO ANALYSIS

The Taguchi notion of quality improvement places emphasis on variance reduction. His approach to variance reduction is widely acknowledged as a very important contribution to statistics and engineering. While others had suggested the use of noise variables before Taguchi, his ideas of linking their use to variance reduction is original and has had a profound impact. Details that report on his analysis ideas can be found in Taguchi (1987).

To understand the rationale behind the Taguchi analysis, one must be reminded that the purpose of parameter design is to find the settings on the control variables that produce a robust product or process. The control variables may be quantitative or qualitative. As a result, the Taguchi analysis does reduce to process optimization, where the performance criterion is the SNR.

The analysis method is simple and is very much linked to the crossed array concept. There is an attempt to **maximize SNR**. The "modeling" that is done essentially relates SNR to the control variables in a "main effects only" scenario. Taguchi rarely considers interactions among the control variables. In fact, many of the designs put forth by Taguchi do not allow estimation of these interactions. Standard ANOVA techniques are used to identify the control factors that impact SNR. In addition, ANOVA is done on $\bar{y}$ in order to ascertain possible adjustment factors. The latter are set to levels that bring the mean to target (in the target is best case), while the factors that effect SNR are set to levels that maximize SNR. This completes the two-step process where it is appropriate. The setting of levels is accomplished by a

pick the winner analysis in which a marginal means approach is accomplished. To gain further insight, consider the following example.

Example 10.1   In an experiment described by Schmidt and Launsby (1990), solder process optimization is accomplished by robust parameter design in a printed circuit board assembly plant. After the parts are inserted into a bare board, the board is put through a wave solder machine which connects all the parts into the circuit. Boards are placed on a conveyor and then put through the following steps: bathing in a flux mixture to remove oxide, preheating to minimize warpage, and soldering. An experiment is designed to determine the conditions which give minimum numbers of solder defects per million joints. The control factors and levels are as follows:

Factor	(-1)	$(+1)$
A, solder pot temperature ($°$ F)	480	510
B, conveyor speed (ft/min)	7.2	10
C, flux density	$0.9°$	$1.0°$
D, preheat temperature ($°$ F)	150	200
E, wave height (in.)	0.5	0.6

In this illustration, three of the noise factors are difficult to control in the process. These are the solder pot temperature, the conveyor speed, and the assembly type. Often in these type of processes, the natural noise factors are **tolerances in the control variables.** In other words, as these factors vary off their nominal values, variability is transmitted to the response. It is known that the control of temperature is within $\pm 5°$ F and that the control of conveyor speed is within ± 0.2 ft/min. Thus, it is conceivable that variability is increased substantially because of the inability to control these two factors at nominal levels. The third noise factor is the assembly type. One of the two types of assembly will be used. Thus, the noise factors are as follows:

Factor	$(+1)$	(-1)
A^*, solder	$5°$	$-5°$
B^*, conveyor	$+0.2$	-0.2
C^*, assembly type	1	2

Both the control array (inner array) and the noise array (outer array) were chosen to be fractional factorials. The inner array is a 2^{5-2} design and the outer array is a 2^{3-1}. The *crossed array* and the response values are as follows:

| Inner Array | | | | | | Outer Array | | | |
A	B	C	D	E	(1)	$a*b*$	$a*c*$	$b*c*$	$(SNR)_s$
1	1	1	-1	-1	194	197	193	275	-46.75
1	1	-1	1	1	136	136	132	136	-42.61
1	-1	1	-1	1	185	261	264	264	-47.81
1	-1	-1	1	-1	47	125	127	42	-39.51
-1	1	1	1	-1	295	216	204	293	-48.15
-1	1	-1	-1	1	234	159	231	157	-45.97
-1	-1	1	1	1	328	326	247	322	-45.76
-1	1	-1	-1	-1	186	187	105	104	-43.59

Analysis

The analysis in this example involves a marginal means modeling of the SNR of Equation (10.1), namely SNR_s, because it is of interest to minimize the number of solder defects. One analytical device used frequently by Taguchi is to depict a **main effects only** analysis through the graphs in Figure 10.5. The mean SNR_s is plotted against levels of each control factor. The means are taken across levels of the other factors. In addition to the mean SN_2 plot, a plot of $\bar{y}$ against the levels of the control variables is also made and appears in Figure 10.6.

It is clear from Figures 10.5 and 10.6 that temperature and flux density are critical control factors that have a profound effect on SNR_s and $\bar{y}$. It also appears that wave height has some influence on SNR_s but little or no effect on $\bar{y}$. Because it is of interest to maximize SNR_s and minimize $\bar{y}$, the following represent suggested operating conditions:

$$\boxed{\begin{array}{l} \text{Solder temperature} = 510° \text{ F} \\ \text{Flux density} = 0.9° \\ \text{Wave height} = 0.5 \text{ in.} \end{array}} \qquad (10.5)$$

The conveyor speed and preheat temperature can be placed at the most economical settings, presumably at low levels. The effect of wave height is marginal compared to the impact of solder temperature and flux density. The analysis suggests that the conditions given by Equations (10.5) are the most **robust conditions**, that is, in the context of this example, those conditions that are most insensitive to changes in the noise variables.

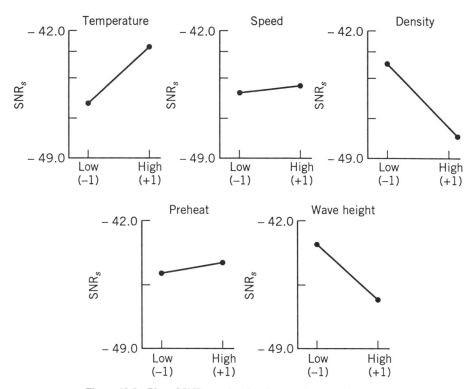

Figure 10.5   Plot of SNR$_s$ against level for each control factor.

Further Comments on the Analysis

In addition to earlier concerns regarding the appropriateness of the SNR, other comments concerning the analysis are in order. The plots in Figures 10.5 and 10.6 are appropriate for selecting optimum (most robust) conditions *if one is willing to assume no interaction among the control variables*. Obviously in many areas of application this is a safe assumption, and certainly Taguchi's success bears this out. However, there are many subject matter areas in which interactions make a huge contribution in any modeling endeavor. This renders the plots of main effect means highly misleading. For example, in Figure 10.5, consider the temperature plot. High positive slope reveals that across levels of the other factors an increase in temperature increases SNRs. But suppose when the density is low (which is an important level to maintain) the effect of temperature is negligible or even negative. Then, there will be little need to use high temperature when one operates with low density.

Many of the designs employed by Taguchi for inner arrays, where control factors reside, do not accommodate interactions. In addition to the apparent difficulty with SNR, this is a source of criticism of methodology suggested by Taguchi. Many of the designs suggested by Taguchi are saturated or near-

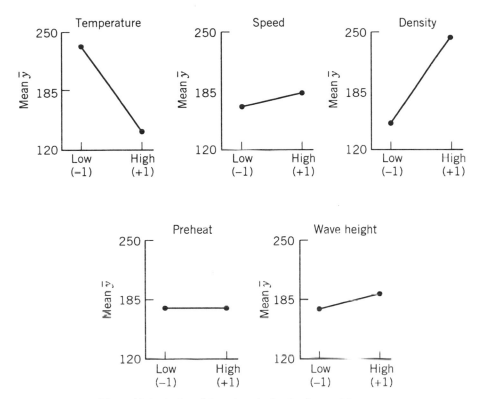

Figure 10.6 A plot of $\bar{y}$ against the levels of control factors.

saturated Plackett–Burman designs and thus do not allow for estimation of interaction among the control variables.

A major component of the methodology of Taguchi's robust design is to use confirmatory trials at the recommended operating conditions. Indeed, this recommendation is quite appropriate for any statistical procedure in which optimum operating conditions are estimated.

Example 10.2 An experiment was conducted in order that a method could be found to assemble an elastometric connector to a nylon tube that would deliver the required pull-off information for use in an automotive engine application. The objective is to maximize pull-off force [see Byrne and Taguchi (1987)]. Four variables were used as control variables. They are interference (A), connector wall thickness (B), insertion depth (C), and percent adhesive in connector pre-dip (D). Three factors were chosen as noise variables because they are difficult to control in the process and make a major contribution to variability in the pull-off force. They are conditioning time (E), conditioning temperature (F), and conditioning relative humidity

(G). The levels are as follows:

Controllable Factors	Levels		
A = Interference	Low	Medium	High
B = Connector wall thickness	Thin	Medium	Thick
C = Insertion depth	Shallow	Medium	Deep
D = Percent adhesive in connector pre-dip	Low	Medium	High

Uncontrollable Factors	Levels	
E = Conditioning time	24 hr	120 hr
F = Conditioning temperature	72° F	150° F
G = Conditioning relative humidity	25%	75%

Note that the control factors are at three levels, while the noise factors are at two levels. The experimental design and measured pull-off force is given in Table 10.1. Notice that the inner array is a $\frac{1}{9}$ of a 3^4 factorial. (Taguchi refers to this as an L_9.) The outer array in the noise variables is a 2^3. The levels for the control variables are -1, 0, and $+1$—representing low, medium, and high, respectively. The usual -1, $+1$ notation is used for low and high as the noise variables. Note also that the design is a crossed array, giving a total of $9 \times 8 = 72$ observations. The SNR$_l$ and $\bar{y}$ values are given. If formal modeling were to be done, then the fraction of the 3^4 could accommodate linear and quadratic terms in each control variable. *However, there are no degrees of freedom left for interaction among the control variables.* The analysis pursued is, once again, the marginal means plots for SNR$_l$ and $\bar{y}$. Figures 10.7 and 10.8 give the marginal means plots for $\bar{y}$ and SNR$_l$, respectively. As

Table 10.1 Inner and Outer Array for Example 10.2

				Outer Array (L_5)										
				E	-1	-1	-1	-1	$+1$	$+1$	$+1$	$+1$		
				F	-1	-1	$+1$	$+1$	-1	-1	$+1$	$+1$		
				G	-1	$+1$	-1	$+1$	-1	$+1$	-1	$+1$		
Inner Array (L_9)													Responses	
Run	A	B	C	D									$\bar{y}$	SNR$_l$
1	-1	-1	-1	-1	15.6	9.5	16.9	19.9	19.6	19.6	20.0	19.1	17.525	24.025
2	-1	0	0	0	15.0	16.2	19.4	19.2	19.7	19.8	24.2	21.9	19.475	25.522
3	-1	$+1$	$+1$	$+1$	16.3	16.7	19.1	15.6	22.6	18.2	23.3	20.4	19.025	25.335
4	0	-1	0	$+1$	18.3	17.4	18.9	18.6	21.0	18.9	23.2	24.7	20.125	25.904
5	0	0	$+1$	-1	19.7	18.6	19.4	25.1	25.6	21.4	27.5	25.3	22.825	26.908
6	0	$+1$	-1	0	16.2	16.3	20.0	19.8	14.7	19.6	22.5	24.7	19.225	25.326
7	$+1$	-1	$+1$	0	16.4	19.1	18.4	23.6	16.8	18.6	24.3	21.6	19.850	25.711
8	$+1$	0	-1	$+1$	14.2	15.6	15.1	16.8	17.8	19.6	23.2	24.2	18.338	24.852
9	$+1$	$+1$	0	-1	16.1	19.9	19.3	17.3	23.1	22.7	22.6	28.6	21.200	26.152

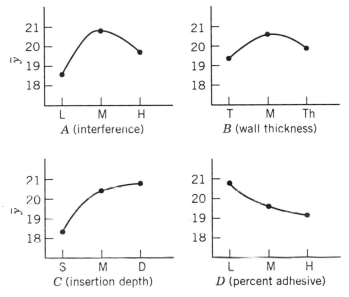

Figure 10.7 Plot of $\bar{y}$ against levels of control variables.

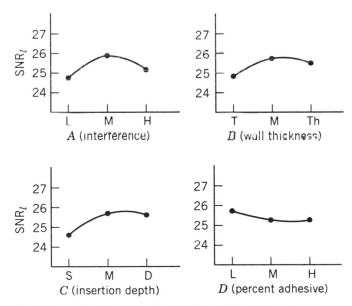

Figure 10.8 Plot of mean SNR_l against level of control variables.

in Example 10.1, the graphs are to be examined and the analysis involves a "pick the winner" approach. The following represent the "optimal" choices of A, B, C, and D according to the plots.

> A:Medium
> B:Medium
> C:Medium or deep
> D:Low

In each case, we have the levels that maximize the mean response and also maximize SNR_1. Again, these choices are made under the assumption that no interaction exists among the control factors.

Actually, when cost and other external factors were taken into account, the researchers decided to use

> A: Medium
> B: Thin
> C: Medium
> D: Low

The "thin" level for B was much less expensive. Neither of the two conditions for A, B, C, and D reported above were in the design. As a result, confirmatory runs were necessary. The authors report that good results were observed. The levels used in the noise variables were E_{Low}, F_{Low}, and G_{Low}.

10.5 FURTHER COMMENTS ON DESIGN AND ANALYSIS IN PARAMETER DESIGN

To this point our emphasis has been placed on experimental design and analysis procedures suggested by Genichi Taguchi. It should be emphasized that despite criticisms put forth by statisticians in the Western World concerning his approach, Taguchi has been quite successful and many practitioners in the United States have made good use of his methods. While there are weaknesses in the methodology illustrated in the two previous examples, there are certainly many real-life situations in which it works well. For an excellent account on Taguchi's contributions the reader is referred to Pignatiello and Ramberg (1991). In the subsections that follow, we will discuss areas where there are weaknesses and will point out situations in which these weaknesses may lead to difficulties. This is designed to motivate an emphasis on response surface methods as an alternative for solving robust parameter design problems.

10.5.1 Experimental Design

The crossed array, or *product array*, begins with two experimental designs, one for the noise variables and one for the control variables. These designs are then *crossed* with each other. The two individual designs are generally quite economical because they are either saturated or near-saturated. However, the crossing of two designs often *does not* produce an economical design. As we will suggest in a subsequent section, there are more economical approaches to designing an experiment for control and noise variables. The difficulty with the crossed array is easily explained through an inspection of the degrees of freedom.

Degrees of Freedom for the Crossed Array
Consider the crossed array produced when A, B, and C are involved in a 2^{3-1} inner array which is crossed with another 2^{3-1} as the outer array involving D, E, and F. The result is 16 observations in which the degree of freedom breakdown is as follows:

Effect	Degrees of Freedom	Effect	Degrees of Freedom
A	1	D	1
B	1	E	1
C	1	F	1
AD	1	CD	1
AE	1	CE	1
AF	1	CF	1
BD	1		
BE	1		
BF	1		

Notice how all the degrees of freedom are for main effects and *control × noise* interactions. No degrees of freedom account for noise × noise interactions or control × control interactions. The lack of estimation of the latter effects represents one of the main difficulties with many crossed arrays. For the sake of economy, each array is taken to be heavily fractionated. However, the total crossed array is not economical and control × control interactions are sacrificed for the sake of an overabundance of control × noise interactions. This is not to imply that control × noise interactions are unimportant. **They are crucial.** However, in many subject matter areas the control × control interactions also represent important effects; and, in the marginal means analysis, the ignored presence of control × control interactions may lead to faulty conclusions. In Example 10.2 we have:

Effect	Degrees of Freedom	Effect	Degrees of Freedom
A	2	E × F	1
B	2	E × G	1
C	2	F × G	1
D	2	E × F × G	1
E	1		
F	1		
G	1		

All other available degrees of freedom (56 of them) are allocated to control × noise interactions.

In sections that follow we suggest economical experimental designs for handling robust design problems in which control × control interactions are available. In addition, these designs are compatible with a response surface approach to the solution of robust parameter design problems.

10.5.2 The Taguchi Analysis

Notwithstanding difficulties discussed earlier with the SNRs, the approach taken by Taguchi and described here can work very well for determining a robust product if control × control interactions are not important. However, the marginal means approach, while placing emphasis on process optimization, does not allow the practitioner to learn more about the process. Alternatives put forth in the panel discussion by Nair et al. (1992) put more emphasis on gaining knowledge about the process. This is done through more rigorous analysis accomplished with more economical experiments. Focus is placed on models that involve control × control interactions and the determining of separate models for the process mean and variance.

10.6 RESPONSE SURFACE ALTERNATIVES FOR PARAMETER DESIGN PROBLEMS

There are various response surface alternatives for solving the parameter design problem. In this section we begin by emphasizing the importance of the control × noise interactions, discuss alternative modeling and analysis strategies, and discuss alternatives to the crossed array experimental designs. With regard to modeling strategies, it should be emphasized that, in general, process optimization may be quite overrated. The more emphasis that is placed on learning about the process, the less important *absolute optimization* becomes. An engineer or scientist can always find (and often do) pragmatic reasons why the suggested optimum conditions cannot be adopted. A thorough study (via a response surface approach) of the control design space will allow nice alternative suggestions for operating conditions.

10.6.1 The Role of the Control × Noise Interaction

We have emphasized the need to model control × control interactions. However, the **control × noise interactions are vitally important**. Indeed, the structure of these interactions determine the nature of nonhomogeneity of process variance that characterizes the parameter design problem. As an illustration consider the case of two control variables and one noise variable with data in a crossed array as follows:

Inner Array		Outer Array		Means
A	B	C		
		−1	+1	
−1	−1	11	15	13.0
−1	1	7	8	7.5
1	−1	10	26	18.0
1	1	10	14	12.0

The response values are given in the table. Consider the plots displaying the control × noise interactions *AC* and *BC* in Figures 10.9 and 10.10. In viewing these interaction plots, one should keep in mind that while *C* is held fixed in the experimental design, it will vary according to some probability distribution during the process. The probability distribution may be discrete

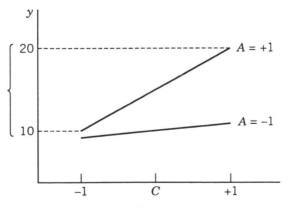

Figure 10.9 *AC* interaction plot.

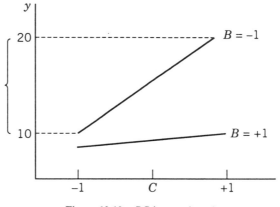

Figure 10.10   *BC* interaction plot.

or continuous. Here it is clear that when we take this variability into account, the picture suggests that $A = +1$ and $B = -1$ will result in **largest product variability**, whereas $A = -1$ and $B = +1$ should result in **smallest product variability**. The brackets on the vertical axes reflect product variability at $A = +1$ and $B = -1$. Now, of course, in this simple case, the conclusion is borne out in the data themselves but it may not be so easy to see in larger problems. In this illustration, we say that A and B are both **dispersion effects**, that is, effects that have influence on the process variance. In addition, both effects are location effects, namely they have an effect on the process mean. In fact, the illustration points out the need to compromise between process mean and process variance. From the plots $A = +1$ and $B = -1$ produces the largest mean response, and if this is a "larger the better" case the optimum conditions will involve simultaneous consideration of the process mean and variance.

It should be apparent to the reader that control × noise interactions display the structure of the process variance and relate which factors are dispersion effects and which are not. Consider a second example with data given by

Inner Array		Outer Array	
A	*B*	*C*	
		-1	$+1$
-1	-1	14.5	22.5
-1	1	19	27.5
1	-1	23.5	32
1	1	28.5	37

Here, the $A \times C$ and $B \times C$ interactions plots are given in Figures 10.11

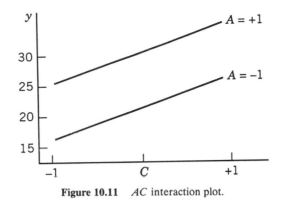

Figure 10.11   *AC* interaction plot.

and 10.12. Here we have parallelism in the two interaction plots. No *AC* factor interaction is exhibited, and thus *A* is *not* a dispersion effect. That is, the process variance is the same at both levels of *A*. No *BC* interaction is exhibited, and thus *B* is *not* a dispersion effect. The factor *A* has a greater location effect than *B*. If this is a "larger the better" case, then the combination *A* = +1 and *B* = +1 is optimal and there is **no robust parameter design problem** because altering *A* and *B* has no impact on variance produced by changing the noise variable *C*. Consider a third example in which the *AC* and *BC* interaction plots appear in Figures 10.13 and 10.14. Figure 10.13 depicts *A* as a dispersion effect with *A* = +1 representing the level of minimum variance. Figure 10.14 suggests that *B* is *not* a dispersion effect. Both *A* and *B* are location effects; and for a "larger the better" scenario, *A* = +1 and *B* = +1 represents optimum conditions. The control × noise interactions can serve as nice diagnostic tools regarding the structure of the process variance. Note that the larger the slopes of the interaction plots, the larger the contributions to the process variance. Here we have discussed the diagnostic nature of the two-factor interactions. However, if

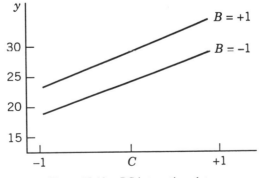

Figure 10.12   *BC* interaction plot.

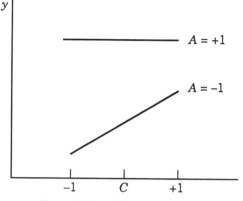

Figure 10.13   *AC* interaction plot.

there are three-factor interactions involving control and noise factors, these too play a role.

Consider the situation of Example 10.2. Here, there are four control factors and from Figure 10.8 factor *A* appears to have an effect on SNR$_l$ and, indeed we learned that *A* medium is the appropriate level. In addition, *A* medium maximizes $\bar{y}$. The two-factor *AG* interaction plot is given in Figure 10.15. This plot implies that *A* medium *minimizes variance produced by changes in relative humidity*. It also reinforces that *A* medium is desirable from the point of view of the mean response.

As a result, it should be clear that the control × noise interaction plot aids in the determination of which factors are dispersion effects, which are location effects, and which levels of the dispersion effects moderate the variance produced by which noise variables. For example, consider a situation in

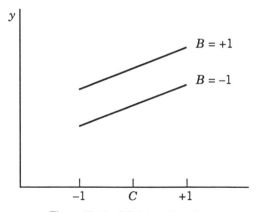

Figure 10.14   *BC* interaction plot.

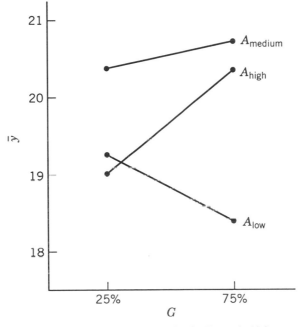

Figure 10.15 AG interaction plot for Example 10.2.

which we have three control factors (A, B, and C) and two noise factors (D and E). Suppose also that interactions AD and BE are found to be significant. In addition, A is found to be insignificant while B and C are significant.

Consider the main effect and interaction plots in Figure 10.16. The plots indicate that C has a strong location effect and B has a mild location effect. Factor B has a huge dispersion effect with minimum process resulting from the use of the -1 level of B. Using the -1 level of **B minimizes that portion of the process variance that results from changes in the noise variable E.** Despite the fact that A interacts with the noise variable D, the use of -1 or $+1$ on A results in the same process variance. However, if A is a continuous variable there may be other levels of A that will produce a horizontal line on the AD interaction plot and thus dramatically reduce the portion of the process variance associated with changes in D. (Note dashed line on AD interaction plot.) Because the factor C does not interact with either noise factor, C *cannot be used to control process variance*. If one were interested in conditions that *minimize mean response*, say, but also result in minimizing process variance, one would choose $B = -1$, $C = +1$, and $A = -1$ or $+1$. However, if A is a continuous variable, the value of A that results in minimum process variance should be sought. This latter idea will be developed in future sections.

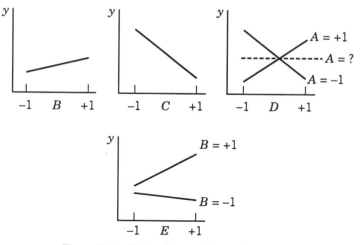

Figure 10.16 Main effect and interaction plots.

10.6.2 Use of the Model Containing Both Control and Noise Variables

From what we have just completed in the previous subsection, it becomes apparent that one may wish to make use of a model that contains main effects and interactions involving both control and noise variables. If the control variables are predominantly continuous variables, this type of model will become a very useful vehicle for the eventual formation of a *dual response* surface approach—that is, a response surface for the process mean and a response surface for the process variance. In what has preceded this section, we have assumed that control and noise factors are fixed effects. Suppose we consider **x** and **z**, the control and noise factors respectively. The modeling of both **x** and **z** in the same model has been called a **response model approach** by Shoemaker et al. (1991). The following subsection illustrates an example of such an approach.

Example 10.3 Etch Rate Example Montgomery (1991b) outlines an experiment in which a nitride etch process on a single wafer plasma etcher is being studied. The response of interest is etch rate and the experimental factors are

A, or x_1: The spacing between the anode and cathode (gap)
B, or x_2: Pressure in the reactor chamber
C, or z_1: Flow rate of the reactant gas
D, or z_2: Power applied to the cathode

Factors A and B are control factors and factors C and D are noise variables because they are quite difficult to control in the process. The factor levels are given by

	x_1	x_2	z_1	z_2
Low (-1)	0.80	450	125	275
High $(+1)$	1.20	550	200	325

Note that all factors are continuous variables.
The design and observed results are as follows:

$A = x_1$	$B = x_2$	$C = z_1$	$D = z_2$	y (etch rate; Å/min)
-1	-1	-1	-1	550
1	-1	-1	-1	669
-1	1	-1	-1	604
1	1	1	1	650
-1	-1	1	-1	633
1	-1	1	-1	642
-1	1	1	-1	601
1	1	1	-1	635
-1	-1	-1	1	1037
1	-1	-1	1	749
-1	1	-1	1	1052
1	1	-1	1	868
-1	-1	1	1	1075
1	-1	1	1	860
-1	1	1	1	1063
1	1	1	1	729

Though the design is a 2^4 factorial array, one can view it as a crossed array in the form $2^2 \times 2^2$, that is, a 2^2 inner array crossed with a 2^2 outer array. But if both the x_i and z_i appear in the same model and the basic response is used (rather than SNR), there is no need to display the design in product array form. An analysis of variance is produced with three- and four-factor interactions pooled into error.

Source	Degrees of Freedom	Mean Square	F
x_1	1	41,310,563	20.28
x_2	1	10.563	
z_1	1	217.563	
z_2	1	374,850.063	183.99
$x_1 x_2$	1	248.063	
$x_1 z_1$	1	2745.063	
$x_1 z_2$	1	99,402.563	48.79
$x_2 z_1$	1	7700.063	3.78
$x_2 z_2$	1	1.563	
$z_1 z_2$	1	18.063	
Error	5	2037.363	

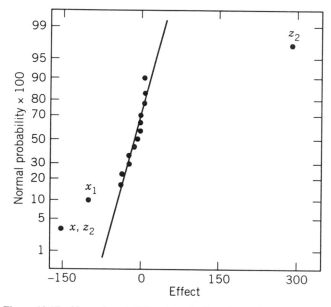

Figure 10.17   Normal probability plot of effects for etch rate example.

A normal probability plot of effects was also used in the analysis, and the result is given in Figure 10.17. Only the effects x_1, z_2, and $x_1 z_2$ are found statistically significant. The interaction $x_2 z_1$ is significant at the $P \cong 0.12$ level. Using only the significant terms the fitted model is given by

$$\hat{y} = 776.0625 - 50.8125 x_1 + 153.0625 z_2 - 76.8125 x_1 z_2 \qquad (10.6)$$

Now, at this point it is of interest to determine from the fitted model any obvious desirable conditions on the design variables. The $x_1 z_2$ interaction plot produces the result in Figure 10.18. Here it is clear that, of the two levels of x_1, the high level, $+1$, is best because it would result in a considerably smaller process variance. However, from Equation (10.6) the main effect term in x_1 would suggest that for maximum mean etch rate, one needs to use the low level, -1, of x_1. As a result there is a tradeoff that should be considered. The tradeoff illustrated in the simple example is derived from an attempt to get high process mean and "robustness" that features low process variance by minimizing variance produced by the random movement of z_2, the power applied to the cathode.

Dual Response Surface Approach for Etch Rate Example
Suppose we continue to focus on the etch rate example. The use of a dual response surface approach in the determination of the most robust product

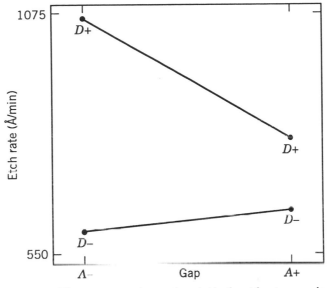

Figure 10.18 The $AD = x_1 z_2$ interaction plot in the etch rate example.

seems to be quite appropriate. Here we define a robust product as one with a high etch rate (use of a process mean response surface) and a relatively low process variance (use of a process variance response surface). Indeed, the examination of these two response surfaces will also often provide insight into what is driving the process mean and variability.

Let us assume that in the process z_2 is a random variable with mean 0 and some variance σ_z^2. Also assume the levels of z_2 are at $\pm \sigma_z$ in coded form. Thus, $\sigma_z = 1$. Now, at this point let us assume that the model is given by

$$y = \beta_0 + \beta_1 x_1 + \gamma_2 z_2 + \delta_{12} x_1 z_2 + \varepsilon \tag{10.7}$$

Allowing that $E(\varepsilon) = 0$ and $E(z_2) = 0$, we have as the process mean model

$$E_{z_2}(y) = \beta_0 + \beta_1 x_1$$

As a result, an estimate of the process mean model is given by

$$\hat{E}_{z_2}(y) = 776.0625 - 50.8125 x_1 \tag{10.8}$$

Now, let us take a variance operator across z_2, that is, consider the operator $\mathrm{Var}_{z_2}(y)$ in the model of Equation (10.7):

$$\mathrm{Var}_{z_2}(y) = \gamma_2^2 \sigma_z^2 + \delta_{12}^2 x_1^2 \sigma_z^2 + 2 \delta_{12} \gamma_2 x_1 \sigma_z^2 + \sigma^2$$

We now replace parameters by estimates from Equation (10.6). As a result,

$$\hat{\text{Var}}_{z_2}(y) = (153.0625)^2 \sigma_z^2 + (76.8125)^2 x_1^2 \sigma_z^2$$

$$- 2(153.00625)(76.1825) x_1 \sigma_z^2 + \hat{\sigma}^2$$

$$= [153.0625 - 76.8125 x_1]^2 \sigma_z^2 + \hat{\sigma}^2 \qquad (10.9)$$

$$= [153.0625 - 76.8125 x_i]^2 + 2037.363 \qquad (10.10)$$

Here we have replaced $\hat{\sigma}^2$ by the error mean square from the analysis and $\sigma_z^2 = 1.0$. Equation (10.10) is a **response surface for the process variance**, written as a quadratic function of the control variable x_1. The tradeoff is readily apparent here. From Equation (10.8), it is clear that the process mean etch rate is maximized for small x_1, say $x_1 = -1$. However, Equation (10.9) suggests that the process variance is minimized for

$$153.0625 - 76.8125 x_1 = 0$$

or

$$x_1 \cong 2$$

But if we stay inside the region of the experiment, we set $x_1 = 1.0$ as the optimum value—that is, a value of 1.20 g for the spacing between the anode and the cathode. As a result, there is no setting for x_1 that maximizes the process mean and, simultaneously, minimizes the process variance. However, a great deal can be learned about the process, and one can have the kind of information that will allow an intelligent choice of x_1 by observing two plots. Consider Figures 10.19 and 10.20. The former shows a plot of the process mean against x_1, and the latter shows a plot of the process *standard deviation* against x_1. Keep in mind that these refer to the mean and standard deviation in etch rate produced by random fluctuation in power applied to the cathode.

As we indicated earlier, the dual response surface analysis produced in this rather simple example is designed to aid the user to achieve a better understanding of the process; that is, it answers the question, How does movement in the spacing between the anode and the cathode impact the mean etch rate and the standard deviation in etch rate? In the "target is best" case, one can use process mean and standard deviation information to create a single number criterion:

$$\hat{E}_z(y - t)^2 = [\hat{E}_z(y) - t]^2 + \hat{\text{Var}}_z y \qquad (10.11)$$

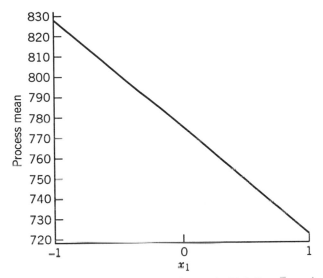

Figure 10.19 Plot of process mean against x_1 for Etch Rate Example.

where t is the desired target. In the present case, however, we are in a "larger the better" scenario and thus one might be interested in quantiles such as

$$\hat{E}_z(y) - 2\hat{\sigma}_z(y) \tag{10.12}$$

In this illustration the value $x_1 = 1$ maximizes the above quantity, which is an

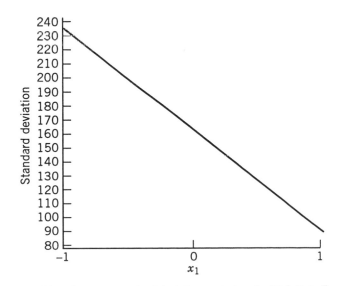

Figure 10.20 Plot of process standard deviation against x_1 for Etch Rate Example.

estimate of a specific lower quantile of the distribution of etch rate. The value $x_1 = 1$ is considered optimal because the movement away from this level results in a faster increase in $\hat{\sigma}_z(y)$ than the decrease experienced in $\hat{E}_{z_2}(y)$. At $x = +1$, the estimated mean etch rate is approximately 726 Å/min and the estimated standard deviation is approximately 88 Å/min.

10.6.3 A Generalization of the Mean and Variance Modeling

As we indicated in the previous subsection, the advantage derived from a dual response surface approach lies in the flexibility it offers. Specifically:

1. It provides an estimate of mean and standard deviation at any location of interest in the control design variables.
2. The engineer or scientist can gain insight regarding the roles of these variables in controlling process mean and variance.
3. It provides a ready source for process optimization via the use of a squared error loss criterion [in Equation (10.11)], or the maximization of an estimated quantile [Equation (10.12)] in the "larger the better" case, or the minimization of $\hat{E}_z(y) + 2\sigma_z(y)$ in the "smaller the better case."
4. Allows the use of constrained optimization.

Lucas (1989), Vining and Myers (1990), Box and Jones (1989), Del Castillo and Montgomery (1993) and others have pointed out the virtues of modeling both process mean and variance. In this subsection we show a generalization of mean and variance modeling. In many examples, the model in control and noise variables may very well involve quadratic terms in the x's, the control variables. Indeed, response surface designs [the central composite design (CCD) or the Box–Behnken design (BBD)] can be of considerable value. As a result, we might consider the **fitted response surface** in x (control variables) and z (noise variables) to be written as

$$\hat{y}(\mathbf{x}, \mathbf{z}) = b_0 + \mathbf{x}'\mathbf{b} + \mathbf{x}'\hat{\mathbf{B}}\mathbf{x} + \mathbf{z}'\mathbf{c} + \mathbf{x}'\hat{\mathbf{\Delta}}\mathbf{z} \qquad (10.13)$$

We are assuming linear main effects in x and z, interactions and quadratic terms in x, and the very important two-factor interactions involving control and noise variables. The fitted model in Equation (10.13) represents a least squares fit of the postulated model

$$y(\mathbf{x}, \mathbf{z}) = \beta_0 + \mathbf{x}'\boldsymbol{\beta} + \mathbf{x}'\mathbf{B}\mathbf{x} + \mathbf{z}'\boldsymbol{\gamma} + \mathbf{x}'\boldsymbol{\Delta}\mathbf{z} + \boldsymbol{\varepsilon} \qquad (10.14)$$

We do not mean to rule out the use of interactions among noise variables or even quadratic terms in noise variables. However, the model in Equation

(10.14) will accommodate many real-life situations and as we shall demonstrate later. The development of mean and variance modeling with the use of (10.13) and (10.14) will provide the machinery for the use of more general models. Let z in Equation (10.14) be $r_z \times 1$ and let x be $r_x \times 1$. The matrix Δ is an $r_x \times r_z$ matrix containing coefficients of xz-type interactions.

Assumptions Underlying Dual Response Development
The ε term in Equation (10.14) plays the same role as the random response error term in any response surface methodology (RSM) model discussed in Chapters 3–9. Assumptions of the type ε_i-i.i.d. $N(0, \sigma)$ are often certainly reasonable. It should be pointed out that the *constant variance* assumption on ε implies that any nonconstancy of variance of the process stems primarily from *inability to control noise variables*. Detection of other sources of heterogeneous variance will be addressed in Section 10.7.

The assumptions associated with the noise variables deserve special attention. As we indicated in Example 10.3, we must assume that the user possesses a reasonable amount of knowledge about the random variable z. It may be assumed that in the experiment the levels of say z_j are centered at the mean that z_j experiences in the process and that the $+1$ levels are at $\mu_{z_j} + \sigma_{z_j}$. As a result, the mean of the z's in coded metric is at 0, and ± 1 levels are one standard deviation from the mean. The latter is not a hard and fast assumption. The experimenter may decide to let ± 1 correspond to $\pm 2\sigma_z$. As a result, we will demonstrate a development of the process variance response surface under the assumption that σ_z is known, and the latter may take on values of 1, $\frac{1}{2}$, $\frac{1}{3}$, and so on, in practice. As a result, let us assume

$$E(z) = 0, \qquad \text{Var}(z) = \sigma_z^2 I_{r_z} \tag{10.15}$$

Equation (10.15) implies the assumption that noise variables are uncorrelated with known variance.

The Mean and Variance Response Surfaces
At this point we shall take expectation and variance operators on the model in Equation (10.14). Because $E(z) = 0$, the expectation operation E_z gives the response surface model

$$E_z[(x, z)] = \beta_0 + x'\beta + x'Bx \tag{10.16}$$

The operator, $\text{Var}_z[y(x, z)]$ plus the error variance around (10.14) gives

$$\text{Var}_z[y(x, z)] = \text{Var}_z[\gamma' + x'\Delta]z + \sigma^2$$

Here the quantity $[\gamma' + x'\Delta] = a'$ is a vector of constants. As a result, we have to evaluate $\text{Var}_z[\mathbf{a'z}]$. Rules for variance operations give

$$\text{Var}_z[\mathbf{a'z}] = \mathbf{a'}\left[\text{Var}(\mathbf{z})\right]\mathbf{a}$$

where $\text{Var}(\mathbf{z})$ is the variance–covariance matrix of $\mathbf{z}$. As a result, we have

$$\text{Var}_z[\boldsymbol{\gamma'} + \mathbf{x'\Delta}]\mathbf{z} = \sigma_z^2[\boldsymbol{\gamma'} + \mathbf{x'\Delta}][\boldsymbol{\gamma'} + \mathbf{x'\Delta}] \text{ and}$$

$$\text{Var}_z[y(\mathbf{x}, \mathbf{z})] = \sigma_z^2(\boldsymbol{\gamma'} + \mathbf{x'\Delta})(\boldsymbol{\gamma'} + \mathbf{x'\Delta})' + \sigma^2$$

$$= \sigma_z^2 \mathbf{l'}(\mathbf{x})\mathbf{l}(\mathbf{x}) + \sigma^2 \tag{10.17}$$

where $\mathbf{l}(\mathbf{x}) = \boldsymbol{\gamma} + \mathbf{\Delta'x}$.

Equation (10.16) is the **response surface for the process mean**, while Equation (10.17) serves as the **response surface for the process variance**. Equation (10.17) deserves special attention. Apart from σ^2, the model error variance, the process variance is simply $\|\mathbf{l}(\mathbf{x})\|^2$, where $\mathbf{l}(\mathbf{x})$ is simply the vector of partial derivatives of $y(\mathbf{x}, \mathbf{z})$ with respect to $\mathbf{z}$. In other words,

$$\mathbf{l}(\mathbf{x}) = \frac{\partial y(\mathbf{x}, \mathbf{z})}{\partial \mathbf{z}} \tag{10.18}$$

The result in Equation (10.17) is very intuitive; the "larger" the vector of derivatives $\partial \mathbf{y}/\partial \mathbf{z}$, the larger the process variance. This relates very nicely to items in Section 10.6.1. The reader should recall that the "flatter" the x–z interaction plots, the smaller the process variance. Clearly, **flat interaction plots result from small values of $\partial y(\mathbf{x}, \mathbf{z}) / \partial \mathbf{z}$**. Also, note the important role of $\mathbf{\Delta}$, the matrix of x–z interaction coefficients, in the process variance response surface. If $\mathbf{\Delta} = 0$, the process variance does not depend on $\mathbf{x}$ and hence one cannot create a robust process by choice of settings of the control variables. In other words, the process variance is constant and thus *there is no robust parameter design problem*. In Equation (10.17) we add the model error variance. In many cases this variance will be negligible compared to the variance that is contributed by variation in the noise variables. In our development here there is an implicit assumption that the noise variables are quantitative in nature. In the case of discrete noise variables the assumption of uncorrelated noise variables is not necessarily appropriate.

Equations (10.16) and (10.17) represent the mean and variance response surfaces developed from the model (or response surface) that contains both control and noise variables. Clearly the estimated response surfaces are obtained by replacing parameters by estimates found in the fitted model in Equation (10.13). As a result, the estimated process mean and variance response surfaces are given by

$$\hat{E}_z[y(\mathbf{x}, \mathbf{z})] = \hat{\mu}_z[y(\mathbf{x}, \mathbf{z})] = b_0 + \mathbf{x}'\mathbf{b} + \mathbf{x}'\hat{\mathbf{B}}\mathbf{x} \qquad (10.19)$$

$$\widehat{\mathrm{Var}}_z[y(\mathbf{x}, \mathbf{z})] = \sigma_z^2 \hat{\mathbf{l}}(\mathbf{x})'\hat{\mathbf{l}}(\mathbf{x}) + \hat{\sigma}^2$$

$$= \sigma_z^2 (\mathbf{c}' + \mathbf{x}'\hat{\boldsymbol{\Delta}})(\mathbf{c}' + \mathbf{x}'\hat{\boldsymbol{\Delta}})' + \hat{\sigma}^2 \qquad (10.20)$$

Here the values $\hat{\sigma}^2$ is the mean squared error in the fitted response surface.

10.6.4 Analysis Procedures Associated with the Two Response Surfaces

We saw a simple special case of a dual response surface analysis in Section 10.6.3. Examples of response surfaces in Equations (10.19) and (10.20) were used in harmony to find suggested operating conditions. Single number criteria can be developed using Equations (10.19) and (10.20). Another approach (particularly useful in the "larger the better" or "smaller the better" case) is to use constrained optimization; that is, select a value of $\hat{\mu}_z[y(\mathbf{x}, \mathbf{z})]$ that one is willing to accept. This does not imply a target on the mean but, say, a value below which one cannot accept. In fact, several of these values may be chosen in order to add flexibility. Call this constraint $\mu_z[y(\mathbf{x}, \mathbf{z})] = m$. One then chooses $\mathbf{x}$ as follows:

$$\underset{\mathbf{x}}{\mathrm{Min}}\ \mathrm{Var}_z[y(\mathbf{x}, \mathbf{z})] \quad \text{subject to} \quad \hat{\mu}_z[y(\mathbf{x}, \mathbf{z})] = m$$

The choice of several values of m may be made in order, once again, to provide flexibility and provide several alternatives for the user.

There are many situations in which focus is placed entirely on the variance response surface. For example, there may be location effects that appear in the mean model but not in the variance model, and they can be used as tuning factors. As a result, the total analysis centers around the variance model. One may be interested in determining $\mathbf{x}$ that minimizes, or makes zero, the process variance in Equation (10.20). To that end, consider the computation of the locus of points $\mathbf{x}_0$ for which

$$\hat{\mathbf{l}}(\mathbf{x}_0) = \mathbf{0} \qquad (10.21)$$

or

$$\mathbf{c} + \hat{\boldsymbol{\Delta}}'\mathbf{x}_0 = \mathbf{0}$$

The condition of Equation (10.21) represents r_z equations in r_x unknowns. If $r_x > r_z$, this result, which produces a zero estimated process variance (apart from the model error variance, σ^2), will result in a line or a plane. If $r_x = r_y$ there will be a single point $\mathbf{x}_0$ which satisfies Equation (10.21). If $r_x < r_z$,

Equation (10.21) may not have a solution. Of course a great deal of interest centers on whether or not the locus of points x_0 passes through the experimental design region in the control variables.

Example 10.4   A scenario illustration involving variance modeling with two control and two noise variables appeared in Nair et al. (1992). The specific model fitted is given by

$$\hat{y} = b_0 + b_1 x_1 + b_2 x_2 + b_{12} x_1 x_2 + b_{11} x_1^2 + b_{22} x_2^2$$

$$+ c_1 z_1 + c_2 z_2 + \hat{\delta}_{11} z_1 x_1 + \hat{\delta}_{21} z_1 x_2 \qquad (10.22)$$

Let us assume that no additional model terms are significant. Now, from what has been discussed in the foregoing, there is some vital diagnostic information here. Both control variables can be used to exert control on variability produced by noise factor z_1. Although there are two noise variables, the variance produced by z_2 will be constant and thus not affected by choice of **x**. Essentially, the use of Equation (10.21) for control of process variance will reduce to only $\partial \hat{y} / \partial z_1$ because in this case

$$\hat{\Delta} = \begin{bmatrix} \hat{\delta}_{11} & 0 \\ \hat{\delta}_{21} & 0 \end{bmatrix}$$

Thus a choice of robust conditions implies a choice of x_1 and x_2 that forces a "flat" $\partial \hat{y} / \partial z_1$. The equation for the process variance $\hat{\mathbf{l}}'(\mathbf{x})\hat{\mathbf{l}}(\mathbf{x}) = \hat{\sigma}_z^2[y(\mathbf{x}, \mathbf{z})]$ becomes (apart from $\hat{\sigma}^2$)

$$\hat{\sigma}_z^2[y(\mathbf{x}, \mathbf{z})] = c_2^2 + (c_1 + \delta_{11} x_1 + \delta_{21} x_2)^2 = c_2^2 + (\partial \hat{y} / \partial z_1)^2$$

Now, suppose $c_1 = \frac{1}{2}$, $\hat{\delta}_{11} = \frac{1}{2}$, and $\hat{\delta}_{21} = -\frac{1}{4}$. Movement away from the design center with x_1 in a negative direction and x_2 in a positive direction will result in a robust product. Consider Figure 10.21. We include in the picture a hypothetical mean model that represents the result of an $\hat{E}_z$ operator on Equation (10.22). We also include the line of minimum $\hat{\sigma}_z^2[y(\mathbf{x}, \mathbf{z})]$. For, say, a target of $\mu_y = 50$, an obvious coordinate in **x** of interest is given by $x_1 = -1$ and $x_2 \cong 0$. If we have a "larger the better" case, tradeoffs and conditions for possible future experiments are evident. The inclusion of contours of constant variance on the same graph will also be useful. This will be illustrated in a future example.

Confidence Regions on Minimum Process Variance Conditions

In the previous illustration we displayed a line of "minimum process variance," where the process variance is the variability produced by the lack of sufficient control of noise variables in the process. At times, what is depicted

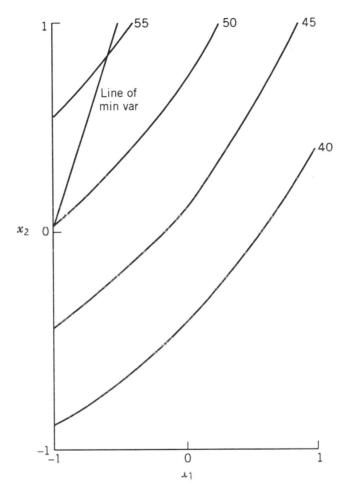

Figure 10.21   Dual response surface analysis for Example 10.4.

as a line in this case may be a plane or even a point in the space of the control variables. If minimizing process variance is important, then some characterization of the precision of the estimates of these conditions may also be crucial. For example, in the previous example, the "line of minimum variance" may represent a set of conditions that are unsatisfactory to the investigator. It would be helpful then to gain some sense of how far away from the line the investigator can move and still maintain "no significant difference" from the minimum variance conditions. The principle here is analogous to that discussed in Chapter 6 where confidence regions were discussed for stationary points in cases where a single response surface is being deployed. Consider, once again, Equation (10.21) with the solution $\mathbf{x}_0$ representing a locus of points or a single point of minimum process variance

Let $\mathbf{t}_0$ represent the "true" condition of minimum variance—that is,

$$\mathbf{l}(\mathbf{t}_0) = \boldsymbol{\gamma} + \boldsymbol{\Delta}'\mathbf{t}_0 = \mathbf{0}$$

One must keep in mind that $\mathbf{c}$ is an unbiased estimate of $\boldsymbol{\gamma}$ and $\hat{\boldsymbol{\Delta}}$ is an unbiased estimate of the matrix $\boldsymbol{\Delta}$. Now, assuming normal errors around the response model containing $\mathbf{x}$ and $\mathbf{z}$ [Equation (10.14)], we have

$$\frac{\left[\hat{\mathbf{l}}(\mathbf{t}_0) - \mathbf{0}\right]'\left[\text{Vâr}\,\hat{\mathbf{l}}(\mathbf{t}_0)\right]^{-1}\left[\hat{\mathbf{l}}(\mathbf{t}_0) - \mathbf{0}\right]}{r_z} \sim F_{r_z, df(E)}$$

where $\hat{\mathbf{l}}(\mathbf{t}_0) = \mathbf{c} + \hat{\boldsymbol{\Delta}}'\mathbf{t}_0$ and $\text{Vâr}\,\hat{\mathbf{l}}(\mathbf{t}_0)$ is the variance–covariance matrix of $\hat{\mathbf{l}}(\mathbf{t}_0)$ with σ^2 replaced by the error mean square σ^2. See Graybill (1976) or Myers and Milton (1991). Here $df(E)$ is the error degrees of freedom for the estimator s^2, the latter obtained from the fitted model of Equation (10.15). As a result the conditions $\mathbf{t}_0$ satisfying the inequality

$$\frac{\left[\hat{\mathbf{l}}(\mathbf{t}_0) - \mathbf{0}\right]'\left[\text{Vâr}\,\hat{\mathbf{l}}(\mathbf{t}_0)\right]^{-1}\left[\hat{\mathbf{l}}(\mathbf{t}_0) - \mathbf{0}\right]}{r_z} \leq F_{1-\alpha, r_z, df(E)} \qquad (10.23)$$

represent the desired $100(1 - \alpha)\%$ confidence region. One should note that r_z in Equation (10.23) denotes the number of noise variables that interact with control variables, and $F_{1-\alpha, r_z, df(E)}$ is the upper percent point of $F_{r_z, df(E)}$.

As an illustration of this confidence region, consider the previous example with

$$\hat{l}(\mathbf{t}_0) = c_1 + \hat{\delta}_{11}t_{1,0} + \hat{\delta}_{21}t_{2,0}$$

The left-hand side of Equation (10.23) reduces to

$$\frac{\left[c_1 + \hat{\delta}_{11}t_{1,0} + \hat{\delta}_{21}t_{2,0}\right]^2}{\left[\begin{array}{c}\text{Var}(c_1) + t_{1,0}^2\,\text{Var}\,\hat{\delta}_{11} + t_{2,0}^2\,\text{Var}\,\hat{\delta}_{21} + 2t_{1,0}\,\text{Cov}(c_1, \hat{\delta}_{11}) \\ + 2t_{1,0}\,\text{Cov}(c_1, \hat{\delta}_{21}) + 2t_{1,0}t_{2,0}\,\text{Cov}(\hat{\delta}_{11}, \hat{\delta}_{21})\end{array}\right]}$$

Now, of course for most reasonable experimental designs the covariance terms in the above equation will vanish. Values of $t_{1,0}$ and $t_{2,0}$ that satisfy the above equation fall inside the confidence region for the "line of location of minimum process variance" of the previous example.

Example 10.5 In the transmission of color TV signals the quality of the decoded signals is determined by the PSNRs (power signal-to-noise ratios in electronics engineering) in the image transmitted. The response here refers to the quality (coarse versus detailed) in the reception of transmitted signals in decibels. The control factors are x_1 (number of tabs in a filter) and x_2 (sampling frequency). These two factors form an inner array, while the noise or outer array involves z_1 (number of bits in an image) and z_2 (voltage applied). The total design is a $3^2 \times 2^2$ crossed array, and the data appear in Table 10.2.

The model to be used is that in equation (10.14). The 3^2 in the inner array allows for a second order model in the control variables. For this specific case, the model can be written out as

$$y(\mathbf{x}, \mathbf{z}) = \beta_0 + \beta_1 x_1 + \beta_2 x_2 + \beta_{11} x_1^2 + \beta_{22} x_2^2 + \beta_{12} x_1 x_2 + \gamma_1 z_1$$

$$+ \gamma_2 z_2 + \delta_{11} x_1 z_1 + \delta_{12} x_1 z_2$$

$$+ \delta_{21} x_2 z_1 + \delta_{22} x_2 z_2 + \varepsilon \tag{10.24}$$

The process variance is determined by taking the variance operator in

Table 10.2 Color TV Image Data for Example 10.5

			Codel Levels			
			-1	0	1	
x_1 (Number of tabs in a filter)			5	13	21 (tabs)	
x_2 (Sampling frequencies)			6.25	9.875	13.5 (MHz)	
z_1 (Number of bits of an image)			256		512 (bits)	
z_2 (Voltage applied)			100		200 (volts)	
Factor			$z_1 \ -1$	-1	1	1
Combination	x_1	x_2	$z_2 \ -1$	1	-1	1
(1)	-1	-1	33.5021	41.2268	25.2683	31.9930
(2)	-1	0	35.8234	38.0689	32.7928	34.0383
(3)	-1	1	33.0773	31.8435	36.2500	34.0162
(4)	0	-1	30.4481	41.2870	15.1493	23.9883
(5)	0	0	34.8679	40.2276	27.7724	31.1321
(6)	0	1	35.2202	37.1008	33.3280	35.2085
(7)	1	-1	21.1553	34.1086	0.7917	15.7450
(8)	1	0	27.6736	38.1477	15.5132	25.9873
(9)	1	1	32.1245	38.1193	26.1673	32.1622

Equation (10.24), namely,

$$
\begin{aligned}
\sigma_z^2[y(\mathbf{x}, \mathbf{z})] &= \big[(\gamma_1^2 + \gamma_2^2 + \delta_{11}^2 x_1^2 + \delta_{12}^2 x_1^2 + \delta_{21}^2 x_2^2 + \delta_{22}^2 x_2^2 \\
&\quad + 2\gamma_1(\delta_{11} x_1 + \delta_{21} x_2) + 2\gamma_2(\delta_{12} x_1 + \delta_{22} x_2) \\
&\quad + 2(\delta_{11}\delta_{21} + \delta_{12}\delta_{22}) x_1 x_2)\big]\sigma_z^2 + \sigma^2 \\
&= \big[(\gamma_1 + \delta_{11} x_1 + \delta_{21} x_2)^2 + (\gamma_2 + \delta_{12} x_1 + \delta_{22} x_2)^2\big]\sigma_z^2 + \sigma^2 \\
&= \Bigg[\bigg(\frac{\partial y(\mathbf{x}, \mathbf{z})}{\partial z_1}\bigg)^2 + \bigg(\frac{\partial y(\mathbf{x}, \mathbf{z})}{\partial z_2}\bigg)^2\Bigg]\sigma_z^2 + \sigma^2 \\
&= \big[(l_1(\mathbf{x}))^2 + (l_2(\mathbf{x}))^2\big]\sigma_z^2 + \sigma^2
\end{aligned}
\tag{10.25}
$$

Note the role of the derivatives $l_1(\mathbf{x})$ and $l_2(\mathbf{x})$ in the process variance function. Again, we seek $\mathbf{x}_0$ that results in minimum process variance. Here we are assuming that the $\pm$ levels of z_1 and z_2 are at $\pm\sigma_z$. **Note that the computation of the location of estimated minimum process variance does not depend on knowledge of σ_z.**

The least squares fit of the model in Equation (10.14) is given by

$$
\begin{aligned}
\hat{y}(\mathbf{x}, \mathbf{z}) = {}& 33.389 - 4.175 x_1 + 3.748 x_2 + 3.348 x_1 x_2 - 2.328 x_1^2 \\
&- 1.867 x_2^2 - 4.076 z_1 + 2.985 z_2 - 2.324 x_1 z_1 \\
&+ 1.932 x_1 z_2 + 3.268 x_2 z_1 - 2.073 x_2 z_2
\end{aligned}
$$

All coefficients are significant and the root error mean square is 0.742 dB. The slopes in the z_1 and z_2 directions are given by

$$
\hat{l}_1(x_1, x_2) = -4.076 - 2.234 x_1 + 3.268 x_2
$$

$$
\hat{l}_2(x_1, x_2) = 2.985 + 1.932 x_1 - 2.073 x_2
$$

Both x_1 and x_2 can be used to exert control on the process variance because both control $\times$ noise interactions are significant.

The point of minimum estimated process variance is found by setting simultaneously, $\hat{l}_1(\mathbf{x}) = 0$ and $\hat{l}_2(\mathbf{x}) = 0$. This gives the result

$$
\mathbf{x}_0 = \begin{bmatrix} x_{1,0} \\ x_{2,0} \end{bmatrix} = \begin{bmatrix} -0.874 \\ 0.625 \end{bmatrix}
$$

which, of course, lies in the experimental design region For the confidence region around this value, we have

$$\text{Vâr}\left[\hat{l}_1(\mathbf{x})\right] = \text{Vâr}\left[\hat{l}_2(\mathbf{x})\right] = 0.0154 + 0.0230x_1^2 + 0.0230x_2^2$$

This result comes from the variances of the $\hat{\delta}_{ij}$ and c_i values. Keep in mind that these estimators are uncorrelated. Thus for the confidence region, Equation (10.23) becomes

$$\frac{\hat{\mathbf{l}}(\mathbf{t})'\left[\text{Var}\,\hat{\mathbf{l}}(\mathbf{t})\right]^{-1}\hat{\mathbf{l}}(\mathbf{t})}{2} \leq F_{2,\,24,\,0.05}$$

which reduces to

$$\frac{(-4.076 - 2.324t_1 + 3.268t_2)^2 + (2.985 + 1.932t_1 - 2.073t_2)^2}{0.0154 + 0.0230(t_1^2 + t_2^2)} \leq 6.8056$$

Any (t_1, t_2) that satisfies the above inequality falls inside the 95% confidence region on the *location of minimum process variance*. Please note Figure 10.22. Two lines $\hat{l}_1(\mathbf{x}) = 0$ and $\hat{l}_2(\mathbf{x}) = 0$ shows intersection at $\mathbf{x}_0 = (-0.874, 0.625)$, the point of estimated minimum process variance. The shaded area displays the confidence region. Also included, however, are contours of constant *estimated mean response*, whose response surface is obtained using the expectation operator on Equation (10.24). We have

$$\hat{E}_z[y(\mathbf{x}, \mathbf{z})] = 33.3889 - 4.1752x_1 + 3.7481x_2$$
$$+ 3.3485x_1x_2 - 2.3277x_1^2 - 1.8670x_2^2$$

Of course the notion of a robust product here implies the choice of (x_1, x_2) that gives **large response with low variance.** Notice that the stationary point for the mean response surface is a point of estimated maximum response. This stationary point, however, does not fall inside the confidence region for the location of minimum variance. The mean response surface is fairly flat in the region of small x_1 and large x_2. As a result, very little is lost on the mean if one operates at $\mathbf{x}_0$, the point of minimum variance. Now, consider Figure 10.23. Here we have superimposed *contours of constant standard deviation.* Here, s is the estimated process standard deviation. Note that the standard deviation is stable in the upper left-hand corner. The estimated standard

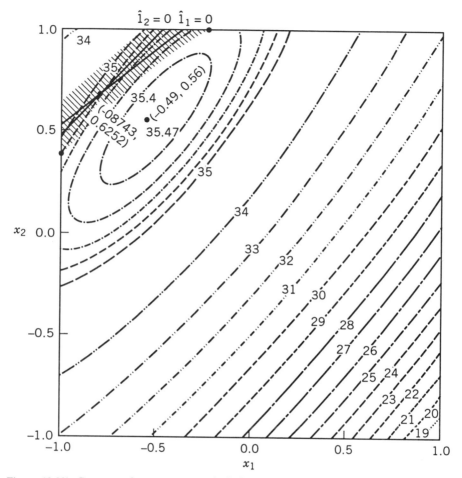

Figure 10.22 Contours of constant mean decibels, point of estimated minimum variance, and 95% confidence region on location of process variance for Example 10.5.

deviation at x_0 is 0.80. Note the standard deviation becomes less stable as one moves away from the 95% confidence region.

From the display given in Figures 10.22 and 10.23, the researcher learns a great deal about the process. The most robust process will clearly be in the area of small x_1 and large x_2. The "trough" created by the shaded area in Figure 10.22 shows considerable flexibility as far as optimum conditions is concerned. At x_0, the estimated process mean is approximately 35 dB and the estimated process standard deviation is 0.8.

In this example the researcher is fortunate that the intersection of $\hat{l}_1(x) = 0$ and $\hat{l}_2(x) = 0$ occurs inside the design region. If this does not occur, then one must use other response surface methods in dealing with the process variance.

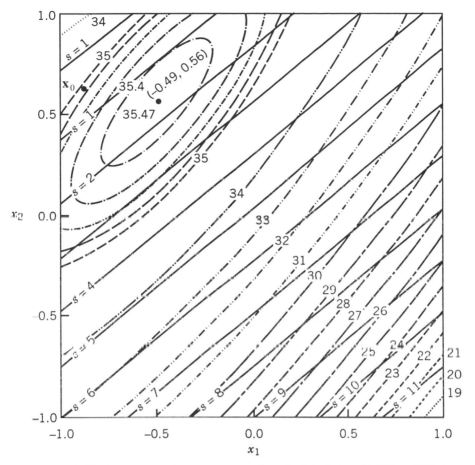

Figure 10.23 Dual response surfaces on mean and standard deviation with Points of Minimum Variance and Maximum Mean (Example 10.5).

10.6.5 Appropriate Estimation of the Process Variance with Applications

In the previous section we have dealt with the use of a mean and variance response surface for learning about the roles of the control and noise design variables in how they influence the process and mean and process variance. We have also discussed the use of these two response surfaces for finding robust conditions on the control variables, where some type of compromise is chosen for accommodating goals on the mean response but achieving a sufficiently small process variance. It can be argued that the two response surfaces can be dealt with in a fashion similar to that displayed in Chapter 6, where we discussed multiple response surface analysis. It is not our intention to recommend a single "best" way of blending together the mean and variance response surfaces.

In many instances the procedure of determining the conditions that produce zero value for the derivatives (i.e., values for which $\hat{l}_1 = 0$, $\hat{l}_2 = 0, \ldots, \hat{l}_{r_z} = 0$) give coordinates that are not practical. There is no assurance that this stationary point will fall inside the region of the design in the control variables. As a result, a ridge analysis is a reasonable approach for handling both the mean and variance response surfaces. This could provide two loci of points, each locus giving constrained optimum conditions, one locus for the mean and one for the variance. In what follows we will supply a few details for the case of the variance response surface. Before we embark on this discussion we will deal with the notion of appropriate estimation of the process variance.

For the case where the noise variables are uncorrelated, the variance response surface is given by Equation (10.25). In a more general setting, say, when the noise variables are correlated and $\text{Var}\,\mathbf{z} = \mathbf{V}$, where $\mathbf{V}$ is a $r_z \times r_z$ variance–covariance matrix, the process variance is

$$\text{Var}_z[y(\mathbf{x}, \mathbf{z})] = \mathbf{l}'(\mathbf{x})\mathbf{V}\mathbf{l}(\mathbf{x}) + \sigma^2$$

where the σ^2 is the error variance for the model of Equation (10.24). Now, if one merely replaces $\mathbf{l}(\mathbf{x})$ by the estimator from the data, and σ^2 by the mean squared error from the fitted model, we have as an estimator of the process variance

$$\hat{\text{Var}}_z[y(\mathbf{x}, \mathbf{z})] = \hat{\mathbf{l}}'(\mathbf{x})\mathbf{V}\hat{\mathbf{l}}(\mathbf{x}) + \text{MSE} \tag{10.26}$$

The result in Equation (10.26) is not an unbiased estimator because $E\hat{\mathbf{l}}'(\mathbf{x})\mathbf{V}\hat{\mathbf{l}}(\mathbf{x}) \neq \mathbf{l}'(\mathbf{x})\mathbf{V}\mathbf{l}(x)$. In fact,

$$E\,\hat{\text{Var}}_z[y(\mathbf{x}, \mathbf{z})] = \mathbf{l}'(\mathbf{x})\mathbf{V}\mathbf{l}(\mathbf{x}) + \sigma^2\,\text{tr}(\mathbf{V}\mathbf{C}) + \sigma^2 \tag{10.27}$$

where the matrix C is simply $\text{Var}\,(\hat{\mathbf{l}}(\mathbf{x}))/\sigma^2$. In other words, $\mathbf{C}$ is the variance covariance of the $\hat{l}$'s, apart from σ^2. Recall the role of this matrix in the confidence region on the location of process variance. See Equation (10.23). For standard experimental designs this matrix will be a diagonal matrix since the c_i and $\hat{\delta}_{jk}$ will all be uncorrelated. However, it should be noted that the diagonal elements, that is, the variances of the $\hat{l}_i(\mathbf{x})$, depend on $\mathbf{x}$. Now, from Equation (10.27) one can work out an unbiased estimator of the process variance, namely,

$$s_z^2[y(\mathbf{x}, \mathbf{z})] = \hat{\mathbf{l}}'(\mathbf{x})\mathbf{V}\hat{\mathbf{l}}(\mathbf{x}) + \text{MSE}(1 - \text{tr}\,\mathbf{V}\mathbf{C}) \tag{10.28}$$

or, for the cases that we have dealing with in this chapter, when the noise variables are uncorrelated with variance all 1.0,

$$s_z^2[y(\mathbf{x}, \mathbf{z})] = \hat{\mathbf{l}}'(x)\hat{l}(\mathbf{x}) + MSE(1 - \text{tr}\,\mathbf{C}) \tag{10.29}$$

The estimator in Equation (10.29) will often differ only slightly from the simpler but biased estimator in Equation (10.26). In fact, we used the latter in examples in previous sections. However, there will be occasions when the difference between the two will not be negligible. This is particularly true when the process variance is being computed on or near the design perimeter because $\text{tr}\,\mathbf{C}$ becomes larger as one approaches the design perimeter. As a result, one cannot ignore $s_z^2[y(\mathbf{x},\mathbf{z})]$ as an estimator of process variance.

In a comparison of the biased estimator of Equation (10.26) and the unbiased estimator of Equation (10.29), it can be shown that unless the error degrees of freedom in the fitting of the original model in Equation (10.24) is 2 or less, the unbiased estimator in Equation (10.29) is preferable. We will not supply the details of that study here. In what follows we discuss the use of the estimated process variance in a ridge analysis-type procedure for the process variance.

Ridge Analysis for Process Variance

The reader should recall from Chapter 6 that the purpose of ridge analysis is to produce a locus of points that are constrained optima on the response. This produces a ridge in the design space which could give rise to future experiments, or it could lead to interesting conditions that are of course confined to the experimental region. Applying the same notion to the process variance, we have

$$\min_{\mathbf{x}} \left[s_z^2 y(\mathbf{x},\mathbf{z}) \right] \quad \text{subject to} \quad \mathbf{x}'\mathbf{x} = R^2 \qquad (10.30)$$

That is, we will develop a locus of points that are each a point of conditionally minimum variance. The use of Lagrange multipliers as in Chapter 6 requires

$$\frac{\partial}{\partial \mathbf{x}} \left[\hat{\mathbf{l}}'(\mathbf{x})\hat{\mathbf{l}}(\mathbf{x}) + \text{MSE}(1 - \text{tr}\,\mathbf{C}) - \lambda(\mathbf{x}'\mathbf{x} - R^2) \right] \qquad (10.31)$$

Now, if the experimental design renders the noise main effects independent of the control $\times$ noise interactions, and the control $\times$ noise interaction independent of each other, the $\mathbf{C}$ matrix reduces to

$$\mathbf{C} = \begin{bmatrix} \mathbf{x}'\mathbf{D}_{11}\mathbf{x} + \text{Var}\,c_1 & & & 0 \\ & \mathbf{x}'\mathbf{D}_{22}\mathbf{x} + \text{Var}\,c_2 & & \\ & & \ddots & \\ 0 & & & \mathbf{x}'\mathbf{D}_{r_z,r_z}\mathbf{x} + \text{Var}\,c_{r_z} \end{bmatrix}$$

where $\mathbf{D}_{jj}$ is an $r_x \times r_x$ matrix containing $\text{Var}\,\hat{\delta}_{ij}/\sigma^2$ on the ith main diagonal. Here $\mathbf{x}' = [x_1, x_2, \ldots, x_{r_x}]$. Thus $\text{tr}\,\mathbf{C}$ becomes

$$\text{tr}(\mathbf{C}) = \sum_{j=1}^{r_z} \mathbf{x}'\mathbf{D}_{jj}\mathbf{x} + \sum_{j=1}^{r_z} c_j$$

Now, we will not supply the details in the differentiation of Equation (10.31). However, partial derivatives in Equation (10.31) set to **0** gives

$$\left[\hat{\Delta} \hat{\Delta}' - \text{MSE} \sum_{j=1}^{r_z} \mathbf{D}_{jj} - \mu \mathbf{I} \right] \mathbf{x} = - \hat{\Delta} \mathbf{c} \qquad (10.32)$$

where $\mathbf{c}' = [c_1, c_2, \ldots, c_{r_z}]$. To ensure that solutions are points of constrained minimum variance, values of μ smaller than the smallest eigenvalue of $\hat{\Delta} \hat{\Delta} - \text{MSE} \sum_{j=1}^{r_z} \mathbf{D}_{jj}$ are to be used in Equation (10.32).

Example 10.6 Consider a situation in which there are two control variables and three noise variables. The design variables x_1, x_2, z_1, z_2, and z_3 are varied in a 23-run variation of a CCD. *Please note that this is not a crossed array*. This type of design is referred to us as a **combined array**; that is, the control and noise variables are combined into one design. The result is considerably more economical than a crossed array for a situation like the one we discuss here. In Section 10.7 we discuss the concept of a combined array in more detail. The design and data are given below:

Observation	x_1	x_2	z_1	z_2	z_3	y
1	1	−1	−1	−1	−1	30.0250
2	−1	1	−1	−1	−1	30.0007
3	−1	−1	1	−1	−1	49.8009
4	−1	−1	−1	1	−1	43.4717
5	−1	−1	−1	−1	1	44.1905
6	1	1	1	−1	−1	31.3911
7	1	1	−1	1	−1	16.0333
8	1	1	−1	−1	1	35.3823
9	1	−1	1	1	−1	30.3383
10	1	−1	1	−1	1	36.3417
11	1	−1	−1	1	1	36.1355
12	−1	1	1	1	−1	30.1289
13	−1	1	1	−1	1	41.3179
14	−1	1	−1	1	1	22.7125
15	−1	−1	1	1	1	43.2415
16	1	1	1	1	1	39.1733
17	−2	0	0	0	0	46.1502
18	2	0	0	0	0	36.0689
19	0	−2	0	0	0	47.3903
20	0	2	0	0	0	31.4659
21	0	0	0	0	0	30.8109
22	0	0	0	0	0	30.7499
23	0	0	0	0	0	30.9655

Note that the design is not a complete CCD. The factorial portion is a $\frac{1}{2}$ fraction of a 2^5, and the axial points do not extend to include axial distances in z_1, z_2, z_3 because we are not fitting quadratic terms in the z's. The fitted regression can be verified as

$$\hat{y} = 30.382 - 2.925x_1 - 4.136x_2 + 2.855x_1x_2 + 2.596x_1^2$$

$$+ 2.175x_2^2 + 2.736z_1 - 2.326z_2 + 2.332z_3$$

$$- 0.278x_1z_1 + 0.893x_1z_2 + 2.574x_1z_3 + 1.999x_2z_1$$

$$- 1.430x_2z_2 + 1.547x_2z_3$$

The important derivatives in the z_1, z_2, and z_3 direction are given by

$$\hat{l}(\mathbf{x}) = \begin{bmatrix} \hat{l}_1 \\ \hat{l}_2 \\ \hat{l}_3 \end{bmatrix} = \begin{bmatrix} 2.736 - 0.278x_1 + 1.999x_2 \\ -2.326 + 0.893x_1 - 1.430x_2 \\ 2.332 + 2.574x_1 + 1.547x_2 \end{bmatrix}$$

The method of ridge analysis requires the minimization of $\hat{l}'\hat{l} - \text{MSE tr}(C)$ subject to the constraint $\sum_{i=1}^{2} x_i^2 = 2$. Here we are assuming that the noise variables are uncorrelated with $\sigma_z^2 = 1.0$ in the coded metric of the noise variables. In this case one needs to solve Equation (10.32) with particular values of μ used in the equation. The matrix $\sum_{j=1}^{r} \mathbf{D}_{jj}$ simply becomes $\frac{3}{16}\mathbf{I}_2$ and MSE $= 0.920$ from the analysis. As a result,

$$\text{MSE} \sum_{j=1}^{2} D_{jj} = \begin{bmatrix} 0.1725 & 0 \\ 0 & 0.1725 \end{bmatrix}$$

while the matrix $\hat{\Delta}$ is the 2×3 matrix containing the control $\times$ noise interaction coefficients in the fitted regression. Eigenvalues of $\hat{\Delta}\hat{\Delta}' - \text{MSE} \sum_{j=1}^{r} \mathbf{D}_{jj}$ turn out to be 9.995 and 5.594. As a result, we replace μ with values smaller than 5.594. It turns out that for a $\mu = -0.486$ the solution to Equation (10.32) falls on a radius $\sqrt{2}$. Indeed, a value $(-0.0156, -1.41405)$ is the point of constrained minimum variance at a radius $\sqrt{2}$. Figure 10.24 displays the result of the ridge analysis. All points on the path in the figure are points of constrained minimum process variance. Obviously future experiments with x_1 near zero and smaller values of x_2 might be considered.

In much of what has preceded we have built mean and variance response surfaces that are generated from a single response surface that is developed from a model with noise and control variables in the same model. In what follows we discuss alternative methods of constructing the variance model.

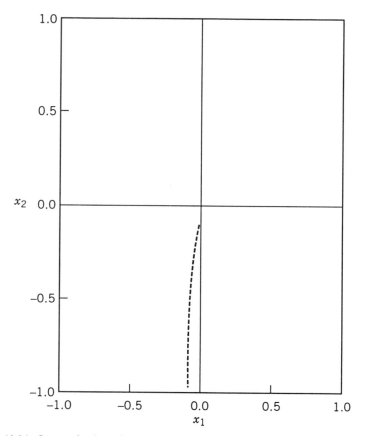

Figure 10.24   Locus of points of minimum constrained process variance for Example 10.6.

10.6.6   The Log Linear Variance Model

Bartlett and Kendall (1946) in a classical paper provided evidence that variance modeling can be accomplished without serious violation of assumptions by using a log linear model of the type

$$\log s_i^2 = \mathbf{x}_i'\boldsymbol{\gamma} + \varepsilon_i^*, \qquad i = 1, 2, \ldots, d \tag{10.33}$$

where s_i^2 is a sample variance obtained from, say, n observations taken at each of d design points. The $\mathbf{x}_i'\boldsymbol{\gamma}$ portion of the model refers to a linear model in a set of design variables. They were able to demonstrate that if the errors are normal around the mean model, i.e., if

$$y_{ij} = \mathbf{x}_i'\boldsymbol{\beta} + \varepsilon_{ij}\sigma_i, \qquad i = 1, 2, \ldots, d; j = 1, 2, \ldots, n$$

where $\varepsilon_{ij} \sim N(0, 1)$ and $\ln \sigma_i^2 = \mathbf{x}_i'\boldsymbol{\gamma}$, then the use of the model in Equation (10.33) leads to approximately normal errors with constant variance. Thus ordinary least squares may be appropriate for modeling variance in a log linear type of procedure. Carroll and Ruppert (1988) discuss further details of variance modeling.

The quality movement of the early 1980s and the influence of Taguchi toward the notion of controlling process variance has resurrected the idea of variance modeling via the log linear approach. Indeed, the crossed array is a natural vehicle where the s_i^2 values are obtained from the data in the outer arrays. For example, in, say, a $3^2 \times 2^2$ crossed array, the sample variances are computed from four observations at each inner array point. Thus we have the following data scenario:

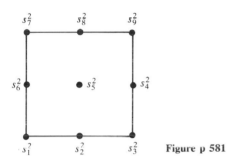

Figure p 581

Thus there are nine sample variances and the resulting

$$\ln s_i^2 = f(x_{1i}, x_{2i}) + \varepsilon_i^* \qquad i - 1, 2, \ldots, 9$$

Here, of course, the $f(x_{1i}, x_{2i})$ could conceivably include a second-order model in the control variables.

Apart from the virtues of homogeneous variance and normality, the use of the log linear variance model also often results in simplicity. The log transformation often reduces the effect of curvature and interaction terms, thereby rendering more simple the total study and selection of candidate regions for robust conditions. One obvious disadvantage is that the design must be a crossed array; and because quality of the variance model is best when the number of observations in the outer array is large, success may be achieved at a relatively high price. For example, one could not use this approach with the CCD in the combined array discussed in the previous section. Additional discussion dealing in the use of crossed arrays and combined arrays will be given in Section 10.7.

Once a log linear variance model is fit, the mean response surfaces model in the control variables can be found in the usual way as discussed in Chapter 6. Thus the dual response surface approach for finding robust or optimum conditions can be carried out. Vining and Myers discuss the approach with illustrations in the panel discussion edited by Nair et al. (1992).

Example 10.7   Consider Example 10.1. Recall that it is of interest to determine conditions on the control variables that produce a low value of response (solder defects per million joints) but also provide consistency (low variability). The data in Example 10.1 were used to develop two models, one for the mean response (standard linear regression model) and one for the variance via the log linear variance model described here. Recall that the crossed array contains a 2_{III}^{5-2} inner array design and a 2_{III}^{3-1} in the outer array. Thus, four outer array points can be used for the sample variances. The fitted models, including significant terms, are given by

$$\hat{y} = 197.175 - 28.525x_1 + 57.975x_2$$

for the mean and

$$\ln(s^2) = 7.4155 - 0.2067x_1 - 0.0309x_2$$

for the variance. Here x_1 is solder pot temperature and $\mathbf{x}_2$ is flux density. Figure 10.25 shows the graphical analysis involving both models. The *mean*

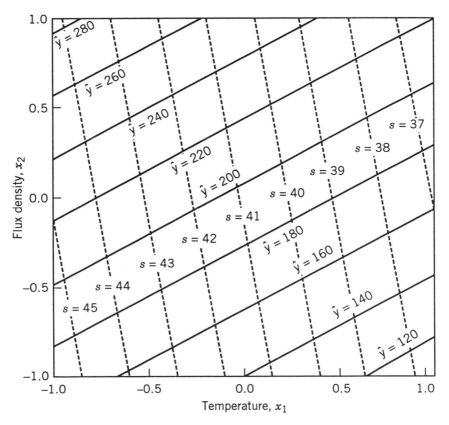

Figure 10.25   Mean and variance models for wave solder from Example 10.7.

and standard deviation are plotted. The latter was found via the operation

$$\text{Process standard deviation} = \left[e^{7.4155 - 0.2067x_1 - 0.0309x_2} \right]^{1/2}$$

As one would expect from the analysis in Example 10.1, high temperature and low flux density minimize mean number of errors and variability simultaneously. However, the graphical dual response surface procedure allows the user to see tradeoffs for better understanding of the process. For the example the user may be dissatisfied with the extreme conditions with solder temperature and flux density, particularly temperature. The graphs provide alternative conditions with resulting mean and variance estimates. It would be of interest to see the results when the variance model is computed from the "variance operator" on a model containing both control and noise variables —that is, the procedure discussed in Sections 10.6.4 and 10.6.5. See Exercise 10.24.

10.7 DESIGNS FOR ROBUST PARAMETER DESIGN; CROSSED AND COMBINED ARRAYS

Thus far in this chapter we have emphasized the use of crossed arrays (or product arrays) in which separate designs are used for the control and noise variables, and the final design is a "crossing" of the inner array (control array) and the outer array (noise array). The impetus for the design construction was provided by Taguchi, who suggested the SNR to provide a summary statistic for the data in the outer array. In the same vein the $\ln s^2$, discussed in the previous section, is a basic response for variance modeling that benefits greatly from this same crossed array design type. However, we have also indicated how, in many instances, the crossed array design can be very costly, with degree of freedom for interaction among the control factors often being sacrificed in favor of large numbers of degrees of freedom for control $\times$ noise interactions. If robust parameter design is to be accomplished, some control $\times$ noise interaction must be exploited.

10.7.1 The Combined Array

The crossed array was clearly born out of the necessity of having considerable information on which to base the SNR. However, if one is attracted to the so-called "response model" approach in which a single model is constructed for both $\mathbf{x}$ and $\mathbf{z}$, with the concomitant development of a mean response surface via $E_z[y(\mathbf{x}, \mathbf{z})]$ and a variance response surface via $\sigma_z^2[y(\mathbf{x}, \mathbf{z})]$, then the crossed array is not needed. For this approach, the so-called **combined array** in which a design is chosen to allow estimability of a reasonable model in $\mathbf{x}$ and $\mathbf{z}$ is often less costly and quite sufficient. The word *combined* implies that separate arrays are not used, but rather one array that considers a combined $\mathbf{x}$ and $\mathbf{z}$ is the appropriate design. The variation of a CCD used in example 10.6 in the process variance ridge analysis is an example of a combined array. In that situation we needed to accommodate a second-order model in the x's as well as x–z interactions and z linear terms. The design chosen was quite appropriate. The concept of a combined array has been proposed by many authors, including Welch et al. (1990), Shoemaker et al. (1991), Lucas (1989), Montgomery (1991a, b), Myers et al. (1992a), and Box and Jones (1989), and in each case their intention was to model **x** and **z** together for some specific purpose.

Combined arrays offer considerably more flexibility in the estimation of effects or regression coefficients as well as savings in run size. The designs that are chosen generally are those that offer the concept of *mixed resolution* [see Lucas (1989)]. That is, because control × noise interactions are so vital in modeling as well as diagnostics, the resolution of the design for these interactions is higher than that for the noise × noise interactions, which are generally considered not as important. The resolution for the control × control interactions may also be higher if they are known to be important. In what follows we consider situations in which models are first order (linear effects) or first order plus interaction in the control variables.

Example 10.8   This example, pointed out in Shoemaker et al. (1991), shows the superiority of the combined array. Suppose there are factors x_1, x_2, x_3 and z_1, z_2, z_3. An obvious crossed array is as follows:

$$2_{\text{III}}^{3-1} \times 2_{\text{III}}^{3-1} \quad \text{(Design 1)}$$

giving a total of 16 runs. This crossed array could be viewed as a combined array with defining relations $I = x_1 x_2 x_3 = z_1 z_2 z_3 = x_1 x_2 x_3 z_1 z_2 z_3$. *Clearly no interactions in the control factors can be estimated.* This is a $\frac{1}{4}$ fraction of a 2^6. *Six main effects are estimable only if no two factor interactions are important.* In addition to six main effects, nine additional degrees of freedom are available and they are all control × noise interactions. These interactions are allowable at the expense of control × control interactions.

As an alternative to design 1, a combined array offers considerably more flexibility. One such design would be a 2^{6-2} with defining relations

$$I = x_1 x_2 x_3 z_1 = z_1 z_2 z_3 = x_1 x_2 x_3 z_2 z_3 \quad \text{(Design 2)}$$

This setup illustrates nicely the concept of mixed resolution. The design is resolution III with regard to noise × noise interactions and resolution IV with regard to other interactions. Indeed, the three control × control interactions can be estimated if one is willing to assume $x_1 z_1, x_2 z_1,$ and $x_3 z_1$ as negligible. In other words, one has the option to "trade" control × noise interactions for control × control interactions if appropriate assumptions are met.

Example 10.9   Consider a situation in which there are four control factors $x_1, x_2, x_3,$ and x_4 and two noise factors z_1 and z_2. Assume it is known that $x_1 x_4, x_2 x_4,$ and $x_3 x_4$ are important. One potential *product array* is the

32-run design

$$2_{IV}^{4-1} \times 2^2 \quad \text{(Design 1)}$$

The 31 available degrees of freedom include all six main effects and 12 two-factor interactions x_1x_2, x_1x_3, x_1x_4, z_1z_2, x_1z_1, x_2z_1, x_3z_1, x_4z_1, x_1z_2, x_2z_2, x_3z_2, and x_4z_2. This represents 18 degrees of freedom. This means that *13 degrees of freedom are being used to estimate higher-order control × noise interactions*. On the o ther hand, a combined array 2_{VI}^{6-1} with 32 runs using

$$I = x_1x_2x_3x_4z_1z_2 \quad \text{(Design 2)}$$

is much more appropriate. This design allows estimation of all six main effects and *all 15 two-factor interactions*. Thus three higher-order interactions in design 1 are being exchanged for three control × control interactions which will likely be more important.

As a "better yet" approach, the combined array to handle the situation in Example 10.9 can be found using computer-generated design algorithms. In this case the model contains six main effects and 12 two-factor interactions. Indeed many designs with less than 32 runs are available. Here we list a 22-run design shown in Shoemaker et al. (1991). The design is given by

$$\mathbf{D}_{\text{design 3}} = \begin{bmatrix} x_1 & x_2 & x_3 & x_4 & z_1 & z_2 \\ 1 & 1 & 1 & 1 & 1 & -1 \\ 1 & -1 & -1 & 1 & 1 & -1 \\ 1 & -1 & -1 & -1 & -1 & -1 \\ 1 & 1 & -1 & 1 & -1 & -1 \\ 1 & 1 & 1 & 1 & -1 & 1 \\ 1 & -1 & 1 & 1 & 1 & 1 \\ -1 & 1 & 1 & 1 & -1 & -1 \\ 1 & -1 & 1 & -1 & -1 & -1 \\ 1 & -1 & 1 & -1 & 1 & -1 \\ -1 & -1 & 1 & -1 & -1 & 1 \\ 1 & 1 & -1 & -1 & 1 & 1 \\ -1 & -1 & -1 & -1 & 1 & 1 \\ -1 & 1 & -1 & -1 & 1 & -1 \\ -1 & 1 & 1 & 1 & -1 & -1 \\ 1 & -1 & -1 & 1 & -1 & 1 \\ -1 & 1 & -1 & 1 & 1 & 1 \\ -1 & -1 & -1 & 1 & -1 & -1 \\ 1 & 1 & 1 & -1 & -1 & 1 \\ -1 & -1 & 1 & 1 & 1 & -1 \\ -1 & 1 & 1 & -1 & 1 & 1 \\ -1 & 1 & -1 & -1 & -1 & 1 \\ -1 & -1 & 1 & 1 & -1 & 1 \end{bmatrix}$$

This design allows efficient estimation of all important terms and leaves three degrees of freedom for lack of fit.

10.7.2 Second-Order Designs for Robust Parameter Design

The flexibility of the combined array and the use of composite designs, BBD and computer-aided designs extend nicely to the use of models that are second order in the control variables and even second order in the noise variables. Here the mixed resolution idea for the CCD is very important. Consider, for example, a crossed array in the case of three control variables and two noise variables. The control array may be a $\frac{1}{3}$ fraction of a 3^3 and the noise array a 2^2 (36 total runs) to accommodate a second-order model in the control variables, the terms z_1, z_2, and z_1z_2, and appropriate control $\times$ noise interactions. A CCD containing the components

(i) 2^{5-1} using $I = x_1x_2x_3z_1z_2$
(ii)   axial points:

$$
\begin{array}{ccccc}
-\alpha & 0 & 0 & 0 & 0 \\
\alpha & 0 & 0 & 0 & 0 \\
0 & -\alpha & 0 & 0 & 0 \\
0 & \alpha & 0 & 0 & 0 \\
0 & 0 & -\alpha & 0 & 0 \\
0 & 0 & \alpha & 0 & 0
\end{array}
$$

(iii)   center runs:

$$
\begin{array}{ccccc}
\mathbf{0} & \mathbf{0} & \mathbf{0} & \mathbf{0} & \mathbf{0}
\end{array}
$$

contains 22 + center runs. This design will estimate x_1, x_2, x_1^2, x_2^2, x_1x_2, z_1, z_2, z_1z_2, and six linear $\times$ linear control $\times$ noise interactions. This leaves seven lack-of-fit degrees of freedom. Again the crossed array often invests in a large number of higher-order control $\times$ noise interactions. The combined array is more efficient.

To better illustrate the idea of mixed resolution, consider a situation in which there are four control variables and three noise variables. Suppose it is important for the response model to contain terms that are second order in the control variables (15 terms including intercept), z_1, z_2, z_3 and 12 control $\times$ noise interactions. (Keep in mind the importance of these interactions in building the variance model.) The CCD with mixed resolution in the factorial portion is quite useful. A 2^{7-2} design is appropriate for the factorial portion. Consider the defining relations

$$I = x_1x_2x_3z_1z_2 = z_1z_2z_3x_4 = x_1x_2x_3x_4z_3$$

This design, then, involves the components

(i) 2^{7-2} with the above defining relations

(ii)   axial points:

x_1	x_2	x_3	x_4	z_1	z_2	z_3
$-\alpha$	0	0	0	0	0	0
α	0	0	0	0	0	0
0	$-\alpha$	0	0	0	0	0
0	α	0	0	0	0	0
0	0	$-\alpha$	0	0	0	0
0	0	α	0	0	0	0
0	0	0	$-\alpha$	0	0	0
0	0	0	α	0	0	0

(iii)   center runs: **0   0   0   0   0   0   0**

As a result, 40 + center runs are used to estimate 30 model terms. Note the careful selection of the defining relations. By definition, the factorial portion is resolution IV. However, it is of higher resolution in the control variables because no two $x_i x_j$ interactions are aliased. In addition, it is effectively a higher-resolution design for the control × noise interactions because no two control × noise interactions are aliased with each other, nor are they aliased with two-factor interactions in the control variables.

10.7.3   Summary Remarks

Because the response surface approach puts so much emphasis on efficient estimation of the appropriate response model in **x** and **z**, with flexibility needed in the model terms selected, the combined array is altogether appropriate. In fact, the response model approach and the combined array are quite compatible. Many times the crossed array will involve an excessive number of experimental runs that are a result of higher-order control × noise interactions.

10.8   DETECTION AND COMPUTATION OF DISPERSION EFFECTS IN HIGHLY FRACTIONATED DESIGNS

In this entire chapter we have dealt with the important notion of variance modeling where the variance is derivable from the inability to control noise variables. However, there are circumstances in which the practitioner is unable to measure noise variables or perhaps he or she is not aware of what the noise variables are. At any rate, there is nonhomogeneous variance from one or more sources and thus there are both locations and dispersion effects. Detection of **dispersion effects**, positive or negative, is often vital, particularly if the experiment is a pilot experiment and directions are sought for future experiments. For example, consider a 2^3 factorial experiment with factors A, B, and C. In addition, suppose we wish to move to a different experimental region via *steepest ascent* or some other sequential procedure. Suppose the calculated effects are as follows:

A: $+7.5$

B: $+4.7$

C: -3.5

Now, these calculated effects are **location effects;** that is, they reflect the change in the mean response as one moves from low to high. In this case, future experiments will involve an increase in A, increase in B, and a decrease in C. However, suppose an increase in A results in a profound *increase in variance* in the response. We say that A is a *dispersion effect* as well as a location effect, and thus we are left with a mean–variance tradeoff if we are to consider moving A to the high level.

In this section the concept of what is mean by a dispersion effect has not changed. In previous sections a dispersion effect was discovered in a factor if it had a certain type of interaction with a noise variable. Here we are assuming that there are not necessarily any noise variables in the problem, and yet there is a need to detect the existence of factors that influence dispersion. Plots and numerical computation of dispersion effects are useful.

10.8.1   The Use of Residuals

The reader should recall from Chapter 2 that much diagnostic information is gained from the use of residuals—that is, the values $e_i = y_i - \hat{y}_i$, where $\hat{y}_i$ is the predicted or fitted value from a regression analysis. Residuals provide considerable information about unexplained variability. For example, when residuals are plotted, an attractive display to the user is a *random scatter around zero*. Patterns in residuals suggest that some assumption that has been made is being violated. The two most prominent violations are:

1. Functional form of model misspecified
2. Nonhomogeneous variance

It is violation 2 that concerns us here. It is of interest to determine if the signal from residual plots reflect a type of pattern that would identifiy certain factors as dispersion effects.

For example, a plot of the residuals against the level of a factor in the case of a two-level design may look as follows:

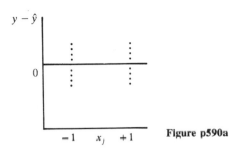

Figure p590a

In this case there is no evidence that x_j is a dispersion effect. On the other hand, suppose the plot of residuals takes the form

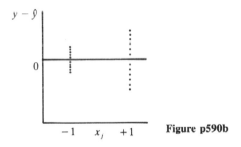

Figure p590b

Now, while one must take considerable care in reading too much into residual plots with small experiments, there is certainly some evidence here that at the high level of x_j, the model error variance is larger than at the low level. While the fitted model here suggests that all three factors are location effects, it appears as if x_j is a dispersion effect, with $x_j = -1$ favored for a consistent product.

Importance of the Fitted Mean Model
It should be understood that the use of residuals, residual plots, and other diagnostic tools for determining dispersion effects is profoundly affected by the fitted mean model because the residuals used are the residuals from the mean model. If the mean model has been handled in a sloppy fashion, residuals should be clouded by model misspecification. Thus the detection of

dispersion effects depends on the presumption that the fitted model for the mean is a good one. In this regard, the detection of dispersion effects will be most successful when there are only a few well-defined effects to be used in the mean model—that is, when there is **sparsity of effects.**

10.8.2   Further Diagnostic Information from Residuals

It should be clear from the contrived example in Section 10.8.1 that one is attempting to determine if the product or process variance depends on the design locations, rather than being homogeneous. In the example, one might conclude from the plots that the error variance is constant except when $x_3 = +1$. When this type of result is detected, it is hoped that the engineer or scientist has, at least in retrospect, some scientific explanation. However, there still remains the issue surrounding what is meant by a "significant change in variance." Can we say with assurance that the plots in Section 10.8.1 indicate that x_3 is a dispersion effect? Are there other diagnostic pieces of information apart from the plots?

An important diagnostic for determining if a specific factor, say factor i, is a dispersion effect is the statistic

$$F_i^* = \ln\left(\frac{s_{i+}^2}{s_{i-}^2}\right) \qquad (10.34)$$

See Box and Meyer (1986). Here s_{i+}^2 is the sample variance of the residuals when factor i is at the high level and s_{i-}^2 is the sample variance of the residuals when factor i is at the low level. Obviously a value close to zero is evidence that the spread in the residuals is not significantly different for the two levels, whereas a value large in magnitude (positive or negative) would be a signal that the variable has a dispersion effect. A large positive value indicates that $\sigma_{i+}^2 > \sigma_{i-}^2$, whereas a large negative effect indicates $\sigma_{i-}^2 > \sigma_{i+}^2$. While there should be no intention of using the F_i^* values in Equation (10.34) as a formal test statistic, it can be a very useful diagnostic because it has been found that under $H_0: \sigma_{i+}^2 = \sigma_{i-}^2$, the F^* values are approximately normally distributed with mean 0 and common variance. This may be exploited as a numerical diagnostic, and the F_i^* values may be studied via normal probability plots.

Example 10.10   In an injection molding experiment it is often important to determine conditions in which shrinkage is minimized. In the present experiment [see Montgomery (1991a)] seven factors were varied in a 2^{7-3} with four center runs. It was felt that there may be curvature in the system. The factors are x_1 (mold temperature), x_2 (screw speed), x_3 (holding time), x_4 (gate size), x_5 (cycle time), x_6 (moisture content), and x_7 (holding

Table 10.3 Injection Molding Data

Standard Order	Actual Run Order	x_1 A	x_2 B	x_3 C	x_4 D	x_5 E($=ABC$)	x_6 F($=BCD$)	x_7 G($=ACD$)	Observed Shrinkage ($\times 10$)
1	8	−	−	−	−	−	−	−	6
2	16	+	−	−	−	+	−	+	10
3	18	−	+	−	−	+	+	−	32
4	17	+	+	−	−	−	+	+	60
5	3	−	−	+	−	+	+	+	4
6	5	+	−	+	−	−	+	−	15
7	10	−	+	+	−	−	−	+	26
8	2	+	+	+	−	+	−	−	60
9	9	−	−	−	+	−	+	+	8
10	15	+	−	−	+	+	+	−	12
11	12	−	+	−	+	+	−	+	34
12	6	+	+	−	+	−	−	−	60
13	13	−	−	+	+	+	−	−	16
14	19	+	−	+	+	−	−	+	5
15	11	−	+	+	+	−	+	−	37
16	1	+	+	+	+	+	+	+	52
17	20	0	0	0	0	0	0	0	25
18	4	0	0	0	0	0	0	0	29
19	14	0	0	0	0	0	0	0	24
20	7	0	0	0	0	0	0	0	27

pressure). The defining relations are

$$x_1x_2x_3x_5 = x_2x_3x_4x_6 = x_1x_3x_4x_7 = I$$

The data appear in Table 10.3. It is of interest to determine settings (within the region of the experiment) that minimizes shrinkage. In addition, any dispersion effect that is found must be dealt with because the variability in the shrinkage must also be taken into account.

The individual location effects of interest were first placed on a normal probability plot to gain some impression of what model terms are important. Table 10.4 gives the reader a clear indication of what effects are estimable. Fifteen effects, including all main effects and two-factor interactions (and aliases), were plotted on the normal probability plot in Figure 10.26. The x_1, x_2, and x_1x_2 effects are found to be important from the plot. All three are positive effects. The x_1x_2 interaction plot in Figure 10.27 is particularly revealing. The use of screw speed, x_2, at the low level minimizes shrinkage in spite of the level of x_1, mold temperature. In addition, if mold temperature is difficult to control in the process, the -1 level of screw spread is also a robust choice. So at this point in the analysis it would seem that $x_2 = -1$ is absolutely necessary.

Table 10.4   A Bias Structure for the 2^{7-3} Design in Table 10.3

$A = BCE = DEF = CDG = BFG$	$AB = CE = FG$
$B = ACE = CDF = DEG = AFG$	$AC = BE = DG$
$C = ABE = BDF = ADG = EFG$	$AD = EF = CG$
$D = BCG = AEF = ACG = BEG$	$AE = BC = DF$
$E = ABC = ADF = BDG = CFG$	$AF = DE = BG$
$F = BCD = ADE = ABG = CEG$	$AG = CD = BG$
$G = ACD = BDE = ABF = CEF$	$BD = CF = EG$
$ABD = CDE = ACF = BEF = BCG = AEG = DEF$	

The next step in the analysis is to deal with possible curvature. Table 10.5 reveals a regression analysis with the regression model

$$\hat{y} = 27.3125 + 6.9375x_1 + 17.8125x_2 + 5.9375x_1x_2 \qquad (10.35)$$

and an analysis of variance that reveals no evidence of curvature. The "curvature" term in the ANOVA is a result of an x_j^2 term being used in the model (only a single degree of freedom). Here it becomes clear that there is no curvature in the system. At this point it would appear the $x_1 = -1$ and $x_2 = -1$ would be proper settings. It turns out that this would reduce mean shrinkage by roughly 10%. But what about variability? The regression equation [Equation (10.35)] was used to compute the residuals. The normal probability plot of residuals in Figure 10.28 does not reveal anything out of the ordinary. However, the residuals were also plotted against levels of the main

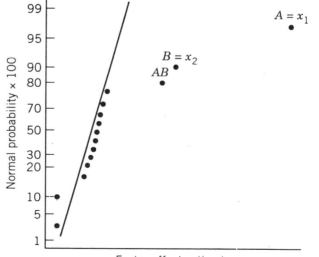

Figure 10.26   Normal probability plot of location effects for Example 10.10.

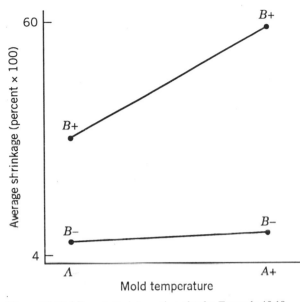

Figure 10.27 $AB = x_1 x_2$ interaction plot for Example 10.10.

effects, and the residual plot involving x_3 was particularly enlightening. It appears in Figure 10.29. It seems clear that variability in the residuals is greater at high holding time than at low holding time. The $s_{(i-)}$ and $s_{(i+)}$ values as well as the F_i^* in Equation (10.34) were computed. They appear, along with the residuals themselves, in Table 10.6. Note that the single F_i^* value that stands out is that of x_3 ($F^* = 2.50$). To gain more diagnostic insight, a normal probability plot of the F^* values appears in Figure 10.30. Clearly, effect x_3 stands out. It would not be unreasonable to say that x_3 has a significant dispersion effect.

It is of interest now to provide a summary picture of mean and variability of the shrinkage response. This is particularly easy to do because there appears to be only three "main players," x_1, x_2, and x_3. The analysis suggests that low x_1, low x_2, and low x_3 are the settings that produce, simultaneously, minimum mean shrinkage and minimum variance in shrinkage. That is, the optimum conditions are

$$x_1 = -1, \qquad x_2 = -1, \qquad x_3 = -1$$

10.8.3   Further Comments Concerning Variance Modeling

In this entire chapter we have focused on the importance of considering process variability. The major sources of process variability often comes from noise variables; and when this is the case, the modeling can be handled

Table 10.5 Regression and ANOVA for Data of Example 10.10

Source of Variation	Sum of Squares	Degrees of Freedom	Mean Square	F	Probability > F
Model	6410.687500	3	2136.8958333	121.645	0.0001
Curvature	3.612500	1	3.6125000	0.206	0.6567
Error	263.500000	15	17.5666667		
Residual	248.750000	12	20.7291667		
Pure error	14.750000	3	4.9166667		
Corrected total	6677.800000	19			

			R-squared	0.9605	
Root mean squared error	4.191261				
Dependent mean	27.312500		Adjusted R-squared	0.9500	
CV	15.35%				

Variable	Parameter Estimate	Degrees of Freedom	Sum of Squares	t for H_0: Parameter $= 0$	Probability > |t|
Intercept	27.312500	1			
A	6.937500	1	7700.062500	6.621	0.0001
B	17.812500	1	5076.562500	17.000	0.0001
AB	5.937500	1	564.062500	5.667	0.0001

522

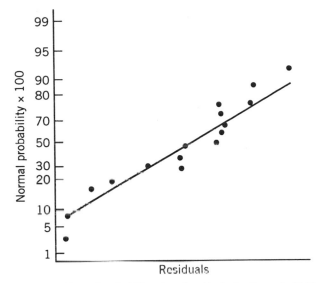

Figure 10.28 Normal probability plot of residuals for Example 10.10.

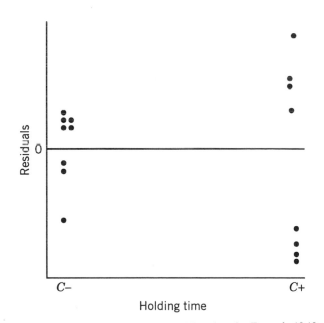

Figure 10.29 Plot of residuals against holding time for Example 10.10.

Table 10.6 Residuals and F^* Values for All Effects for Example 10.10

x_1	x_2	x_1x_2	x_3	x_1x_3	x_2x_3	x_5	x_4	x_1x_4	x_2x_4	$x_1x_2x_4$	x_3x_4	x_7	x_6	x_4x_5	Residuals
−	−	+	−	+	+	−	−	+	+	−	+	−	−	+	−2.50
+	−	−	−	−	+	+	−	−	+	+	+	+	−	−	−0.50
−	+	−	−	+	−	+	−	+	−	+	+	−	+	−	+0.25
+	+	+	−	−	−	−	−	−	−	−	+	+	+	+	2.00
−	−	+	+	−	−	−	−	+	+	−	−	+	+	−	−4.50
+	−	−	+	+	−	+	−	−	+	+	−	−	−	+	4.50
−	+	−	+	−	+	+	−	+	−	+	−	+	−	+	−6.26
+	+	+	+	+	+	−	−	−	−	−	−	−	+	−	2.00
−	−	+	−	+	+	+	+	−	−	+	−	+	+	+	+0.50
+	−	−	−	−	+	−	+	+	−	−	−	−	+	−	1.50
−	+	−	−	+	−	−	+	−	+	−	−	+	−	+	1.75
+	+	+	−	−	−	+	+	+	+	+	−	−	−	−	2.00
−	−	+	+	−	−	+	+	−	−	+	+	−	−	+	7.50
+	−	−	+	+	−	−	+	+	−	−	+	+	+	−	−5.50
−	+	−	+	−	+	−	+	−	+	−	+	−	+	−	4.75
+	+	+	+	+	+	+	+	+	+	+	+	+	+	+	−6.00
S_{i-} 4.59	4.41	4.10	1.63	4.52	4.33	4.25	3.59	2.75	4.41	3.51	3.50	3.65	3.12	3.40	
S_{i+} 3.80	4.01	4.33	5.70	3.68	3.85	4.17	4.64	3.39	4.01	4.72	3.12	3.50	3.88	4.87	
F_i^* −0.38	−0.18	0.11	2.50	−0.41	−0.24	−0.03	0.51	0.42	−0.18	0.52	0.51	0.23	−0.30	0.72	

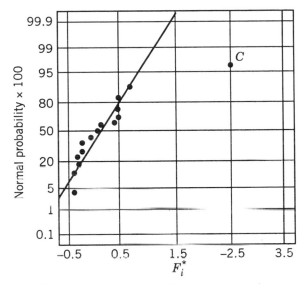

Figure 10.30 Normal probability plots of dispersion effects F_i^* for Example 10.10.

conveniently through the use of a combined array and the resulting response surface in **x** and **z** and response surfaces that result for both mean and variance. Another source of variance modeling is the log linear model, where $\ln(s^2)$ becomes the response. In this context, the s^2 was viewed as the sample variance associated with the outer array in a crossed array, but this need not be the case. The s^2 may very well come from *replicated observations* in, say, a two-level design; in other words, there may be no noise variables in the design and yet a nonconstant variance may be evident through replication (n runs at each of d design points) and the use of the model

$$\ln s_i^2 = f(\mathbf{x}_i) + \varepsilon_i, \qquad i = 1, 2, \ldots, d \qquad (10.36)$$

Additional levels of variance modeling and/or the computation of dispersion effects can be accomplished for cases as in the Example 10.10. The experiment contains a large number of factors, and because of cost constraints the design is a highly fractionated two-level design with no replication. Here, formal variance modeling is not necessarily a good practice due to lack of sufficient information on variability. However, the use of residuals can highlight dispersion effects as in the example. The reader should be aware of the notion of retrieval of variance information and dispersion effects at different levels of experimentation and analysis. In addition, more discussion is needed regarding different methods of calculating dispersion effects.

What Is a Dispersion Effect? (Case of Replicated Observations)

Box and Meyer (1986) and Nair and Pregibon (1986) discuss the notion of dispersion effects for various types of scenarios. Consider a situation in which the design is a two-level factorial or fraction and n observations appear at each data point. The fitted model may be a log linear model with

$$\ln(s_i^2) = \gamma_0 + \gamma_1 x_{i1} + \gamma_2 x_{i2} + \cdots + \gamma_k x_{ik} + \varepsilon_i \qquad i = 1, 2, \ldots, d \quad (10.37)$$

In this case, the replication variance is more reliable as process variance information than that obtained from residuals. In the latter case, the reliability of the variance information depends to a great extent on the quality of the mean model that created the residuals.

The least squares estimators of the γ_i in Equation (10.37) are derived from

$$\hat{\boldsymbol{\gamma}} = (\mathbf{X}'\mathbf{X})^{-1}\mathbf{X}'\mathbf{e} \qquad (10.38)$$

where $\mathbf{e}$ is a vector of log s^2 values. Because the rows of $\mathbf{X}'$ contain $+1$ and -1 values depending on whether the specific factor is high or low and because $(\mathbf{X}'\mathbf{X})^{-1} = (1/d)\mathbf{I}$, the parameter estimates for the model of Equation (10.37) become

$$\hat{\gamma}_0 = \overline{\ln s^2}$$

$$\hat{\gamma}_i = \left[\sum_{j=i+} \ln s_j^2 - \sum_{j=i-} \ln s_j^2 \right] \Big/ d, \qquad i = 1, 2, \ldots, k \quad (10.39)$$

Here $\sum_{j=i+}$ implies that the sum is taken over the $\ln s^2$ values for which the ith variable is at the $+1$ level, while $\sum_{j=i-}$ implies that sum is over the -1 level. In addition, $(\overline{\ln s^2})$ is the average of the $\ln s^2$ values over the experiment. One could, of course, extend what we know about the relationship between effects and regression coefficients for the mean model. Thus the **computed dispersion effect** for the ith variable can be written

$$(\text{Dispersion effect})_i = \left[\left(\overline{\ln s^2}\right)_{i+} - \left(\overline{\ln s^2}\right)_{i-} \right] \qquad i = 1, 2, \ldots, k \quad (10.40)$$

where, obviously $(\overline{\ln s^2})_{i+}$ is the average $\ln s^2$ value when variable i is at the high level and $(\overline{\ln s^2})_{i-}$ is the same for variable i at the lower level. Now, what is done with those dispersion effects or regression coefficients? Normal probability plots can certainly be used to highlight important dispersion effects. The normality assumption with constant variance for these effects is certainly justified. We had suggested earlier that work by Bartlett and Kendall (1946) and others indicate that $\ln s^2$ is approximately normal, and the averaging process strengthens the assumption. Of course, it is quite likely that in most applications where replications are involved, the number of design points and, hence, number of variables will not be large. As a result, there may not be many effects being studied (includng interactions) so the normal probability plot may not contain a very large "sample size." However, t- or F-tests in which lack of fit is used as the "error" for testing main effect dispersion effects will certainly supply useful information.

In addition to the dispersion effects given in Equation (10.40), another representing a slight variation has been suggested. Rather than summing the log variances, the alternative dispersion effects takes sums first—that is,

$$(\text{Dispersion effect*})_i = \ln\left[\sum_{j=i+} s_j^2\right] - \ln\left[\sum_{j-i} s_j^2\right]$$

$$= \ln\left[\frac{\overline{s_{i+}^2}}{\overline{s_{i-}^2}}\right] \qquad (10.41)$$

This, of course, is closer in concept to the F^* dispersion effect discussed earlier for unreplicated experiments.

The Unreplicated Case

We have already discussed the F^* dispersion effect calculation and illustrated with an extensive case study in injection molding. However, it may be of interest to determine the genesis of this dispersion effect which was developed by Box and Meyer (1986). Consider first an experiment with a two-level unreplicated factorial. Consider the mean model of N observations

$$y_i = \mathbf{x}_i'\boldsymbol{\beta} + \varepsilon_i^*\sigma_i, \qquad i = 1, 2, \ldots, N \qquad (10.42)$$

where the ε_i^* are independent $N(0, 1)$ and

$$\ln \sigma_i^2 = \mathbf{x}_i'\boldsymbol{\gamma} \qquad (10.43)$$

Now, clearly the standard deviation moves as $\mathbf{x}$ moves. The model in Equation (10.43) need not imply that all the variables in the mean model are in the variance model. In fact, the mean model often will contain interactions (or even main effects) not contained in the variance model. We noticed that in our injection molding example.

Consider now the distribution of an "error" in the mean model—that is, the distribution of

$$\varepsilon_i = y_i - \mathbf{x}_i'\boldsymbol{\beta}$$

We know that

$$\varepsilon_i^2 \sim \sigma_i^2\chi_1^2$$

and thus

$$\ln \varepsilon_i^2 = \ln \sigma_i^2 + \ln \chi_1^2 \tag{10.44}$$

Thus, one could view Equation (10.44) as a "model" for the log of the ith squared residual. However, please keep in mind that we are assuming in Equation (10.44) that *the mean is known*—that is, ε_i is not an observed residual $y_i - \hat{y}_i$ but rather $y_i - \mathbf{x}_i'\boldsymbol{\beta}$. As a result, one can view an *approximate* model of the log squared residual as

$$\ln e_i^2 \cong \ln \sigma_i^2 + \ln \chi_1^2$$

with $\ln \chi^2$ representing an approximately normal random variable with mean approximately zero and constant variance. From Equation (10.44) we have

$$\ln e_i^2 = \mathbf{x}_i'\boldsymbol{\gamma} + \tilde{\varepsilon}_i \tag{10.45}$$

(it can be shown that $\operatorname{Var} \ln \chi_\nu^2 \cong 2/\nu$) where $\tilde{\varepsilon}_i$ has approximately mean 0 and variance 2. At this point one should consider least squares estimation in Equation (10.45). The parameter estimates from which dispersion effects can be computed are given by

$$\tilde{\gamma}_0 = \overline{\ln e^2}$$

$$\hat{\gamma}_i = \left[\sum_{j=i+} \ln e_j^2 - \sum_{j=i-} \ln e_j^2 \right] \Big/ N$$

As a result, an obvious dispersion effect will be $2\hat{\gamma}_i$, which is given by

$$(\text{Dispersion effect})_i = \left[\overline{\ln e^2} \right]_{i+} - \left[\overline{\ln e^2} \right]_{i-} \tag{10.46}$$

where $\overline{\ln e^2}$ implies the average of the log of the squared residual. Again *normal probability plotting*, once a mean model of high quality has been chosen, is certainly a reasonable diagnostic procedure. In fact, dispersion effects will be approximately normal with mean zero (assuming a zero dispersion effect) and constant variance. Thus, any dispersion effect off the straight line is diagnosed as having mean not zero and, thus, an important dispersion effect.

From the foregoing it should be evident to the reader why F_i^* is also a reasonable dispersion effect. As in the case with replicated factorials, the F_i^* in Equation (10.34) is much like that in Equation (10.46) except that the $\sum_j$ operation occurs before taking logs. In this regard, another possible dispersion effect is given by

$$F_i = \left[\ln \sum_{j=i+} e_j^2 - \ln \sum_{j=1-} e_j^2 \right] \Big/ N = \ln \left[\frac{\sum_{j=i+} e_j^2}{\sum_{j=i-} e_j^2} \right]$$

Of course this is motivated through the use of the ε_j rather than the e_j. It is reasonable to correct the e_j for their respective means so one is using estimates of σ_{i+}^2 and σ_{i-}^2. Keep in mind that the extence of a nonzero γ_i in the model of Equation (10.43) presumes that the variances at $x_i = +1$ and $x_i = -1$ are distinct. As a result, a useful dispersion effect is given by

$$F_i^* = \ln\left[\frac{s_{i+}^2}{s_{i-}^2}\right]$$

which was introduced without motivation in Equation (10.34). It is important for the user to understand that the practical use of dispersion effects in unreplicated factorials or fractional factorials should include diagnostic plots (e.g., normal probability plots) and *not* formal significance testing.

EXERCISES

10.1 Consider Exercise 3.10 in Chapter 3. It is of interest to maximize the etch rate. As a result, it is a "larger the better" scenario. Suppose that C, gas flow, and D, power applied to the cathode, are difficult to control in the process.

    (a) Treating C and D as noise variables, illustrate, with a drawing, that this design is a crossed array.

    (b) Analyze the data and use the standard "larger the better" SNR. Do pick the winner analysis and thus determine optimum values of A, anode–cathode gap, and B, pressure in the reactor chamber.

10.2 Consider Example 4.2 in Chapter 4. Suppose, in this process for manufacture of integrated circuits, that the temperature is very difficult to control. There is some concern over variability in wafer resistivity due to uncontrolled variability in temperature. Consider the other factors as control variables.

    (a) From the analysis given in Chapter 4, can any of the control variables be used to exert some influence over this variability? Explain.

    (b) It is of interest to maximize wafer resistivity and still minimize variability produced by changes in temperature. Can this be done? Explan.

10.3 Consider Exercise 10.2.

    (a) Give an estimate of the process variance as a function of x_1, implant dose.

    (b) What assumptions were made in constructing this variance function? Discuss the variance–mean tradeoff here. If we use $\hat{\mu} - 2\sigma$ as a criterion, give "optimal" values of implant dose and time.

10.4 Consider Exercise 10.3. Because E, furnace position, and D, oxide thickness, are unimportant, reduce the design to a 2^3 factorial with duplicate runs. Using temperature as a single noise variable, determine the optimal values of implant dose and time. Use a Taguchi analysis with the appropriate SNR ratio for a "larger the better" situation.

10.5 Consider Exercise 3.9. Use the concept of dispersion effects to determine if any factors have impact on process variance. Comment on your results.

10.6 Consider Example 10.1 dealing in solder process optimization.
  (a) Construct and edit a response model involving all control and noise variables.
  (b) Generate an estimated mean model as functions of the control variables.
  (c) Generate an estimated process variance function as a function of the control variables.

10.7 Consider Example 10.2. The design is criticized for not allowing interaction among the control variables to be studied.
  (a) What design would be a good candidate to replace the crossed array in this example? Do not allow your design to admit a run size any greater than the design listed in the example. Use a design that allows quadratic effects in the control factors and allows construction of a response model that can be used to generate the process mean and variance models.
  (b) Explain why your design is better than that used in the example.

10.8 Consider the filtration rate data in Example 3.2. Obviously the filtration rate should be maximized. Suppose, however, that temperature must be treated as a noise variable.
  (a) What factors can be used to control the variability in filtration rate due to random fluctuation in temperature in the range $[-1, +1]$?
  (b) What levels (high or low) of formaldehyde and stir rate result in minimum process variance? (Study interaction plots.)

10.9 Consider Exercise 10.8.
  (a) Again, consider temperature as a noise variable. Use the fitted regression model as a response model to produce a variance model assuming that one design unit in temperature is $2\sigma_z$,

where σ_z is the standard deviation of temperature in the process.

(b) What levels of C and D result in a minimum process variance?

(c) Discuss the mean-variance tradeoff here. Give a quantitative discussion. What are reasonable levels of C and D?

10.10 Consider the design used in Example 10.3. Suppose all six two-factor interaction effects are found to be important. Suppose, in addition, that A and B are control variables and C and D are noise variables. Suppose $\sigma_{z_1} = \sigma_{z_2} = 1.0$.

(a) Write out the form of the mean model.

(b) Write out the form of the variance model.

10.11 Consider an experimental situation with $A(x_1)$, $B(x_2)$, and $C(z)$—that is, two control variables and a single noise variable. Suppose the response model that is fit is given by

$$\hat{y} = b_0 + b_1 x_1 + b_2 x_2 + cz + b_{12} x_1 x_2 + b_{1z} x_1 z + b_{2z} x_2 z + b_{12z} x_1 x_2 z$$

Suppose all terms are significant in the analysis.

(a) Write out the form of the mean model.

(b) Write out the form of the variance model.

(c) What assumptions are being made in the construction of the models in (a) and (b)?

10.12 Suppose a study is conducted with three control variables and two noise variables. A $\frac{1}{2}$ fraction of a 2^5 factorial is used as the experimental design. Only the main effects and the $x_1 z_1$, $x_2 z_1$, and $x_2 z_2$ interactions are found to be important. The interaction plots are as follows:

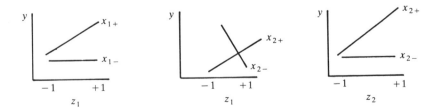

The main effect plots are as follows:

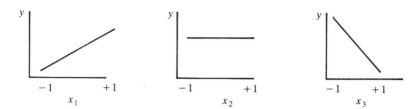

The purpose of the experiment is to determine condition on the control variables that *minimize mean response.*

(a)   Give approximate conditions on x_1, x_2, and x_3 that result in minimum mean response.

(b)   Give approximate conditions on x_1, x_2, and x_3 that give approximate minimum process variance.

(c)   Is there a tradeoff between optimum conditions? Explain.

10.13 Consider Exercise 10.12. Write out the form of the variance model.

10.14 Consider the etch rate data in Example 10.3. Equation (10.10) leads to conditions for minimum process variance as

$$x_1 = \frac{153.0625}{76.8125}$$

The constraint induced by the experimental region forced the recommendation that one should operate at $x_1 = 1.0$.

(a)   Find a confidence region on the location (in x_1) of minimum process variance. Does the region cover $x_1 = 1.0$?

(b)   Comment on your findings in (a).

10.15 Consider Exercise 3.9. Investigate the data set for dispersion effects by fitting a log variance model. Comment.

10.16 Suppose a crossed array of the type

$$2_{III}^{6-3} \times 2_{III}^{3-1}$$

is being proposed. Suppose you wish to use two levels on each variable but you wish to have access to more two-factor interactions among control variables. In addition, you wish to fit a response model involving control and noise variables. Suggest an alternative design.

10.17 Consider a situation with three control variables and three noise variables. In order to construct an appropriate response model, the design should contain linear main effects, two-factor interactions among the control variables, and two-factor interactions between control and noise variables.

(a) What is the minimum number of design points required?

(b) Give a 22-point design that is efficient for fitting this response function.

(c) Give an efficient 25-point design.

10.18 There are situations in which two-factor interactions involving the noise variables are important, though response models involving these interactions are not emphasized in this text. Consider the fitted model

$$\hat{y} = b_0 + b_1 x_1 + b_2 x_2 + b_{12} x_1 x_2 + c_1 z_1 + c_2 z_2 + c_{12} z_1 z_2$$

Assume that σ_{z_1} and σ_{z_2} are both 1.0. In addition, assume that z_1 and z_2 are uncorrelated in the process.

(a) Using the notation in the above fitted model, give an estimate of the mean model.

(b) Give an estimate of the variance model.

(c) Are there any additional assumptions made in (b)?

10.19 Consider Exercise 10.17 again. Suppose it is known that factor A (control factor) is very likely to interact with noise variables D and E. All control variables A, B, and C are likely to be involved in two-factor interactions. Design a combined array that allows for an appropriate robust parameter design. Use 16 observations with a regular fraction of a 2^6. Give appropriate defining relations.

10.20 Consider the data in Example 6.6 with the use of the conversion response. Suppose, in fact, that time and catalyst are control variables and temperature is a noise variable. Use the fitted model in Section 6.5.3 to produce the following:

(a) A process mean model

(b) A process variance model

What assumptions did you make in developing the above models?

10.21 As we indicated in the text, the CCD can be very useful. Suppose there are four control variables and four noise variables. Give a CCD that is a combined array and allows for estimation of a second-order model in the control variables, all control $\times$ noise interactions and noise main effect terms. This design should contain 64 factorial points, 8 axial points, plus center runs. Use $\alpha = 1.0$.

10.22 Use a computer-generated design package to produce a design to accommodate the constraints in Exercise 10.21. Use a 3^8 factorial for candidate points. Use 45 total design points.

10.23 Consider a situation with the three control and two noise variables to accommodate a model that is second order in the control variables and involves linear noise variable terms as well as all control × noise interactions.

(a) Give an appropriate CCD ($\alpha = 1.0$) containing 22 points plus center runs.

(b) Give an appropriate BBD.

10.24 Consider Example 10.1 in Section 10.4.

(a) Fit an appropriate response model.

(b) Use the response model to generate a model for the process mean and a model for the process variance.

(c) Compare your results with those in Section 10.4.

CHAPTER 11

Experiments with Mixtures

11.1 INTRODUCTION

In the response surface examples discussed previously, the levels chosen for any factor in the experimental design are independent of the levels chosen for the other factors. For example, suppose that there are three variables—stirring rate, reaction time, and temperature—that affect the yield of a chemical process. The levels of temperature may be chosen independently of the levels of reaction time and stirring rate, and consequently we may think of the region of experimentation as either a cube or a sphere. The experimental design will consist of an appropriate set of points over this cuboidal or spherical region (such as a factorial design or a central composite design).

A **mixture experiment** is a special type of response surface experiment in which the factors are the ingredients or components of a mixture, and the response is a function of the proportions of each ingredient. These proportionate amounts of each ingredient are typically measured by weight, by volume, by mole ratio, and so forth.

For example, consider chemical etching of semiconductor wafers. The process consists of placing the wafers in a solution composed of several acids —say nitric acid, hydrochloric acid, and phosphoric acid—and observing the etch rate. In such a process, we wish to determine if a combination of these acids is more effective in etching the wafers than any of the three acids by themselves. Because the volume of the etching chamber is fixed, an experiment might consist of testing various combinations of these acids, where all mixes or blends have the same volume. If x_1, x_2, and x_3 represent the proportions by volume of the three acids, then some of the blends that might be of interest are as follows:

Blend 1:	$x_1 = 0.3,$	$x_2 = 0.3,$	$x_3 = 0.4$
Blend 2:	$x_1 = 0.2,$	$x_2 = 0.5,$	$x_3 = 0.3$
Blend 3:	$x_1 = 0.5,$	$x_2 = 0.5,$	$x_3 = 0.0$
Blend 4:	$x_1 = 1.0,$	$x_2 = 0.0,$	$x_3 = 0.0$

Blends 1 and 2 are examples of **complete** mixtures; that is, they are made up of all three of the acids. Blend 3 is a **binary** blend, consisting of two of the three acids, and blend 4 is a mixture that is made up of 100% by volume of only one of the three acids. Each of these blends is called a *mixture*. Notice that in this example $x_1 + x_2 + x_3 = 1$; and because of this constraint, the levels of the factors cannot be chosen independently. For instance, in blend 1 above as soon as we indicate that acid 1 makes up 30% of the mixture by volume ($x_1 = 0.3$) and acid 2 makes up 30% of the mixture by volume ($x_2 = 0.3$), it is immediately obvious that acid 3 must make up 40% of the mixture by volume (that is, $x_3 = 0.4$ because $x_1 = x_2 = 0.3$, and $x_1 + x_2 + x_3 = 1.0$).

In general, suppose that the mixture consists of q ingredients or components, and let x_i represent the proportion of the ith ingredient in the mixture. Then in light of the above discussion we must require that

$$x_i \geq 0, \qquad i = 1, 2, \ldots, q \tag{11.1}$$

and

$$\sum_{i=1}^{q} x_i = x_1 + x_2 + \cdots + x_q = 1 \tag{11.2}$$

The constraint in Equation (11.2) makes the levels of the factors x_i nonindependent, and this makes mixture experiments different from the usual response surface experiments we have discussed previously.

The constraints in Equations (11.1) and (11.2) are shown graphically in Figure 11.1 for $q = 2$ and $q = 3$ components. For two components, the feasible factor space for the mixture experiment includes all values of the two components for which $x_1 + x_2 = 1$, which is the line segment shown in Figure 11.1(*a*). With three components, the feasible space for the mixture experiment is the triangle in Figure 11.1(*b*) that has vertices corresponding to **pure blends**—that is, mixtures that are made up of 100% of a single ingredient and edges that are **binary blends**. In general, the experimental region for a mixture problem with q components is a **simplex**, which is a regularly sided figure with q vertices in $(q - 1)$ dimensions.

The coordinate system for mixture proportions is a **simplex coordinate system**. For example, with $q = 3$ components, the experimental region is shown in Figure 11.2. Each of the three vertices in the equilateral triangle corresponds to a pure blend, and each of the three sides of the triangle represents a mixture that has none of one of the three components (the component labeled on the opposite vertex). The nine grid lines in each direction mark off 10% increments in the respective components. Interior points in the triangle represent mixtures in which all three ingredients are present at nonzero proportionate amounts. The centroid of the triangle

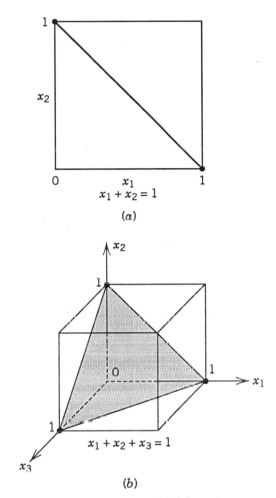

Figure 11.1 Constrained factor space for mixtures with (a) $q = 2$ components and (b) $q = 3$ components.

corresponds to the mixture with equal proportions $x_1 = \frac{1}{3}$, $x_2 = \frac{1}{3}$, $x_3 = \frac{1}{3}$ of all ingredients.

Applications of mixture experiments are found in many areas. A common application is **product formulation**, in which a product is formed by mixing several ingredients together. Examples include (a) formulation of gasoline by combining several refinery products and other types of chemicals such as anti-icers, (b) formulation of soaps, shampoos, and detergents in the consumer products industry, and (c) formulation of products such as cake mixes and beverages in the foods industry. In addition to product formulation, mixture experiments arise frequently in process engineering and development where some property of a final product depends on a mixture of several

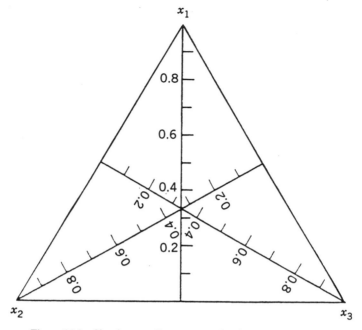

Figure 11.2   Simplex coordinate system for three components.

ingredients interacting with the product at a particular process stage. An example is the wafer etching process described above. Other examples include gas or plasma etching in the semiconductor industry, many joining processes such as soldering, and chemical milling or machining.

In this chapter we discuss experimental design, data analysis, and model-building techniques for the original mixture experiment defined by Equations (11.1) and (11.2). In this problem, the entire simplex region is investigated and the only design variables are the mixture components $x_1, x_2, \ldots, x_q$. There are many variations of the original mixture problem. One of these is the addition of upper and lower bounds on some of the component proportions. These upper or lower bounds occur because the problem may require that at least a certain minimum proportion of an ingredient be present in the blend, or that an ingredient cannot exceed a specified maximum proportion. In some mixture problems there are **process variables** $z_1, z_2, \ldots, z_p$ in addition to the mixture ingredients. For example, in the wafer etching example above, the mixture ingredients x_1, x_2, and x_3 are the proportions of the three acids, and two process variables could be the temperature of the etching solution (z_1) and the agitation rate (z_2) of the wafers. Including process variables in mixture experiments usually involves constructing designs that utilize different levels of the process variables via a factorial design in combination with the mixture variables. Finally, in some mixture problems the experimenter may be interested in not only the formulation of the mixture but the amount of the mixture that is applied to the experimental units. For example, the opacity of a coating is a function of the ingredients from which the coating is formulated, but it also depends upon the thickness of the coating surface when it is applied. Some of these variations of the original mixture experiment will be discussed in Chapter 12.

11.2 SIMPLEX DESIGNS AND CANONICAL MIXTURE POLYNOMIALS

The primary differences between a standard response surface experiment and a mixture experiment are that (1) a special type of design must be used and (2) the form of the mixture polynomial is slightly different from the standard polynomials used in response surface methodology (RSM). In this section we introduce designs that allow an appropriate response surface model to be fit over the entire mixture space. Because the mixture space is a simplex, all design points must be at the vertices, on the edges or faces, or in the interior of a simplex. Much of the work in this area was originated by Scheffé (1958, 1959, 1965). Cornell (1990) is an excellent and very complete reference on the subject.

11.2.1 Simplex Lattice Designs

A simplex lattice is just a uniformly spaced set of points on a simplex. A $\{q, m\}$ simplex lattice design for q components consists of points defined by the following coordinate settings: The proportions taken on by each component are the $m + 1$ equally spaced values from 0 to 1,

$$x_i = 0, \frac{1}{m}, \frac{2}{m}, \dots, 1, \qquad i = 1, 2, \dots, q \tag{11.3}$$

and all possible combinations (mixtures) of the proportions from this equation are used. As an example, let $q = 3$ and $m = 2$; consequently,

$$x_i = 0, \tfrac{1}{2}, 1, \qquad i = 1, 2, 3$$

and the simplex lattice consists of the following six points:

$$(x_1, x_2, x_3) = (1,0,0), (0,1,0), (0,0,1), (\tfrac{1}{2}, \tfrac{1}{2}, 0), (\tfrac{1}{2}, 0, \tfrac{1}{2}), (0, \tfrac{1}{2}, \tfrac{1}{2})$$

This is a $\{3, 2\}$ simplex lattice design, and it is shown in Figure 11.3(a). The three vertices $(1,0,0)$, $(0,1,0)$, and $(0,0,1)$ are pure blends, and the points $(\tfrac{1}{2}, \tfrac{1}{2}, 0)$, $(\tfrac{1}{2}, 0, \tfrac{1}{2})$, and $(0, \tfrac{1}{2}, \tfrac{1}{2})$ are binary blends or two-component mixtures located at the midpoints of the three edges of the triangle. These binary

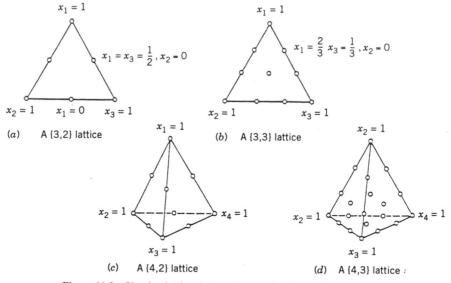

Figure 11.3 Simplex lattice designs for $q = 3$ and $q = 4$ components.

blends are made up of equal parts of two of the three ingredients. Figure 11.3 also shows the $\{3, 3\}$, the $\{4, 2\}$, and the $\{4, 3\}$ simplex lattice designs. As we will subsequently see, the notation "$\{q, m\}$" implies a simplex lattice design in q components that will support a mixture polynomial of degree m. In general, the number of points in a $\{q, m\}$ simplex lattice design is

$$N = \frac{(q + m - 1)!}{m!\,(q - 1)!} \tag{11.4}$$

Now consider the form of the polynomial model that we might fit to data from a mixture experiment. A first-order model is

$$E(y) = \beta_0 + \sum_{i=1}^{q} \beta_i x_i \tag{11.5}$$

However, because of the restriction $x_1 + x_2 + \cdots + x_q = 1$ we know that the parameters $\beta_0, \beta_1, \ldots, \beta_q$ are not unique. We could, of course, make the substitution

$$x_q = 1 - \sum_{i=1}^{q-1} x_i$$

in Equation (11.5), removing the dependency among the x_i terms and producing unique estimates of the parameters $\beta_0, \beta_1, \ldots, \beta_{q-1}$. Although this will work mathematically, it is not the preferred approach, because it obscures the effect of the qth component because the term $\beta_q x_q$ is not included in the equation. Thus, in a sense, we are sacrificing information on component q.

An alternative approach is to multiply some of the terms in the original response surface polynomial by the identity $x_1 + x_2 + \cdots + x_q = 1$ and then simplify. For example, consider Equation (11.5) and multiply the β_0 term by $x_1 + x_2 + \cdots + x_q = 1$, yielding

$$E(y) = \beta_0(x_1 + x_2 + \cdots + x_q) + \sum_{i=1}^{q} \beta_i x_i$$

$$= \sum_{i=1}^{q} \beta_i^* x_i \qquad (11.6)$$

where $\beta_i^* = \beta_0 + \beta_i$. This is called the **canonical form** of the first-order mixture model. In general, the canonical forms of the mixture models (with the asterisks removed from the parameters) are as follows:

Linear:

$$E(y) = \sum_{i=1}^{q} \beta_i x_i \qquad (11.7)$$

Quadratic:

$$E(y) = \sum_{i=1}^{q} \beta_i x_i + \sum \sum_{i<j} \beta_{ij} x_i x_j \qquad (11.8)$$

Full Cubic:

$$E(y) = \sum_{i=1}^{q} \beta_i x_i + \sum \sum_{i<j} \beta_{ij} x_i x_j + \sum \sum_{i<j} \delta_{ij} x_i x_j (x_i - x_j)$$

$$+ \sum \sum \sum_{i<j<k}^{q} \beta_{ijk} x_i x_j x_k \qquad (11.9)$$

Special Cubic:

$$E(y) = \sum_{i=1}^{q} \beta_i x_i + \sum \sum_{i<j}^{q} \beta_{ij} x_i x_j + \sum \sum \sum_{i<j<k}^{q} \beta_{ijk} x_i x_j x_k \qquad (11.10)$$

The terms in the canonical mixture polynomials have simple interpretations. Geometrically, in Equations (11.7) through (11.10), the parameter β_i represents the expected response to the pure mixture $x_i = 1$, $x_j = 0$, $j \neq i$,

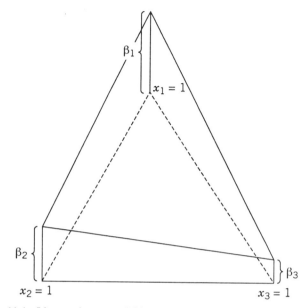

Figure 11.4 Linear mixture model in three components with $\beta_1 > \beta_2 > \beta_3$.

and is the height of the mixture surface at the vertex $x_i = 1$. The portion of each polynomial given by

$$\sum_{i=1}^{q} \beta_i x_i$$

is called the **linear blending** portion. When blending is strictly additive, then Equation (11.7) is an appropriate model. Figure 11.4 presents the case where $q = 3$ and $\beta_1 > \beta_2 > \beta_3$.

Quadratic blending is illustrated in Figure 11.5(a). Notice that the quadratic term $\beta_{12}x_1x_2$ represents the excess response from the quadratic model $E(y) = \beta_1 x_1 + \beta_2 x_2 + \beta_{12}x_1x_2$ over the linear model. This is often called the *synergism (or antagonism) due to nonlinear blending*. For example, if large positive values of y are desired and if β_{12} is positive, synergistic blending is occurring, whereas if β_{12} is negative, antagonistic blending is occurring. In the cubic model, terms such as $\delta_{12}x_1x_2(x_1 - x_2)$ enable one to model both synergistic and antagonistic blending along the x_1–x_2 edge; see Figure 11.5(b). A cubic term such as $\beta_{123}x_1x_2x_3$ accounts for ternary blending among the three components in the interior of the simplex.

It is also helpful to consider the way in which the individual terms in a mixture model contribute to the shape of the response surface. A linear term such as $\beta_1 x_1$ only contributes to the model when $x_1 > 0$; and the maximum contribution occurs at $x_1 = 1$, in which case the maximum effect contributed

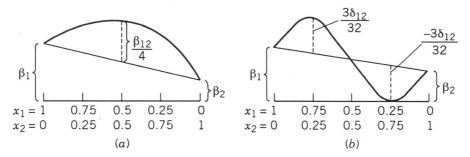

Figure 11.5 Nonlinear blending. (*a*) Quadratic blending with $\beta_{ij} > 0$. (*b*) Cubic blending with $\delta_{12} > 0$.

by x_1 is β_1. The quadratic term $\beta_{12}x_1x_2$ contributes to the model at every point in the simplex where $x_1 > 0$ and $x_2 > 0$. The maximum contribution occurs at the edge joining the vertices x_1 and x_2 and is at the point $x_1 = x_2 = \frac{1}{2}$. The maximum contribution to the model from this term is $\beta_{12}/4$. A cubic term such as $\beta_{123}x_1x_2x_3$ contributes to the model at every point for which $x_1 > 0$, $x_2 > 0$, and $x_3 > 0$ (in the interior of the simplex), and the maximum contribution is of magnitude $\beta_{123}/27$ at the point $x_1 = x_2 = x_3 = \frac{1}{3}$.

Example 11.1 A Three-Component Mixture   Cornell (1990) describes a mixture experiment in which three components—polyethylene (x_1), polystyrene (x_2), and polypropylene (x_3)—were blended to form fiber that will be spun into yarn for draperies. The product developers were interested in only the pure and binary blends of these three materials. The response variable of interest is yarn elongation in kilograms of force applied. A $\{3, 2\}$ simplex lattice design is used to study the product. The design and the observed responses are shown in Table 11.1. Notice that replicate observations are run, with two replicates at each of the pure blends and three

Table 11.1   The $\{3, 2\}$ Simplex Lattice Design for the Yarn Elongation Problem

Design Point	Component Proportions			Observed Elongation Values	Average Elongation Value (y)
	x_1	x_2	x_3		
1	1	0	0	11.0, 12.4	11.7
2	$\frac{1}{2}$	$\frac{1}{2}$	0	15.0, 14.8, 16.1	15.3
3	0	1	0	8.8, 10.0	9.4
4	0	$\frac{1}{2}$	$\frac{1}{2}$	10.0, 9.7, 11.8	10.5
5	0	0	1	16.8, 16.0	16.4
6	$\frac{1}{2}$	0	$\frac{1}{2}$	17.7, 16.4, 16.6	16.9

replicates at each of the binary blends. The error standard deviation can be estimated from these replicate observations as $\hat{\sigma} = 0.85$.

Table 11.2 shows output from the *Design-Expert* computer program, which was used to analyze the yarn elongation data. This program fits both the linear and quadratic mixture model to the data, and the upper panel of Table 11.2 shows the sequential F-tests for the linear blending terms and the quadratic terms. The F-test for the linear terms is significant, and the F-test for the contribution of the quadratic terms over the linear terms is also significant, indicating that a quadratic mixture model should be used. The middle panel of Table 11.2 shows a lack-of-fit test for the linear mixture model. The value of $F = 32.32$ is very large, indicating that a linear mixture model is inadequate. Finally, the bottom panel of Table 11.2 presents several informative summary statistics for the linear and quadratic models. These statistics indicate that the quadratic model is superior to the linear model.

The F-test for the linear terms in the upper panel of Table 11.2 warrants some discussion. This is a test of the hypothesis that there is no linear blending occurring in the mixture. If there is no linear blending, then the

Table 11.2 Design-Expert Output for the Yarn Elongation Data

Sequential Model Sum of Squares					
Source	Sum of Squares	Degrees of Freedom	Mean Square	F-Value	Probability $> F$
Mean	2750.0	1	2750.0		
Linear	57.6	2	28.8	4.477	0.0353
Quadratic	70.7	3	23.6	32.32	0.0001
Residual	6.6	9	0.7		
Total	2884.9	15			

Lack-of-Fit Tests					
Model	Sum of Squares	Degrees of Freedom	Mean Square	F-Value	Probability $> F$
Linear	70.7	3	23.6	32.32	0.0001
Quadratic	0.0	0			
Pure Error	6.6	9	0.7		

ANOVA Summary Statistics of Model Fit					
Source	Unaliased Terms	Residual Degrees of Freedom	Root Mean Squared Error	R-Squared	Adjusted R-Squared	Press
Linear	3	12	2.54	0.4273	0.3319	120.81
Quadratic	6	9	0.85	0.9514	0.9243	18.30

response surface is a level plane above the simplex region; that is, $\beta_1 = \beta_2 = \beta_3 = \beta$. Expressed formally, the hypotheses to be tested are

$$H_0: \qquad \beta_1 = \beta_2 = \beta_3 = \beta$$
$$H_1: \qquad \text{At least one equality is false}$$

Because from Table 11.2 the test statistic is $F_{\text{linear}} = 4.477$ with a P-value of

Table 11.3 Quadratic Mixture Model for the Yarn Elongation Data

ANOVA for Quadratic Model

Source	Sum of Squares	Degrees of Freedom	Mean Square	F-Value	Probability $> F$
Model	128.3	5	25.66	35.20	0.0001
Residual	6.6	9	0.73		
Corrected total	134.9	14			
Root mean squared error	0.85		R-squared	0.9514	
Dependent mean	13.54		Adjusted R-squared	0.9243	

Independent Variable	Coefficient Estimate	Degrees of Freedom	Standard Error	t for H_0 Coefficient = 0	Probability $> \lvert t \rvert$
x_1	11.70	1	0.60	Not applicable	
x_2	9.40	1	0.60	Not applicable	
x_3	16.40	1	0.60	Not applicable	
$x_1 x_2$	19.00	1	2.61	7.285	0.0001
$x_1 x_3$	11.40	1	2.61	4.371	0.0018
$x_2 x_3$	-9.60	1	2.61	-3.681	0.0051

Observation	Actual Value	Predicted Value	Residual	h_{ii}	Studentized Residual	Cook's Distance	R-Student
1	11.00	11.70	-0.70	0.500	-1.160	0.224	-1.185
2	12.40	11.70	0.70	0.500	1.160	0.224	1.185
3	15.00	15.30	-0.30	0.333	-0.430	0.015	-0.410
4	14.80	15.30	-0.50	0.333	-0.717	0.043	-0.696
5	16.10	15.30	0.80	0.333	1.148	0.110	1.171
6	17.70	16.90	0.80	0.333	1.148	0.110	1.171
7	16.60	16.90	-0.30	0.333	-0.430	0.015	-0.410
8	16.40	16.90	-0.50	0.333	-0.717	0.043	-0.696
9	8.80	9.40	-0.60	0.500	-0.994	0.165	-0.993
10	10.00	9.40	0.60	0.500	0.994	0.165	0.993
11	10.00	10.50	-0.50	0.333	-0.717	0.043	-0.696
12	9.70	10.50	-0.80	0.333	-1.148	0.110	-1.171
13	11.80	10.50	1.30	0.333	1.865	0.290	2.245
14	16.80	16.40	0.40	0.500	0.663	0.073	0.641
15	16.00	16.40	-0.40	0.500	-0.663	0.073	-0.641

$P = 0.0353$, we would reject H_0 and conclude that there is linear blending in the system.

Table 11.3 summarizes the quadratic model fit. The analysis of variance for the quadratic model, shown in the upper panel of Table 11.3, tests the hypothesis that the response surface is a level plane above the simplex region. Formally, this hypothesis is

$$H_0: \quad \beta_1 = \beta_2 = \beta_3 = \beta, \quad \beta_{12} = \beta_{13} = \beta_{23} = 0$$

$$H_1: \quad \text{At least one equality is false}$$

Because the test statistic is $F = 35.2$ (with a P-value of $P = 0.0001$), this hypothesis is rejected. Notice that t-tests are provided for the quadratic terms on the model because tests of the hypotheses regarding each of these individual terms (i.e., $H_0: \beta_{ij} = 0$) are meaningful. However, because similar tests on the linear blending terms are not meaningful ($H_0: \beta_i = 0$ does not test for linear blending from component i), these t-tests are not displayed.

The fitted quadratic mixture model is

$$\hat{y} = 11.7x_1 + 9.4x_2 + 16.4x_3 + 19.0x_1x_2 + 11.4x_1x_3 - 9.6x_2x_3$$

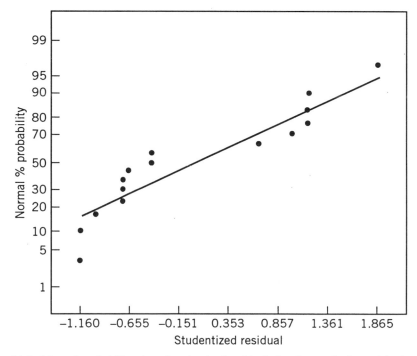

Figure 11.6 Normal probability plot of studentized residuals for the quadratic model of yarn elongation.

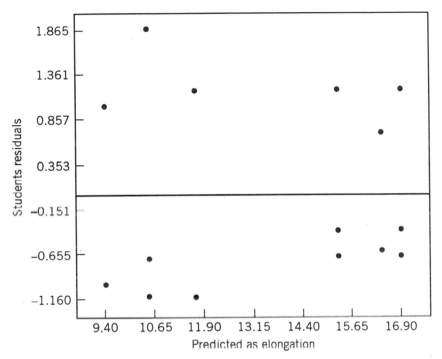

Figure 11.7 Plot of studentized residuals versus predicted yarn elongation.

Because $b_3 > b_1 > b_2$, we would conclude that component 3 (polypropylene) produces yarn with the highest elongation. Furthermore, because b_{12} and b_{13} are positive, blending components 1 and 2 or components 1 and 3 produce higher elongation values than would be expected just by averaging the elongations of the pure blends. This is an example of "synergistic" blending effects. Components 2 and 3 have antagonistic blending effects because b_{23} is negative.

Figures 11.6 and 11.7 present a normal probability plot of the studentized residuals from the quadratic model. These plots are satisfactory. In general, we recommend analyzing studentized residuals from mixture experiments (in contrast to the ordinary least squares residuals) because the points in mixture designs can have substantial differences in their leverage values. Studentized residuals account for leverage through the term $(1 - h_{ii})$ that appears in the denominator [see Equation (2.50) in Chapter 2].

Figure 11.8 presents a contour plot of elongation and a three-dimensional response surface plot. These plots are helpful in interpreting the results. From examining the figure, we note that if maximum elongation is desired, a blend of components 1 and 3 should be chosen consisting of about 80% component 3 and 20% component 1.

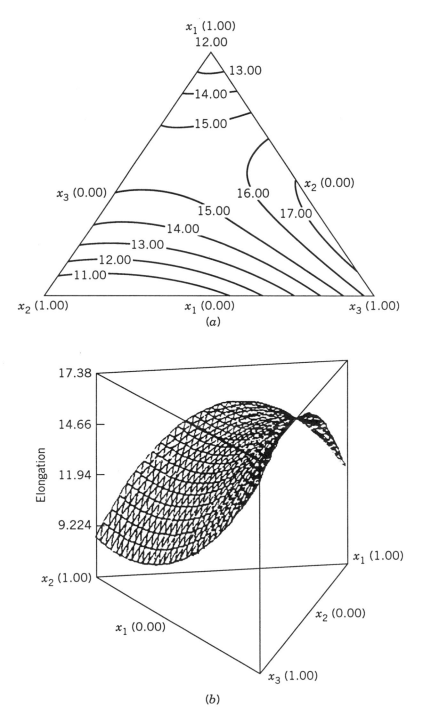

Figure 11.8 (*a*) Contour plot of predicted yarn elongation (*b*) Three-dimensional surface plot.

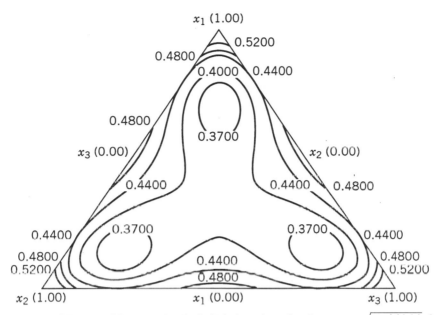

Figure 11.9 Contours of constant standard deviation of predicted response $\sqrt{Var\left[\hat{y}(\mathbf{x})\right]}$ for the quadratic mixture model for the yarn elongation data.

Figure 11.9 presents the contours of constant standard deviation of predicted yarn longation ($\sqrt{Var\left[\hat{y}(\mathbf{x})\right]}$) over the complete mixture space. Notice that the quality of prediction is better, in a variance sense, at points in the interior of the simplex than on the edges or at the vertices. However, one should be very careful about making predictions at points in the interior, because no experimental trials were performed there. The $\{3, 2\}$ simplex lattice is a **boundary point design**; that is, all of the design points are either binary mixtures or pure mixtures. In this application the engineers were not interested in the ternary blends, so the lack of interior points is not a major concern. When one is interested in prediction in the interior, it is highly desirable to augment simplex-type designs with interior design points. We will return to this again in Section 11.2.3.

11.2.2 The Simplex-Centroid Design and Its Associated Polynomial

A q-component simplex-centroid design consists of $2^q - 1$ distinct design points. These design points are the q permutations of $(1, 0, 0, \ldots, 0)$ or single-component blends, the $\binom{q}{2}$ permutations of $(\frac{1}{2}, \frac{1}{2}, 0, \ldots, 0)$ or all binary mixtures, the $\binom{q}{3}$ permutations of $(\frac{1}{3}, \frac{1}{3}, \frac{1}{3}, 0, \ldots, 0)$, and so forth, and the overall centroid $(1/q, 1/q, \ldots, 1/q)$. Figure 11.10 shows the simplex-centroid designs for $q = 3$ and $q = 4$ components. Note that the design points are

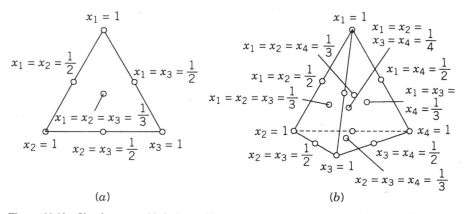

Figure 11.10 Simplex-centroid designs with three and four components. (*a*) $q = 3$. (*b*) $q = 4$.

located at the centroid of the $(q - 1)$-dimensional simplex and at the centroids of all the lower-dimensional simplices contained within the $(q - 1)$-dimensional simplex.

The design points in the simplex-centroid design will support the polynomial

$$E(y) = \sum_{i=1}^{q} \beta_i x_i + \sum_{i<}^{q} \sum_{j} \beta_{ij} x_i x_j + \sum_{i<}^{q} \sum_{j<} \sum_{k} \beta_{ijk} x_i x_j x_k$$

$$+ \cdots + \beta_{12\ldots q} x_1 x_2 \ldots x_q \qquad (11.11)$$

which is a qth-order mixture polynomial. For $q = 3$ components, this model is

$$E(y) = \beta_1 x_1 + \beta_2 x_2 + \beta_3 x_3 + \beta_{12} x_1 x_2 + \beta_{13} x_1 x_3$$

$$+ \beta_{23} x_2 x_3 + \beta_{123} x_1 x_2 x_3$$

which is the special cubic polynomial from Equation (11.10). For $q = 4$ components, the model is

$$E(y) = \sum_{i=1}^{4} \beta_i x_i + \sum_{i<}^{4} \sum_{j} \beta_{ij} x_i x_j + \sum_{i<}^{4} \sum_{j<} \sum_{k} \beta_{ijk} x_i x_j x_k$$

$$+ \beta_{1234} x_1 x_2 x_3 x_4$$

or the special cubic model with an additional quartic term. Because they are relatively efficient designs for fitting the special cubic model, simplex-centroid designs are often used when the experimenter thinks that some cubic terms may be necessary in the final model.

11.2.3 Augmentation of Simplex Designs with Axial Runs

The standard simplex-lattice and simplex-centroid designs are **boundary point designs**; that is, with the exception of the overall centroid, all the design points are on the boundaries of the simplex. For example, consider the {3, 3} lattice design shown in Figure 11.3(*b*). This design has 10 points: the three vertices, the six points that are the thirds of the edges, and the overall centroid. Thus there are three points that tell us something about the pure blends, six points that provide information about binary blends, and only one point that tells us anything about complete mixtures. We could say that the distribution of information about these different types of mixtures is 3:6:1.

If one is interested in making predictions about the properties of complete mixtures, it would be highly desirable to have more runs in the interior of the simplex. We recommend augmenting the usual simplex designs with **axial runs** and the overall centroid (if the centroid is not already a design point).

We define the **axis of component i** as the line or ray extending from the base point $x_i = 0$, $x_j = 1/(q - 1)$ for all $j \neq i$ to the opposite vertex where $x_i = 1$, $x_j = 0$ for all $j \neq i$. Figure 11.11 shows the axes for components 1, 2, and 3 in a three-component system. Notice that the base point will always lie at the centroid of the $(q - 2)$-dimensional boundary of the simplex that is opposite the vertex $x_i = 1$, $x_j = 0$ for all $j \neq i$. [The boundary is sometimes called a $(q - 2)$-flat.] The length of the component axis is one unit. **Axial points** are positioned along the component axes a distance Δ from the

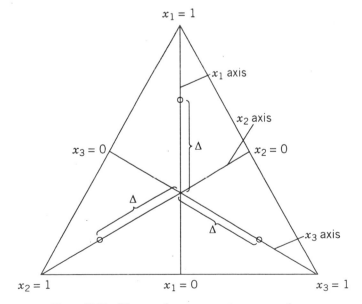

Figure 11.11 The axes for components x_1, x_2, and x_3.

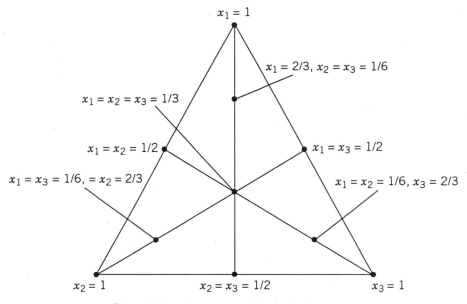

Figure 11.12   An augmented simplex-lattice design.

centroid. The maximum value for Δ is $(q - 1)/q$. We recommend that axial runs be placed midway between the centroid of the simplex and each vertex so that $\Delta = (q - 1)/2q$. Sometimes these points are called **axial check blends**, because a fairly common practice is to exclude them when fitting the preliminary mixture model, and then use the responses at these axial points to check the adequacy of the fit of the preliminary model.

Figure 11.12 shows the $\{3, 2\}$ simplex-lattice design augmented with the axial points. This design has 10 points, with four of these points in the interior of the simplex. Thus the distribution of information in this augmented simplex lattice is $3 : 3 : 4$. Contrast this to the 10-point $\{3, 3\}$ simplex lattice, for which the distribution of information is $3 : 6 : 1$. Cornell (1986) presents an extensive comparison of these two designs. He notes that the $\{3, 3\}$ simplex lattice will support fitting the full cubic model, while the augmented simplex lattice will not; however, the augmented simplex lattice will allow the experimenter to fit the special cubic model or to add special quadratic terms such as $\beta_{1233} x_1 x_2 x_3^2$ to the quadratic model. The augmented simplex lattice is superior for studying the response of complete mixtures in the sense that it can detect and model curvature in the interior of the triangle that cannot be accounted for by the terms in the full cubic model. The augmented simplex lattice has more power for detecting lack of fit than does the $\{3, 3\}$ lattice. This is particularly useful when the experimenter is unsure about the proper model to use and also plans to sequentially build a model by starting with a simple polynomial (perhaps first order), test the model for

Table 11.4   Augmented Simplex-Lattice Design for the Etch Rate Experiment

Design Point	x_1 (Acid A)	x_2 (Acid B)	x_3 (Acid C)	Etch Rate $y(\text{Å}/\text{m})$
1	1	0	0	540, 560
2	0	1	0	330, 350
3	0	0	1	295, 260
4	$\frac{1}{2}$	$\frac{1}{2}$	0	610
5	0	$\frac{1}{2}$	$\frac{1}{2}$	330
6	$\frac{1}{2}$	0	$\frac{1}{2}$	425
7	$\frac{2}{3}$	$\frac{1}{6}$	$\frac{1}{6}$	710
8	$\frac{1}{6}$	$\frac{2}{3}$	$\frac{1}{6}$	640
9	$\frac{1}{6}$	$\frac{1}{6}$	$\frac{2}{3}$	460
10	$\frac{1}{3}$	$\frac{1}{3}$	$\frac{1}{3}$	800, 850

Table 11.5   Sequential Model Fitting for the Etch Rate Data

Sequential Model Sum of Squares

Source	Sum of Squares	Degrees of Freedom	Mean Square	F-Value	Probability $>F$
Mean	3,661,828.6	1	3,661,828.6		
Linear	133,755.0	2	66,877.5	2.13	0.165
Quadratic	229,364.9	3	76,455.0	5.29	0.027
Special cubic	107,877.8	1	107,877.8	96.52	20.001
Full cubic	4,240.9	2	2,120.4	2.96	0.142
Residual	3,582.9	5	716.5		
Total	4,127,850.0	14			

Lack of Fit Tests

Model	Sum of Squares	Degrees of Freedom	Mean Square	F-Value	Probability $>F$
Linear	342,803.9	7	48,972.0	86.58	0.001
Quadratic	113,439.1	4	28,359.8	50.14	0.001
Special cubic	5,561.3	3	1,853.8	3.28	0.141
Full cubic	1,320.4	1	1,320.4	2.33	0.201
Pure Error	2,262.5	4	565.6		

ANOVA Summary Statistics of Models Fit

Source	Unaliased Terms	Residual Degrees of Freedom	Root Mean Squared Error	R-Squared	Adjusted R-Squared	PRESS
Linear	3	11	177.1	0.2793	0.1483	516,539.9
Quadratic	6	8	120.3	0.7584	0.6073	812,042.5
Special cubic	7	7	33.4	0.9837	0.9697	80,312.0
Full cubic	9	5	26.8	0.9925	0.9805	186,374.0

lack of fit, then augment the model with higher-order terms, test the new model for lack of fit, and so forth.

Example 11.2 Wet chemical etching is often performed on the backs of silicon wafers prior to metalization in the semiconductor industry. The

Table 11.6 Special Cubic Model for the Etch Rate Experiment

Source	Sum of Squares	Degrees of Freedom	Mean Square	F Value	Probability $> F$
Model	470997.6	6	78499.6	70.23	< 0.001
Residual	7823.8	7	1117.7		
Lack of fit	5561.3	3	1853.8	3.28	0.141
Pure error	2262.5	4	565.6		
Corrected total	478821.4	13			

Root mean squared error	33.4		R-Squared	0.9837	PRESS = 80,312.0
Dependent Mean	511.4		Adj R-Squared	0.9697	
Coefficient of variation.	6.54%		Pred R-Squared	0.8323	

Component	Coefficient Estimate	Degrees of Freedom	Standard Error	t for H_0 Coefficient $= 0$	Probability $> \|t\|$	VIF
x_1	550.2	1	23.2	Not applicable		1.55
x_2	344.7	1	23.2	Not applicable		1.55
x_3	268.3	1	23.2	Not applicable		1.55
$x_1 x_2$	689.5	1	146.5	4.71	0.002	2.16
$x_1 x_3$	−9.0	1	146.5	−0.06	0.953	2.16
$x_2 x_3$	58.1	1	146.5	0.40	0.703	2.16
$x_1 x_2 x_3$	9243.3	1	940.9	9.82	< 0.001	2.99

Observation	Actual Value	Predicted Value	Residual	h_{ii}	Student Residual	Cook's Distance	R-student
1	540.0	550.2	−10.2	0.483	−0.42	0.024	−0.40
2	560.0	550.2	9.8	0.483	0.41	0.022	0.38
3	330.0	344.7	−14.7	0.483	−0.61	0.050	−0.58
4	350.0	344.7	5.3	0.483	0.22	0.006	0.20
5	295.0	268.3	26.7	0.483	1.11	0.164	1.13
6	260.0	268.3	−8.3	0.483	−0.34	0.016	−0.32
7	610.0	619.8	−9.8	0.912	−0.99	1.453	−0.99
8	425.0	407.0	18.0	0.912	1.81	4.862	2.31
9	330.0	321.0	9.0	0.912	0.90	1.204	0.89
10	800.0	812.2	−12.2	0.375	−0.46	0.018	−0.43
11	850.0	812.2	37.8	0.375	1.43	0.176	1.58
12	710.0	717.4	−7.4	0.206	−0.25	0.002	−0.23
13	640.0	620.2	19.8	0.206	0.66	0.016	0.64
14	460.0	523.8	−63.8	0.206	−2.14	0.170	−3.38

etching solution is a mixture of three different acids: A, B, and C. The experimenters wish to study how the composition of this mixture affects the etch rate. The experimenters assumed that a quadratic mixture model would be appropriate, and they selected the augmented simplex-lattice design shown in Figure 11.12 for use in this study. They decided to replicate the vertices and the overall centroid so that an internal estimate of error could be obtained. Thus the complete experimental design, shown in Table 11.4 along with the response, consists of 14 runs.

Table 11.5 presents the results of fitting linear, quadratic, special cubic, and full cubic models to the data obtained in this experiment (the computations were performed using *Design-Expert*). The sequential F-tests in this table indicate that the contribution of the quadratic terms to the model (over the linear terms) is significant, as is the contribution of the special cubic term $\beta_{123}x_1x_2x_3$ (over the linear and quadratic terms). Furthermore, the quadratic model displays significant lack of fit, whereas the special cubic model does not; and, in addition, the special cubic model has a smaller error mean square, a larger adjusted R^2, and a smaller value of the PRESS statistic than does the quadratic model. Terms in the cubic model are aliased and should be ignored. On the basis of this analysis, the experimenters selected the special cubic model.

portion of this table is the overall analysis of variance for this model. The

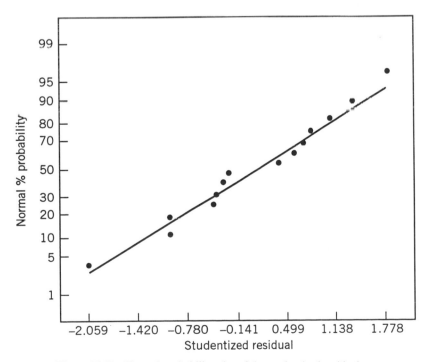

Figure 11.13 Normal probability plot of the studentized residuals.

F-test for the model is used to test the hypotheses

$$H_0: \quad \beta_1 = \beta_2 = \beta_3 = \beta, \quad \beta_{12} = \beta_{13} = \beta_{23} = \beta_{123} = 0$$

$$H_1: \quad \text{At least one equality is false}$$

Because $F_0 = 70.23$ $(P = < 0.001)$, the null hypothesis is rejected. The t-statistics on the individual model terms indicate that two of the quadratic terms, $x_1 x_3$ and $x_2 x_3$, could be deleted; however, the experimenters decided to retain them in order to keep the model hierarchial. The F-statistic for lack of fit is testing for contributions of terms beyond the special cubic. This design could support three special quartic terms. The lack-of-fit test indicates that these terms are unnecessary. The diagnostic information at the bottom of Table 11.6 is generally satisfactory. The centers-of-edges points in this design have relatively large leverage values (0.912), and this is responsible in part for the two large values of Cook's distance (1.453, 1.204, and 4.862) reported in the table.

The fitted special cubic model is

$$\hat{y} = 550.2x_1 + 344.7x_2 + 268.3x_3 + 689.5x_1x_2 - 9.0x_1x_3$$
$$+ 58.1x_2x_3 + 9243.3x_1x_2x_3$$

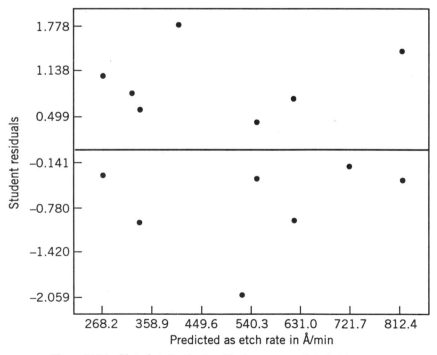

Figure 11.14 Plot of studentized residuals versus predicted etch rate.

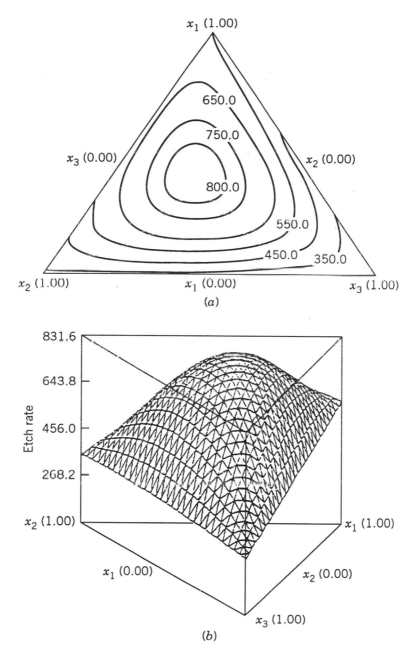

Figure 11.15 Etch rate response surface. (*a*) Contour plot. (*b*) Three-dimensional surface plot.

Figures 11.13 and 11.14 present the normal probability plot of the studentized residuals and a plot of the studentized residuals versus the predicted values, respectively, for this model. The plots are satisfactory, so we conclude that the special cubic model is adequate to describe the etch rate response surface.

Figure 11.15 presents plots of the etch rate response surface. In this process it is important to obtain an etch rate that is at least 750 Å/ min. The contour plot in Figure 11.15(a) indicates that there are several formulations that will meet that requirement.

We would like to emphasize an important point from Example 11.2. Notice that the experimenters originally intended to fit a quadratic mixture model; however, after the experiment had been performed, it was clear that a higher-order model (the special cubic) would be required to adequately model the response. This happens frequently in mixture experiments: The order of the required model is higher than the one initially planned for by the experimenters. If the experimenter designs the experiment without some additional runs to measure lack of fit (and fit higher-order terms, if necessary) and the proposed model is inadequate, then he or she is in trouble. For that reason, we recommend that the experimenter should, whenever possible, make 8–10 additional runs beyond the minimum required to fit the model, with about half of these runs chosen to be replicates of some points in the design and the others chosen as new distinct points so that the lack of fit of the model can be investigated. In this example, the tentative model was the quadratic, which in three components is a six-parameter model. The use of the augmented simplex lattice automatically provided four additional distinct runs. Then the experimenters decided to replicate four runs to give an internal estimate of error. Whenever it is possible to do so, we strongly recommend this design strategy.

11.3 RESPONSE TRACE PLOTS

The response trace is a plot of the estimated response values as we move away from the centroid of the simplex and along the component axes. More generally, we may think of the centroid as a "reference mixture" with component proportions given by $r_i = 1/q_i$, $i = 1, 2, \ldots, q$ (in Chapter 12 we will discuss other choices for the proportions in the reference blend). Consider the ith component and suppose we move away from the centroid by changing the proportion of this component by an amount Δ_i (note that we could make Δ_i either positive or negative). Along the ith axis as the value of x_i either increases or decreases, the values of the other component proportions x_j, $j \neq i$, either decrease or increase, but the relative proportions for these other components remains the same.

To illustrate, consider a $q = 3$ component mixture, so that $r_i = \frac{1}{3}$, $i = 1, 2, 3$. Suppose that component 1 is increased by $\Delta_1 = \frac{1}{6}$ so that now $x_1 = r_1 + \Delta_1 = \frac{1}{3} + \frac{1}{6} = \frac{3}{6} = \frac{1}{2}$. The new proportions for components 2 and 3 are $x_2 = x_3 = \frac{1}{4}$. Thus the ratio of x_2 to x_3 is identical at the centroid $x_2/x_3 = (\frac{1}{3})/(\frac{1}{3}) = 1$ and at the new point along the x_1-axis, because at this point $x_2/x_3 = (\frac{1}{4})/(\frac{1}{4}) = 1$.

We may give a general equation for these results. If x_i is changed by an amount Δ_i from the reference mixture, then the new proportions are

$$x_i = r_i + \Delta_i$$

$$x_j = r_j - \frac{\Delta_i r_j}{(1 - r_i)}, \qquad j \neq i \tag{11.12}$$

To construct a response surface trace plot, we would choose some number of blends along each component axis, calculate the coordinates of each mixture using Equation (11.12), and then substitute these coordinates into the fitted mixture model to obtain the predicted values of the response. Then

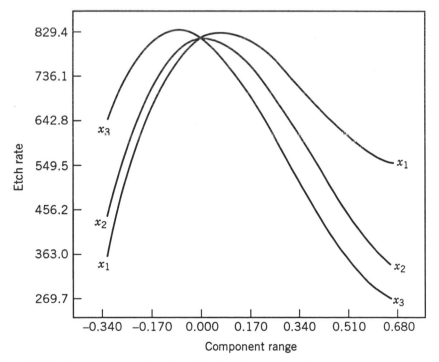

Figure 11.16 Response trace plot of predicted etch rate along the component axes x_1, x_2, and x_3.

these predicted responses are plotted against the changes made in the x_i, $i = 1, 2, \ldots, q$. There will be q of these plots, and it is customary to plot them on the same graph.

To illustrate, Figure 11.16 is the response trace plot for the etch rate experiment described in Example 11.2. The vertical axis is the predicted etch rate, and the horizontal axis is the incremental change Δ_i made in each component. The reference mixture, which is the overall centroid, is shown as the point 0.000 on the horizontal axis. Notice the strong nonlinear effect of all three acids on the etch rate. Also, the etch rate increases slightly and then decreases as we move along the x_1 axis from the centroid toward the x_1-vertex, and etch rate increases as we move away from the centroid along the x_3-axis toward the base point $x_3 = 0$, $x_1 = x_2 = \frac{1}{2}$.

In this example, etch rate is very sensitive to changes in all three component proportions. If one or more of the response traces is a horizontal line, this indicates that these components have little effect on the response; that is, we have discovered ingredients that are *inactive*. This can have important economic consequences.

11.4 REPARAMETERIZING CANONICAL MIXTURE MODELS TO CONTAIN A CONSTANT TERM (β_0)

It is occasionally convenient to reparameterize the canonical mixture polynomial so that it contains a constant term, or intercept, β_0. This can be easily done by simply deleting one of the $\beta_i x_i$ terms, say $\beta_q x_q$. Thus, instead of the linear mixture model

$$E(y) = \beta_1 x_1 + \beta_2 x_2 + \cdots + \beta_q x_q \qquad (11.13)$$

we would write

$$E(y) = \beta_0 + \beta_1^* x_1 + \beta_2^* x_2 + \cdots + \beta_{q-1}^* x_{q-1} \qquad (11.14)$$

and instead of the quadratic model

$$E(y) = \sum_{i=1}^{q} \beta_i x_i + \sum_{i<j}^{q} \beta_{ij} x_i x_j \qquad (11.15)$$

we would write

$$E(y) = \beta_0 + \sum_{i=1}^{q-1} \beta_i^* x_i + \sum_{i<}^{q} \sum_{j} \beta_{ij} x_i x_j \qquad (11.16)$$

Note that only the parameters in the linear blending portion of the canonical model are affected. Specifically, the intercept β_0 takes on the role of the parameter β_q corresponding to the deleted component x_q, and the remaining parameters β_i^* represent the differences $\beta_i^* = \beta_i - \beta_q$, $i = 1, 2, \ldots,$ $q - 1$.

We will use the quadratic model for yarn elongation in Example 11.1 to illustrate. The fitted canonical model was

$$\hat{y} = 11.7x_1 + 9.4x_2 + 16.4x_3 + 19.0x_1x_2 + 11.4x_1x_3 - 9.6x_2x_3$$

Now fit the model in Equation (11.14) with $q = 3$ by the method of least squares. This results in

$$\hat{y} = 16.4 - 4.7x_1 - 7.0x_2 + 19x_1x_2 + 11.4x_1x_3 - 9.6x_2x_3$$

That is,

$$b_0 = 16.4, \qquad b_1^* = -4.7, \qquad b_2^* = -7.0$$

Notice that

$$b_0 = b_3 = 16.4$$

$$b_1^* = b_1 - b_3 = 11.7 - 16.4 = -4.7$$

and

$$b_2^* = b_2 - b_3 = 9.4 - 16.4 = -7.0$$

and that the estimates of the quadratic terms are the same in both models.

This reparameterization of the canonical polynomial can be useful in analyzing mixture data with standard multiple regression computer software. Most regression software packages will fit a model with the intercept term suppressed. This option will produce the correct estimates of the β's in a canonical mixture model, but it will result in an incorrect value for the model or regression sum of squares, because the no-intercept option does not correct this sum of squares for the overall mean. As a result, the F-test, R^2, and the adjusted R^2 will be incorrect. Reparameterizing the model as we have discussed will result in correct output for the regression or model sum of squares, R^2, and the adjusted R^2. Then the correct values of the linear blending coefficients can be calculated from

$$b_i = b_i^* + b_0, \quad i = 1, 2, \ldots, q - 1$$
$$b_q = b_0 \tag{11.17}$$

EXERCISES

11.1 Three different motor fuels can be blended to form a gasoline. The miles per gallon (MPG) performance of this gasoline is the response variable. An augmented simplex-lattice design was used to study the blending properties of these three fuels, and the results are shown in the table below:

Design Point	Component Proportion			MPG
	x_1	x_2	x_3	y
1	1	0	0	24.5, 25.1
2	0	1	0	24.8, 23.9
3	0	0	1	22.7, 23.6
4	$\frac{1}{2}$	$\frac{1}{2}$	0	25.1
5	$\frac{1}{2}$	0	$\frac{1}{2}$	24.3
6	0	$\frac{1}{2}$	$\frac{1}{2}$	23.5
7	$\frac{1}{3}$	$\frac{1}{3}$	$\frac{1}{3}$	24.8, 24.1
8	$\frac{2}{3}$	$\frac{1}{6}$	$\frac{1}{6}$	24.2
9	$\frac{1}{6}$	$\frac{2}{3}$	$\frac{1}{6}$	23.9
10	$\frac{1}{6}$	$\frac{1}{6}$	$\frac{2}{3}$	23.7

(a) Fit a linear mixture model to these data. Is the model adequate?

(b) Fit a quadratic mixture model to these data. Is this model an improvement over the linear model?

(c) What model would you use to predict the MPG performance of the gasoline obtained by blending these three fuels?

(d) Plot the response surface contours for the model in part (c). What blend would you use to maximize mileage performance?

11.2 A mixture experiment was performed to determine if an artificial sweetener could be developed for a soft drink beverage that would minimize the aftertaste. Four different sweeteners were evaluated, and a simplex-centroid design with $q = 4$ components was used to evaluate the different blends. The response variable is aftertaste, with low values being desirable. The data are shown below:

Design Point	Component Proportions				Aftertaste
	x_1	x_2	x_3	x_4	y
1	1	0	0	0	19
2	0	1	0	0	8
3	0	0	1	0	15
4	0	0	0	1	10
5	$\frac{1}{2}$	$\frac{1}{2}$	0	0	13
6	$\frac{1}{2}$	0	$\frac{1}{2}$	0	16
7	$\frac{1}{2}$	0	0	$\frac{1}{2}$	12
8	0	$\frac{1}{2}$	$\frac{1}{2}$	0	11
9	0	$\frac{1}{2}$	0	$\frac{1}{2}$	5
10	0	0	$\frac{1}{2}$	$\frac{1}{2}$	10
11	$\frac{1}{3}$	$\frac{1}{3}$	$\frac{1}{3}$	0	14
12	$\frac{1}{3}$	$\frac{1}{3}$	0	$\frac{1}{3}$	10
13	$\frac{1}{3}$	0	$\frac{1}{3}$	$\frac{1}{3}$	14
14	0	$\frac{1}{3}$	$\frac{1}{3}$	$\frac{1}{3}$	8
15	$\frac{1}{4}$	$\frac{1}{4}$	$\frac{1}{4}$	$\frac{1}{4}$	12

(a) Fit an appropriate mixture model to these data. Test the model for adequacy.

(b) Construct the response trace plot for this experiment. Provide a practical interpretation of this plot.

(c) Construct contour plots of predicted aftertaste that would be useful to a decision-maker in choosing a blend of these four sweeteners that minimizes the aftertaste. What blend would you recommend?

11.3 Consider the linear mixture model from Exercise 11.1(a). Reparametize this model to the form

$$E(y) = \beta_0 + \beta_1^* x_1 + \beta_2^* x_2$$

and fit this model using least squares. Verify that this model can be converted to the linear mixture model you found in Exercise 11.1(a) using the method discussed in Section 11.4.

11.4 A thickening agent is added to a liquid soap product to control the viscosity. This agent usually makes up 10% of the soap by weight. There are three different salt compounds that are blended to make up the thickener. The product formulators have decided to investigate the combination of these ingredients using a mixture experiment. Let x_1^*, x_2^*, and x_3^* represent the actual proportions of the three salt compounds, with $x_1^* + x_2^* + x_3^* = 0.1$. Notice that if we let $x_i = x_i^*/0.1$, then the design points can be expressed as proportions $0 \leq x_i \leq 1$ with the standard mixture constraint $x_1 + x_2 + x_3 = 1$. The experiments decided to use an augmented simplex-lattice design to study the viscosity of the thickener. The design and the viscosity data are shown below. The centroid was replicated three times.

Design Point	Component Proportions			Viscosity
	x_1	x_2	x_3	
1	1	0	0	410
2	0	1	0	880
3	0	0	1	1445
4	$\frac{1}{2}$	$\frac{1}{2}$	0	683
5	$\frac{1}{2}$	0	$\frac{1}{2}$	465
6	0	$\frac{1}{2}$	$\frac{1}{2}$	456
7	$\frac{2}{3}$	$\frac{1}{6}$	$\frac{1}{6}$	354
8	$\frac{1}{6}$	$\frac{2}{3}$	$\frac{1}{6}$	521
9	$\frac{1}{6}$	$\frac{1}{6}$	$\frac{2}{3}$	700
10	$\frac{1}{3}$	$\frac{1}{3}$	$\frac{1}{3}$	465
11	$\frac{1}{3}$	$\frac{1}{3}$	$\frac{1}{3}$	392
12	$\frac{1}{3}$	$\frac{1}{3}$	$\frac{1}{3}$	428

(a) Fit linear and quadratic mixture models to these data. Test both models for lack of fit. Does either model seem adequate, based on this test?

(b) Select an appropriate mixture model for the viscosity data. Construct appropriate residual plots to check for model adequacy.

(c) Construct and interpret a response trace plot for this experiment.

(d) Construct a response surface contour plot for viscosity.

(e) Suppose that a desirable value for viscosity of this thickening agent is 900 centipoise. Is there a formulation that would produce the desired viscosity?

11.5 One of the authors (DCM) is an owner of a vineyard and winery in Newberg, Oregon. The partners that operate this business are developing a new red wine that is a blend of three different varieties of grapes; Pinot Noir-Pommard, Pinot Noir-Wadenswil, and Gamay Noir. A three-component mixture design was used to study the blends of these three ingredients. The design is shown in the table below, where x_1 = proportion of Pinot Noir-Pommard, x_2 = proportion of Pinot Noir-Wadenswil, and x_3 = proportion of Gamay Noir in the product. Each of the 10 blends was tasted by a panel of experts and ranked on the basis of the task-test results, with rank 1 being the best, rank 2 the next best, and so forth.

Design Point	Component Proportions			Task-Test Ranks			
	x_1	x_2	x_3	DCM	HPN	DLB	JPN
1	1	0	0	2	3	3	2
2	0	1	0	1	2	1	3
3	0	0	1	4	4	5	6
4	$\frac{1}{2}$	$\frac{1}{2}$	0	3	1	2	1
5	$\frac{1}{2}$	0	$\frac{1}{2}$	5	5	4	4
6	0	$\frac{1}{2}$	$\frac{1}{2}$	6	7	8	5
7	$\frac{2}{3}$	$\frac{1}{6}$	$\frac{1}{6}$	7	6	9	8
8	$\frac{1}{6}$	$\frac{2}{3}$	$\frac{1}{6}$	8	8	6	7
9	$\frac{1}{6}$	$\frac{1}{6}$	$\frac{2}{3}$	10	9	10	9
10	$\frac{1}{3}$	$\frac{1}{3}$	$\frac{1}{3}$	9	10	7	10

(a) Fit linear and quadratic mixture models to the averages of the rank data. Test both of these models for lack of fit. Does either model seem adequate?

(b) Select an appropriate mixture model for the average rank response. Construct appropriate residual plots to check for model adequacy.

(c) Construct and interpret a response trace plot for this experiment.

(d) Construct a response surface contour plot for the average rank response. What blends of the three varieties of grapes would you recommend?

11.6 Reconsider the sweetener experiment in Exercise 11.2. Suppose that the experimenters are interested in a second response, the cost of the sweetener. For each of the 15 formulations of the product tested in the simplex-centroid design, the cost of the mixture of ingredients used in the formula and a manufacturing cost estimate are combined. For the design points, listed in the same order as in the original table in Exercise 11.2, these costs are as follows:

Design Point	Cost	Design Point	Cost	Design Point	Cost
1	4	6	8	11	12
2	15	7	16	12	18
3	6	8	13	13	15
4	25	9	18	14	16
5	10	10	15	15	15

(a) Build an appropriate mixture model for the cost response.

(b) Construct a response trace plot for the cost response. Interpret this graph.

(c) Construct a response surface contour plot of the cost response. What practical advice could you give a decision-maker based on this graph?

(d) Consider both the response surface for cost and the aftertaste response surface developed in Exercise 11.2. Suppose that your objective is to select a blend of ingredients so that the sweetener has low aftertaste while keeping the cost of the product below 12. What blend of sweeteners would you recommend?

11.7 **Continuation of Exercise 11.6** Reconsider the cost response surface from Exercise 11.6. Suppose that your objective was to minimize the cost of the sweetener while keeping the aftertaste below 13. What formulation of ingredients would you recommend for this product?

11.8 Consider the second-order polynomial in two variables:

$$E(y) = \beta_0 + \beta_1 x_1 + \beta_2 x_2 + \beta_{12} x_1 x_2 + \beta_{11} x_1^2 + \beta_{22} x_2^2$$

(a) Show that if the term β_0 is replaced by $\beta_0(x_1 + x_2)$, if x_1^2 is replaced by $x_1(1 - x_2)$, and if x_2^2 is replaced by $x_2(1 - x_1)$, the canonical form of the quadratic mixture model results.

(b) Why are the substitutions described in (a) required in a mixture experiment?

11.9 When observations are collected only at the points of a $\{q, m\}$ simplex lattice, the estimates of the parameters in the canonical polynomial are simple functions of the observed responses. Consider the $\{3, 2\}$ lattice, let $\bar{y}_i$ be the average of n_i observations at each vertex $x_i = 1$, $x_j = 0$ for $j \neq i$, and let $\bar{y}_{ij}$ be the average of n_{ij} observations at the edge midpoints or 50:50 binary blends $x_i = x_j = \frac{1}{2}$, $x_k = 0$ for $i < j < k = 1, 2, 3$.

(a) Show that by substituting $\bar{y}_i$ and $\bar{y}_{ij}$ for the expected response at each of the six design points, the following equations result:

$$\bar{y}_1 = \beta_1, \qquad \bar{y}_2 = \beta_2, \qquad \bar{y}_3 = \beta_3$$

$$\bar{y}_{12} = \tfrac{1}{2}\beta_1 + \tfrac{1}{2}\beta_2 + \tfrac{1}{4}\beta_{12}$$

$$\bar{y}_{13} = \tfrac{1}{2}\beta_1 + \tfrac{1}{2}\beta_3 + \tfrac{1}{4}\beta_{13}$$

$$\bar{y}_{23} = \tfrac{1}{2}\beta_2 + \tfrac{1}{2}\beta_3 + \tfrac{1}{4}\beta_{23}$$

(b) Show that the solution to the equations given in part (a) are as follows:

$$b_1 = \bar{y}_1, \qquad b_2 = \bar{y}_2, \qquad b_3 = \bar{y}_3$$

$$b_{12} = 4\bar{y}_{12} - 2(\bar{y}_1 + y_2)$$

$$b_{13} = 4\bar{y}_{13} - 2(\bar{y}_1 + \bar{y}_3)$$

$$b_{23} = 4\bar{y}_{23} - 2(\bar{y}_2 + \bar{y}_3)$$

Notice that the constants 4 and 2 in the last three equations do not depend on the number of replicates at each design point, but arise instead from the $x_i = x_j = \frac{1}{2}$ values at each edge midpoint.

(c) Use the equations derived in part (b) above to calculate the parameter estimates for the yarn elongation data in Example 11.1. Verify that the results obtained agree with the least squares estimates of these parameters obtained in Example 11.1.

11.10 Consider the quadratic mixture model

$$y = \beta_1 x_1 + \beta_2 x_2 + \beta_3 x_3 + \beta_{12} x_1 x_2 + \beta_{12} x_1 x_3 + \beta_{23} x_2 x_3 + \varepsilon$$

where ε is the usual NID $(0, \hat{\sigma})$ random error term. Suppose that the $\{3, 2\}$ lattice design as described in Exercise 11.8 above is run, and the resulting data are used to estimate the coefficients in the quadratic model.

(a) Show that

$$E(b_i) = \beta_i$$

and that

$$E(b_{ij}) = \beta_{ij}$$

(b) Show that

$$\text{Var}(b_i) = \frac{\sigma^2}{n_i}, \qquad i = 1, 2, 3$$

and

$$\text{Var}(b_{ij}) = \frac{16\sigma^2}{n_{ij}} + \frac{4\sigma^2}{n_i} + \frac{4\sigma^2}{n_j}, \qquad i < j = 2, 3$$

(c) Show that

$$\text{Cov}(b_i, b_j) = 0, \qquad i \neq j$$

$$\text{Cov}(b_i, b_{ij}) = \frac{-2\sigma^2}{n_i}$$

$$\text{Cov}(b_{ij}, b_{ik}) = \frac{4\sigma^2}{n_i}, \qquad j \neq k$$

11.11 Consider the simplex-centroid design for three components. Let $\bar{y}_i$, $\bar{y}_{ij}$, and $\bar{y}_{123}$ be the averages of n_i observations at the three vertices, n_{ij} observations at the edge midpoints, and n_{123} observations at the design centroid. The model of interest is

$$E(y) = \sum_{i=1}^{3} \beta_i x_i + \sum_{i<}^{3} \sum_{j} \beta_{ij} x_i x_j + \beta_{123} x_1 x_2 x_3$$

(a) Show that by substituting $\bar{y}_i$, $\bar{y}_{ij}$, and $\bar{y}_{123}$ for the expected response at each of the seven design points, the following equations result:

$$\bar{y}_i = \beta_i, \qquad i = 1, 2, 3$$

$$\bar{y}_{ij} = \tfrac{1}{2}(\beta_i + \beta_j) + \tfrac{1}{4}\beta_{ij}, \qquad i < j = 2, 3$$

$$\bar{y}_{123} = \tfrac{1}{3}(\beta_1 + \beta_2 + \beta_3) + \tfrac{1}{9}(\beta_{12} + \beta_{13} + \beta_{23}) + \tfrac{1}{27}\beta_{123}$$

(b) Show that the solution to the equation given in part (a) are

$$b_i = \bar{y}_i, \qquad i = 1, 2, 3$$

$$b_{ij} = 4\bar{y}_y - 2(\bar{y}_i + \bar{y}_j), \qquad i < j = 2, 3$$

$$b_{123} = 27\bar{y}_{123} - 12(\bar{y}_{12} + \bar{y}_{13} + \bar{y}_{23}) + 9(\bar{y}_1 + \bar{y}_2 + \bar{y}_3)$$

(c) What effect does the observation at the centroid have on the estimates of the linear and quadratic terms in this model?

11.12 Consider the special cubic model

$$\hat{y} = \sum_{i=1}^{3} b_i x_i + \sum_{i<}\sum_{j} b_{ij} x_i x_j + b_{123} x_1 x_2 x_3$$

(a) How large must the b_{ij} term be if the term $b_{12} x_i x_j$ is to contribute as much to the model at the edge midpoint as the linear term $b_i x_i$?

(b) How large must the cubic term b_{123} be if the contribution of the term $b_{123} x_1 x_2 x_3$ to the model at the centroid is as large as the contribution of the quadratic term $b_{ij} x_i x_j$?

11.13 Consider the quadratic model in two components:

$$E(y) = \beta_1 x_1 + \beta_2 x_2 + \beta_{12} x_1 x_2$$

(a) Suppose that four observations are collected at the following points:

$$y_1 \text{ at } x_1 = 1, x_2 = 0$$
$$y_2 \text{ at } x_1 = 0, x_2 = 1$$
$$y_3 \text{ and } y_4 \text{ at } x_1 = x_2 = \tfrac{1}{2}$$

Derive the estimates of the model coefficients for this design.

(b) Suppose that four observations are collected at the following points:

$$y_1 \text{ at } x_1 = 1, x_2 = 0$$
$$y_2 \text{ at } x_1 = 0, x_2 = 1$$
$$y_3 \text{ at } x_1 = \tfrac{1}{3}, x_0 = \tfrac{2}{3}$$
$$y_4 \text{ at } x_1 = \tfrac{2}{3}, x_1 = \tfrac{1}{3}$$

Derive the estimates of the model coefficients for this design.

(c) Which of these two designs would you prefer, if your objective was to find minimum variance estimates of the coefficients in the quadratic model?

CHAPTER 12

Other Mixture Design and Analysis Techniques

In Chapter 11 we focused on mixture experiments for exploring the complete simplex region. We introduced simplex-lattice and simplex-centroid designs that would enable us to fit response surface models over this region. In many mixture experiments there are restrictions on the component proportions that prevent us from exploring the entire simplex region. Frequently these restrictions take the form of lower and/or upper bounds on the component proportions. This problem is discussed in this chapter. In addition, we also briefly discuss several other variations of the mixture problem, including the use of ratios of components as the design variables, the inclusion of process variables in mixture experiments, and screening techniques to identify the most important components in the mixture.

12.1 CONSTRAINTS ON THE COMPONENT PROPORTIONS

In many mixture experiments there are constraints on the component proportions. These are often upper- and/or lower-bound constraints of the form $L_i \leq x_i \leq U_i$, $i = 1, 2, \ldots, q$, where L_i is the lower bound for the ith component and U_i is the upper bound for the ith component. Essentially, L_i represents a minimum proportion of the ith component that must be present in the mixture and U_i represents a maximum proportion of the ith component that must be present in the mixture. The general form of the constrained mixture problem is

$$x_1 + x_2 + \cdots + x_q = 1$$

$$L_i \leq x_i \leq U_i, \qquad i = 1, 2, \ldots, q \tag{12.1}$$

where $L_i \geq 0$ and $U_i \leq 1$ for $i = 1, 2, \ldots, q$.

The effect of the upper- and lower-bound restriction in Equation (12.1) is to limit the feasible space for the mixture experiment to a subregion of the simplex. Therefore, experimental designs for constrained mixture problems must be limited to the feasible subregion of the simplex. In this section, we present design and analysis techniques for these constrained mixture spaces.

12.1.1 Lower-Bound Constraints on the Component Proportions

We now consider the case where only the lower bounds in Equation (12.1) are imposed, so that the constrained mixture problem becomes

$$x_1 + x_2 + \cdots + x_q = 1$$

$$L_i \le x_i \le 1, \qquad i = 1, 2, \ldots, q \qquad (12.2)$$

To illustrate the effect that lower bounds have on the feasible mixture design space, consider a three-component example with constraints

$$0.3 \le x_1, \qquad 0.4 \le x_2, \quad \text{and} \quad 0.1 \le x_3$$

Figure 12.1 shows the feasible mixture space after these constraints are imposed. Notice that the imposition of lower bounds does not affect the shape of the feasible mixture space; it is still a simplex. In general, this will always be the case; if only lower bounds are imposed on any of the component proportions, the feasible region for the mixture experiment will be a smaller simplex inscribed inside the original (or unconstrained) region. If all lower bounds are equal ($L_1 = L_2 = \cdots = L_q$), the centroid of the original simplex will also be the centroid of the smaller inscribed simplex.

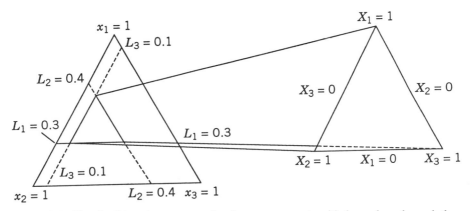

Figure 12.1 The feasible mixture space for three components with lower bounds, and the redefinition of the region as a simplex in the L-pseudocomponents. [Adapted from Cornell (1990), with permission of the publisher.]

Because the feasible experimental region is still a simplex, it seems reasonable to define a new set of components that will take on the values 0 to 1 over the feasible region. This will make design construction and model fitting easier over the constrained region. These redefined components are called **L-pseudocomponents**, or sometimes just **pseudocomponents**. The pseudocomponents X_i are defined using the following transformation:

$$X_i = \frac{x_i - L_i}{1 - L} \tag{12.3}$$

where

$$L = \sum_{i=1}^{q} L_i < 1$$

is the sum of all of the lower bounds. To illustrate, we may use the situation described and shown graphically in Figure 12.1. The pseudocomponents are

$$X_1 = \frac{x_1 - 0.3}{0.2}, \qquad X_2 = \frac{x_2 - 0.4}{0.2}, \qquad X_3 = \frac{x_3 - 0.1}{0.2}$$

because $L_1 = 0.3$, $L_2 = 0.4$, $L_3 = 0.1$, $L = 0.3 + 0.4 + 0.1 = 0.8$, and $1 - L = 0.2$. The factor space in terms of pseudocomponents is shown on the right-hand side of Figure 12.1. The orientation of the L-pseudocomponent simplex is the same as the original simplex, and the height of the L-pseudocomponent simplex is $1 - L$.

Constructing a design in the L-pseudocomponents is easy. One simply specifies the design points in the X_i and then converts them to the settings in the original components using

$$x_i = L_i + (1 - L)X_i \tag{12.4}$$

To illustrate, suppose that we decide to use a simplex-centroid design for our experiment. Table 12.1 shows the design points in the pseudocomponents, along with the corresponding setting for the original components.

The original component settings in Table 12.1 were calculated by using Equation (12.4). This becomes

$$x_1 = 0.3 + 0.2X_1$$

$$x_2 = 0.4 + 0.2X_2$$

$$x_3 = 0.1 + 0.2X_3$$

To illustrate, consider the point $X_1 = \frac{1}{2}$, $X_2 = \frac{1}{2}$, and $X_3 = 0$. In terms of

Table 12.1   Pseudocomponent Settings and Original Component Settings, Simplex-Centroid Design

Pseudocomponents			Original Components		
X_1	X_2	X_3	x_1	x_2	x_3
1	0	0	0.5	0.4	0.1
0	1	0	0.3	0.6	0.1
0	0	1	0.3	0.4	0.3
$\frac{1}{2}$	$\frac{1}{2}$	0	0.4	0.5	0.1
$\frac{1}{2}$	0	$\frac{1}{2}$	0.4	0.4	0.2
0	$\frac{1}{2}$	$\frac{1}{2}$	0.3	0.5	0.2
$\frac{1}{3}$	$\frac{1}{3}$	$\frac{1}{3}$	0.3667	0.4667	0.1666

the original components, this point becomes

$$x_1 = 0.3 + 0.2(\tfrac{1}{2}) = 0.4$$

$$x_2 = 0.4 + 0.2(\tfrac{1}{2}) = 0.5$$

$$x_3 = 0.1 + 0.2(0) = 0.1$$

as shown in Table 12.1.

We recommend using pseudocomponents to fit the mixture model. The reason for this is that constrained design spaces usually have moderately high to high levels of multicollinearity or ill conditioning (see Appendix 2 for more details), and this can have serious impact on the least squares estimators of regression coefficients in that the variances of these coefficient estimators are inflated. In general, a mixture model in the pseudocomponents will have lower levels of multicollinearity than will the same model in the original component proportions. For more discussion of this point, see Montgomery and Voth (1994) and St. John (1984). It is, of course, a relatively simple matter to convert a model in the pseudocomponents into the corresponding model in the actual components.

Example 12.1   We will use a three-component example from Montgomery and Voth (1994) to illustrate fitting and interpreting a model with pseudo-components. The example includes blending fuel (x_1), oxidizer (x_2), and binder (x_3) together to form a propellant used in aircrew escape systems. As in most mixture experiments, several responses are of interest, including both physical and mechanical properties of the propellant; however, we will concentrate on three responses. The first is burning rate (y_1, cm/sec), a critical characteristic if the escape system is to perform satisfactorily. Burning rate is measured by testing several samples of propellant from the same run and reporting the average observed burning rate. The second response is the standard deviation of the observed burning rates at each run (y_2, cm/sec); low variability is desirable, implying consistent performance across different escape systems assembled from propellant grains from the same batch. The third response is a manufacturability index (y_3) that reflects the cost and difficulty associated with producing a particular mixture. The constraints for this mixture problem are

$$x_1 + x_2 + x_3 = 0.9$$

$$0.30 \leq x_1$$

$$0.20 \leq x_2$$

$$0.20 \leq x_3$$

Notice that the three components must make up 90% of the mixture. We will model this system in terms of pseudocomponents, say

$$X_i = \frac{x_i - L_i}{0.9 - \sum\limits_{i=1}^{3} L_i}, \qquad i = 1, 2, 3 \qquad (12.5)$$

This transformation yields pseudocomponents X_i such that $0 \leq X_i \leq 1$, $i = 1, 2, 3$, and $X_1 + X_2 + X_3 = 1$. Because all the constraints in this problem are lower-bound constraints, the experimental region is a simplex. To draw this region on the usual coordinate system, we must remember that the range of each x_i on the graph is from 0 to 1. Therefore, the constraints that should be plotted on the graph are $0.3/0.9 \leq x_1$, $0.2/0.9 \leq x_2$, and $0.2/0.9 \leq x_3$ or

$$0.3333 \leq x_1$$

$$0.2222 \leq x_2$$

$$0.2222 \leq x_3$$

This constrained region is shown in Figure 12.2.

   The experimenters decided that a quadratic mixture model would likely be adequate to describe the relationship between the responses and the three component proportions. They selected a 10-point design consisting of the simplex-centroid augmented with the interior axial points located midway between the centroid and the opposed vertex. They decided to replicate each vertex twice and the centroid three times, resulting in the final 15-run design shown in Table 12.2. The relationship between the settings in the pseudo-

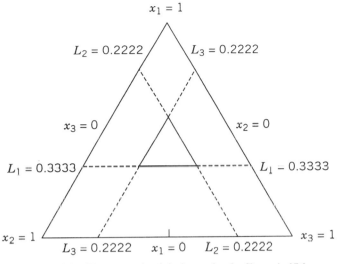

Figure 12.2 The constrained design region for Example 12.1.

components and the actual component settings is

$$x_i = L_i + X_i\left(0.9 - \sum_{i=1}^{3} L_i\right)$$

To illustrate for the first design point in Table 12.2 where

$$X_1 = 1, \qquad X_2 = X_3 = 0$$

we have

$$x_1 = 0.3 + 1(0.2) = 0.5$$
$$x_2 = 0.2 + 0(0.2) = 0.2$$
$$x_3 = 0.2 + 0(0.2) = 0.2$$

Table 12.3 shows the results of sequentially fitting the linear, quadratic, and special cubic models to the data in Table 12.2, using the mean burning rate response y_1. Based on the lack-of-fit test and the small P-value associated with the special cubic term, we would recommend the special cubic model. Note also that the value of the PRESS statistic.

$$\text{PRESS} = \sum_{i=1}^{n} e_{(i)}^2$$

is smallest for the special cubic model. Recall that $e_{(i)}$ in PRESS is the

Table 12.2 Data for Propellant Example

Observation	Pseudocomponents			Original Components			Burning Rate y_1	Standard Deviation of Burning Rate y_2	Manufacturability Index y_3
	X_1	X_2	X_3	x_1 (fuel)	x_2 (oxidizer)	x_3 (binder)			
1	1.0	0.0	0.0	0.5	0.2	0.3	32.5	4.1	32
2	1.0	0.0	0.0	0.5	0.2	0.3	37.9	3.7	25
3	0.5	0.5	0.0	0.4	0.3	0.3	44.0	6.8	20
4	0.5	0.0	0.5	0.4	0.2	0.4	63.2	4.7	18
5	0.0	1.0	0.0	0.3	0.4	0.3	54.5	8.9	18
6	0.0	1.0	0.0	0.3	0.4	0.3	32.5	9.2	21
7	0.0	0.5	0.5	0.3	0.3	0.4	94.0	4.5	17
8	0.0	0.0	1.0	0.3	0.2	0.5	64.0	14.0	14
9	0.0	0.0	1.0	0.3	0.2	0.5	78.5	13.0	16
10	0.666666666	0.166666666	0.166666666	0.433333333	0.233333333	0.333333333	67.1	3.5	20
11	0.166666666	0.666666666	0.166666666	0.333333333	0.333333333	0.333333333	73.0	5.2	22
12	0.166666666	0.166666666	0.666666666	0.333333333	0.233333333	0.433333333	87.5	7.0	17
13	0.333333333	0.333333333	0.333333333	0.366666666	0.266666666	0.366666666	112.5	4.6	19
14	0.333333333	0.333333333	0.333333333	0.366666666	0.266666666	0.366666666	98.5	3.5	20
15	0.333333333	0.333333333	0.333333333	0.366666666	0.266666666	0.366666666	103.6	3.0	18

Table 12.3 Analysis of Variance Summary and Lack-of-Fit Tests

ANOVA Summary of Models Fit

Source	Sum of Squares	Degrees of Freedom	Mean Square	F-Value	Probability $> F$	R^2	Adjusted R^2	PRESS
Intercept	72,564.993	1	72,564.993					
Linear	2,395.909	2	1,197.954	2.007	0.1770	0.2507	0.1258	11,323.250
Quadratic	5,486.852	3	1,828.951	9.827	0.0034	0.8247	0.7274	8,284.901
Special cubic	1,063.615	1	1,063.615	13.92	0.0058	0.9360	0.8881	3,296.975
Error	611.402	8	76.425					
Corrected total	9,557.778	14						

Lack-of-Fit Tests

Model	Sum of Squares	DF	Mean Square	F-Value	Probability $> F$
Linear	6,699.757	7	957.108	10.36	0.0102
Quadratic	1,212.905	4	303.226	3.281	0.1123
Special cubic	149.290	3	49.763	0.5384	0.6763
Pure error	462.112	5	92.422		

residual obtained when y_i is predicted with a model based on all observations except the ith. As noted in Chapter 2, regression models with small values of PRESS are usually good prediction equations.

In judging the relative size of PRESS, it is often useful to compute the R^2 for prediction based on PRESS

$$R^2_{\text{Prediction}} = 1 - \frac{\text{PRESS}}{SS_{\text{Total}}}$$

This statistic can be thought of as a measure of the capability of the model to predict new data, and it can be compared to the usual R^2 and R^2-adjusted statistics. For the special cubic model we have

$$R^2_{\text{Prediction}} = 1 - \frac{3296.975}{9555.777} = 0.6550$$

This is somewhat smaller than the other two R^2-values (see Table 12.4).

Table 12.4 contains a summary of the special cubic model. The fitted model in the pseudocomponents is

$$\hat{y}_1 = 35.49X_1 + 42.78X_2 + 70.36X_3 + 16.02X_1X_2$$
$$+ 36.33X_1X_3 + 136.82X_2X_3 + 854.98X_1X_2X_3$$

By substituting the definitions of the pseudocomponents into this equation, we may obtain a model for mean burning rate in terms of the original (actual) components. However, it is usually not necessary to do this, because modern computer software will plot the response surface contours in either the pseudocomponents or the actual components.

From Table 12.4, notice that in the final model there are three values of the hat diagonals h_{ii} that are large (0.91); these are associated with the centers-of-edges design points. This has occurred because the h_{ii} are both model- and design-dependent, and the final model here is of higher-order than the one that was initially contemplated. This happens frequently in mixture experiments and often has undesirable consequences. For example, in this situation, the extremely large value of h_{ii} associated with design point 7 is accompanied by a moderately large studentized residual, resulting in a large value of Cook's distance measure. Furthermore, if we examine the PRESS residuals in Table 12.4, we note that design point 7 has a relative large PRESS residual; in fact, this observation accounts for about 40% of the PRESS statistic. This is why $R^2_{\text{Prediction}}$ for this model seemed low. Therefore, design point 7 exerts disproportionate influence on the model. We will investigate this in more detail subsequently.

Figure 12.3 presents a normal probability plot of the studentized residuals from the special cubic model. We recommend plotting studentized residuals from mixture models because the highly nonuniform distribution of leverage

Table 12.4 Results of Fitting the Special Cubic Model for the Mean Burning Rate Response

Variable	Parameter Estimate	Degrees of Freedom	Standard Error	t for H_0 Parameter $= 0$	Probability $> \lvert t \rvert$
x_1	35.49456	1	6.07214	5.845	0.0004
x_2	42.77552	1	6.07214	7.045	0.0001
x_3	70.36123	1	6.07214	11.59	0.0001
$x_1 x_2$	16.02049	1	38.29236	0.4184	0.6867
$x_1 x_3$	36.33478	1	38.29236	0.9489	0.3705
$x_2 x_3$	136.82049	1	38.29236	3.573	0.0073
$x_1 x_2 x_3$	854.98182	1	229.18322	3.731	0.0058

Standard error of mean $= 2.25721$

Observation	Actual Value	Predicted Value	Residual	h_u	Studentized Residual	Cook's Distance	PRESS Residual
1	32.50000	35.49456	−2.99456	0.482	−0.476	0.030	5.78100
2	37.90000	35.49456	2.40544	0.482	0.382	0.019	4.64371
3	44.00000	43.14016	0.85984	0.910	0.327	0.154	9.55378
4	63.20000	62.01159	1.18841	0.910	0.452	0.294	13.2046
5	54.50000	42.77552	11.72448	0.482	1.864	0.463	22.63413
6	32.50000	42.77552	−10.27552	0.482	−1.634	0.355	−19.83691
7	94.00000	90.77350	3.22650	0.910	1.227	2.165	35.85006
8	64.00000	70.36123	−6.36123	0.482	−1.011	0.136	−12.28037
9	78.50000	70.36123	8.13877	0.482	0.886	0.223	15.71191
10	67.10000	67.96999	−0.86999	0.186	−0.110	0.000	−1.06878
11	73.00000	79.98427	−6.98427	0.186	−0.886	0.026	−8.56018
12	87.50000	95.46999	−7.96999	0.786	−1.011	0.033	−9.79114
13	112.50000	102.22929	10.27071	0.273	1.378	0.102	14.12752
14	98.50000	102.22929	−3.72929	0.273	−0.500	0.013	−5.12970
15	103.60000	102.22929	1.37071	0.273	0.184	0.002	1.88543

in many mixture designs results in residuals that may have very different variances. Thus, in a normal probability plot, the standardized residuals should lie approximately along a straight line. Figure 12.4 presents a contour plot for the mean burning rate response.

Table 12.5 summarizes the model-building activity for the other two responses, standard deviation of burning rate (y_2), and manufacturability index (y_3). In both cases, we selected the model with minimum PRESS, resulting in a quadratic model for the standard deviation of burning rate and a linear model for the manufacturability index. The values of $R^2_{\text{Prediction}}$ for those models, also shown in Table 12.5, are reasonably close to the ordinary R^2-statistic for the selected model. Plots of the studentized residuals for both models were satisfactory, and the contour plots are shown in Figures 12.5 and 12.6.

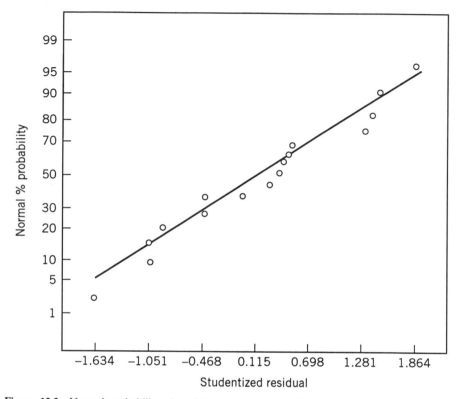

Figure 12.3   Normal probability plot of the studentized residuals for the mean burning rate response.

As in most mixture designs, here the experimenters are interested in formulating a product that satisfies several customer requirement. Specifically, the mean burning rate must excced 95 cm/ sec, the standard deviation of burning rate should be small (less than 4.5 cm/ sec, say), and the manufacturability index should not exceed 20. Figure 12.7 presents a plot of the design region with these constraints superimposed. This graph identifies the feasible mixtures that satisfy all three response constraints. Obviously, there are many product formulations that will be satisfactory. The final choice between formulations could now be made using other supplemental criteria. For example, the conclusions here are based on predicted mean responses, and no measure of uncertainty in these predictions was evaluated. One could use confidence or prediction intervals to account for the uncertainty. Alternatively, the effects of tightening the constraints can be investigated. Figure 12.8 illustrates the effects of requiring the standard deviation of burning rate to be less than 4 cm/ sec and the manufacturability index to be less than 19. This new set of constraints has greatly reduced the feasible mixture space, and it is approaching the practical limits of reduction in both

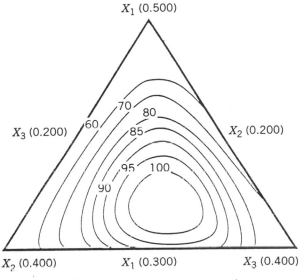

Figure 12.4   Mean burning rate contour plot.

variability and cost for this product if the burning rate requirement is to be met.

Finally, the effect of the influential point (design point number 7) may be investigated. A simple way to do this is to withhold the point, refit the model (or models) in question, and examine the changes that occur. The results of this analysis are shown in Figure 12.9 which shows the feasible space for the mixture based on the new models. Comparing Figures 12.7 and 12.9 indicates that point 7 is moderately influential, because the new feasible region in

Table 12.5   Mixture Models for Standard Deviation of Burning Rate (y_2) and Manufacturability Index (y_3)

Model Form	y_2		y_3	
	R^2	PRESS	R^2	PRESS
Linear	0.4686	150.60	0.7942	39.83
Quadratic	0.9830	7.15	0.8222	69.50
Special cubic	0.9844	13.47	0.8337	140.51

$$\hat{y}_1 = 3.88159X_1 + 9.03873X_2 + 13.63397X_3$$
$$- 0.19048X_1X_2 - 16.61905X_1X_3$$
$$- 27.67619X_2X_3$$
$$R^2_{\text{Prediction}} = 0.9575$$
$$\hat{y}_3 = 23.13333X_1 + 19.73333X_2 + 14.73333X_3$$
$$R^2_{\text{prediction}} = 0.6456$$

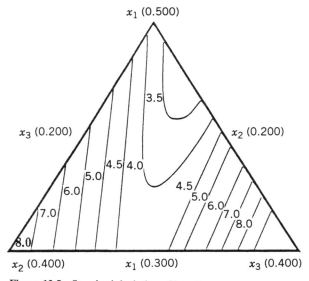

Figure 12.5   Standard deviation of burning rate contour plot.

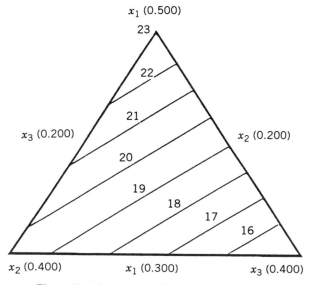

Figure 12.6   Manufacturability index contour plot.

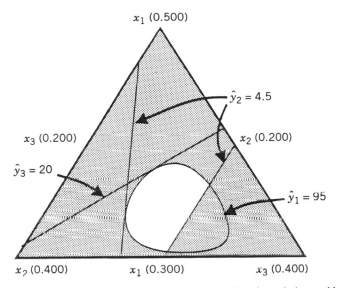

Figure 12.7 Region of an optimum solution to the propellant formulation problem.

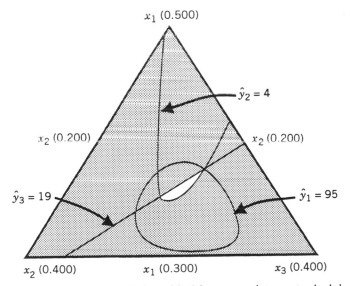

Figure 12.8 Region of an optimum solution with tighter constraints on standard deviation of burning rate and manufacturability index.

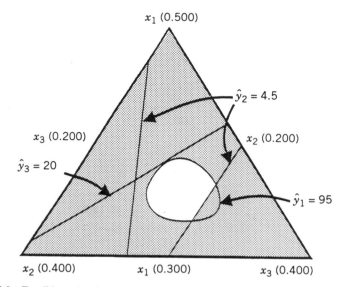

Figure 12.9 Feasible region for the propellant formulation with design point 7 withheld.

Figure 12.9 is considerably smaller than its counterpart in Figure 12.7 and shifted toward the X_1 vertex. However, there are still many mixtures that fall into the feasible spaces on both plots, and it would be prudent to select the final formulation from one of those. It may be appropriate to consider either robust fitting methods or bounded-influence regression (see Appendix 3) for mixture models when the distribution of leverage is very nonuniform throughout the design space, because these methods will downweight influential observations in comparison to least squares.

12.1.2   Upper-Bound Constraints on Component Proportions

Sometimes only upper-bound constraints of the form $x_i \le U_i$ are placed on the component proportions. In these cases, a simple modification of the simplex-lattice design consists of replacing the restricted components with mixtures consisting of combinations of the restricted components and predetermined combinations of the unrestricted components. This approach is discussed and illustrated in Cornell (1990). However, as is obvious from inspection of Figure 12.10 where a single upper-bound constraint $x_1 \le U_1$ has been imposed, the presence of upper bounds can lead to an experimental region that is not a simplex. In such cases, computer-generated designs (perhaps based on the D-optimal criterion) are logical design alternatives. The points labeled A through K in Figure 12.10 would be logical candidate points for a D-optimal algorithm to consider. Note that points A, C, G, and I are vertices of the constrained region, B, D, F and H are centers of edges,

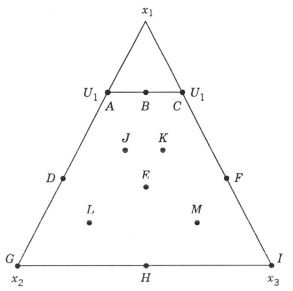

Figure 12.10 A three-component mixture problem with an upper-bound constraint $x_1 \le U_1$.

E is the overall centroid, and J, K, L, and M are points that lie midway between the centroid and the vertices of the constrained region (these are somewhat analogous to the axial check blends often added to simplex-type designs).

In some cases, upper-bound constraints lead to a design space that is a simplex. For example, consider the three-component problem

$$x_1 \le 0.4, \qquad x_2 \le 0.5, \quad \text{and} \quad x_3 \le 0.3$$

Is shown in Figure 12.11(a). Notice that the feasible region for this mixture experiment is an inverted simplex. Obviously a standard simplex-type design can be defined over this region. Figure 12.11(b) shows the feasible region for the three-component problem:

$$x_1 \le 0.7, \qquad x_2 \le 0.5, \quad \text{and} \quad x_3 \le 0.8$$

In this case, the feasible experimental region is not a simplex, and, once again, a computer-generated design would be a logical choice.

In general, when upper-bound constraints $x_1 \le U_1$ are present, the feasible region will be an inverted simplex that lies entirely within the original or unconstrained simplex if and only if

$$\sum_{i=1}^{q} U_i - U_{\min} \le 1 \qquad (12.6)$$

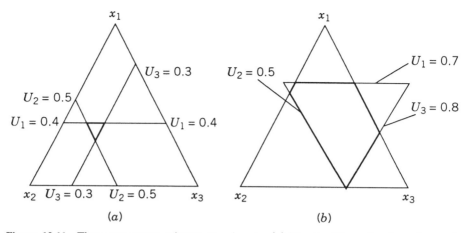

Figure 12.11 Three-component mixture experiments. (a) The feasible region is an inverted simplex. (b) The feasible region is not a simplex.

where U_{min} is the smallest of the upper bounds U_i. For example, for the situation in Figure 12.11(a) we have

$$\sum_{i=1}^{3} U_i = 0.4 + 0.5 + 0.3 = 1.2$$

and $U_{min} = 0.3$, so

$$\sum_{i=1}^{3} U_i - U_{min} = 1.2 - 0.3 = 0.9 < 1$$

and the feasible region is an inverted simplex. On the other hand, in Figure 12.11(b) we have

$$\sum_{i=1}^{3} U_i = 0.7 + 0.5 + 0.8 = 2.0$$

and $U_{min} = 0.5$, so

$$\sum_{i=1}^{3} U_i - U_{min} = 2.0 - 0.5 = 1.5 > 1$$

and the feasible region is not a simplex.

When two or more of the component proportions are constrained by upper bounds, Crosier (1984) suggests defining upper-bound pseudocomponents, or U-pseudocomponents:

$$u_i = \frac{U_i - x_i}{\sum\limits_{i=1}^{q} U_i - 1}, \qquad i = 1, 2, \ldots, q \qquad (12.7)$$

where

$$\sum_{i=1}^{q} U_i > 1$$

The region of the U-pseudocomponents is an inverted simplex; and when this inverted simplex lies entirely within the original simplex, it is easy to set up a standard simplex-type design in the U-pseudocomponents. The relationship between the settings in the original component proportions and the design points in the U-pseudocomponents u_i is

$$x_i = U_i - \left(\sum_{i=1}^{q} U_i - 1 \right) u_i \qquad (12.8)$$

One then typically fits a model in the U-pseudocomponents. However, because the orientation of this design space is inverted, one must be careful in interpreting the coefficients in the fitted model when making inferences about the fitted surface in the original components.

12.1.3 Active Upper- and Lower-Bound Constraints

We now consider the case where there are both lower and upper bounds on the component proportions. In these cases, the feasible mixture region is no longer a simplex. To illustrate, suppose that we wish to formulate a shampoo in terms of three component proportions $x_1 =$ lauryl sulfate, $x_2 =$ cocamide, and $x_3 =$ lauramide, and such that

$$x_1 + x_2 + x_3 = 0.50$$

$$0.20 \leq x_1 \leq 0.30$$

$$0.07 \leq x_2 \leq 0.10$$

$$0.13 \leq x_3 \leq 0.20$$

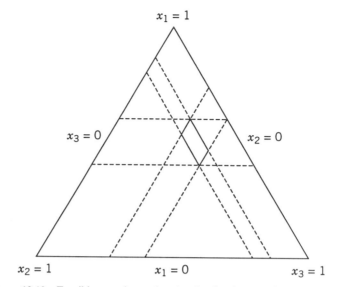

Figure 12.12 Feasible experimental region for the shampoo foam experiment.

The remaining 50% of the mixture consists of water, perfume, and coloring agents. The response variable of interest is the foam height of the shampoo.

Figure 12.12 shows the feasible region for this shampoo foam experiment. Remember that in constructing this graph, the components must be scaled so that $x_1 + x_2 + x_3 = 1$, so the upper and lower bounds on this graph were obtained by dividing the original upper and lower bounds by 0.5. Notice that the region is not a simplex; therefore, standard simplex-type mixture designs cannot be used.

An another example, consider the flare experiment described by McLean and Anderson (1966). The objective of the experiment was to determine the formulation of a railroad flare so as to maximize the illumination level. The flare consists of four ingredients: x_1 = magnesium, x_2 = sodium nitrate, x_3 = strontium nitrate, and x_4 = binder. The constraints on the four component proportions were

$$x_1 + x_2 + x_3 + x_4 = 1$$

$$0.40 \le x_1 \le 0.60$$

$$0.10 \le x_2 \le 0.50$$

$$0.10 \le x_3 \le 0.50$$

$$0.03 \le x_4 \le 0.08$$

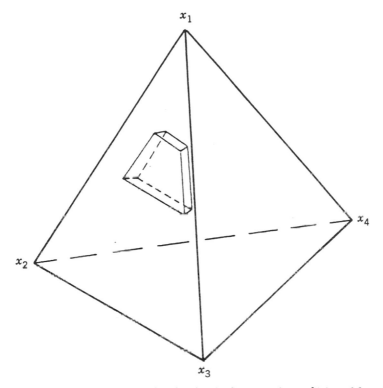

Figure 12.13 Constrained experimental region for the flare experiment. [Adapted from Cornell (1990), with permission of the publisher.]

The constrained experimental region is shown in Figure 12.13. Notice that the feasible region for this experiment is not a simplex.

In general, when both upper- and lower-bound constraints are active in a mixture experiment, the feasible design space is an irregular hyperpolytope, and, usually, some type of computer-generated design should be used in the experiment. We now discuss three popular criteria for constructing these designs.

Extreme Vertices Designs

The extreme vertices of the constrained region are formed by the combinations of the upper- and lower-bound constraints. In the shampoo foam study ($q = 3$ components) there are four extreme vertices, and in the flare problem ($q = 4$ components) there are eight extreme vertices. McLean and Anderson (1966) proposed using these points as the basis of a design, along with a subset of the remaining centroids or points that are in the center of the edges, faces, and so forth, including the overall centroid of the region.

Table 12.6 Extreme Vertices Design for McLean and Anderson's Flare Experiment

Run[a]	x_1	x_2	x_3	x_4	y
1	0.4000	0.4700	0.1000	0.0300	145
2	0.4000	0.1000	0.4700	0.0300	75
3	0.6000	0.2700	0.1000	0.0300	220
4	0.6000	0.1000	0.2700	0.0300	195
5	0.4000	0.4200	0.1000	0.0800	230
6	0.4000	0.1000	0.4200	0.0800	180
7	0.6000	0.2200	0.1000	0.0800	350
8	0.6000	0.1000	0.2200	0.0800	300
9	0.4000	0.2725	0.2725	0.0550	190
10	0.5000	0.1000	0.3450	0.0550	220
11	0.5000	0.3450	0.1000	0.0550	260
12	0.5000	0.2350	0.2350	0.0300	260
13	0.6000	0.1725	0.1725	0.0550	310
14	0.5000	0.2100	0.2100	0.0800	410
15	0.5000	0.2225	0.2225	0.0550	425

[a]Vertices are runs 1–8, constraint plane centroids are runs 9–14, and the overall centroid is run 15.

McLean and Anderson (1966) demonstrated this strategy for the flare problem assuming that a quadratic model would be adequate to describe the response. They set up a 15-point design consisting of the eight vertices, the six constraint plane centroids, and the overall centroid. This design, along with the response variable, is shown in Table 12.6.

While there are computer software programs that implement the extreme vertices design strategy, we feel that the other two approaches described below are generally superior. We recommend that one of these strategies be used in practice.

D-Optimal Designs

The *D*-optimal criterion (or other alphabetic optimality criteria) can be used to select points for a mixture design in a constrained region. Recall that this particular criterion selects design points from a list of candidate points so that the variances of the model regression coefficients are minimized. From a practical viewpoint, the points are selected so that the determinant of the **X'X** matrix is maximized.

For mixture experiments, these *D*-optimal design algorithms require (1) a set of reasonable candidate points from which to select the design points, (2) a convenient method for actually identifying the coordinates of these points in the constrained design space, and (3) a systematic procedure or set of rules for selecting the points. The set of candidate points to use should depend upon the order of the model the experimenter wishes to fit. Based on our practical experience, we would recommend the following:

1. **Linear Model.** The candidate points should include the vertices of the region, the edge centers, the overall centroid, and the axial points that are located halfway between the overall centroid and the vertices.
2. **Quadratic Model.** The candidate points should include the vertices, the edge centers, the constraint plane centroids, the overall centroid, and the axial points.
3. **Cubic or Special Cubic Model.** The candidate points should include the vertices, the thirds of edges, the constraint plane centroids, the overall centroid, and the axial points.

There are a variety of algorithms for finding the coordinates of the vertices of a constrained region. These include the extreme vertices algorithm of McLean and Anderson (1966), the XVERT algorithm of Snee and Marquardt (1974), the modification of this by Nigam et al. (1983), called the XVERT1 algorithm, and an algorithm described by Cornell (1990) based on the U-pseudocomponents. The coordinates of the other candidate points can be expressed as linear combinations of the coordinates of the extreme vertices. Piepel (1988) presents the CONAEV algorithm for finding the centers of edge points and all other centroids.

Once an appropriate set of candidate points has been identified, a D-optimal point selection algorithm can be used to select the points to be used in the design. In the following examples, we will illustrate one such algorithm.

Example 12.2 Consider the shampoo foam experiment described previously. Recall that

$$x_1 + x_2 + x_3 = 0.5$$

$$0.2 \leq x_1 \leq 0.3$$

$$0.07 \leq x_2 \leq 0.10$$

$$0.13 \leq x_3 \leq 0.20$$

Figure 12.12 shows the feasible region for this experiment (recall that in Figure 12.12 the component proportions are defined as $x_i/0.5$ so that the region can be drawn on the usual graph paper). We will construct a D-optimal design for this problem, assuming that the experimenter is planning to fit a quadratic model. We will specify as our candidate design points the four vertices of the region, the four edge centers, the overall centroid, and the four axial runs that lie midway between the centroid and the vertices of the constrained region. Table 12.7 lists these design points using the actual

Table 12.7 Candidate Design Points for the Shampoo Foam Experiment, Example 12.2

	Vertices					*Axial Check Points*		
1	Pseudo	1.0000	0.0000	0.0000	9	Pseudo	0.7500	0.0750	0.1750
	Real	0.6000	0.1400	0.2600		Real	0.5500	0.1550	0.2950
	Actual	0.3000	0.0700	0.1300		Actual	0.2750	0.0775	0.1475
2	Pseudo	0.7000	0.3000	0.0000	10	Pseudo	0.6000	0.2250	0.1750
	Real	0.5400	0.2000	0.2600		Real	0.5200	0.1850	0.2950
	Actual	0.2700	0.1000	0.1300		Actual	0.2600	0.0925	0.1475
3	Pseudo	0.0000	0.3000	0.7000	11	Pseudo	0.2500	0.2250	0.5250
	Real	0.4000	0.2000	0.4000		Real	0.4500	0.1850	0.3650
	Actual	0.2000	0.1000	0.2000		Actual	0.2250	0.0925	0.1825
4	Pseudo	0.3000	0.0000	0.7000	12	Pseudo	0.4000	0.0750	0.5250
	Real	0.4600	0.1400	0.4000		Real	0.4800	0.1550	0.3650
	Actual	0.2300	0.0700	0.2000		Actual	0.2400	0.0775	0.1825

	Edge Centers					*Overall Centroid*		
5	Pseudo	0.8500	0.1500	0.0000	13	Pseudo	0.5000	0.1500	0.3500
	Real	0.5700	0.1700	0.2600		Real	0.5000	0.1700	0.3300
	Actual	0.2850	0.0850	0.1300		Actual	0.2500	0.0850	0.1650
6	Pseudo	0.6500	0.0000	0.3500					
	Real	0.5300	0.1400	0.3300					
	Actual	0.2650	0.0700	0.1650					
7	Pseudo	0.3500	0.3000	0.3500					
	Real	0.4700	0.2000	0.3300					
	Actual	0.2350	0.1000	0.1650					
8	Pseudo	0.1500	0.1500	0.7000					
	Real	0.4300	0.1700	0.4000					
	Actual	0.2150	0.0850	0.2000					

component proportions, the proportions defined by $x_i/0.5$ (called "real proportions" in Table 12.7), and the L-pseudocomponents. Figure 12.14 shows the candidate points in terms of the L-pseudocomponents.

At least six of the 13 points in Table 12.7 must be selected in order to fit the quadratic model. We would recommend that several additional runs be used in the design so that an estimate of error can be obtained and model adequacy can be checked. In general, we recommend a minimum of 7–10 additional runs, with about half of these runs used as replicates and the other half chosen as distinct design points so that lack of fit of the model can be checked. See Snee (1985) and Montgomery and Voth (1994) for more discussion of these recommendations.

The experimenters decided to select 7 additional runs, allocated as 4 replicate runs and 3 distinct points. Along with the 6 runs required to fit the model, this results in a design with 13 runs. The D-optimal design was generated using the routine in the *Design-Expert* software package. This design is shown in Table 12.8. The "DSN ID" column indicates which

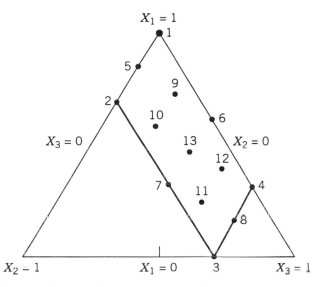

Figure 12.14 Feasible design region using the L-pseudocomponents for the shampoo foam height experiment in Example 12.2.

candidate points were selected, and that these points were also replicated. The remaining distinct design points are three edge centers (DSN ID points 5, 6, and 7), and two of the axial blends (points 9 and 12).

Table 12.9 shows the details of model fitting. Based on the relatively small value of PRESS, the adjusted R^2, and the P-value for the special cubic term, the experimenters selected the special cubic model as the best model for

Table 12.8   D-Optimal Design for the Shampoo Foam Experiment, Example 12.2

Observation	DSN ID	Run Order	X_1	X_2	X_3	y_1 Height (mm)
1	1	11	1.000	0.0000	0.0000	152.0
2	1	12	1.000	0.0000	0.0000	140.0
3	2	3	0.700	0.3000	0.0000	150.0
4	2	6	0.700	0.3000	0.0000	145.0
5	3	5	0.000	0.3000	0.7000	141.0
6	3	2	0.000	0.3000	0.7000	138.0
7	4	10	0.300	0.0000	0.7000	153.0
8	4	4	0.300	0.0000	0.7000	147.0
9	5	8	0.850	0.1500	0.0000	165.0
10	6	7	0.650	0.0000	0.3500	170.0
11	7	1	0.350	0.3000	0.3500	148.0
12	9	13	0.750	0.0750	0.1750	175.0
13	12	9	0.400	0.0750	0.5250	163.0

Table 12.9   Model Fitting for the Shampoo Foam Experiment, Example 12.2

Sequential Model Sum of Squares

Source	Sum of Squares	Degrees of Freedom	Mean Square	F-Value	Probability $> F$
Mean	303,705.3	1	303,705.3		
Linear	377.4	2	188.7	1.438	0.2825
Quadratic	1008.0	3	336.0	7.730	0.0126
Special cubic	153.4	1	153.4	6.103	0.0484
Residual	150.8	6	25.1		
Lack of fit	(43.8)	2	21.9	0.8193	0.5033
Pure error	(107.0)	4	26.8		
Total	305,395.0	13			

Lack-of-fit tests

Model	Sum of Squares	Degrees of Freedom	Mean Square	F-Value	Probability $> F$
Linear	1205.3	6	200.9	7.509	0.0357
Quadratic	197.3	3	65.8	2.458	0.2026
Special cubic	43.8	2	21.9	0.8193	0.5033
Pure Error	107.0	4	26.8		

ANOVA Summary Statistics of Models Fit

Source	Unaliased Terms	Residual Degrees of Freedom	Root Mean Squared Error	R-Squared	Adjusted R-Squared	PRESS
Linear	3	10	11.46	0.2234	0.0681	2054.97
Quadratic	6	7	6.59	0.8199	0.6913	1035.39
Special cubic	7	6	5.01	0.9107	0.8215	657.08

foam height. In terms of the pseudocomponents, this model is

$$\hat{y} = 146.74 X_1 - 370.32 X_2 + 94.90 X_3$$

$$+ 745.43 X_1 X_2 + 183.21 X_1 X_3 + 876.64 X_2 X_3 - 524.99 X_1 X_2 X_3$$

The analysis of variance for this model is shown in Table 12.10. Notice that all of the second-order terms and the special cubic term are statistically significant (their P-values are all less than 0.05). Residual plots from the model were satisfactory, and thus the special cubic equation above is the final model.

Figure 12.15 shows the shampoo foam height response surface, both as a contour plot and as a three-dimensional surface plot. The experimenters' objective was to formulate a product with foam height in excess of 170 mm.

Table 12.10   ANOVA for the Special Cubic Model, Shampoo Foam Experiment

Source	Sum of Squares	Degrees of Freedom	Mean Square	F-Value	Probability $> F$
Model	1538.9	6	256.48	10.20	0.0062
Residual	150.8	6	25.14		
Lack of fit	(43.8)	2	21.92	0.8193	0.5033
Pure error	(107.0)	4	26.75		
Corrected total	1689.7	12			

Root mean squared error	5.01	R-squared	0.9107
Dependent mean	152.85	Adjusted R-squared	0.8215
Coefficient of Variation	3.28%		

Independent Variable	Coefficient Estimate	Degrees of Freedom	Standard Error	t for H_0: Coefficient $= 0$	Probability $> \|t\|$
X_1	146.74	1	3.49	Not applicable	
X_2	-370.32	1	130.14	Not applicable	
X_3	94.90	1	15.01	Not applicable	
X_1X_2	745.43	1	185.13	4.026	0.0069
X_1X_3	183.21	1	43.18	4.243	0.0054
X_2X_3	876.64	1	187.33	4.680	0.0034
$X_1X_2X_3$	-524.99	1	212.51	-2.470	0.0484

From inspection of the response surface plots we note that there are many formulations that would satisfy this objective.

Figure 12.16 is the response surface trace plot. In this plot, the centroid of the constrained region is the reference mixture, and the response trace is computed along a line connecting the centroid to the original vertex $x_i = 1$. This is often called "*Cox's direction*" (see Section 12.4 for more discussion). All three components in this trace plot exhibit strong nonlinear effects. Because all of the higher-order terms in the model were significant, this result is not unexpected.

Experimenters often select D-optimal designs because the concept of minimizing the variance of the regression coefficients is intuitively pleasing. However, as we have noted before, this does not necessarily guarantee that the variance of the predicted response across the region of interest is well-behaved. In fact, the variance of predicted response may be very poorly behaved. Figure 12.17 shows contours of constant standard deviation of predicted foam height from the shampoo experiment. Notice the very irregular contours of constant standard deviation of predicted response. There are some areas in which the model predicts reasonably well, and other areas relatively nearby where the model predicts poorly. In particular, the model

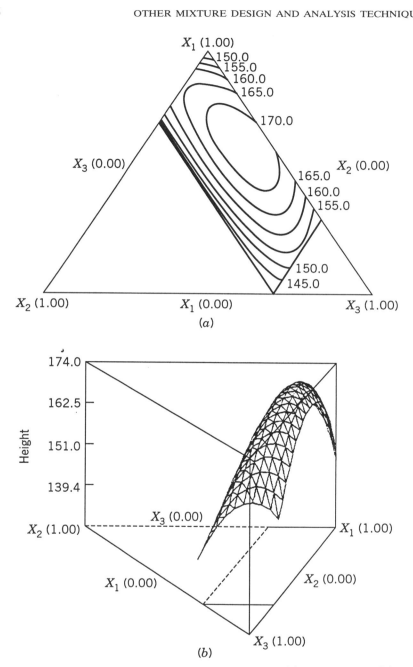

Figure 12.15 The shampoo foam height response surface. (*a*) Contour plot. (*b*) Three-dimensional surface plot.

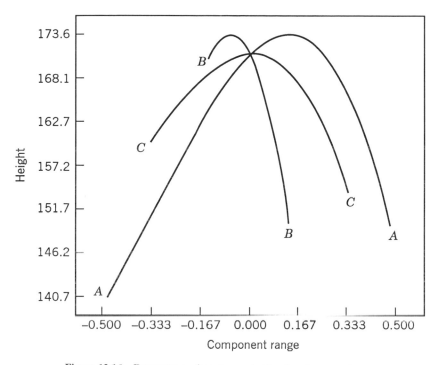

Figure 12.16 Response surface trace plot for the shampoo foam experiment.

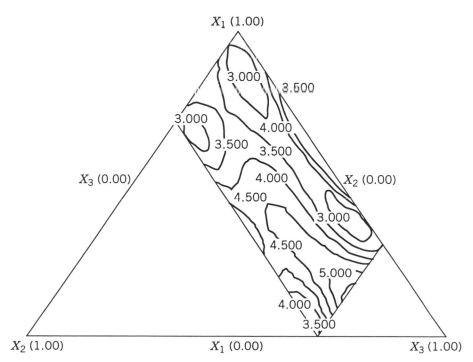

Figure 12.17 Contours of constant standard deviation of predicted response for the shampoo foam experiment.

Table 12.11 A 10-Point *D*-Optimal Design for the Shampoo Foam Experiment

DSN ID Selected	Type of Point
1	Vertex
2	Vertex
3	Vertex
4	Vertex
4	Vertex
5	Edge center
6	Edge center
7	Edge center
8	Edge center
13	Overall centroid

does not predict as well as one might wish near the center of the region. This is due, in part, to the tendency of the *D*-optimal criterion to "load up" the vertices of the region with design points (recall that each vertex is replicated twice in our design).

Another problem with *D*-optimal point selection is that the results obtained are often heavily dependent on the size of the design—that is, the number of runs requested. To illustrate, suppose we select a *D*-optimal design with 10 runs for the shampoo foam experiment, assuming that a quadratic model is of interest. Table 12.11 shows the design given by *Design-Expert*. Notice that the algorithm selects all four vertices, all four centers of edges, and the overall centroid, and it replicates one of the vertices. This is a very different design than the one obtained in Example 12.2, and we do not think that it is as good as the 13-point design that was actually run. The 10-point design simply doesn't have enough replicate runs to give a reasonable estimate of error.

As another example of constructing *D*-optimal designs for a constrained mixture problem, consider the flare problem of McLean and Anderson (1966), described earlier. Suppose that the experimenter initially entertains a quadratic mixture model for the illumination response. Table 12.12 lists the 8 vertices, the 12 edge centers, the 6 constraint plane centroids, the 8 axial check blends, and the overall centroid for the constrained region. Because the quadratic model has 10 parameters, a reasonable design should have between 17 and 20 runs; some of the additional runs are allocated to replicates, and the others are distinct design points for checking lack of fit.

Panel A of Table 12.13 shows a 19-point *D*-optimal design for the flare problem constructed using *Design-Expert*. We forced the algorithm to select five replicate runs and four distinct runs for testing lack of fit, in addition to the 10 distinct runs required to fit the model. Notice that the algorithm selected six of the eight vertices, five of the eight edge centers, two of the six

Table 12.12   Candidate Design Points for the Flare Experiment

Vertices					
1	Pseudo	0.0000	1.0000	0.0000	0.0000
	Actual	0.4000	0.4700	0.1000	0.0300
2	Pseudo	0.0000	0.0000	1.0000	0.0000
	Actual	0.4000	0.1000	0.4700	0.0300
3	Pseudo	0.5405	0.4595	0.0000	0.0000
	Actual	0.6000	0.2700	0.1000	0.0300
4	Pseudo	0.5405	0.0000	0.4595	0.0000
	Actual	0.6000	0.1000	0.2700	0.0300
5	Pseudo	0.0000	0.8919	0.0000	0.1081
	Actual	0.4000	0.4300	0.1000	0.0700
6	Pseudo	0.0000	0.0000	0.8919	0.1081
	Actual	0.4000	0.1000	0.4300	0.0700
7	Pseudo	0.5405	0.3514	0.0000	0.1081
	Actual	0.6000	0.2300	0.1000	0.0700
8	Pseudo	0.5405	0.0000	0.3514	0.1081
	Actual	0.6000	0.1000	0.2300	0.0700
Edge Centers					
9	Pseudo	0.0000	0.5000	0.5000	0.0000
	Actual	0.4000	0.2850	0.2850	0.0300
10	Pseudo	0.2703	0.7297	0.0000	0.0000
	Actual	0.5000	0.3700	0.1000	0.0300
11	Pseudo	0.0000	0.9459	0.0000	0.0541
	Actual	0.4000	0.4500	0.1000	0.0500
12	Pseudo	0.2703	0.0000	0.7297	0.0000
	Actual	0.5000	0.1000	0.3700	0.0300
13	Pseudo	0.0000	0.0000	0.9459	0.0541
	Actual	0.4000	0.1000	0.4500	0.0500
14	Pseudo	0.5405	0.2297	0.2297	0.0000
	Actual	0.6000	0.1850	0.1850	0.0300
15	Pseudo	0.5405	0.4054	0.0000	0.0541
	Actual	0.6000	0.2500	0.1000	0.0500
16	Pseudo	0.5405	0.0000	0.4054	0.0541
	Actual	0.6000	0.1000	0.2500	0.0500
17	Pseudo	0.0000	0.4459	0.4459	0.1081
	Actual	0.4000	0.2650	0.2650	0.0700
18	Pseudo	0.2703	0.6216	0.0000	0.1081
	Actual	0.5000	0.3300	0.1000	0.0700
19	Pseudo	0.2703	0.0000	0.6216	0.1081
	Actual	0.5000	0.1000	0.3300	0.0700
20	Pseudo	0.5405	0.1757	0.1757	0.1081
	Actual	0.6000	0.1650	0.1650	0.0700

Table 12.12 (*Continued*)

		Constraint Plane Centroids			
21	Pseudo	0.0000	0.4730	0.4730	0.0541
	Actual	0.4000	0.2750	0.2750	0.0500
22	Pseudo	0.2703	0.0000	0.6757	0.0541
	Actual	0.5000	0.1000	0.3500	0.0500
23	Pseudo	0.2703	0.6757	0.0000	0.0541
	Actual	0.5000	0.3500	0.1000	0.0500
24	Pseudo	0.2703	0.3649	0.3649	0.0000
	Actual	0.5000	0.2350	0.2350	0.0300
25	Pseudo	0.5405	0.2027	0.2027	0.0541
	Actual	0.6000	0.1750	0.1750	0.0500
26	Pseudo	0.2703	0.3108	0.3108	0.1081
	Actual	0.5000	0.2150	0.2150	0.0700
		Axial Check Points			
27	Pseudo	0.1351	0.6689	0.1689	0.0270
	Actual	0.4500	0.3475	0.1625	0.0400
28	Pseudo	0.1351	0.1689	0.6689	0.0270
	Actual	0.4500	0.1625	0.3475	0.0400
29	Pseudo	0.4054	0.3986	0.1689	0.0270
	Actual	0.5500	0.2475	0.1625	0.0400
30	Pseudo	0.4054	0.1689	0.3986	0.0270
	Actual	0.5500	0.1625	0.2475	0.0400
31	Pseudo	0.1351	0.6149	0.1689	0.0811
	Actual	0.4500	0.3275	0.1625	0.0600
32	Pseudo	0.1351	0.6149	0.1689	0.0811
	Actual	0.4500	0.1625	0.3275	0.0600
33	Pseudo	0.4054	0.3446	0.1689	0.0811
	Actual	0.5500	0.2275	0.1625	0.0600
34	Pseudo	0.4054	0.1689	0.3446	0.0811
	Actual	0.5500	0.1625	0.2275	0.0600
		Overall Centroid			
35	Pseudo	0.2703	0.3378	0.3378	0.0541
	Actual	0.5000	0.2250	0.2250	0.0500

constraint plane centroids, and one of the eight axial check blends. The replicates were run at four vertices and one edge center. For comparison, Panel B of Table 12.13 contains a 15-point *D*-optimal design, where no restrictions were placed on how the points were selected. Notice that this design contains all eight vertices, five edge centers, and two constraint plane centroids.

We prefer the design in Panel A because it contains (a) enough runs to fit higher-order terms should the quadratic model be inadequate (a relatively

Table 12.13   Two *D*-Optimal Designs for the Flare Experiment

A		B		
A 19-Point Design, with Five Replicates Required[a]		A 15-Point Design[a]		
1, 1	11	1	6	11
2, 2	14	2	7	15
4	15	3	8	18
5	18	4	9	21
6, 6	22	5	10	22
8, 8	26			
9, 9	27			

[a]The design point identification numbers refer to the points listed in Table 12.12.

common occurrence in mixture experiments) and (b) replicate runs so that an estimate of error can be made. Obviously, many other designs could be generated for this problem. In actual practice, we often generate several *D*-optimal designs for a problem, varying the number of replicate runs and the total number of runs on each trial. In this way, the experimenter can more clearly see any benefit associated with making a few additional runs, as well as get some idea of the total number of runs that should be made. Another strategy that we have used is to generate a *D*-optimal design with a few runs beyond the minimum number (but no required replicates), and then select a few points in this design for replication. For example, we might start with the 15-point design for the flare problem in Panel B of Table 12.13 and then replicate four or five of those points. We usually replicate design points that have higher leverage values whenever possible, because this generally produces smaller variance of prediction in the neighborhood of these high-leverage points, and it minimizes the impact of unusual observations or outliers that might occur at these points. For an example of this strategy applied to the flare problem, see Montgomery and Voth (1994).

Distance-Based Designs
The distance point selection criterion attempts to spread the design points out uniformly over the feasible region. The algorithm for selecting points is very simple: Start with the point that is as close as possible to a vertex of the unconstrained region, and then add to this the point for which the Euclidean distance to the first point is a maximum. All subsequent points are added similarly; that is, choose the point for which the minimum Euclidean distance to the other points in the design is maximum. Obviously, the distance criterion requires a grid of candidate points. We usually recommend the same candidate points that would be used for the *D*-optimal criterion.

Table 12.14   Distance-Based Designs for the Shampoo Foam Experiment

A		B	
A 13-Point Design with Four Replicates Required[a]		A 10-Point Design[a]	
1, 1	9	1	6
2, 2	10	2	9
3	11	3	10
4, 4	13	4	11
5, 5		5	13

[a]The design point identification numbers refer to the points listed in Table 12.7.

To illustrate the types of designs produced by the distance criterion, consider the two distance-based designs for the shampoo foam experiment shown in Table 12.14. Panel A of this table contains a 13-point design. In constructing this design, we forced the algorithm to replicate four runs. Panel B of Table 12.14 contains a 10-point design in which we did not require any replication.

The 13-point design uses all four vertices, one of the edge centers, three of the axial points, and the overall centroid. Three of the vertices and the edge center are replicated. In general, the distance criterion will usually force more points into the interior of the feasible region than will the D-optimal criterion. As an illustration, compare this design with the 13-point D-optimal design in Table 12.8. Notice that the distance criterion placed four runs in the interior, whereas the D-optimal criterion placed only two runs in the interior. The 10-point distance design (Panel B of Table 12.14) also has four interior runs, whereas the 10-point D-optimal design (Table 12.11) has only one. Many experiments consider this "uniform" spacing of design points over the region of interest a useful feature of the distance criterion.

Table 12.15 presents two designs for the flare problem that further illustrate the distance criterion. A 19-point design in which five replicate points were forced into the design is in Panel A, and a 15-point design is shown in Panel B. The 19-point design contains five vertices, four edge centers, four axial runs, and the overall centroid; one of the vertices and the four edge centers are replicated. This is very different from the 19-run D-optimal design for this problem, which selected more of the vertices, edge centers, and constraint plane centroids. The 15-point distance design has six vertices, four edge centers, four axial runs, and the overall centroid. In contrast, the 15-point D-optimal design has all eight vertices, five edge centers, and two constraint plane centroids. The tendency of the distance criterion to put points in the interior of the region is clearly evident.

Generally, the distance criterion first chooses points to evenly cover the boundary of the region, and then it adds interior points only when these points are further from the points already in the design than other nonin-

Table 12.15 Distance-Based Designs for the Flare Problem

A			B		
A 19-Point Design with Five Replicates Required[a]			A 15-Point Design[a]		
1	9, 9	27	1	6	27
2	18, 18	28	2	9	28
3	19, 19	29	3	18	29
4	20, 20	30	4	19	30
5, 5		35	5	20	35

[a]The design point identification numbers refer to the points listed in Table 12.12.

cluded boundary points. As the number of variables increases, the likelihood of selecting interior points in a design with a reasonable number of points goes down, and thus for four or more components we usually prefer the D-optimal criterion. We have encountered cases where the distance criterion selects many low-dimensional centroids or axial check blends and relatively few vertices. Such designs would generally not be preferable to a D-optimal design from a variance viewpoint. Consequently, we recommend examining plots of the variance of prediction over the region of interest as an aid in design selection.

12.1.4 Multicomponent Constraints

Sometimes in addition to upper and/or lower bounds on the individual component proportions, there are other linear constraints, such as

$$C_j \le A_{1j}x_1 + A_{2j}x_2 + \cdots + A_{qj}x_q \le D_j \tag{12.9}$$

for $j = 1, 2, \ldots, m$. These types of constraints are called **multicomponent constraints**. Some of the constants A_{ij} may be zero, so that in general, not all components are forced to appear in each constraint.

When multicomponent constraints are present, they can change the shape of the feasible region for the experiment. That is, they can change the number of vertices, edges, and so on, that define the feasible space. To illustrate, consider the following example:

$$x_1 + x_2 + x_3 = 1$$
$$0.1 \le x_1$$
$$x_2 \le 0.8$$
$$x_3 \le 0.7$$
$$0.4 \le x_1 + 0.8x_2$$

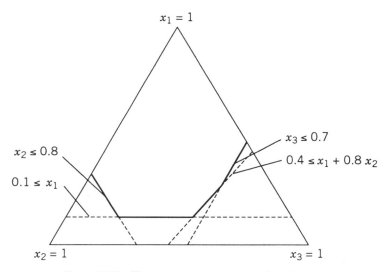

Figure 12.18   Three-component constrained region.

Figure 12.18 shows the feasible region for this design. Notice that the effect of the multicomponent constraint is to "cut off" part of the region that is defined by the upper- and lower-bound constraints. As a result, both the number of vertices and edges in the feasible region are increased by one.

Not all computer software for mixture experiments can handle multicomponent constraints. Piepel (1988) has published an algorithm called CON-VRT, which can locate the extreme vertices of the region when multicomponent constraints are active. His paper appeared in the Computer Programs Column of the April 1988 issue of the *Journal of Quality Technology*. Design-Expert is a commercially available code that can handle multicomponent constraints.

12.2   MIXTURE EXPERIMENTS USING RATIOS OF COMPONENTS

In some mixture experiments, the experimenter may wish to work with ratios of the mixture components instead of the original component proportions. For example, in formulating a detergent, it may be convenient or natural for the chemist to think of the ratio of an emulsifier to water rather than the actual proportion of each ingredient in a blend. In such an experiment, it might be natural to think of all the ingredients as a ratio of the ingredient to water.

There are usually several ways to set up the ratios. For example, if there are three components, then one way to define the ratio variables is

$$r_1 = \frac{x_1}{x_3}, \qquad r_2 = \frac{x_2}{x_3}$$

Another possibility is

$$r_1 = \frac{x_2}{x_1}, \qquad r_2 = \frac{x_2}{x_3}$$

In general, each ratio r_1 and r_2 must contain at least one of the components used in the other ratio. This holds true in general: If there are q components, then we may define $q - 1$ ratios $r_1, r_2, \ldots, r_{q-1}$, and each ratio should contain at least one of the components used in at least one of the other ratios in the set. The $q - 1$ ratio variables are **independent**, consequently any type of standard response surface polynomial can be fit in the ratios $r_1, r_2, \ldots, r_{q-1}$. Obviously, standard response surface designs can be used in these situations.

Example 12.3   Cornell (1990) presents a three-component example of blending refinery products to form gasoline. The response variable of interest is octane number. The ratios used are

$$r_1 = \frac{X_3}{X_1}, \qquad r_2 = \frac{X_3}{X_2}$$

and the process currently operates at $r_1 = 1$, $r_2 = 2$ at a composition of $x_1 = 0.4$, $x_2 = 0.2$, and $x_3 = 0.4$. The region of interest is constrained by

$$0.2 \le x_1 \le 0.6$$
$$0.1 \le x_2 \le 0.4$$
$$0.2 \le x_3 \le 0.6$$

The feasible region is shown in Figure 12.19.

Suppose that the experimenter wishes to fit the quadratic model

$$y = \beta_0 + \beta_1 r_1 + \beta_2 r_2 + \beta_{12} r_1 r_2 + \beta_{11} r_1^2 + \beta_{22} r_2^2 + \varepsilon$$

in the ratio variables. Any type of response surface design in the two variables r_1 and r_2 could be used. Cornell chose values of the ratios of $r_1 = (0.5, 1.0, 1.5)$ and $r_2 = (1.0, 2.0, 3.0)$ and used a 3^2 design. The rays representing these ratios are shown in Figure 12.19. The nine design points are at the intersection of the six rays. Table 12.16 shows the design in both the ratios and the original component proportions. The settings of the original components in terms of the ratios (which are required to run the experiment) were found by solving

$$r_1 = \frac{x_3}{x_1}, \qquad r_2 = \frac{x_3}{x_2}, \qquad x_1 + x_2 + x_3 = 1$$

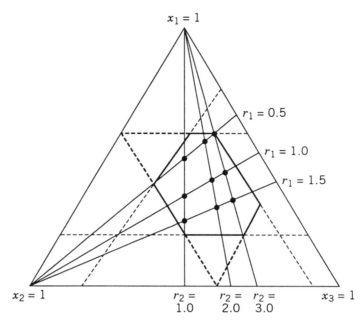

Figure 12.19 Feasible design region for Example 12.3, along with design points in the ratio variables.

for x_1, x_2, and x_3, yielding

$$x_1 = \frac{r_2}{r_1 + r_1 r_2 + r_2}$$

$$x_2 = \frac{r_1 r_2}{r_1 + r_1 r_2 + r_2}$$

$$x_3 = \frac{r_1}{r_1 + r_1 r_2 + r_2}$$

Table 12.16 Octane Data for Example 12.3

Run	$r = x_3/x_1$	$r = x_3/x_2$	x_1	x_2	x_3	Octane Number, y
1	0.5	1.0	0.50	0.25	0.25	82.8
2	1.0	1.0	0.33	0.33	0.33	87.3
3	1.5	1.0	0.25	0.37	0.37	84.9
4	0.5	2.0	0.57	0.14	0.29	85.6
5	1.0	2.0	0.40	0.20	0.40	93.0
6	1.5	2.0	0.31	0.23	0.46	89.3
7	0.5	3.0	0.60	0.10	0.30	87.2
8	1.0	3.0	0.43	0.14	0.43	89.6
9	1.5	3.0	0.33	0.17	0.50	90.6

Table 12.17   Analysis of Variance for the Quadratic Model in Example 12.3

Source	Sum of Squares	Degrees of Freedom	Mean Square	F-Value	Probability $> F$
Model	71.04	5	14.208	5.205	0.1025
Residual	8.19	3	2.730		
Corrected total	79.23	8			

Root mean square error	1.652		R-squared 0.8966		
Dependent mean	87.811		Adjusted R-squared 0.7244		
Coefficient of Variation	1.88%				

Independent Variable	Coefficient Estimate	Degrees of Freedom	Standard Error	t for H_0 Coefficient $= 0$	Probability $> \lvert t \rvert$
Intercept	91.456	1	1.231	74.27	
$A = z_1$	1.533	1	0.674	2.273	0.1076
$B = z_2$	2.067	1	0.674	3.064	0.0548
A^2	−3.233	1	1.168	−2.768	0.0697
B^2	−2.233	1	1.168	−1.912	0.1519
AB	0.325	1	0.826	0.3934	0.7203

Table 12.17 presents the analysis of variance summary for the quadratic model. The model coefficients in this table are given in terms of "coded" orthogonal variables z_1 and z_2, where

$$z_1 = \frac{r_1 - 1.0}{0.5}, \qquad z_2 = \frac{r_2 - 2.0}{1.0}$$

In terms of these variables, the model is

$$\hat{y} = 91.456 + 1.533z_1 + 2.067z_2 + 0.325z_1z_2 - 3.233z_1^2 - 2.233z_2^2$$

The corresponding model in terms of the ratios r_1 and r_2 is

$$\hat{y} = 63.689 + 27.633r_1 + 10.350r_2 + 0.650r_1r_2 - 12.933r_1^2 - 2.233r_2^2$$

Figure 12.20 is the octane contour plot in terms of the ratios r_1 and r_2. If the experimenter wants to blend gasoline with an octane number of 89, we see from this plot that several blends will satisfy this objective. For example, $r_1 = 1.0$ and $r_2 = 1.33$ is a feasible set of ratios. In terms of the original

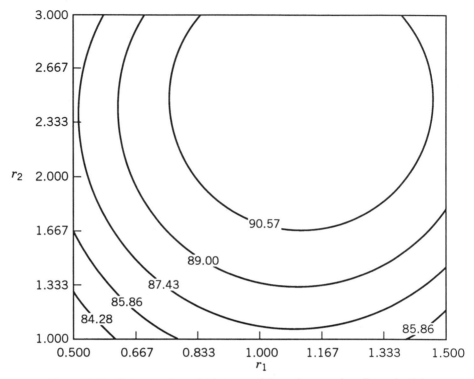

Figure 12.20   Octane contour plot in terms of the ratios r_1 and r_2, Example 12.3.

components, this point is

$$x_1 = \frac{r_2}{r_1 + r_1 r_2 + r_2} = \frac{1.33}{1.0 + (1.0)1.33 + 1.33} = 0.36$$

$$x_2 = \frac{r_1 r_2}{r_1 + r_1 r_2 + r_2} + \frac{(1.0)1.33}{1.0 + (1.0)1.33 + 1.33} = 0.36$$

$$x_3 = \frac{r_1}{r_1 + r_1 r_2 + r_2} = \frac{1.0}{1.0 + (1.0)1.33 + 1.33} = 0.28$$

The ratio variables approach is a very useful one, and we recommend it in cases where it is convenient or natural for the experimenters to work in terms of these variables. The approach does have two potential disadvantages. First, the interpretation is in terms of a function of the original component proportions (the ratios), and this may not be a natural framework for the experimenter to work in. The second disadvantage relates to how the design in terms of the ratio variables fits into the feasible mixture space. By referring to Figure 12.19, we see that the feasible mixture space for Example 12.3 is actually larger than the region of experimentation for the 3^2 design in the ratios r_1 and r_2. If it is important to predict the response over this entire region, then the 3^2 design would not be as a good a choice as a D-optimal design in the original component proportions. See Piepel and Cornell (1994) for other comments on the use of ratios.

12.3 PROCESS VARIABLES IN MIXTURE EXPERIMENTS

Process variables are factors in an experiment that are not mixture components but which could affect the blending properties of the mixture ingredients. For example, the effectiveness of an etching solution (measured as an etch rate, say) is not only a function of the proportions of the three acids that are combined to form the mixture, but also depends on the temperature of the solution and the agitation rate.

In general, there may be q mixture components $x_1, x_2, \ldots, x_q$ and p process variables $z_1, z_2, \ldots, z_p$. The etching example above has $q = 3$ mixture components and $p = 2$ process variables.

There are two approaches to including process variables in mixture experiments. The first approach involves transforming the mixture variables into $(q - 1)$-independent variables and then treating all $p + q - 1$ factors in a conventional (nonmixture) experiment. Using ratios of the components is one transformation of the mixture variables that is widely used in practice. To illustrate, consider the etching solution described above, and let the ratios be defined as

$$r_1 = \frac{x_1}{x_3}, \qquad r_2 = \frac{x_2}{x_3}$$

The process variables are $z_1 =$ temperature and $z_2 =$ agitation rate. Then a simple model relating these factors is

$$E(y) = \beta_0 + \beta_1 r_1 + \beta_2 r_2 + \alpha_1 z_1 + \alpha_2 z_2$$
$$+ \delta_{11} r_1 z_1 + \delta_{12} r_1 z_2 + \delta_{21} r_2 z_1 + \delta_{22} r_2 z_2$$
$$+ \beta_{12} r_1 r_2 + \alpha_{12} z_1 z_2$$

In this model, we may obtain information about the linear and nonlinear blending properties of the mixture components through the parameters β_1, β_2, and β_{12}; information about the main effects and interaction of the process variables through the parameters α_1, α_2, and α_{12}; and information on the dependency between the mixture components and the process variables through the parameters δ_{ij}. In addition to the ratios approach, there are

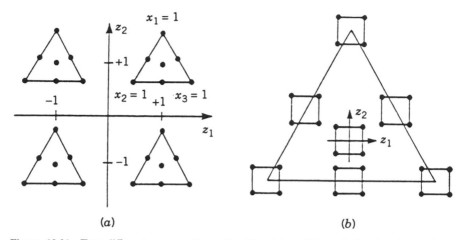

(a)   (b)

Figure 12.21 Two different presentations of a 28-point combined design for three mixture components and two process variables. (a) The seven-point simplex-centroid constructed at each of the $2^2 = 4$ points of the factorial arrangement. (b) A complete 2^2 constructed at each of the seven points of the simplex-centroid design. [From Cornell (1990), with permission of the publisher.]

other transformations that could be used to convert the mixture components into a set of $(q - 1)$-independent mixture-related variables. For a good discussion of these transformations, see Cornell (1990).

The second approach is to directly model the mixture components. This involves setting up a mixture design at each treatment combination of the factorial experiment used for the process variables. Alternatively, we can think of this as setting up a factorial in the process variables at each point in the mixture design. These two viewpoints of the design problem are shown in Figure 12.21 for the etching problem described above, assuming that a 2^2 factorial is used for the process variables and a simplex-centroid design is used for the three mixture components. The combined design has 28 points, and can be used to collect response data for fitting a combined model that includes both the mixture and the process variables. If we wish to fit a quadratic model in the mixture variables and a model with both main effects and the two-factor interaction in the process variables, then the combined model can be written as

$$
E(y) = \sum_{i=1}^{3} \beta_i x_i + \sum_{i}^{3} \sum_{<j} \beta_{ij} x_i x_j
$$

$$
+ \sum_{k=1}^{2} \left[\sum_{i=1}^{3} \alpha_{ik} x_i + \sum_{i}^{3} \sum_{<j} \alpha_{ijk} x_i x_j \right] z_k
$$

$$
+ \left[\sum_{i=1}^{3} \delta_{i12} x_i + \sum_{i}^{3} \sum_{<j} \delta_{ij12} x_i x_j \right] z_1 z_2 \qquad (12.10)
$$

This model has 24 parameters. Fortunately, they have relatively straightforward interpretations. The first six terms on the right-hand side of Equation (12.10) are the linear and nonlinear blending portion of the model. They do not depend on the settings of the process variables. All remaining parameters measure the effects of the process variables on the linear and nonlinear blending properties of the mixture components.

The usual approach to analyzing such a model is to fit the complete combined model and then test hypotheses on individual parameters or groups of parameters using the "extra sum of squares" method. This complete combined model would be very useful for predicting the response of a particular mixture formulation at combinations of the process variables that are not factorial design points. To predict at design points in the factorial space we would recommend fitting individual mixture models to the mixture experiments at each factorial point. Contour plotting of the response in terms of the mixture components at each point in the factorial array is then a reasonably effective way to interpret the response surface.

As the number of mixture components and process variables increase, the size of these combined designs becomes very large. Several authors have suggested fractionating these combined designs, although there are no "obvious" best choices of the fractional design. This is a significant problem when there are constraints on the mixture components, and the mixture space is a hyperpolytope. The D-optimal criterion is often used for generating a design for these problems. In order to produce valid models, we caution the experimenter to be sure that there are sufficient runs in the mixture part of the experiment to support the mixture polynomial at each treatment combination in the factorial design. That is, if the combined design has too many runs, we would prefer to fractionate the factorial part of the design. Then one should fit individual mixture models to the data obtained at each different treatment combination in the process variables and use these fitted models to generate response surface contour plots for analysis. We feel that this is a safer approach to prediction and optimization than direct interpretation of the combined model. For more discussion of mixture experiments with process variables, see Cornell (1990), Piepel and Cornell (1994), and Cornell (1995).

12.4 SCREENING MIXTURE COMPONENTS

In some mixture experiments there are a large number of components, and the initial objective of the experimenter is to screen these components to identify the ones that are most important. In fact, any time there are six or more components ($q \geq 6$) the experimenter should consider a **screening experiment** to reduce the number of components.

Screening is often done using the first-order mixture model

$$E(y) = \beta_1 x_1 + \beta_2 x_2 + \cdots + \beta_q x_q \qquad (12.11)$$

If the experimental region is a simplex, it is usually a good idea to make the ranges of the components as similar as possible, because then the relative effects of the components can be assessed by ranking the ratios of the parameter estimates β_1, $i = 1, 2, \ldots, q$ relative to their standard errors. Obviously, important components will have large values of $b_i/\mathrm{se}(b_i)$. If the effects of two or more components are equal, then the proportions of the individual components are summed and their sum is considered to be a "new" component.

An **axial design** in which the design points are placed only on the component axes is often recommended for screening mixture components. The simplest type of axial design has points positioned equidistant from the centroid toward each vertex. Generally, as the number of components increases, the greater the "spread" of these axial points should be in order to increase the precision of the parameter estimates. If the vertices of the simplex are complete mixtures, as will be the case if only upper-bound constraints are active, we recommend placing the axial runs at the vertices. On the other hand, if the vertices are pure blends, some experimenters prefer to run the axial check blends as the axial design, because every run in the design will be a complete mixture.

It is, of course, still possible to examine the ratios $b_i/\mathrm{se}(b_i)$ in situations where the constraints form a region of experimentation that is not a simplex. We usually suggest a D-optimal design for the linear model [Equation (12.11)] in these cases.

To screen out unimportant components, it is necessary to measure the effects of the individual components. In general, we define the effect of component i as the change in the expected response from a change in the proportion of component i while holding the relative proportions of the other components constant. The largest change that we can make in component i is from 0 to 1. To keep the proportions of the other components constant, this change must be made along the x_i-axis. The estimate of the response when $x_i = 0$ is

$$\hat{y} = \sum_{j \neq i}^{q} b_j [1/(q - 1)]$$

$$= (q - 1)^{-1} \sum_{j \neq i}^{q} b_j$$

Similarly, at the vertex where $x_i = 1$, the predicted response is

$$\hat{y} = b_i$$

If the first-order model is adequate, the **effect of component** i is just the difference in these two predicted responses, or

$$\text{Effect of component } i = b_i - (q-1)^{-1} \sum_{\substack{j \neq i}}^{q} b_j, \qquad i = 1, 2, \ldots, q \quad (12.12)$$

This equation can be used for model reduction. If the true surface is linear and

$$b_i = (q-1)^{-1} \sum_{\substack{j \neq i}}^{q} b_j$$

then the effect of component i is zero. Then the term $\beta_i x_i$ can be removed from the model and component x_i is not considered further in the experiment.

When the feasible region is constrained, then often we find that the ranges of the individual components are not equal. This renders the estimate of an effect in Equation (12.12) less useful. A simple way to remedy this situation is to **adjust** the component effect estimate as follows:

$$\text{Adjusted effect of component } i = R_i \left[b_i - (q-1)^{-1} \sum_{\substack{j \neq i}}^{q} b_j \right], \quad i = 1, 2, \ldots, q$$

$$(12.13)$$

where $R_i = U_i - L_i$ is the range of component i. Notice that the adjustment essentially weights the component effect by the range of the ith component. Piepel (1982) and Cornell (1990, Section 5.9.2) also discuss measuring component effects in constrained regions.

The response trace plots introduced in Chapter 11 (Section 11.3) are also very useful in screening mixture components. If one (or more) of the curves on the response trace plot is a horizontal line, then we have detected an **inactive ingredient**. In constrained regions that are not simplexes, care must be taken in defining both the reference mixture and the direction for measuring the component effect. Generally, the centroid of the constrained region is chosen as the reference mixture, and the effect of the ith component is measured along a line connecting this centroid to the vertex $x_i = 1$. This direction was introduced by Cox (1971) in presenting new mixture models as alternatives to those proposed by Scheffe. Consequently, this is sometimes called *Cox's direction*. The construction of the response surface trace plot for constrained regions is then identical to the procedure discussed in Section 11.3 for a simplex region of experimentation. The trace plot in Figure 12.16 (the shampoo foam experiment, Example 12.2) uses Cox's direction. Piepel (1982) introduced another direction that could be used to construct trace plots. His direction is the line extending from a reference mixture (usually the centroid of the constrained region) to the vertices of the L-pseudocomponent region. Figure 12.22 illustrates and compares these two directions for the shampoo foam experiment. These two directions are very closely related, and usually there is little difference in the trace plots.

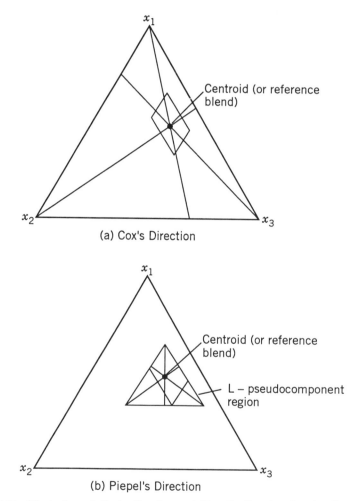

Figure 12.22 Illustration of Cox's direction and Piepel's direction for the Shampoo Foam Experiment, Example 12.1.

EXERCISES

12.1 Suppose that you have fit a second-order mixture model using L-pseudocomponents for $q = 3$. The fitted model is

$$\hat{y} = b_1 X_1 + b_2 X_2 + b_3 X_3 + b_{12} X_1 X_2 + b_{13} X_1 X_3 + b_{23} X_2 X_3$$

You wish to convert this to a model in the original components, say

$$\hat{y} = a_1 x_1 + a_2 x_2 + a_3 x_3 + a_{12} x_1 x_2 + a_{13} x_1 x_3 + a_{23} x_2 x_3$$

Show that the a's can be expressed in the terms of the b's as follows:

$$a_1 = \frac{b_{12} L_2 (L_1 - 1) + b_{13} L_3 (L_1 - 1) + b_{23} L_2 L_3}{(1 - L)^2} + \frac{b_1 - \sum_{i=1}^{3} b_i L_i}{1 - L}$$

$$a_2 = \frac{b_{12} L_1 (L_2 - 1) + b_{13} L_1 L_3 + b_{23} L_3 (L_2 - 1)}{(1 - L)^2} + \frac{b_2 - \sum_{i=1}^{3} b_i L_i}{1 - L}$$

$$a_3 = \frac{b_{12} L_1 L_2 + b_{13} L_1 (L_3 - 1) + b_{23} L_2 (L_3 - 1)}{(1 - L)^2} + \frac{b_3 - \sum_{i=1}^{3} b_i L_i}{1 - L}$$

$$a_{ij} = b_{ij} / (1 - L)^2, \qquad i, j = 1, 2, 3; i < j$$

12.2 An article in *Industrial Quality Control* ("Experiments with Mixtures of Components Having Lower Bounds," by I. S. Kurotori, Vol. 22, 1966, pp. 592–596) describes an experiment to investigate the formulation of a rocket propellant. The propellent consists of fuel (x_1), oxidizer (x_2), and binder (x_3). The restrictions on the component proportions are $0.20 \le x_1$, $0.40 \le x_2$, and $0.20 \le x_1 \le 0.40$, with $x_1 + x_2 + x_3 = 1$. The response variable of interest is elasticity.

(a) Draw a graph of the feasible region.

(b) The data collected by Kurotori are shown below:

Design Point	x_1	x_2	x_3	X_1	X_2	X_3	Elasticity
1	0.40	0.40	0.40	1	0	0	2650
2	0.20	0.60	0.20	0	1	0	2450
3	0.20	0.40	0.40	0	0	1	2350
4	0.30	0.50	0.20	0.50	0.50	0	2950
5	0.30	0.40	0.30	0.50	0	0.50	2750
6	0.20	0.50	0.30	0	0.50	0.50	2400
7	0.27	0.46	0.27	0.333	0.33	0.333	3000
8	0.33	0.43	0.23	0.667	0.166	0.166	2980
9	0.23	0.53	0.23	0.166	0.667	0.166	2770
10	0.23	0.43	0.33	0.166	0.166	0.667	2690

What type of experimental design has been used? Critique the choice of design.

(c) Fit an appropriate model relating elasticity to the component proportions (use the pseudocomponents).

(d) Convert the model found in part (c) to an equivalent model in the original component proportions.

(e) Plot the contours of constant elasticity and interpret the fitted response surface.

12.3 Koons and Wilt (1985) describe an experiment involving the formulation of acrylonitrile-butadiene styrene (ABS), a material used to make plastic pipe. ABS is normally made from two materials: grafted polybutadiene (x_1 = graft) and styrene-acrylonitrile (x_2 = SAN). Mixture experiments were performed to determine whether or not coal-tar pitch (x_3 = pitch) could be added to the SAN–graft combination and form a pipe that would have physical properties which met ASTM specifications. The motive was to reduce pipe cost and utilize a by-product produced elsewhere in the plant. Experience indicated that the following constraints (on weight) should be used:

$$0.45 < x_1 < 0.70$$

$$0.30 < x_2 < 0.55$$

$$0.0 < x_3 < 0.25$$

The physical properties of interest for the ABS pipe are: Izod impact strength (y_1, ft-lb/ in.), deflection temperature under load (y_2, °F) and yield strength (y_3, psi).

(a) Graph the feasible region for this experiment.

(b) The response data obtained in the experiment performed by Koons and Wilt are shown below:

Run	x_1	x_2	x_3	X_1	X_2	X_3	y_1	y_2	y_3
1	0.700	0.300	0.000	1	0	0	7.4	205	4350
2	0.450	0.550	0.000	0	1	0	6.1	213	6080
3	0.450	0.300	0.250	0	0	1	0.8	183	5065
4	0.575	0.425	0.000	0.50	0.50	0	7.3	204	5280
5	0.575	0.300	0.125	0.50	0	0.50	3.9	188	5080
6	0.450	0.425	0.125	0	0.50	0.50	2.1	191	5740
7	0.535	0.380	0.085	0.34	0.32	0.34	4.4	197	5410
8	0.600	0.350	0.050	0.60	0.20	0.20	6.1	192	5035
9	0.500	0.450	0.050	0.20	0.60	0.20	4.9	202	5615
10	0.500	0.350	0.150	0.20	0.20	0.60	2.5	186	5385

(c) What type of experimental design has been used? Comment on the experimenter's choice of design.

(d) Build appropriate response surface models for all three responses. Build these models using the pseudocomponents.

(e) Convert the models found in part (d) to equivalent models in the actual component proportions.

(f) Construct response surface contour plots for the three response variables y_1, y_2, and y_3.

(g) In order to produce acceptable pipe from this modified ABS material, the following response constraints must be satisfied: $y_1 > 1$ ft-lb/ in, $y_2 > 190°F$, and $y_3 > 5000$ psi. Is there a formulation of the modified ABS material that will satisfy these specifications?

(h) If the objective is to include as much pitch in the formulation as possible and still satisfy the response specifications in part (g), what formulation would you recommend?

12.4 The following constraints are imposed on three mixture components:

$$0.15 \le x_1, \qquad 0.25 \le x_2, \qquad 0.10 \le x_3$$

$$x_1 + x_2 + x_3 = 1$$

(a) Draw a graph of the feasible region.

(b) Suppose that you could only afford to perform eight runs, and your objective is to fit a first-order mixture model. What runs would you recommend?

(c) Suppose that your objective is to fit a quadratic mixture model. If you can afford to perform 12 runs, what blends would you include in the design?

(d) List the points in the design from part (c) in terms of the L-pseudocomponents.

12.5 Consider the data from McLean and Anderson's flare experiment, shown in Table 12.6.

(a) Convert the component setting in the extreme vertices design in this table to L-pseudocomponents.

(b) Fit a quadratic mixture model to the illumination response, using the L-pseudocomponents. Comment on the adequacy of this model.

(c) Add the special cubic terms to the model found in part (b) above. Does the addition of these terms improve the model?

(d) Construct response surface contour plots in terms of components x_1, x_2, and x_3 for various levels of x_4. Interpret the fitted surface.

12.6 The table shown below presents the actual values of the observed response from the shampoo foam height experiment in Example 12.2, the predicted response from the quadratic model, the residuals, the values of h_{ii}, and the studentized residuals.

Run	Actual Value	Predicted Value	Residual	h_{ii}	Studentized Residual
1	152.00	146.74	5.26	0.484	1.461
2	140.00	146.74	− 6.74	0.484	− 1.872
3	150.00	148.16	1.84	0.466	0.501
4	145.00	148.16	− 3.16	0.466	− 0.863
5	141.00	139.43	1.57	0.499	0.443
6	138.00	139.43	− 1.43	0.499	− 0.403
7	153.00	148.93	4.07	0.473	1.119
8	147.00	148.93	− 1.93	0.473	− 0.529
9	165.00	164.22	0.78	0.617	0.250
10	170.00	170.28	− 0.28	0.874	− 0.156
11	148.00	146.95	1.05	0.909	0.698
12	175.00	171.20	3.80	0.346	0.936
13	163.00	167.83	− 4.83	0.409	− 1.254

(a) Construct a normal probability plot of the studentized residuals from this experiment. Interpret the plot.

(b) Plot the studentized residuals versus the predicted values of foam height. What conclusions can you show about model adequacy?

(c) Two points in the design (a) have relative high leverage ($h_{10,\,10} = 0.874$, $h_{11,\,11} = 0.909$). Where are these points located in the feasible space? Discuss the practical implications of having design points with high average.

12.7 A three-component mixture experiment is subject to the following constraints on the component proportions:

$$0.3 \leq x_1 \leq 0.6, \qquad 0.1 \leq x_2 \leq 0.5, \qquad 0.5 \leq x_3 \leq 0.45$$

$$x_1 + x_2 + x_3 = 1$$

(a) Graph the feasible region for this experiment.

(b) Suppose that the experimenter wishes to fit a quadratic model to the response. Identify a set of candidate points that the experimenter should consider in the selected design.

(c) Find a D-optimal design for this experiment with $n = 14$ runs, assuming that four of these runs must be replicates.

12.8 Consider the constrained mixture problem in Exercise 12.7.

(a) Find a 10-point design for this experiment using the distance criterion, assuming that you are not required to replicate any points.

(b) Find a 14-point distance-criterion design for this experiment, assuming that four of these points must be replicates.

12.9 Consider the three-component mixture experiment from Exercise 12.7. Suppose the experimenter wishes to fit a linear mixture model to the response and is willing to make $n = 8$ runs.

(a) Construct a D-optimal design for this problem.

(b) Construct a distance-criterion design for this problem.

(c) Compare and contrast the designs found in parts (a) and (b). Which design would you recommend?

12.10 A four-component mixture experiment is subject to the following constraints on the component proportions:

$$0.8 \leq x_1, \qquad 0.05 \leq x_2 \leq 0.15, \qquad 0.02 \leq x_3 \leq 0.10, \qquad x_4 \leq 0.05$$

$$x_1 + x_2 + x_3 + x_4 = 1$$

(a) Suppose that the experimenter wishes to fit a quadratic model to the response Identify a set of candidate points that the experimenter should consider in the selected design.

(b) Find a D-optimal design for this experiment with $n = 15$ runs, assuming that you are not required to replicate any points.

(c) Find a D-optimal design for this experiment with $n = 18$ runs, assuming that four of these runs must be replicates.

12.11 Consider the constrained mixture problem in Exercise 12.10.

(a) Find a 15-point design for this experiment using the distance criterion, assuming that you are not required to replicate any points.

(b) Find an 18-point design for this experiment using the distance criterion. Assume that four runs must be replicates.

12.12 Consider the four-component mixture experiment from Exercise 12.10. Suppose that the experimenter wishes to fit a linear mixture model to the response and is willing to make $n = 10$ runs.

(a) Construct a D-optimal design for this problem.

(b) Construct a distance-based design for this problem.

(c) Compare and contrast the designs in parts (a) and (b). Which design would you recommend?

12.13 Heinsman and Montgomery (1995) describe a four-component mixture experiment used to optimize the formulation of a household detergent. The constraints on the component proportions are:

$$0.5 \le x_1 \le 1, \qquad 0 \le x_2 \le 0.5, \qquad 0 \le x_3 \le 0.5, \qquad 0 \le x_4 \le 0.05$$

$$x_1 + x_2 + x_3 + x_4 = 1$$

The authors measured four responses: y_1 = product life (in lather units), y_2 = soil pellets, y_3 = foam height, and y_4 = total foam. The last three responses reflect the grease-cutting ability of the product. All four responses should be as large as possible. The data from the experiment they ran is shown in the following table. The design used by the authors was generated by selecting some points with the distance criterion and the remaining points with the D-optimal criterion.

(a) Build response surface models for all four responses.

(b) Plot contours of each response and interpret the fitted surfaces.

(c) What formulation of these ingredients would you recommend?

The Experimental Design for Exercise 12.13

Observation	x_1	x_2	x_3	x_4	X_1	X_2	X_3	X_4	y_1 (Lather Units)	y_2 (Pellets)	y_3 (Foam Height)	y_4 (Total Foam)
1	1.000	0.0000	0.0000	0.00000	1.000	0.000	0.000	0.0000	7.170	7.00	95.0	559.0
2	0.500	0.5000	0.0000	0.00000	0.000	1.000	0.000	0.0000	2.680	20.00	92.0	1320.0
3	0.500	0.0000	0.5000	0.00000	0.000	0.000	1.000	0.0000	3.080	3.00	44.0	275.0
4	0.950	0.0000	0.0000	0.05000	0.900	0.000	0.000	0.1000	6.990	7.00	73.0	508.0
5	0.500	0.4500	0.0000	0.05000	0.000	0.900	0.000	0.1000	2.920	20.00	105.0	1436.0
6	0.500	0.0000	0.4500	0.05000	0.000	0.000	0.900	0.1000	2.890	5.00	45.0	371.0
7	0.750	0.2500	0.0000	0.00000	0.500	0.500	0.000	0.0000	4.830	20.00	88.0	12.1
8	0.750	0.0000	0.2500	0.00000	0.500	0.000	0.500	0.0000	3.850	8.00	53.0	510.0
9	0.500	0.2500	0.2500	0.00000	0.000	0.500	0.500	0.0000	3.130	20.00	70.0	1123.0
10	0.725	0.2250	0.0000	0.05000	0.450	0.450	0.000	0.1000	4.430	20.00	80.0	1196.0
11	0.725	0.0000	0.2250	0.05000	0.450	0.000	0.450	0.1000	3.600	8.00	75.0	581.0
12	0.650	0.1500	0.1500	0.05000	0.300	0.300	0.300	0.1000	3.750	20.00	58.0	1061.0
13	0.650	0.1500	0.1500	0.05000	0.300	0.300	0.300	0.1000	3.260	20.00	59.0	1087.0
14	0.829	0.0792	0.0792	0.01250	0.658	0.158	0.158	0.0250	5.390	8.00	65.0	546.0
15	0.658	0.1583	0.1583	0.02500	0.317	0.317	0.317	0.0500	4.310	20.00	55.0	1069.0
16	0.579	0.3292	0.0792	0.01250	0.158	0.658	0.158	0.0250	2.640	20.00	80.0	1310.0
17	0.579	0.0792	0.3292	0.01250	0.158	0.158	0.658	0.0250	3.560	20.00	57.0	1011.0
18	0.804	0.0792	0.0792	0.03750	0.608	0.158	0.158	0.0750	5.230	20.00	68.0	1039.0
19	0.579	0.3042	0.0792	0.03750	0.158	0.608	0.158	0.0750	3.220	20.00	76.0	1192.0
20	0.579	0.0792	0.3042	0.03750	0.158	0.158	0.608	0.0750	3.520	20.00	59.0	1087.0

12.14 Cornell (1990) describes an experiment to optimize the texture of fish patties made from three types of fish: mullet (x_1), sheepshead (x_2), and croaker (x_3). The texture is measured by the amount of force required to break the patty. A single process variable, cooking temperature, is included in the experiment; this factor is run at two levels, 375°F and 425°F (actually, Cornell used three process variables in this experiment, but to keep things simple, we will ignore the other two). The results of this experiment are shown below:

Mullet (x_1)	Sheepshead (x_2)	Croaker (x_3)	Cooking Temperature (°F)	Breaking Force (grams $\times 10^{-3}$)
1	0	0	375	1.84
1	0	0	425	2.86
0	1	0	375	1.51
0	1	0	425	1.60
0	0	1	375	0.67
0	0	1	425	1.10
$\frac{1}{2}$	$\frac{1}{2}$	0	375	1.29
$\frac{1}{2}$	$\frac{1}{2}$	0	425	1.53
$\frac{1}{2}$	0	$\frac{1}{2}$	375	1.42
$\frac{1}{2}$	0	$\frac{1}{2}$	425	1.81
0	$\frac{1}{2}$	$\frac{1}{2}$	375	1.116
0	$\frac{1}{2}$	$\frac{1}{2}$	425	1.50
$\frac{1}{3}$	$\frac{1}{3}$	$\frac{1}{3}$	375	1.59
$\frac{1}{3}$	$\frac{1}{3}$	$\frac{1}{3}$	425	1.68

(a) Fit a special cubic model to the texture response at the high temperature level only. Construct a contour plot of the response surface and interpret the results.

(b) Fit a special cubic model to the texture response at the low temperature level only. Construct a contour plot of the response surface and interpret the results.

(c) Fit a combined model to the texture response that is a special cubic in the mixture components and where the process variable is defined as $z = (T - 400)/25$, where T is the temperature in degrees Fahrenheit. Is there a simple relationship between the parameter estimates in the combined model and the parameter estimates in the two individual model built at the low and high temperature levels? Does this relationship seem reasonable?

12.15 Consider a three-component mixture experiment with the following constraints are the mixture proportions:

$$x_1 \leq 0.5, \qquad x_2 \leq 0.6, \qquad x_3 \leq 0.6$$
$$x_1 + x_2 + x_3 = 1$$

(a) Graph the feasible region for this experiment.

(b) Suppose that the experimenter considers a quadratic model to be adequate. What type of experimental design would you recommend? Construct the design.

12.16 Consider the following first-order mixture model in seven components:

$$\hat{y} = 35x_1 + 85x_2 + 140x_3 + 77x_4 + 90x_5 + 100x_6 + 120x_7$$

(a) Assuming that all components were varied over approximately the same ranges, calculate the effects of each component.

(b) Is there any indication that some components can be dropped from further consideration?

12.17 Consider a mixture experiment with the following constraints on the component proportions:

$$x_1 \leq 0.2, \qquad x_2 \leq 0.6, \qquad x_3 \leq 0.15,$$
$$x_4 \leq 0.65, \qquad x_5 \leq 0.15, \qquad x_6 \leq 0.75$$
$$x_1 + x_2 + x_3 + x_4 + x_5 + x_6 = 1$$

(a) Suggest a screening design for this mixture experiment.

(b) Suppose that you have obtained the following first-order model for the response.

$$\hat{y} = 20x_1 + 30x_2 + 10x_3 + 16x_4 + 45x_5 + 80x_6$$

Calculate the adjusted effects of each component. Interpret the results of this analysis.

(a) Suggest a screening design for this mixture experiment.

(b) Suppose that you have obtained the following first-order model for the response.

$$\hat{y} = 20x_1 + 30x_2 + 10x_3 + 16x_4 + 45x_5 + 80x_6$$

Calculate the adjusted effects of each component. Interpret the results of this analysis.

CHAPTER 13

Continuous Process Improvement with Evolutionary Operation

13.1 INTRODUCTION

Response surface methodology is often applied to pilot plant operations or in a process development environment by research and development personnel. When it is applied to a full-scale production process, it is usually only done once (or relatively infrequently) because the experimental procedure is relatively elaborate. However, conditions that were optimum for the pilot plant may not be optimum for the full-scale process. The pilot plant may produce a small amount of product per day, whereas the full-scale process will produce much larger quantities. This "scale-up" of the pilot plant to the full-scale production process usually results in distortion of the optimum conditions. Furthermore, actual process equipment may differ in many respects from the pilot or prototype production process. Even if the full-scale plant begins operation at the optimum, it will eventually "drift" away from that point because of variations in raw materials, environmental changes, and operating personnel.

Box (1957) proposed **evolutionary operation** (EVOP) as a method for continuous monitoring and improvement of a full-scale process with the objective of moving the operating conditions toward the optimum or following a "drift." EVOP does not require large or sudden changes in operating conditions that might disrupt production. It was proposed as a method of routine plant operation that is carried out by manufacturing or operating personnel with minimum involvement of the engineering or development staff.

EVOP consists of systematically introducing small changes in the levels of the process variables under consideration. Usually, a 2^k design is employed to do this. The changes in the variables are relatively small, so that serious disturbances in yield, quality, or product characteristics will not occur, yet they must be large enough for potential improvements in process performance to eventually be discovered. Data are collected on the response variables of interest at each point of the 2^k design. When one observation has been taken at each design point, a cycle is said to have been completed. The effects and interactions of the process variables are then computed. Eventually, after several cycles, the effect of one or more process variables or their interactions may appear to have a significant effect on the response. At this point, a decision may be made to change the basic operating conditions to improve the response. When improved conditions have been detected, a **phase** is said to have been completed.

In testing the significance of process variables and interactions, an estimate of experimental error is required. In the original version of EVOP proposed by Box, this error estimate is calculated from the cycle data using a range method. Also, the 2^k design is usually centered about the current best operating conditions. By comparing the response at this point with the 2^k points in the factorial portion, we may check on curvature or, as it is sometimes called, **change in mean** (CIM). If the process is really centered at the maximum, say, then the response at the center should be significantly greater than the response at the 2^k peripheral points.

In theory, EVOP can be applied to k process variables. In practice, only two or three variables are usually considered. In the next section, we will give a two-variable example of the original version of EVOP, as proposed by Box (1957). Box and Draper (1969) give a detailed discussion of the three-variable case, including necessary forms and worksheets. Then we will discuss how EVOP can be implemented using modern computer software. Finally, we will discuss a variation of EVOP based on the simplex design, and we will give some advice about the practical implementation of EVOP.

13.2   AN EXAMPLE OF EVOP

We will illustrate EVOP using a chemical process whose yield is a function of temperature (x_1) and reaction time (x_2). The current operating conditions are $x_1 = 150°C$ and $x_2 = 30$ min. The EVOP procedure uses the 2^2 design plus the center point shown in Figure 13.1. Notice that each of the five points in the design are numbered. The cycle is completed by running each design point in numerical order $(1, 2, 3, 4, 5)$. This run order is used because it is easy for operating personnel to remember. Furthermore, if there are time or other nuisance factor effects, this run order confounds these effects with blocks. The yields in the first cycle are shown in Figure 13.1.

The yields from the first cycle are entered in the EVOP calculation sheet, shown in Table 13.1. At the end of the first cycle, no estimate of the standard deviation can be made. The effects and interaction for temperature and pressure are calculated in the usual manner for a 2^2 design.

A second cycle is then run and the yield data are entered in another EVOP calculation sheet, shown in Table 13.2. At the end of the second cycle, the experimental error can be estimated and the estimates of the effects can be compared to approximate 95% (two standard deviation) limits. Note that "range," shown on the right-hand side of the worksheet refers to the range of

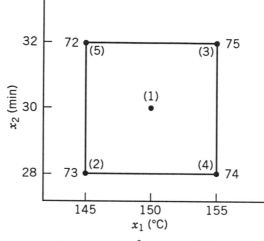

Figure 13.1 A 2^2 design for EVOP.

Table 13.1   EVOP Calculation Sheet, $n = 1$

5	3			
.1		Cycle: $n = 1$	Phase: 1	
		Response: Yield	Date: 3/27/94	
2	4			

Operating Conditions	Calculation of Averages					Calculation of Standard Deviation
	(1)	(2)	(3)	(4)	(5)	
(i) Previous cycle sum						Previous sum $S =$
(ii) Previous cycle average						Previous average $\bar{S} =$
(iii) New observations	74	73	75	74	72	New $S =$ range $\times$ $f_{5,n} =$
(iv) Differences [(iii) − (ii)]						Range of (iv) $=$
(v) New sums [(i) + (iii)]	74	73	75	74	72	New sum $S =$
(vi) New averages [$\bar{y}_i = $ (v)/n]	74	73	75	74	72	New average $\bar{S} =$ $\dfrac{\text{New sum } S}{n-1}$

Calculation of Effects	Calculation of Error Limits

Temperature effect

$$= \tfrac{1}{2}(\bar{y}_3 + \bar{y}_4 - \bar{y}_2 - \bar{y}_5) = 2.00$$

For new average $\dfrac{2}{\sqrt{n}}\bar{S} =$

Time Effect

$$= \tfrac{1}{2}(\bar{y}_3 + \bar{y}_3 - \bar{y}_2 - \bar{y}_4) = 0.00$$

For new effects $\dfrac{2}{\sqrt{n}}\bar{S} =$

Interaction effect

$$= \tfrac{1}{2}(\bar{y}_2 + \bar{y}_3 - \bar{y}_4 - \bar{y}_5) = 1.00$$

Change-in-mean effect

$$\tfrac{1}{5}(\bar{y}_2 + \bar{y}_3 + \bar{y}_4 + \bar{y}_5 - 4\bar{y}_1) = -0.40$$

For change in mean $\dfrac{1.78}{\sqrt{n}}\bar{S} =$

Table 13.2   EVOP Calculation Sheet, $n = 2$

5	3
	.1
2	4

Cycle: $n = 2$ Phase: 1
Response Yield   Date: 3/27/94

Operation Conditions	Calculation of Averages					Calculation of Standard Deviation
	(1)	(2)	(3)	(4)	(5)	
(i) Previous cycle sum	74	73	75	74	72	Previous sum $S =$
(ii) Previous cycle average	74	73	75	74	72	Previous average $\bar{S} =$
(iii) New observations	72	71	76	75	73	New $S =$ range $\times$
						$f_{5,n} = (3.6)(0.3) = 0.90$
(iv) Differences [(iii) − (ii)]	−2	2	−1	1	1	Range of (iv) = 3.0
(v) New sums [(i) + (iii)]	146	144	151	149	145	New sum $S = 0.90$
(vi) New Averages	73	72	75.5	74.5	72.5	New average $\bar{S} =$
$[\bar{y}_i - (v)/n]$						New sum S
						$\dfrac{\text{New sum } S}{n-1} = 0.90$

Calculation of Effects	Calculation of Error Limits
Temperature effect	For new average
$= \frac{1}{3}(\bar{y}_3 + y_4 - y_2 - y_5) = 2.75$	$\dfrac{2}{\sqrt{n}}\bar{S} = 1.27$
Time effect	For new effects
$= \frac{1}{2}(\bar{y}_3 + \bar{y}_5 - \bar{y}_2 - \bar{y}_4) = 0.75$	$\dfrac{2}{\sqrt{n}}\bar{S} = 1.27$
Interaction	
effect $= \frac{1}{2}(\bar{y}_2 + \bar{y}_3 - \bar{y}_4 - \bar{y}_5) = 0.25$	
Change-in-mean effect	For change in mean
$= \frac{1}{5}(\bar{y}_2 + \bar{y}_3 + \bar{y}_4 + \bar{y}_5 - 4\bar{y}_1) = 0.50$	$\dfrac{1.78}{\sqrt{n}}\bar{S} = 1.13$

the differences in row (iv); thus the range is $+1.0 - (-2.0) = 3.0$. This range is converted into s, an estimate of the process standard deviation, by multiplying the range times the factor $f_{5,n} = f_{5,2} = 0.30$ from Table 13.3. The estimate of the standard deviation from each cycle is averaged with the standard deviation estimate from previous cycles through the calculation

$$\text{New average } \bar{s} = \frac{\text{new sum } s}{n-1}$$

Table 13.3   Values of $f_{k,n}$

$n =$	2	3	4	5	6	7	8	9	10
$k = 5$	0.30	0.35	0.37	0.38	0.39	0.40	0.40	0.40	0.41
9	0.24	0.27	0.29	0.30	0.31	0.31	0.31	0.32	0.32
10	0.23	0.26	0.28	0.29	0.30	0.30	0.30	0.31	0.31

This new average $\bar{s}$ is then used in the calculation of the error limits in the bottom half of the worksheet.

Notice that at the end of the second cycle the temperature effect exceeds its error limit. This is equivalent to the effect estimate differing from zero by at least two standard deviations, so a change in operating conditions is warranted. Because the temperature effect is positive, we should increase temperature in order to increase yield. Therefore, a reasonable strategy would be to begin a new EVOP phase around the point $x_1 = 155°C$ and $x_2 = 30$ min.

An important aspect of EVOP is feeding the information generated back to the process operators and supervisors. This is accomplished by a prominently displayed EVOP information board. The information board for this example at the end of cycle 2 is shown in Figure 13.2.

Most of the quantities on the two-variable EVOP worksheet follow directly from the analysis of the 2^k factorial design. For example, the variance of any effect estimate, such as time $= \frac{1}{2}(\bar{y}_3 + \bar{y}_5 - \bar{y}_2 - \bar{y}_4)$, is found as follows:

$$\text{Var(Time)} = \text{Var}\left[\frac{1}{2}(\bar{y}_3 + \bar{y}_5 - \bar{y}_2 - \bar{y}_4)\right] = \frac{1}{4}\left(\frac{\sigma^2}{n} + \frac{\sigma^2}{n} + \frac{\sigma^2}{n} + \frac{\sigma^2}{n}\right)$$

$$= \frac{1}{4}\left(\frac{4\sigma^2}{n}\right) = \frac{\sigma^2}{n}$$

where σ^2 is the variance of the individual observation (y). Thus, two standard deviation (corresponding to 95%) error limits on any effect would be $\pm 2\sigma/\sqrt{n}$. The variance of the change in mean is

$$\text{Var(CIM)} = \text{Var}\left[\frac{1}{5}(\bar{y}_2 + \bar{y}_3 + \bar{y}_4 + \bar{y}_5 - 4\bar{y}_1)\right]$$

$$= \frac{1}{25}\left(\frac{4\sigma^2}{n} + \frac{16\sigma^2}{n}\right) = \left(\frac{20}{25}\right)\frac{\sigma^2}{n}$$

Thus, two standard deviation error limits on the CIM are

$$\pm\left(2\sqrt{20/25}\right)\sigma/\sqrt{n} = \pm 1.78\sigma/\sqrt{n}.$$

In the worksheet, σ is replaced by its estimate s.

The standard deviation σ is estimated by the range method. Let $y_i(n)$ denote the observation at the ith design point in cycle n and $\bar{y}_i(n)$ denote the corresponding average of $y_i(j)$, after cycle n ($j = 1, 2, \ldots, n$). The quantities in row (iv) of the EVOP worksheet are the differences $y_i(n) - \bar{y}_i(n - 1)$. The

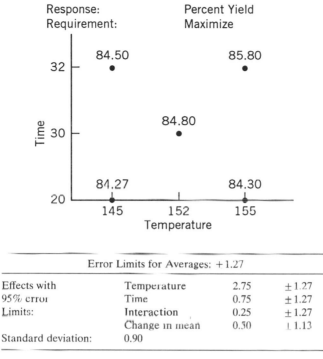

Response: Percent Yield
Requirement: Maximize

Figure 13.2   EVOP information board-cycle 1.

variance of these differences is

$$\text{Var}[y_i(n) - \bar{y}_i(n - 1)] \equiv \sigma_D^2 = \sigma^2\left(1 + \frac{1}{n - 1}\right) = \sigma^2\frac{n}{n - 1}$$

The range of the differences, say R_D, is related to the estimate of the standard deviation of the differences by $\sigma_D = R_D/d_2$. The factor d_2 (which is widely used in quality control work) depends on the number of observations used in computing R_D. Now $R_D/d_2 = \sigma\sqrt{n/(n - 1)}$, so

$$\hat{\sigma} = \sqrt{\frac{n - 1}{n}}\frac{R_D}{d_2} = f_{k,n}R_D = s$$

can be used to estimate the standard deviation of the observations, where k denotes the number of points used in the design. For a 2^2 design with one center point we have $k = 5$, and for a 2^3 design with one center point we have $k = 9$. Values of $f_{k,n}$ are given in Table 13.3.

13.3   EVOP USING COMPUTER SOFTWARE

As originally proposed, EVOP was implemented with manual calculations using a worksheet format, as illustrated in the previous section. A more modern approach would be to implement EVOP using the computer. Spreadsheet software could be easily developed for this purpose. Alternatively, one could use any software program for the analysis of 2^k factorial designs to implement EVOP. To illustrate, we will show how the popular

```
                       DESIGN — EASE ANALYSIS
================================================================
Response: yield; File = No File     Run on 03/27/94 at 14:15:10
================================================================
Var        VARIABLE         Units      -1 LEVEL     +1 LEVEL
A          temp             degC       145.000      155.000
B          time             min         28.000       32.000

                                      STANDARDIZED    SUM OF
                VARIABLE    COEFFICIENT   EFFECT      SQUARES
    OVERALL AVERAGE         73.6000
                      A      1.0000       2.0000     4.000000
                      B      0.0000       0.0000     0.000000
                     AB      0.5000       1.0000     1.000000
          CENTER POINT      0.5000                   0.200000

Computations done for Factorial
================================================================
Model selected for Factorial:
================================================================
Results of Factorial Model Fitting

                    ANOVA for Selected Model
               SUM OF                 MEAN        F
  SOURCE       SQUARES      DF        SQUARE     VALUE     PROB > F
  MODEL        5.000000     3        1.666667
  CURVATURE    0.200000     1        0.200000
  RESIDUAL     0.000000     0
  COR TOTAL    5.200000     4

  ROOT MSE                            R-SQUARED    1.0000
  DEP MEAN     73.600000
```

Final Equation in Terms of Uncoded Variables:

        yield =
                    268.500000
            −   1.300000 * temp
            −   7.500000 * time
            +   0.050000 * temp * time

Figure 13.3   Design-Ease output after $n = 1$ cycles.

microcomputer program *Design-Ease* can be used to perform the EVOP calculations using the two cycles of data from the example in Section 13.2.

Figure 13.3 shows the output from *Design-Ease* after the end of the first EVOP cycle. Notice that the effects estimates for the main effects of temperature and time and the interaction effect estimates agree with those given in the EVOP worksheet Table 13.1. However, the CIM effect does not agree with the center point effect. The center point coefficient in Figure 13.3 is calculated as

$$\text{center point effect} = \bar{y}_1 - \tfrac{1}{4}\left(\bar{y}_2 + \bar{y}_3 + \bar{y}_4 + \bar{y}_5\right)$$

$$= 74 - \tfrac{1}{4}(73 + 74 + 75 + 72)$$

$$= 74 - 73.5$$

$$- 0.5$$

while the CIM is

$$\text{CIM} = \tfrac{1}{5}\left(\bar{y}_2 + \bar{y}_3 + \bar{y}_4 + \bar{y}_5 - 4\bar{y}_1\right)$$

$$= \tfrac{1}{5}[73 + 74 + 75 + 72 - 4(74)]$$

$$= \tfrac{1}{5}(294 - 296)$$

$$= -0.4$$

However, it is obvious that both quantities provide an estimate of curvature in the true response function. Figure 13.4 is a square plot of the data from *Design-Ease* at the end of the first cycle.

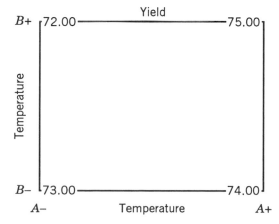

Figure 13.4 A square plot of the data from cycle 1 (from Design-Ease).

```
Response: yield; File = No File      Run on 03/27/94 at 14:20:07
```

Var	VARIABLE	Units	−1 LEVEL	+1 LEVEL
A	temp	degC	145.000	155.000
B	time	min	28.000	32.000

VARIABLE	COEFFICIENT	STANDARDIZED EFFECT	SUM OF SQUARES
OVERALL AVERAGE	73.5000		
A	1.3750	2.7500	15.12500
B	0.3750	0.7500	1.12500
AB	0.1250	0.2500	0.12500
CENTER POINT	−0.6250		0.62500

Computations done for Factorial

Model selected for Factorial:

Results of Factorial Model Fitting

ANOVA for Selected Model

SOURCE	SUM OF SQUARES	DF	MEAN SQUARE	F VALUE	PROB > F
MODEL	16.37500	3	5.45833	4.962	0.0585
CURVATURE	0.62500	1	0.62500	0.5682	0.4849
RESIDUAL	5.50000	5	1.10000		
PURE ERROR	5.50000	5	1.10000		
COR TOTAL	22.50000	9			

ROOT MSE	1.048809	R-SQUARED	0.7486
DEP MEAN	73.500000	ADJ R-SQUARED	0.5977
C.V.	1.43%		

Predicted Residual Sum of Squares (PRESS) = 22.00000

VARIABLE	COEFFICIENT ESTIMATE	DF	STANDARD ERROR	t FOR HO COEFFICIENT=0	PROB>|t|
INTERCEPT	73.625000	1	0.370810		
A	1.375000	1	0.370810	3.708	0.0139
B	0.375000	1	0.370810	1.011	0.3583
AB	0.125000	1	0.370810	0.3371	0.7497
CENTER POINT	−0.625000	1	0.829156	−0.7538	0.4849

Figure 13.5 Design-ease output after $n = 2$ cycles.

```
Final Equation in Terms of Coded Variables:
        yield =
                    73.500000
            +          1.375000 * A
            +          0.375000 * B
            +          0.125000 * A * B

Final Equation in Terms of Uncoded Variables:
        yield =
                    83.000000
            —          0.100000 * temp
            —          1.687500 * time
            +          0.012500 * temp * time
```

Obs Ord	ACTUAL VALUE	PREDICTED VALUE	RESIDUAL	LEVER	STUDENT RESID	COOK'S DIST	t VALUE	RUN Ord
1	73.0000	72.0000	1.0000	0.500	1.348	0.364	1.512	3
2	71.0000	72.0000	−1.0000	0.500	−1.348	0.364	−1.512	2
3	74.0000	74.5000	−0.5000	0.500	−0.674	0.091	−0.632	8
4	75.0000	74.5000	0.5000	0.500	0.674	0.091	0.632	7
5	72.0000	72.5000	−0.5000	0.500	−0.674	0.091	−0.632	4
6	73.0000	72.5000	0.5000	0.500	0.674	0.091	0.632	10
7	75.0000	75.5000	−0.5000	0.500	−0.674	0.091	−0.632	6
8	76.0000	75.5000	0.5000	0.500	0.674	0.091	0.632	1
9	74.0000	73.0000	1.0000	0.500	1.348	0.364	1.512	9
10	72.0000	73.0000	−1.0000	0.500	−1.348	0.364	−1.512	5

Figure 13.5 (*Continued*).

Figure 13.5 presents the output from *Design-Ease* after the completion of the second cycle. The factorial effect estimates agree with those obtained from the tabular worksheet version of EVOP in Table 13.2. However, as noted previously, the test for curvature is performed as in a standard 2^2 factorial design with center points. In the analysis of variance portion of Figure 13.5 there is a formal statistical test for curvature (the P-value is 0.4849, so there is no indication of curvature). Also this computer program uses a t-statistic to test the significance of main effects and interactions, whereas the tabular EVOP essentially uses confidence intervals. We observe that the main effect of factor A = temperature is significant (the P-value for the t-test is 0.0139), so the conclusions from this analysis would agree with those from the tabular EVOP; that is, the temperature variable should be adjusted in the positive direction.

The computer software uses the error mean square from the analysis of variance to estimate σ^2. The process standard deviation is estimated as the square root of this quantity, or $\tilde{\sigma} = \sqrt{MS_E} = \sqrt{5.5} = 1.049$. This estimate is

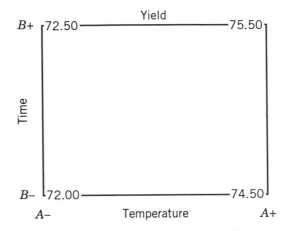

Figure 13.6   A square plot of the data after cycle $n = 2$ (from Design-Ease).

slightly different than the error estimate obtained in the tabular EVOP ($s = 0.90$). However, the tabular EVOP procedure estimates σ using a range method, and we would generally prefer the analysis of variance approach. Figure 13.6 shows the square plot of average responses at the end of the second cycle.

13.4   SIMPLEX EVOP

The experimental design usually employed in EVOP is a 2^k factorial augmented by a center point. An alternative EVOP procedure has been suggested by Spendley et al. (1962). Their scheme is based on the *simplex*, which we recall from Chapter 6 is an orthogonal first-order experimental design, requiring only one more observation than the number of variables under investigation. Thus, if k variables are being studied, then the number of trials in the design $n = k + 1$. The n observations are taken at the vertices of a regular-sided simplex, which for $k = 2$ is an equilateral triangle and for $k = 3$ is a tetrahedron. The design matrix **D** for a simplex of arbitrary orientation may be constructed from the last k column of $n^{1/2}\mathbf{O}$, where **O** is any ($n \times n$) orthogonal matrix having elements in the first column equal. The design points are the rows of the **D** matrix. The jth row of **D** will be denoted vectorially by $\mathbf{d}'_j$. The advantage of this design relative to a factorial is that fewer trials are required.

To apply this technique in a two-factor or a three-factor EVOP would require three and four periods, respectively, to observe the process response (yield, say) at the design points. Then the simplex EVOP procedure would adjust the process variables according to the following rules (assuming that the objective is to maximize the response).

1. Denote by y_i the response at the ith design point, $i = 1, 2, \ldots, n$. Let the minimum value of the response occur at design point $\mathbf{d}'_j$. Form a new simplex by deleting $\mathbf{d}'_j$ from $\mathbf{D}$ and substituting the new design point,

$$\mathbf{d}'^*_j = 2n^{-1}(\mathbf{d}'_1 + \mathbf{d}'_2 + \cdots + \mathbf{d}'_{j-1} + \mathbf{d}'_{j+1} + \cdots + \mathbf{d}'_n) - \mathbf{d}'_j \quad (13.1)$$

Run the process for the next period using the factor levels for $x_1, x_2, \ldots, x_n$ that are the elements of $\mathbf{d}^*_j$.

2. Apply rule 1 unless a design point has occurred in n successive simplexes without being eliminated. Should this situation arise for the ith design point, discard y_i and run the process during the next period using the factor levels in $\mathbf{d}'_i$. Then apply rule 1.

3. Should y_i be the minimum response in the mth simplex and y_{i*} be the minimum yield in the $(m + 1)$th simplex, do not return to the mth design. Instead of oscillating, move from the $(m + 1)$th design by discarding the second largest absolute current error.

We have described these rules for the case of maximizing the response. To minimize the response, replace the work "minimum" with "maximum" in the above rules.

Figure 13.7 shows how a simplex EVOP scheme can systematically move a process from a relative poor starting point to a much improved estimate of the optimum. We note also that the simplex can be used effectively in some

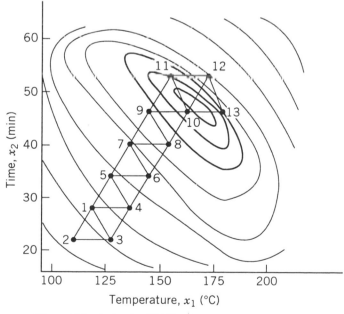

Figure 13.7 A simplex EVOP scheme for $k = 2$ variables.

situations as a type of "automatic" steepest ascent procedure. It has also been used as a mathematical optimization algorithm.

While some authors have advocated the sequential use of the simplex as a replacement for the more conventional factorial-based EVOP (indeed, some have suggested using the simplex as a replacement for variable screening and steepest ascent), we are not generally in favor of this. The simplex design does not provide direct information about interaction effects. In fact, because the simplex is a resolution III design, main effects and two-factor interactions are aliased. Thus the design may not provide information useful in building up process knowledge. Furthermore, when there is a moderate level of noise (error) associated with the response, then the sequential simplex may behave erratically. Replicate runs could be made at each vertex to counteract the effect of noise, but this increases the resources required to run the simplex, and its potential advantage in this regard to conventional factorials would be diminished.

13.5   SOME PRACTICAL ADVICE ABOUT USING EVOP

Our experience in using EVOP has led to several observations and suggestions about its practical implementation that may prove helpful. In this section, we share some of these ideas.

Usually, some care needs to be taken to ensure that a reasonable list of candidate variables are available for EVOP analysis and experimentation. We recommend starting with two (or perhaps three) variables that operating personnel think are the most important, but if several cycles occur (say five to eight, or so) and no significant effects emerge, then new variables should be introduced into the design (or new levels of the old variables tried) and a new EVOP phase started. It is particularly important to keep an open mind when identifying candidate variables. Often we have found that a variable was not realized to be important simply because it had never been changed.

Sometimes a process experiences relative large run-to-run variation, and this is used as an "excuse" for not using EVOP (or any statistically designed experiments, for that matter). This is certainly not a valid argument, because each cycle of the EVOP design is a replicate, and replication is a very effective noise-reduction technique. By building up information over several cycles and working with effect estimates that are based on averages of the responses at each design point, quite often large and important effects can be discovered even in noisy processes. It is also possible to discover that some variable settings result in less variability than do others (indeed, you could use the range or standard deviation of the observation at each design point as a second response). In noisy processes, sometimes more cycles may be required for the important effects to emerge, but the presence of process variability is not a deterrent to the use of designed experiments—it's the reason that statistically based designs *must* be used.

We have encountered some objection to the use of EVOP on the grounds that it is in violation of some of the principles of statistical process control

(SPC). Specifically, SPC encourages operating personnel to leave the process alone so long as it is "statistical control," and this is directly opposed to the EVOP procedure which introduces changes in some process variables almost continuously. We must remember that the objectives of these two procedures are different. EVOP is concerned with optimization, including following a process that drifts over time. SPC, on the other hand, is concerned with the detection and elimination of isolatable, external upsets in the process (called *assignable causes*) that may increase process variability or shift the process off-target. It may not be desirable to monitor and tightly control the process around the wrong target. Furthermore, the process knowledge generated through any designed experiment may lead to improvements in operating performance far faster and more efficiently than the use of SPC methods alone.

Finally, we give a few remarks on training and education in EVOP methods. We recommend that education concerning EVOP be included as part of a basic course in experimental design fundamentals. This course should be offered to process engineers, quality engineers, development personnel, researchers, and other technical professionals We have conducted courses such as this for many years. Usually they are from 3 to 5 days in length. Shorter versions of this course (1–2 days) should be given to key management personnel, so that they will understand the basic notions of designed experiments as well as the EVOP strategy. Plant operating personnel can usually be trained to run and evaluate EVOP schemes in about one-half day. We have found that training and education are critical to the successful implementation of both EVOP (in particular), and experimental design methods (in general).

EXERCISES

13.1 Yield during four cycles of a chemical process is shown in the following table. The variables are percent concentration (x_1) at levels 50, 55, and 60 g/liter and temperature (x_2) at 240°C, 242°C, and 244°C. Analyze by EVOP worksheet methods.

| Cycle | Conditions | | | | |
	(1)	(2)	(3)	(4)	(5)
1	80.7	79.8	80.2	84.2	77.5
2	79.1	82.8	82.5	84.6	78.3
3	76.6	79.1	79.0	82.3	81.1
4	80.5	79.8	84.5	81.0	80.1

13.2 Reconsider the data from Exercise 13.1. Use a standard 2^2 factorial design analysis procedure to evaluate the data from each cycle. Compare your results with the worksheet method in Exercise 13.1.

638

13.3 The elasticity of a polymer is being investigated using a two-variable EVOP scheme. The variables are reaction time x_1 at levels 30, 35, and 40 min and catalyst feed rate at 20, 22, and 24 g/min. Data for five cycles are shown in the following table. Analyze by EVOP worksheet methods.

Cycle	Conditions				
	(1)	(2)	(3)	(4)	(5)
1	17.0	16.3	17.8	17.1	16.5
2	16.7	16.1	17.1	17.0	16.8
3	17.5	16.4	16.9	16.9	16.9
4	16.9	15.9	17.3	16.9	16.3
5	16.8	16.1	17.6	17.4	16.2

13.4 Reconsider the data from Exercise 13.3. Use a standard 2^k factorial design analysis procedure to evaluate the data from each cycle. Compare your result with the worksheet method in Exercise 13.3.

13.5 Design an EVOP calculation sheet for three variables. Assume that a single center point is used. Derive the two standard error limits for the effects and for the change in mean.

13.6 Consider a two-variable EVOP scheme. Suppose that the center point could be easily replicated in each cycle. In what order would you recommend that the six runs be made? Justify your answer. Would this affect how your estimate of error is obtained?

13.7 Box and Draper (1969) describe a two-variable EVOP application to a plastic film extrusion process. The two variables are extrusion rate (x_1) and additive percent (x_2). There are three responses: y_1 = tear resistance, y_2 = gloss, and y_3 = opacity. The objective is to maximize tear resistance, maintain gloss between 8 and 10 units, and keep opacity below 7. Data for five cycles follow. Analyze by EVOP worksheet methods and draw conclusions.

Point 1			Point 2			Point 3			Point 4			Point 5		
y_1	y_2	y_3	y_1	y_2	y_3	y_1	y_2	y_3	y_1	y_2	y_3	y_1	y_2	y_3
7.0	10.1	6.2	6.3	10.0	1.9	7.8	9.7	3.4	7.1	9.4	2.1	6.5	9.7	4.0
6.7	8.8	2.2	6.1	10.6	2.3	7.5	8.0	3.0	7.0	8.9	1.8	6.8	9.1	1.0
7.0	10.0	4.1	6.4	9.6	2.8	7.1	9.0	2.9	7.5	8.9	1.7	6.9	9.5	3.4
6.5	7.3	4.0	6.4	10.0	3.7	7.3	7.9	3.4	6.9	8.9	1.8	6.6	10.1	2.8
6.4	9.7	3.5	6.9	10.1	2.7	7.2	9.5	6.7	6.6	8.3	0.0	6.9	9.5	8.2

13.8 Reconsider the data in Exercise 13.7. Use a standard 2^k factorial design analysis procedure to evaluate the data from each cycle. Compare your results with those from the worksheet approach used in Exercise 13.7.

13.9 Reconsider the situation in Exercise 13.7. Show how the desirability function approach to multiple response optimization discussed in Chapter 6 could be applied in this situation. What are the advantages and disadvantages of this method in the EVOP application environment?

13.10 Set up a simplex EVOP scheme for two variables. Verify that Equation (13.1) in the text "reflects" the simplex when one of the design points is deleted.

APPENDIX 1

Variable Selection and Model-Building in Regression

In response surface work it is customary to fit the "full" model corresponding to the situation at hand. That is, in steepest ascent we usually fit the "full" first-order model, and in the analysis of a second-order model we usually fit the "full" quadratic. Similarly, in the analysis of mixture data, the experimenter usually fits the full Scheffé polynomial.

An experimenter may encounter situations where the full model might not be appropriate; that is, a model based on a **subset** of the regressors in the full model may be superior. **Variable selection** or **model-building** techniques may be used to identify the "best" subset of regressors to include in a regression model.

In this appendix we give a brief presentation of regression model-building and variable selection methods, and we illustrate their application to a response surface problem. Throughout, we assume that there are K candidate regressors denoted $x_1, x_2, \ldots, x_K$ and a single response variable y. All models will have an intercept term β_0, so that the full model has $K + 1$ parameters. We also assume that all variables are correctly and completely specified (for example, $x_3 = 1/x_1$, or $x_3 = x_1^2$, etc.). For a more general discussion of model-building or variable selection, see Montgomery and Peck (1992) or Myers (1990).

A.1.1 BIAS ARISING FROM AN INADEQUATE MODEL

We will write the model with all K regressors as

$$\mathbf{y} = \mathbf{X}\boldsymbol{\beta} + \boldsymbol{\varepsilon} \tag{A.1.1}$$

Suppose that r variables are deleted from the model in Equation (A.1.1), so that the number of variables that are retained is $p = K + 1 - r$. Because the subset model contains an intercept, it uses $p - 1$ of the original regressors.

Now we can write Equation (A.1.1) as

$$y = X_p\beta_p + X_r\beta_r + \varepsilon \qquad (A.1.2)$$

where X has been partitioned into X_p, an $n \times p$ matrix whose columns represent the intercept and the $p - 1$ regressors in the subset model, and X_r, an $n \times r$ matrix whose columns represent the r variables that are deleted from the full model. Let β be partitioned similarly into β_p and β_r. For the full model, the least squares estimate of β is just

$$b = (X'X)^{-1}x'y$$

and for the subset model

$$y = X_p\beta_p + \varepsilon$$

the least squares estimate of β_p is

$$b_p = (X'_pX_p)^{-1}X'_py \qquad (A.1.3)$$

The expected value of b_p is

$$
\begin{aligned}
E(b_p) &= E\left[(X'_pX)^{-1}X'_py\right] \\
&= (X'_pX)^{-1}X'_pE(y) \\
&= (X'_pX_p)^{-1}X'_pE(X_p\beta_p + X_r\beta_r + \varepsilon) \\
&= (X'_pX_p)^{-1}X'_p(X_p\beta_p + X_r\beta_r) \\
&= \beta_p + (X'_pX_p)^{-1}X'_pX_r\beta_r \\
&= \beta_p + A\beta_r \qquad (A.1.4)
\end{aligned}
$$

The $A = (X'_pX_p)^{-1}X'_pX_r$ is sometimes called the alias matrix and was discussed in chapter 9. From examining Equation (A.1.4) it is clear that b_p, the estimate of the regression coefficient in the subset model β_p, is biased unless the term $A\beta_p$ is zero. Now if the deleted variables in X_r are unimportant, the regression coefficients in β_r will be zero, and consequently $A\beta_r$ will be zero. So if all of the important variables are retained in the subset model, an unbiased estimate b_p will result. Also, if the deleted variables X_r are orthogonal to the retained variables X_p so that $X'_pX_r = 0$, then b_p is unbiased.

We can also show that deleting unimportant variables from the model improves the precision of estimation for the parameter estimates of the retained variables. This is also true for the variance of the predicted response. Therefore, there is a strong motivation for correctly specifying the regression model: Leaving out important regressors introduces bias into the parameter estimates, while including unimportant variables weakens the prediction or estimation capability of the model.

A.1.2 PROCEDURES FOR VARIABLE SELECTIONS

We will now present several of the more widely used methods for selecting the appropriate subset of variables for a regression model. We will also discuss and illustrate several of the criteria that are typically used to decide which subset of the candidate regressors leads to the best model.

All Possible Regressions

This procedure requires that the analyst fit all the regression equations involving one-candidate regressors two-candidate regressors, and so on. These equations are evaluated according to some suitable criterion and the "best" regression model selected. If we assume that the intercept term β_0 is included in all equations, then there are K candidate regressors and there are 2^K total equations to be estimated and examined. For example, if $K = 4$, then there are $2^4 = 16$ possible equations, whereas if $K = 10$, then there are $2^{10} = 1024$ possible regression equations. Clearly the number of equations to be examined increases rapidly as the number of candidate regressors increases. Prior to the development of efficient computer codes, generating all possible regressions was impractical for problems involving more than a few regressors. The availability of high-speed computers has led to the development of several very efficient algorithms for all possible regressions.

    Example A.1.1 Table A.1.1 presents the results of running a rotatable central composite design (CCD) on a process used to make a polymer additive for motor oil. The response variable of interest is the average

Table A.1.1 Factors and Response for Example A.1.1

Time ξ_1	Catalyst ξ_2	Time x_1	Catalyst x_2	$y\ (M_n)$
30	4	−1	−1	2320
40	4	1	−1	2925
30	6	−1	1	2340
40	6	1	1	2000
27.93	5	−1.414	0	3180
42.07	5	1.414	0	2925
35	3.586	0	−1.414	1930
35	6.414	0	1.414	1860
35	5	0	0	2980
35	5	0	0	3075
35	5	0	0	2790
35	5	0	0	2850
35	5	0	0	2910

molecular weight (or the M_n), and the two process variables are reaction time in minutes and catalyst addition rate. The table shows the design in terms of both the "natural" variables and the usual coded variables.

The results of fitting the full quadratic model to the molecular weight response using the Design-Expert software product is shown in Table A.1.2. The quadratic model was selected here because the F-statistic for the quadratic terms (over the contribution of the linear terms) was large ($F = 27.89$ in Table A.1.2), and because the linear model displayed strong lack of fit. However, the quadratic model is not satisfactory. Note that two terms have relatively small t-statistics (the P-values for the linear and quadratic terms in "Time" are 0.844 and 0.483, respectively) and that the PRESS statistic for the cubic model in Table A.1.2 is smaller than PRESS for the quadratic model. Of course, the CCD does not contain enough points to fit a full cubic model (this is the meaning of the warning message in Table A.1.2), but one can fit the pure cubic terms. There is some indication here that a subset model might be more appropriate than the full quadratic.

We will use the all-possible-regressions procedure to identify a model. We will restrict the candidate variables for the model to those in the full quadratic polynomial and require that all models obey the principal of hierarchy. A model is said to be hierarchial if the presence of higher-order terms (such as interaction and second-order terms) requires the inclusion of all lower-order terms contained within those of higher order. For example, this would require the inclusion of both main effects if a two-factor interaction term was in the model. Many regression model-builders believe that hierarchy is a reasonable model-building practice when fitting polynomials.

Table A.1.3 presents the all-possible-regressions results. The values of the residual sum of squares, the residual mean square, R^2, R^2_{adj}, and PRESS are given for each model. Table A.1.3 also shows the value of the C_p statistic for each subset model. The C_p statistic is a measure of the total mean squared error for the p-term regression model. We define the total standardized mean squared error for a regression model with p terms as

$$\Gamma_p = \frac{1}{\sigma^2} \sum_{i=1}^{n} \left[E(y_i) - E(\hat{y}_i) \right]^2$$

$$= \frac{1}{\sigma^2} \left[\sum_{i=1}^{n} \{ E(y_i) - E(\hat{y}_i) \}^2 + \sum_{i=1}^{n} V(\hat{y}_i) \right]$$

$$= \frac{1}{\sigma^2} \left[(\text{bias})^2 + \text{variance} \right]$$

We use the mean squared error from the full $K + 1$ term model as an estimate of σ^2; that is, $\hat{\sigma}^2 = MS_E$ (full model). An estimator of Γ_p is the C_p

Table A.1.2 Design-Expert Output for the Full Quadratic Model, Example A.1.1

Factor	Units	−1 Level	+1 Level
A—time	Minutes	30.000	40.000
B—catalyst	Rate	4.000	6.000

WARNING: The cubic model is aliased!

Sequential Model Sum of Squares

Source	Sum of Squares	Degrees of Freedom	Mean Square	F-Value	Probability > F
Mean	89,368,248.1	1	89,368,248.1		
Linear	127,157.9	2	63,578.9	0.26	0.779
Quadratic	2,289,864.9	3	763,288.3	27.89	< 0.001
Cubic	130,117.1	2	65,058.6	5.29	0.058
Residual	61,487.0	5	12,297.4		
Total	91,976,875.0	13			

Lack-of-Fit Tests

Model	Sum of Squares	Degrees of Freedom	Mean Square	F-Value	Probability > F
Linear	2,431,949.0	6	405,324.8	32.74	0.002
Quadratic	142,084.2	3	47,361.4	3.83	0.114
Cubic	11,967.0	1	11,967.0	0.97	0.381
Pure error	49,520.0	4	12,380.0		

Model Summary Statistics

Source	Root Mean Squared Error	R-Squared	Adjusted R-Squared	Predicted R-Squared	PRESS
Linear	498.1	0.0487	−0.1415	−0.9028	4963701.9
Quadratic	165.4	0.9265	0.8741	0.5830	1087751.3
Cubic	110.9	0.9764	0.9434	0.6767	843265.4

ANOVA for Quadratic Model

Source	Sum of Squares	Degrees of Freedom	Mean Square	F-Value	Probability > F
Model	2,417,022.8	5	483,404.6	17.66	< 0.001
Residual	191,604.2	7	27,372.0		
Lack of fit	(142,084.2)	(3)	47,361.4	3.83	0.114
Pure error	(49,520.0)	(4)	12,380.0		
Corrected total	2,608,627.0	12			

Root mean squared error	165.4		R-squared	0.9265
Dep Mean	2621.9		Adjusted R-squared	0.8741
CV	6.31%		Predicted R-squared	0.5830

Predicted residual sum of squares (PRESS) = 1087751.3

Table A.1.2 (*Continued*)

Factor	Coefficient Estimate	Degrees of Freedom	Standard Error	t for H_0 Coefficient $= 0$	Probability $> |t|$	VIF
Intercept	2921.0	1	74.0	39.48		
A—time	−11.9	1	58.5	−0.20	0.844	1.00
B—catalyst	−125.5	1	58.5	−2.15	0.069	1.00
A^2	46.4	1	62.7	0.74	0.483	1.02
B^2	−532.5	1	62.7	−8.49	< 0.001	1.02
AB	−236.3	1	82.7	−2.86	0.024	1.00

Final Equation in Terms of Coded Factors
$$M_n =$$
$$2921.0$$
$$- \quad 11.9 * A$$
$$- \quad 125.5 * B$$
$$+ \quad 46.4 * A^2$$
$$- \quad 532.5 * B^2$$
$$- \quad 236.3 * AB$$

Final Equation in Terms of Actual Factors
$$M_n =$$
$$- \; 15{,}674.7$$
$$+ \; 103.89 * \text{Time}$$
$$+ \quad 6853.3 * \text{Catalyst}$$
$$+ \qquad 1.8567 * \text{Time}^2$$
$$- \qquad 532.51 * \text{Catalyst}^2$$
$$- \qquad 47.250 * \text{Time} * \text{Catalyst}$$

Table A.1.3 All Possible Regressions Results, Example A.1.1

Terms in Model	SS Residual	MS Residual	R^2	R^2_{adj}	PRESS	C_p
x_1	2,607,485.0	237,044.1	0.0004	−0.0904	3,728,849.2	86.26
x_2	2,482,610.9	225,691.9	0.0483	−0.0382	4,240,155.7	81.70
x_1, x_2	2,481,469.0	248,146.9	0.0487	−0.1415	4,963,701.9	83.66
$x_1, x_2, x_1 x_2$	2,258,212.8	250,912.5	0.1343	−0.1542	5,379,306.7	77.50
x_1, x_2, x_1^2	2,386,620.7	265,180.1	0.0851	−0.2199	5,736,756.9	82.19
$x_1, x_2, x_1 x_2, x_1^2$	2,163,364.5	270,420.6	0.1707	−0.2440	6,580,579.4	76.04
x_1, x_2, x_2^2	429,842.7	47,760.3	0.8352	0.7803	1,123,679.6	10.70
$x_1, x_2, x_1 x_2, x_2^2$	206,586.5	25,823.3	0.9208	0.8812	853,249.0	4.55
$x_1, x_2, x_1 x_2, x_1^2, x_2^2$	191,604.2	27,372.0	0.9265	0.8741	1,087,751.3	6.00
x_1, x_1^2	2,512,636.7	251,263.7	0.0368	−0.1558	4,399,694.1	84.80
x_2, x_2^2	430,984.6	43,098.5	0.8343	0.8017	864,737.5	8.75

statistic

$$C_p = \frac{SS_E(p)}{\hat{\sigma}^2} - n + 2p \qquad \text{(A.1.5)}$$

where $SS_E(p)$ is the error sum of squares for the p-term model. If the p-term model has negligible bias, then it can be shown that

$$E(C_p|\text{zero bias}) = p$$

Therefore, the values of C_p for each regression model under consideration should be evaluated relative to p. The regression equations that have substantial bias will have values of C_p that are greater than p. One then chooses as the "best" regression equation either a model with minimum C_p or a model with a slightly larger C_p that does not contain as much bias (i.e., $C_p \cong p$) as the minimum.

The selection of the appropriate subset model is usually based on the summary statistics given in Table A.1.1. Note that the subset model containing the terms x_1, x_2, x_1x_2, and x_2^2 has the smallest residual mean square, the largest adjusted R^2, the smallest value of PRESS, and $C_p = C_5 = 4.55$, which is just less than $p = 5$, so this equation contains little bias.

Table A.1.4 shows the Design-Expert output for this reduced quadratic model. This model is likely to be a better predictor of molecular weight than the full quadratic. Figure A.1.1 shows the contour plot and three-dimensional response surface plot from this reduced quadratic model. The objective of the engineer was to develop a process capable of producing a product with M_n of 2900. Obviously, there are several values of time and catalyst addition produce satisfactory results.

Figure A.1.2 shows the contours of constant standard deviation of predicted response for the reduced quadratic model. Notice that the contours are not circular, even though a rotatable design was used. This has occurred, of course, because a reduced quadratic model was fit instead of the full quadratic. The property of rotatability is both design- and model-dependent.

Stepwise Regression Methods

Because evaluating all possible regressions can be burdensome computationally, various methods have been developed for evaluating only a small number of subset regression models by either adding or deleting regressors one at a time. These methods are generally referred to as *stepwise-type procedures*. They can be classified into three broad categories: (1) forward selection, (2) backward elimination, and (3) stepwise regression, which is a popular combination of procedures 1 and 2. We now briefly describe and illustrate these procedures.

Table A.1.4   Design-Expert Output for the Reduced Quadratic Model in Example A.1.1

Source	Sum of Squares	Degrees of Freedom	Mean Square	F-Value	Probability $> F$
Model	2,402,040.4	4	600,510.1	23.25	< 0.001
Residual	206,586.5	8	25,823.3		
Lack of fit	(157,066.5)	(4)	39,266.6	3.17	0.145
Pure error	(49,520.0)	(4)	12,380.0		
Corrected total	2,608,626.9	12			

Root mean squared error	160.7		R-squared	0.9208
Dependent mean	2,621.9		Adjusted R-squared	0.8812
CV	6.13%		Predicted R-squared	0.6729

Predicted residual sum of squares (PRESS) = 853249.0

Factor	Coefficient Estimate	Degrees of Freedom	Standard Error	t for H_0 Coefficient = 0	Probability $> \lvert t \rvert$	VIF
Intercept	2953.3	1	58.0	50.88		
A—time	−11.9	1	56.8	−0.21	0.839	1.00
B—catalyst	−125.5	1	56.8	−2.21	0.058	1.00
B^2	−538.6	1	60.4	−8.91	< 0.001	1.00
AB	−236.3	1	80.3	−2.94	0.019	1.00

Final Equation in Terms of Coded Factors

$M_n =$

$$2953.3$$
$$- \quad 11.9 * A$$
$$- \quad 125.5 * B$$
$$- \quad 538.6 * B_2$$
$$- \quad 236.3 * AB$$

Final Equation in Terms of Actual Factors

$M_n =$

$$- 18068.1$$
$$+ \quad 233.86 * \text{Time}$$
$$+ \quad 6913.8 * \text{Catalyst}$$
$$- \quad 538.55 * \text{Catalyst}^2$$
$$- \quad 47.250 * \text{Time} * \text{Catalyst}$$

Observation	Actual Value	Predicted Value	Residual	Lever	Studentized Residual	Cook's Distance	R-Student	Run Order
1	2320.0	2316.0	4.0	0.598	0.04	0.000	0.04	7
2	2925.0	2764.6	160.4	0.598	1.57	0.737	1.77	2
3	2340.0	2537.4	−197.4	0.598	−1.94	1.116	−2.49	12
4	2000.0	2041.0	−41.0	0.598	−0.40	0.048	−0.38	4
5	3180.0	2970.2	209.8	0.380	1.66	0.338	1.92	11
6	2925.0	2936.4	−11.4	0.380	−0.09	0.001	−0.08	5
7	1930.0	2054.0	−124.0	0.598	−1.22	0.440	−1.26	1
8	1860.0	1699.0	161.0	0.598	1.58	0.741	1.78	10
9	2980.0	2953.3	26.7	0.130	0.18	0.001	0.17	9
10	3075.0	2953.3	121.7	0.130	0.81	0.020	0.79	13
11	2790.0	2953.3	−163.3	0.130	−1.09	0.036	−1.10	3
12	2850.0	2953.3	−103.3	0.130	−0.69	0.014	−0.66	8
13	2910.0	2953.3	−43.3	0.130	−0.29	0.003	−0.27	6

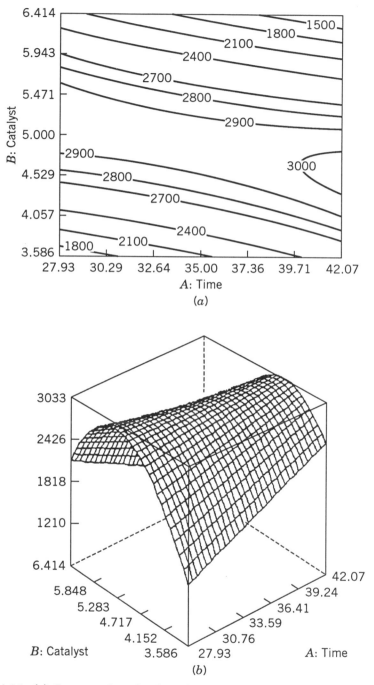

Figure A.1.1 (*a*) Contours of predicted number average molecular weight, Example A.1.1. (*b*) Response surface plot.

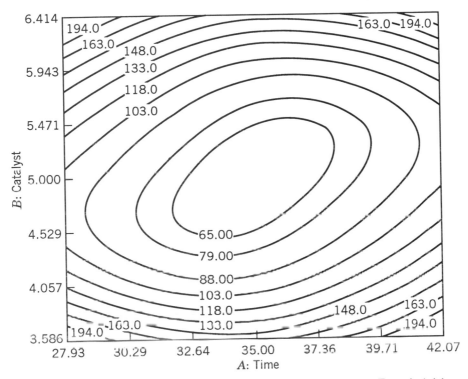

Figure A.1.2   Contours of constant standard deviation of predicted response, Example A.1.1.

Forward Selection

This procedure begins with the assumption that there are no regressors in the model other than the intercept. An effort is made to find an optimal subset by inserting regressors into the model one at a time. The first regressor selected for entry into the equation is the one that has the largest simple correlation with the response variable y. Suppose that this regressor is x_1. This is also the regressor that will produce the largest value of the F-statistic for testing significance of regression. This regressor is entered if the F-statistic exceeds a preselected F-value, say F_{IN} (or F-to-enter). The second regressor chosen for entry is the one that now has the largest correlation with y after adjusting for the effect of the first regressor entered (x_1) on y. We refer to these correlations as *partial correlations*. They are the simple correlations between the residuals from the regression $\hat{y} = \hat{\beta}_0 + \hat{\beta}_1 x_1$ and the residuals from the regressions of the other candidate regressors on x_1, say $\hat{x}_j = \hat{\alpha}_{0j} + \hat{\alpha}_{1j} x_1$, $j = 2, 3, \ldots, K$.

Suppose that at step 2 the regressor with the highest partial correlation with y is x_2. This implies that the largest partial F-statistic is

$$F = \frac{SS_R(x_2|x_1)}{MS_E(x_1, x_2)}$$

Table A.1.5   Design-Expert Output for Forward Selection

Factor	Units	−1 Level	+1 Level
A—time	Minutes	30.000	40.000
B—catalyst	Rate	4.000	6.000

Forward Selection Regression

Alpha to enter $= 5.00 \times 10^{-2}$

Factor	Coefficient Estimate	t for H_0 Coefficient $= 0$	Probability $> |t|$	R-Squared	Mean Squared Error
Forced intercept	2622				
A	−11.95	−0.07	0.947		
B	−125.5	−0.71	0.492	4.87×10^{-2}	2.48×10^{5}
Entered B^2 intercept	2953				
A	−11.95	−0.15	0.881		
B	−125.5	−1.62	0.139		
B^2	−538.6	−6.55	< 0.001	0.8352	4.78×10^{4}
Entered AB intercept	2953				
A	−11.95	−0.21	0.839		
B	−125.5	−2.21	0.058		
B^2	−538.6	−8.91	< 0.001		
AB	−236.3	−2.94	0.019	0.9208	2.58×10^{4}

If this F value exceeds F_{IN}, then x_2 is added to the model. In general, at each step the regressor having the highest partial correlation with y (or equivalently the largest partial F-statistic given the other regressors already in the model) is added to the model if its partial F-statistic exceeds the preselected entry level F_{IN}. The procedure terminates either when the partial F-statistic at a particular step does not exceed F_{IN} or when the last candidate regressor is added to the model.

Example A.1.2   Table A.1.5 shows the results of applying forward selection to the experimental data from Example A.1.1. We used the forward selection option in Design-Expert, which selects the cutoff value F_{IN} by choosing a type I error rate α. We used the default value of $\alpha = 0.05$. We also forced the program to include both of the main effects, namely, time and catalyst. Notice that the final model is identical to the model found using all possible regressions.

Backward Elimination

Forward selection begins with no regressors in the model and attempts to insert variables until a suitable model is obtained. Backward elimination attempts to find a good model by working in the opposite direction. That is, we begin with a model that includes all K candidate regressors. Then the partial F-statistic (or a t-statistic, which is equivalent) is computed for each regressor as if it were the last variable to enter the model. The smallest of these partial F-statistics is compared with a preselected value, F_{OUT} (or F-to-remove) for example; and if the smallest partial F value is less than F_{OUT}, that regressor is removed from the model. Now a regression model with $K - 1$ regressors is fit, the partial F-statistics for this new model calculated, and the procedure repeated. The backward elimination algorithm terminates when the smallest partial F-value is not less than the preselected cutoff value F_{OUT}.

Backward elimination is often a very good variable selection procedure. It is particularly favored by analysts who like to see the effect of including all the candidate regressors, just so that nothing "obvious" will be missed

Example A.1.3   Table A.1.6 presents the results of applying backward elimination to the data of Example A.1.1 using Design-Expert. This program uses the t-statistic as a guide to eliminating variables, which is, of course, equivalent to the partial F-statistic. We used the default value of $\alpha = 0.05$ and forced the final model to include both main effects. We also started with the reduced cubic model, which includes the pure cubic terms in time and catalyst addition rate, referred to as factors A and B, respectively, in the computer output. Once again, notice that the final model is identical to the all-possible-regressions final model.

Stepwise Regression

The two procedures described above suggest a number of possible combinations. One of the most popular is the stepwise regression algorithm. This is a modification of forward selection in which at each step all regressors entered into the model previously are reassessed via their partial F- or t-statistics. A regressor added at an earlier step may now be redundant because of the relationship between it and regressors now in the equation. If the partial F-statistic for a variable is less than F_{OUT}, that variable is dropped from the model.

Stepwise regression requires two cutoff values, F_{IN} and F_{OUT}. Some analysts prefer to choose $F_{IN} = F_{OUT}$, although this is not necessary. Sometimes we choose $F_{IN} > F_{OUT}$, making it relatively more difficult to add a regressor than to delete one.

Example A.1.4   Table A.1.7 presents the results of applying stepwise regression as implemented in Design-Expert to the data of Example A.1.1. This program uses t-statistics for selecting regressors, and we used the default value of $\alpha = 0.10$ for both entering and removing variables. We also specified all of the terms in the quadratic model plus the cubic terms in time

Table A.1.6   Design-Expert Output for Backward Elimination

Factor	Units	−1 Level	+1 Level
A—time	Minutes	30.000	40.000
B—catalyst	Rate	4.000	6.000

Backward Elimination Regression

Alpha to exit = 5.00×10^{-2}

Factor	Coefficient Estimate	t for H_0 Coefficient = 0	Probability $> \lvert t \rvert$	R-Squared	Mean Squared Error
Forced					
intercept	2622				
A	−11.95	−0.07	0.947		
B	−125.5	−0.71	0.492	4.87×10^{-2}	2.48×10^5
Removed A^2					
intercept	2953				
A	222.8	1.76	0.128		
B	−427.9	−3.39	0.015		
B^2	−538.6	−12.69	< 0.001		
AB	−236.3	−4.19	0.006		
A^3	−156.5	−1.96	0.098		
B^3	201.6	2.52	0.045	0.9707	1.27×10^4
Removed A^3					
intercept	2953				
A	−11.95	−0.25	0.808		
B	−427.9	−2.86	0.024		
B^2	−538.6	−10.70	< 0.001		
AB	−236.3	−3.53	0.010		
B^3	201.6	2.13	0.071	0.9519	1.79×10^4
Removed B^3					
intercept	2953				
A	−11.95	−0.21	0.839		
B	−125.5	−2.21	0.058		
B^2	−538.6	−8.91	< 0.001		
AB	−236.3	−2.94	0.019	0.9208	2.58×10^4

and catalyst addition rate as candidate regressors. We forced in both main effects and required a hierarchial model. These decisions essentially forced all three candidate terms into the model. The quadratic term in time was added in the final step to ensure model hierarchy.

General Comments on Stepwise-Type Procedures

The stepwise regression algorithms described above have been criticized on various grounds, the most common being that none of the procedures generally guarantees that the best subset regression model of any size will be identified. Furthermore, because all the stepwise-type procedures terminate with one final equation, inexperienced analysts may conclude that they have found a model that is in some sense optimal. Part of the problem is that it is

Table A.1.7 Design-Expert Stepwise Regression Output

Factor	Units	−1 Level	+1 Level
A—time	Minutes	30.000	40.000
B—catalyst	Rate	4.000	6.000

Stepwise Regression

Alpha to enter = 0.1000
Alpha to exit = 0.1000

Factor	Coefficient Estimate	t for H_0 Coefficient = 0	Probability $> |t|$	R-Squared	Mean Squared Error
Forced					
intercept	2622				
A	−11.95	−0.07	0.947		
B	−125.5	−0.71	0.492	4.87×10^2	2.48×10^5
Entered B_2					
intercept	2953				
A	−11.95	−0.15	0.881		
B	−125.5	−1.62	0.139		
B_2	−538.6	−6.55	< 0.001	0.8352	4.78×10^4
Entered AB					
intercept	2953				
A	−11.95	−0.21	0.839		
B	−125.5	−2.21	0.058		
B_2	−538.6	−8.91	< 0.001		
AB	−236.3	−2.94	0.019	0.9208	2.58×10^4
Entered B_3					
intercept	2953				
A	−11.95	−0.25	0.808		
B	−427.9	−2.86	0.024		
B_2	−538.6	−10.70	< 0.001		
AB	−236.3	−3.53	0.010		
B_3	201.6	2.13	0.071	0.9519	1.79×10^4
Entered A_3					
intercept	2953				
A	222.8	1.76	0.128		
B	−427.9	−3.39	0.015		
B_2	−538.6	−12.69	< 0.001		
AB	−236.3	−4.19	0.006		
B_3	201.6	2.52	0.045		
A_3	−156.5	−1.96	0.098	0.9707	1.27×10^4

ANOVA for Reduced Cubic Model (Hierarchy Corrected)

Source	Sum of Squares	Degrees of Freedom	Mean Square	F-Value	Probability $> F$
Model	2,547,139.9	7	363,877.1	29.59	< 0.001
Residual	61,487.0	5	12,297.4		
Lack of fit	(11,967.0)	(1)	11,967.0	0.97	0.381
Pure error	(49,520.0)	(4)	12,380.0		
Corrected total	2,608,626.9	12			

Root mean squared error	110.9	R-squared	0.9764
Dependent mean	2,621.9	Adjusted R-squared	0.9434
CV	4.23%	Predicted R-squared	0.6767

Predicted residual sum of squares (PRESS) = 843265.4

Table A.1.7 (*Continued*)

Factor	Coefficient Estimate	Degrees of Freedom	Standard Error	t for H_0 Coefficient $= 0$	Probability $> |t|$	VIF
Intercept	2921.0	1	49.6	58.90		
A—time	222.8	1	124.0	1.80	0.132	10.0
B—catalyst	−427.9	1	124.0	−3.45	0.018	10.0
A_2	46.4	1	42.1	1.10	0.320	1.02
B_2	−532.5	1	42.1	−12.66	< 0.001	1.02
AB	−236.3	1	55.4	−4.26	0.008	1.00
A_3	−156.5	1	78.5	−1.99	0.103	10.0
B_3	201.6	1	78.5	2.57	0.050	10.0

Final Equation in Terms of Coded Factors

$$M_n =$$

$$
\begin{aligned}
& 2921.0 \\
+\ & 222.8 * A \\
-\ & 427.9 * B \\
+\ & 46.4 * A^2 \\
-\ & 532.5 * B^2 \\
-\ & 236.3 * AB \\
-\ & 156.5 * A_3 \\
+\ & 201.6 * B_3
\end{aligned}
$$

Final Equation in Terms of Actual Factors

$$M_n =$$

$$
\begin{aligned}
& 12{,}676.1 \\
-\ & 4450.7 * \text{Time} \\
+\ & 21672 * \text{Catalyst} \\
+\ & 133.33 * \text{Time}^2 \\
-\ & 3556.8 * \text{Catalyst}^2 \\
-\ & 47.250 * \text{Time} * \text{Catalyst} \\
-\ & 1.2521 * \text{Time}^3 \\
+\ & 201.62 * \text{Catalyst}^3
\end{aligned}
$$

Observation	Actual Value	Predicted Value	Residual	Lever	Student Residual	Cook's Distance	R-Student	Run Order
1	2320.0	2358.7	−38.7	0.875	−0.99	0.852	−0.98	7
2	2925.0	2963.7	−38.7	0.875	−0.99	0.852	−0.98	2
3	2340.0	2378.7	−38.7	0.875	−0.99	0.852	−0.98	12
4	2000.0	2038.7	−38.7	0.875	−0.99	0.852	−0.98	4
5	3180.0	3141.3	38.7	0.875	0.99	0.851	0.98	11
6	2925.0	2886.3	38.7	0.875	0.99	0.851	0.98	5
7	1930.0	1891.3	38.7	0.875	0.99	0.851	0.98	1
8	1860.0	1821.3	38.7	0.875	0.99	0.851	0.98	10
9	2980.0	2921.0	59.0	0.200	0.59	0.011	0.55	9
10	3075.0	2921.0	154.0	0.200	1.55	0.075	1.93	13
11	2790.0	2921.0	−131.0	0.200	−1.32	0.055	−1.46	3
12	2850.0	2921.0	−71.0	0.200	−0.72	0.016	−0.68	8
13	2910.0	2921.0	−11.0	0.200	−0.11	0.000	−0.10	6

likely that there is not one best subset model, but that there are several equally good ones.

The analyst should also keep in mind that the order in which the regressors enter or leave the model does not necessarily imply an order of importance to the variables. It is not unusual to find that a regressor inserted into the model early in the procedure becomes negligible at a subsequent step. For example, suppose that forward selection chooses x_4 (say) as the first regressor to enter. However, when x_2 (say) is added at a subsequent step, x_4 is no longer required because of high positive correlation between x_2 and x_4. This is a general problem with the forward selection procedure. Once a regressor has been added, it cannot be removed at a later step.

Note that forward selection, backward elimination, and stepwise regression do not necessarily lead to the same choice of final model. The correlation between the regressors affect the order of entry and removal. Some users have recommended that all the procedures be applied in the hopes of either seeing some agreement or learning something about the structure of the data that might be overlooked by using only one selection procedure. Furthermore, there is not necessarily any agreement between any of the stepwise-type procedures and all possible regressions.

For these reasons, stepwise-type variable selection procedures should be used with caution. Our own preference is for the stepwise regression algorithm followed by backward elimination. The backward elimination algorithm is often less adversely affected by the correlative structure of the regressors than is forward selection. We have also found it helpful to run the problem several times with different choices of F_{IN} and F_{OUT} (or different values of α). This will often allow the analyst to generate several different models for evaluation.

APPENDIX 2

Multicollinearity and Biased Estimation in Regression

In some response surface and mixture experiments, there can be one or more near-linear dependencies among the regressor variables in the model. Regression model-builders refer to this as **multicollinearity** among the regressors. Multicollinearity can have serious effects on the estimates of the model parameters and on the general applicability of the final model. In this appendix, we give a brief introduction to the multicollinearity problem along with biased estimation, one of the parameter estimation techniques useful in dealing with multicollinearity.

The effects of multicollinearity may be easily demonstrated. Consider a regression model with two regressor variables x_1 and x_2, and suppose that x_1 and x_2 have been "standardized" by subtracting the average of that variable from each observation and dividing by the square root of the corrected sum of squares. This unit length scaling, as it is called, results in the $\mathbf{X}'\mathbf{X}$ matrix having the form of a correlation matrix; that is, the main diagonals are 1 and the off-diagonals are the simple correlation between regressor x_i and regressor x_j. The model is

$$y_i = \beta_0 + \beta_1 x_{i1} + \beta_2 x_{i2} + \varepsilon_i, \qquad i = 1, 2, \ldots, n$$

Now if the response is also centered, then the estimate of β_0 is zero. The $(\mathbf{X}'\mathbf{X})^{-1}$ matrix for this model is

$$\mathbf{C} = (\mathbf{X}'\mathbf{X})^{-1} = \begin{bmatrix} 1/\left(1 - r_{12}^2\right) & -r_{12}/\left(1 - r_{12}^2\right) \\ -r_{12}/\left(1 - r_{12}^2\right) & 1/\left(1 - r_{12}^2\right) \end{bmatrix}$$

where r_{12} is the simple correlation between x_1 and x_2.

Now, if multicollinearity is present, x_1 and x_2 are highly correlated, and $|r_{12}| \to 1$. In such a situation, the variances and covariances of the regression coefficients become very large, because $V(b_j) = C_{jj}\sigma^2 \to \infty$ as $|r_{12}| \to 1$, and

$\text{Cov}(b_1, b_2) = C_{12}\sigma^2 \to \pm\infty$ depending on whether $r_{12} \to \pm 1$. The large variances for b_j imply that the regression coefficients are very poorly estimated. Note that the effect of multicollinearity is to introduce a near-linear dependency in the columns of the $\mathbf{X}$ matrix. As $r_{12} \to 1$ or -1, this linear dependency becomes exact.

Similar problems occur when multicollinearity is present and there are more than two regressor variables. In general, the diagonal elements of the matrix $\mathbf{C} = (\mathbf{X'X})^{-1}$ can be written as

$$C_{jj} - \frac{1}{\left(1 - R_j^2\right)}, \qquad j = 1, 2, \ldots, k \qquad (A.2.1)$$

where R_j^2 is the coefficient of multiple determination resulting from regressing x_j on the other $k - 1$ regressor variables. Clearly, the stronger the linear dependency of x_j on the remaining regressor variables (and hence the stronger the multicollinearity), the larger the value of R_j^2 will be. We say that the variance of b_j is "inflated" by the quantity $(1 - R_j^2)^{-1}$. Consequently, we usually call

$$\text{VIF}(b_j) - \frac{1}{\left(1 - R_j^2\right)}, \qquad j = 1, 2, \ldots, k \qquad (A.2.2)$$

the **variance inflation factor** for b_j. Note that these factors are the main diagonal elements of the inverse of the correlation matrix. They are an important measure of the extent to which multicollinearity is present.

Although the estimates of the regression coefficients are very imprecise when multicollinearity is present, the fitted model may still be useful. For example, suppose we wish to predict new observations. If these predictions are required in the region of the x-space where the multicollinearity is in effect, then often satisfactory results will be obtained because while individual β_j may be poorly estimated, the function $\sum_{j=1}^{k}\beta_j x_{ij}$ may be estimated quite well. On the other hand, if the prediction of new observations requires extrapolation, then generally we would expect to obtain poor results. Successful extrapolation usually requires good estimates of the individual model parameters.

Multicollinearity arises for several reasons. It will occur when the analyst collects the data such that a constraint of the form $\sum_{j=1}^{k}a_j x_j = \mathbf{0}$ holds among the columns of the $\mathbf{X}$ matrix (the a_j are constants, not all zero). For example, if four regressor variables are the components of a mixture, then such a constraint will always exist because the sum of the component proportions is always constant. Usually, however, these constraints do not hold exactly, and the analyst does not know that they exist.

There are several ways to detect the presence of multicollinearity. Some of the more important of these will be briefly discussed.

1. The variance inflation factors, defined in Equation (A.2.2), are very useful measures of multicollinearity. The larger the variance inflation factor, the more severe the multicollinearity. Some authors have suggested that if any variance inflation factors exceed 10, then multicollinearity is a problem. Other authors consider this value too liberal and suggest that the variance inflation factors should not exceed 4 or 5.

2. The determinant of the $X'X$ matrix in correlation form may also be used as a measure of multicollinearity. The value of this determinant can range between 0 and 1. When the value of the determinant is 1, the columns of the X matrix are orthogonal (i.e., there is no intercorrelation between the regressor variables), and when the value is 0, there is an exact linear dependency among the columns of X. The smaller the value of the determinant, the greater the degree of multicollinearity.

3. The eigenvalues or characteristic roots of the $X'X$ matrix in correlation form provide a measure of multicollinearity. The eigenvalues of $X'X$ are the roots of the equation

$$|X'X - \lambda I| = 0$$

One or more eigenvalues near zero implies that multicollinearity is present. If λ_{max} and λ_{min} denote the largest and smallest eigenvalues of $X'X$, then the *condition number* $\kappa = \lambda_{max}/\lambda_{min}$ can also be used as a measure of multicollinearity. The larger the value of the condition number, the greater the degree of multicollinearity. Generally, if the condition number $\kappa = \lambda_{max}/\lambda_{min}$ is less than 100, there is little problem with multicollinearity.

4. Sometimes inspection of the individual elements of the correlation matrix can be helpful in detecting multicollinearity. If an element $|r_{ij}|$ is close to one, then x_i and x_j may be strongly multicollinear. However, when more than two regressor variables are involved in a multicollinear fashion, the individual r_{ij} are not necessarily large. Thus, this method will not always enable us to detect the presence of multicollinearity.

5. If the F-test for significance of regression is significant, but tests on the individual regression coefficients are not significant, then multicollinearity may be present.

Several remedial measures have been proposed for resolving the problem of multicollinearity. Augmenting the data with new observations specifically designed to break up the approximate linear dependencies that currently exist is often suggested. However, sometimes this is impossible for economic reasons, or because of the physical constraints that relate the x_j. Another possibility is to delete certain terms from the model. This suffers from the disadvantage of discarding the information contained in the deleted terms.

Because multicollinearity primarily affects the stability of the regression coefficients, it would seem that estimating these parameters by some method that is less sensitive to multicollinearity than ordinary least squares would be helpful. **Ridge regression** is one of several methods that have been suggested for this. In ridge regression, the regression coefficients estimates are obtained by solving

$$\mathbf{b}^*(\theta) = (\mathbf{X}'\mathbf{X} + \theta\mathbf{I})^{-1}\mathbf{X}'\mathbf{y} \qquad\qquad (A.2.3)$$

where $\theta \geq 0$ is a constant. Generally, values of θ in the interval $0 \leq \theta \leq 1$ are appropriate. The ridge estimator $\mathbf{b}^*(\theta)$ is not an unbiased estimator of $\boldsymbol{\beta}$, as is the ordinary least squares estimator $\mathbf{b}$, but the mean squared error of $\mathbf{b}^*(\theta)$ will be smaller than the mean square error of $\mathbf{b}$. Thus ridge regression seeks to find a set of regression coefficients that is more "stable," in the sense of having a small mean square error. Because multicollinearity usually results in ordinary least squares estimators that may have extremely large variances, ridge regression is suitable for situations where the multicollinearity problem exists.

To obtain the ridge regression estimator from Equation (A.2.3), one must specify a value for the constant θ. Generally, there is an "optimum" θ for any problem, but the simplest approach is to solve Equation (A.2.3) for several values of θ in the interval $0 \leq \theta \leq 1$. Then a plot of the values of $\mathbf{b}^*(\theta)$ against θ is constructed. This display is called the *ridge trace*. The appropriate value of θ is chosen subjectively by inspection of the ridge trace. Typically, a value for θ is chosen such that relatively stable parameter estimates are obtained. In general, the variance of $\mathbf{b}^*(\theta)$ is a decreasing function of θ, while the squared bias $[\boldsymbol{\beta} - \mathbf{b}^*(\theta)]^2$ is an increasing function of θ. Choosing the value of θ involves trading off these two properties of $\mathbf{b}^*(\theta)$. See Marquardt and Snee (1975) for a good discussion of the practical aspects of ridge regression. Montgomery and Peck (1992) and Myers (1990) discuss biased estimation in detail.

Multicollinearity is usually not a big problem in a well-designed and well-executed response surface experiment. For example, with three factors and a rotatable central composite design (CCD), the largest variance inflation factor is 1.018. That is, the design is nearly orthogonal. However, a poorly designed or poorly executed response surface experiment could have substantial problems with multicollinearity, as we will see in the next example.

Mixture experiments may often have substantial levels of multicollinearity. Unconstrained mixture experiments where lattice-type designs are used present no serious problems, nor do spaces constrained by lower bounds, if the L-pseudocomponents are used. However, in mixture problems with both upper and lower bounds where the constrained region is a hyperpolytope, the multicollinearity problem can be significant. You should always fit mixture models in these constrained spaces in terms of the pseudocomponents, because the variance inflation factors are much smaller (and hence the

Table A.2.1   Acetylene Data for Example A.2.1

Observations, i	Conversion of n-Heptane to Acetylene (%)	Reactor Temperature (°C)	Ratio of H_2 to n-Heptane (Mole ratio)	Contact Time (sec)
1	49.0	1300	7.5	0.0120
2	50.2	1300	9.0	0.0120
3	50.5	1300	11.0	0.0115
4	48.5	1300	13.5	0.0130
5	47.5	1300	17.0	0.0135
6	44.5	1300	23.0	0.0120
7	28.0	1200	5.3	0.0400
8	31.5	1200	7.5	0.0380
9	34.5	1200	11.0	0.0320
10	35.0	1200	13.5	0.0260
11	38.0	1200	17.0	0.0340
12	38.5	1200	23.0	0.0410
13	15.0	1100	5.3	0.0840
14	17.0	1100	7.5	0.0980
15	20.5	1100	11.0	0.0920
16	29.5	1100	17.0	0.0860

multicollinearity problem less severe) for the pseudocomponent version of the model. For more details, see St. John (1984) and Montgomery and Voth (1994).

Example A.2.1   Table A.2.1 presents data concerning the percentage of conversion of n-heptane to acetylene and three explanatory variables [Himmelblau (1971), Kunugi, Tamura, and Naito (1961), and Marquardt and Snee (1975)]. This is chemical process data for which a full quadratic response surface in all three variables will be considered as an appropriate model. A plot of contact time versus reactor temperature is shown in Figure A.2.1. Clearly these two regressors are highly correlated, so there are potential multicollinearity problems in these data.

The full quadratic model for the acetylene data is

$$P = \gamma_0 + \gamma_1 T + \gamma_2 H + \gamma_3 C + \gamma_{12} TH + \gamma_{13} TC + \gamma_{23} HC$$
$$+ \gamma_{11} T^2 + \gamma_{22} H^2 + \gamma_{33} C^2 + \varepsilon$$

where

$$P = \text{Percentage of conversion}$$

$$T = \frac{\text{Temperature} - 1212.50}{80.623}$$

$$H = \frac{H_2/(n - \text{heptane}) - 12.44}{5.662}$$

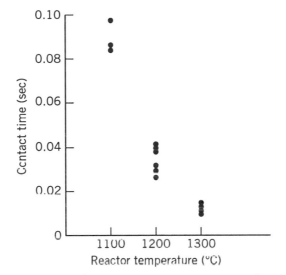

Figure A.2.1 Contact time versus reactor temperature, acetylene data.

and

$$C = \frac{\text{Contact time} - 0.0403}{0.03164}$$

Each of the original regressors has been centered and scaled by subtracting the average (centering) and dividing by the standard deviation. The squared and cross-product terms are generated from the scaled linear terms. The least squares fit is

$$\hat{P} = 35.897 + 4.019T + 2.781H - 8.031C - 6.457TH - 26.982TC$$

$$- 3.768HC - 12.524T^2 - 0.972H^2 - 11.594C^2$$

The summary statistics for this model are displayed in Table A.2.2.

This table also shows the variance inflation factors for this model. Because there are several VIFs that exceed 10, we conclude that multicollinearity is a serious problem in this model.

Figure A.2.2 presents a contour plot of the predicted response in the contact time–reactor temperature space, with mole ratio of $H_2 = 12.44$. The predicted conversions at many locations in this region are negative, an obvious impossibility. It seems that the least squares model fits the data reasonably well but predicts conversion over the region of interest very poorly. A likely cause of this is multicollinearity.

To obtain the ridge solution for the acetylene data, we must solve the equations $(\mathbf{X'X} + \theta\mathbf{I})\mathbf{b}^*(\theta) = \mathbf{X'y}$ for several values of $0 \leq \theta \leq 1$, with $\mathbf{X'X}$

Table A.2.2 Summary Statistics for the Least Squares Acetylene Model

Term	Regression Coefficient	Standard Error	t_0	VIF
Intercept	35.8971	1.0903	32.93	
T	4.0187	4.5012	0.89	374
H	2.7811	0.3074	9.05	1.74
C	-8.0311	6.0657	-1.32	679.11
TH	-6.4568	1.4660	-4.40	31.03
TC	-26.9818	21.0224	-1.28	6565.91
HC	-3.7683	1.6554	-2.28	36.50
T^2	-12.5237	12.3239	-1.02	1762.58
C^2	-11.5943	7.7070	-1.50	1158.13
H^2	-0.9721	0.3746	-2.60	3.17

$MS_E = 0.8126$, $R^2 = 0.998$, $F_0 = 289.72$.
When the response is standardized, $MS_E = 0.00038$ for the least squares model.

and $\mathbf{X'y}$ in correlation form. The standardized data, using the unit length scaling, are shown in Table A.2.3. The ridge trace is shown in Figure A.2.3, and the ridge coefficients for several values of θ are listed in Table A.2.4. This table also shows the residual mean square and R^2 for each ridge model. Notice that as θ increases, MS_E increases and R^2 decreases. The ridge trace illustrates the instability of the least squares solution, because

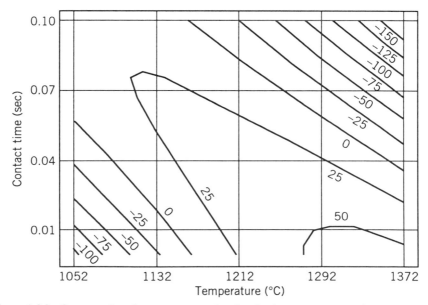

Figure A.2.2 Contour plot of percent conversion for the least squares acetylene model with mole ratio of $H_2 = 12.44$.

Table A.2.3 Standardized Acetylene Data

Observation i	y	x_1	x_2	x_3	$x_1 x_2$	$x_1 x_3$	$x_2 x_3$	x_1^2	x_2^2	x_3^2
1	.27979	.28022	-.22544	-.23106	-.33766	-.02085	.30952	.07829	-.04116	-.03452
2	.03583	.28022	-.15704	-.23106	-.25371	-.02085	.23659	.07829	-.13270	-.03452
3	.31234	.28022	-.06584	-.23514	-.14179	-.02579	.14058	.07829	-.20378	-.02735
4	.26894	.28022	.04817	-.22290	.00189	-.01098	.01960	.07829	-.21070	-.04847
5	.24724	.28022	.20777	-.21882	.19398	-.00605	-.14065	.07829	-.06745	-.05526
6	.18214	.28022	.48139	-.23106	.52974	-.02085	-.44415	.07829	.59324	-.03452
7	.17590	-.04003	-.32577	-.00255	-.00413	.25895	.07300	-.29746	.15239	-.23548
8	-.09995	-.04003	-.22544	-.01887	-.02171	.26177	.08884	-.29746	-.04116	-.23418
9	-.03486	-.04003	-.06584	-.06784	-.04970	.27023	.08985	-.29746	-.20378	-.21822
10	-.02401	-.04003	.04817	-.11680	-.06968	.27869	.04328	-.29746	-.21070	-.18419
11	.04109	-.04003	.20777	-.05152	-.09766	.26741	.01996	-.29746	-.06745	-.22554
12	.05194	-.04003	.48139	.00561	-.14563	.25754	.08202	-.29746	.59329	-.23538
13	.45800	-.36029	-.32577	.35653	.45252	-.29615	-.46678	.32879	.15239	.24374
14	.41460	-.36029	-.22544	.47078	.29423	-.47384	-.42042	.32879	-.04116	.60000
15	-.33865	-.36029	-.06584	.42187	.04240	-.39769	-.05859	.32879	-.20378	.43527
16	-.14335	-.36029	.20777	.37285	-.38950	-.32153	-.42738	.32879	-.06745	.28861

[a]The standardized data was constructed from the centered and scaled form of the original data in Table A.2.1.

Table A.2.4 Coefficients at Various Values of θ

θ	0.000	0.001	0.002	0.004	0.008	0.016	0.032	0.064	0.128	0.256	0.512
b_1	.3377	.6770	.6653	.6362	.6003	.5672	.5392	.5122	.4806	.4379	.3784
b_2	.2337	.2242	.2222	.2199	.2173	.2148	.2117	.2066	.1971	.1807	.1554
b_3	−.6749	−.2129	−.2284	−.2671	−.3134	−.3515	−.3735	−.3800	−.3724	−.3500	−.3108
b_{12}	−.4799	−.4479	−.4258	−.3913	−.3437	−.2879	−.2329	−.1862	−.1508	−.1249	−.1044
b_{13}	−2.0344	−.2774	−.1887	−.1350	−.1017	−.0809	−.0675	−.0570	−.0454	−.0299	−.0092
b_{23}	−.2675	−.2173	−.1920	−.1535	−.1019	−.0433	.0123	.0562	.0849	.0985	.0991
b_{11}	−.8346	.0643	.1035	.1214	.1262	.1254	.1249	.1258	.1230	.1097	.0827
b_{22}	−.0904	−.0732	−.0682	−.0621	−.0558	−.0509	−.0481	−.0464	−.0444	−.0406	−.0341
b_{33}	−1.0015	−.2451	−.1853	−.1313	−.0825	−.0455	−.0267	−.0251	−.0339	−.0464	−.0586
MS_E	.00038	.00047	.00049	.00054	.00062	.00074	.00094	.00127	.00206	.00425	.01002
R^2	.998	.997	.997	.997	.996	.996	.994	.992	.988	.975	.940

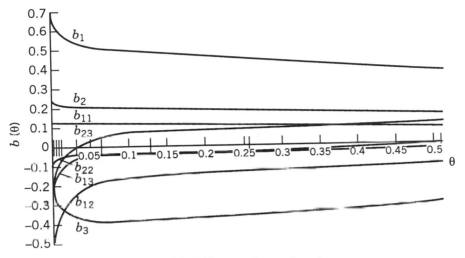

Figure A.2.3   Ridge trace for acetylene data.

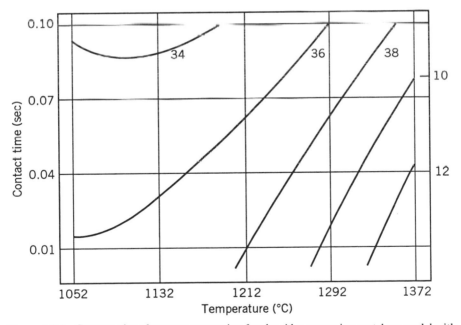

Figure A.2.4   Contour plot of percent conversion for the ridge regression acetylene model with mole ratio of $H_2 = 12.44$.

there are large changes in the regression coefficients for small values of θ. However, the coefficients stabilize rapidly as θ increases.

Judgment is required to interpret the ridge trace and select an appropriate value of θ. We want to choose θ large enough to provide stable coefficients, but not necessarily large because this introduces additional bias and increases the residual mean square. From Figure A.2.3 we see that reasonable coefficient stability is achieved in the region $0.0008 < \theta < 0.064$, without a severe increase in the residual mean square (or loss in R^2). If we choose $\theta = 0.032$, the ridge regression model is

$$\hat{y} = 0.5392 x_1 + 0.2117 x_2 - 0.3735 x_3 - 0.2329 x_1 x_2 - 0.0675 x_1 x_3$$
$$+ 0.0123 x_2 x_3 + 0.1249 x_1^2 - 0.0481 x_2^2 - 0.0267 x_3^2$$

Note that in this model the estimates of the coefficients β_{13}, β_{11}, and β_{23} are considerably smaller than the least squares estimates and the original negative estimates of β_{23} and β_{11} are now positive.

Figure A.2.4 shows a contour plot for the ridge model. Comparing Figure A.2.2 and A.2.4, we note the ridge model gives much more realistic predictions over the region of interest than does least squares. Thus we conclude that the ridge regression approach in this example has produced a model that is superior to the original least squares fit.

APPENDIX 3

Robust Regression

A.3.1 THE NEED FOR ROBUST ESTIMATION

When the observations $\mathbf{y}$ in the linear regression model $\mathbf{y} = \mathbf{X}\boldsymbol{\beta} + \boldsymbol{\varepsilon}$ are normally distributed, the method of least squares works well in the sense that it produces an estimate of $\boldsymbol{\beta}$ that has good statistical properties. However, when the observations follow some nonnormal distribution, particularly one that has longer or heavier tails than the normal, the method of least squares may not be appropriate. Heavy-tailed distributions usually generate outliers, and these outliers may have a strong influence on the least squares estimate. In effect, outliers "pull" the least squares fit too much in their direction, and consequently the identification of these outliers is difficult because their residuals have been made artificially small. Skillful residual analysis coupled with the use of techniques for identifying influential observations can help the analyst discover these problems. However, the successful use of these diagnostic procedures often requires abilities beyond those of the average analyst.

A number of authors have proposed *robust regression* procedures designed to dampen the effect of observations that would be highly influential if least squares were used. That is, a robust procedure tends to leave the residuals associated with outliers large, thereby making the identification of influential points much easier. In addition to insensitivity to outliers, a robust estimation procedure should be nearly as efficient as least squares when the underlying distribution is normal. For some basic references in robust regression, see Montgomery and Peck (1992), Rousseeuw (1984), and Rousseeuw and Leroy (1987).

To motivate the discussion and to demonstrate why it may be desirable to use an alternative to least squares when the observations are nonnormal, consider the simple linear regression model

$$y_i = \beta_0 + \beta_1 x_i + \varepsilon_i, \qquad i = 1, 2, \ldots, n \qquad (A.3.1)$$

where the errors are independent random variables that follow the double

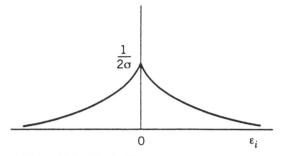

Figure A.3.1   The double exponential distribution.

exponential distribution

$$f(\varepsilon_i) = \frac{1}{2\sigma} e^{-|\varepsilon_i|/\sigma}, \qquad -\infty < \varepsilon_i < \infty \qquad (A.3.2)$$

The double exponential distribution is shown in Figure A.3.1. The distribution is more pointed in the middle than the normal and tails off to zero as $|\varepsilon_i|$ goes to infinity. However, since the density function goes to zero as $e^{-|\varepsilon_i|}$ goes to zero and the normal density function goes to zero as $e^{-\varepsilon_i^2}$ goes to zero, we see that the double exponential distribution has heavier tails than the normal.

We will use the method of maximum likelihood to estimate β_0 and β_1. The likelihood function is

$$L(\beta_0, \beta_1) = \prod_{i=1}^{n} \frac{1}{2\sigma} e^{-|\varepsilon_i|/\sigma} = \frac{1}{(2\sigma)^n} \exp\left(-\sum_{i=1}^{n} |\varepsilon_i|/\sigma\right) \quad (A.3.3)$$

Therefore maximizing the likelihood function would involve minimizing $\sum_{i=1}^{n} |\varepsilon_i|$, the sum of the absolute errors. It can be shown that the method of maximum likelihood applied to the regression model with normal errors leads to the least squares criterion. Thus the assumption of an error distribution with heavier tails than the normal implies that the method of least squares is no longer an optimal estimation technique. Note that the absolute error criterion would weight outliers far less severely than would least squares. Minimizing the sum of the absolute errors is often called the L_1-norm regression problem. Least squares is the L_2-norm regression problem.

A.3.2   *M*-ESTIMATORS

We have noted that the L_1-norm regression problem arises naturally from the maximum likelihood approach with double exponential errors. In general, we may define a class of robust estimators that minimize a function ρ of the residuals—for example,

$$\min_{\beta} \sum_{i=1}^{n} \rho(e_i) = \min_{\beta} \sum_{i=1}^{n} \rho(y_i - \mathbf{x}_i'\boldsymbol{\beta}) \qquad (A.3.4)$$

where $\mathbf{x}_i'$ denotes the ith row of $\mathbf{X}$. An estimator of this type is called an M-estimator, where M stands for maximum likelihood. That is, the function ρ is related to the likelihood function for an appropriate choice of the error distribution. For example, if the method of least squares is used (implying that the error distribution is normal), then $\rho(z) = \frac{1}{2}z^2$, $-\infty < z < \infty$. The M-estimator is not necessarily scale invariant [i.e., if the residuals $y_i - \mathbf{x}'\boldsymbol{\beta}$ were multiplied by a constant, the new solution to Equation (A.3.4) might not be the same as the old one]. To obtain a scale-invariant version of this estimator, we usually solve

$$\min_{\beta} \sum_{i=1}^{n} \rho\left(\frac{e_i}{s}\right) = \min_{\beta} \sum_{i=1}^{n} \rho\left(\frac{y_i - \mathbf{x}_i'\boldsymbol{\beta}}{s}\right) \qquad (A.3.5)$$

where s is a robust estimate of the standard deviation. A popular choice for s is

$$s = \text{median}|e_i - \text{median}(e_i)| / 0.6745$$

The constant 0.6745 makes s an approximately unbiased estimator of σ if n is large and the error distribution is normal.

To minimize Equation (A.3.5), equate the first partial derivatives of ρ with respect to β_j $(j = 0, 1, \ldots, k)$ equal to zero, yielding a necessary condition for a minimum. This gives the system of $p = k + 1$ equations

$$\sum_{i=1}^{n} x_{ij}\psi\left(\frac{y_i - \mathbf{x}_i'\boldsymbol{\beta}}{s}\right) = 0, \qquad j = 0, 1, \ldots, k \qquad (A.3.6)$$

where $\psi = \rho'$ and x_{ij} is the ith observation on the jth regressor and $x_{i0} = 1$. In general, the ψ function is nonlinear and Equation (A.3.6) must be solved by iterative methods. While several nonlinear optimization techniques could be employed, **iteratively reweighted least squares** is most widely used. This approach is usually attributed to Beaton and Tukey (1974). To use iteratively reweighted least squares, suppose that an initial estimate $\mathbf{b}_0$ is available. Then write the $p = k + 1$ Equations (A.3.6) as

$$\sum_{i=1}^{n} x_{ij}\psi\left(\frac{y_i - \mathbf{x}_i'\boldsymbol{\beta}}{s}\right)$$

$$= \sum_{i=1}^{n} \frac{x_{ij}\{\psi[(y_i - \mathbf{x}_i'\boldsymbol{\beta})/s]/(y_i - \mathbf{x}_i'\boldsymbol{\beta})/s\}(y_i - \mathbf{x}_i'\boldsymbol{\beta})}{s} = 0$$

$$j = 0, 1, \ldots, k \quad (A.3.7)$$

and

$$\sum_{i=1}^{n} x_{ij} w_{i0}(y_i - \mathbf{x}'_i \boldsymbol{\beta}) = 0, \qquad j = 0, 1, \ldots, k \qquad (A.3.8)$$

where

$$w_{i0} = \begin{cases} \dfrac{\psi[(y_i - \mathbf{x}'_i \mathbf{b}_0)/s]}{(y_i - \mathbf{x}'_i \mathbf{b}_0)/s} & \text{if } y_i \neq \mathbf{x}'_i \mathbf{b}_0 \\ 1 & \text{if } y_i = \mathbf{x}'_i \mathbf{b}_0 \end{cases} \qquad (A.3.9)$$

In matrix notation, Equation (A.3.8) becomes

$$\mathbf{X}'\mathbf{W}_0 \mathbf{X} \boldsymbol{\beta} = \mathbf{X}'\mathbf{W}_0 \mathbf{y} \qquad (A.3.10)$$

where $\mathbf{W}_0$ is an $n \times n$ diagonal matrix of "weights" with diagonal elements $w_{10}, w_{20}, \ldots, w_{n0}$ given by Equation (A.3.9). Equation (A.3.10) are called the **weighted least squares normal equations**. Consequently the one-step estimator is

$$\mathbf{b}_1 = (\mathbf{X}'\mathbf{W}_0 \mathbf{X})^{-1} \mathbf{X}'\mathbf{W}_0 \mathbf{y} \qquad (A.3.11)$$

At the next step we recompute new weights $\mathbf{W}_1$ from Equation (A.3.9) but using $\mathbf{b}_1$ instead of $\mathbf{b}_0$. Then a new $\mathbf{b}_2$ is computed via weighted least squares. Usually only a few iterations are required to achieve convergence. The iteratively reweighted least squares procedure requires only a standard weighted least squares computer program.

A number of popular robust criterion functions are shown in Table A.3.1. The behavior of these ρ-functions and their corresponding ψ-functions are illustrated in Figures A.3.2 and A.3.3, respectively. Robust regression procedures can be classified by the behavior of their ψ-function. The ψ-function controls the weight given to each residual and (apart from a constant of proportionality) is sometimes called the *influence function*. For example, the ψ-function for least squares is unbounded, and thus least squares tends to be nonrobust when used with data arising from a heavy-tailed distribution. The Huber t-function [Huber (1964)] has a monotone ψ-function and does not weight large residuals as heavily as least squares. The last three influence functions actually redescend as the residual becomes larger. Ramsay's E_a-function [see Ramsay (1977)] is a soft redescender; that is, the ψ-function is asymptotic to zero for large $|z|$. Andrew's wave function and Hampel's 17A function [see Andrews et al. (1972) and Andrews (1974)] are hard redescenders; that is, the ψ-function equals zero for sufficiently large $|z|$. We should note that the ρ-functions associated with the redescending ψ-functions are nonconvex, and this in theory can cause convergence problems in the iterative estimation procedure. However, this is not a common occur-

Table A.3.1 Robust Criterion Functions

Criterion	$\rho(z)$	$\psi(z)$	$w(z)$	Range
Least squares	$\frac{1}{2}z^2$	z	1.0	$|z| < \infty$
Huber's t-function	$\frac{1}{2}z^2$	z	1.0	$|z| \leq t$
$t = 2$	$|z|t - \frac{1}{2}t^2$	$t\,\text{sign}(z)$	$\dfrac{t}{|z|}$	$|z| > t$
Ramsay's E_u-function $a = 0.3$	$a^{-2}[1 - \exp(-a|z|)$ $\cdot(1 + a|z|)]$	$z\exp(-a|z|)$	$\exp(-a|z|)$	$|z| < \infty$
Andrew's wave function	$a[1 - \cos(z/a)]$	$\sin(z/a)$	$\dfrac{\sin(z/a)}{z/a}$	$|z| \leq a\pi$
$a = 1.339$	$2a$	0	0	$|z| > a\pi$
Hampel's 17A function	$\frac{1}{2}z^2$	z	1.0	$|z| \leq a$
$u = 1.7$ $b = 3.4$ $c = 8.5$	$a|z| - \frac{1}{2}a^2$	$u\,\text{sign}(z)$	$a/|z|$	$a < |z| \leq b$
	$\dfrac{a(c|z| - \frac{1}{2}z^2)}{c - b}$	$\dfrac{u\,\text{sign}(z)(c - |z|)}{c - b}$ $-(7/6)a^2$	$\dfrac{a(c - |z|)}{|z|(c - b)}$	$b < |z| \leq c$
	$a(b + c - a)$	0	0	$|z| < c$

rence. Furthermore, each of the robust criterion functions requires the analyst to specify certain "tuning constants" for the ψ-functions. We have shown typical values of these turning constants in Table A.3.1.

Several authors [Andrews (1974), Hogg (1979), and Hocking (1978)] have noted that the starting value $\mathbf{b}_0$ used in robust estimation must be chosen carefully. Using the least squares solution can disguise the high-leverage points. The L_1-norm estimates would be good choice of starting values.

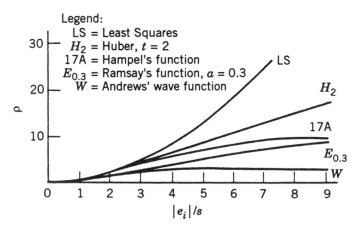

Figure A.3.2 Robust criterion functions.

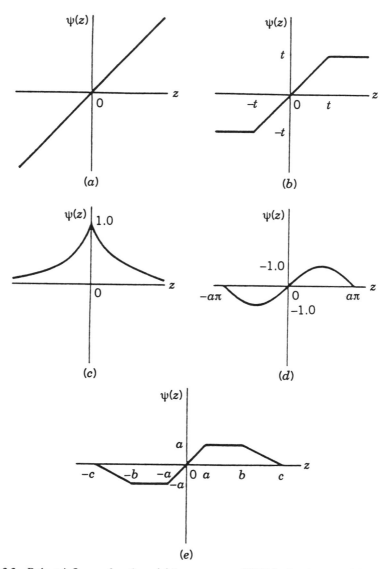

Figure A.3.3   Robust influence functions: (*a*) Least squares; (*b*) Huber's *t*-function; (*c*) Ramsay's E_a-function; (*d*) Andrew's wave function; (*e*) Hempel's 17A function.

Andrews (1974) and Dutter (1977) also suggest procedures for choosing the starting values.

At present it is difficult to give strong recommendations concerning the error structure of the final robust regression estimates **b**. Determining the covariance matrix of **b** is important if we are to construct confidence intervals or make other model inferences. Huber (1973) has shown that asymptotically

b has an approximate normal distribution with covariance matrix:

$$\sigma^2 \frac{E[\psi^2(\varepsilon/\sigma)]}{\{E[\psi'(\varepsilon/\sigma)]\}^2}(\mathbf{X'X})^{-1}$$

Therefore a reasonable approximation for the covariance matrix of **b** is

$$\frac{(ns)^2}{n-p} \frac{\sum_{i=1}^{n} \psi^2[(y_i - \mathbf{x}_i'\mathbf{b})/s]}{\left(\sum_{i=1}^{n} \psi'[(y_i - \mathbf{x}_i'\mathbf{b})/s]\right)^2}(\mathbf{X'X})^{-1}$$

The weighted least squares computer program also produces an estimate of the covariance matrix

$$\frac{\sum_{i=1}^{n} w_i(y_i - \mathbf{x}_i'\mathbf{b})^2}{n-p}(\mathbf{X'WX})^{-1}$$

Other suggestions are in Welsch (1975) and Hill (1979). There is no general agreement about which approximation to the covariance matrix **b** is best. Both Welsch and Hill note that these covariance matrix estimates perform poorly for **X** matrices that have outliers. Ill-conditioning (multicollinearity) also distorts robust regression estimation estimates. However, there are indications that in many cases we can make approximate inferences about **b** using procedures similar to the usual normal theory.

Robust regression methods have much to offer the data analyst. They can be extremely helpful in locating outliers and highly influential observations. Whenever a least squares analysis is performed, it would be useful to perform a robust fit also. If the results of the two procedures are in substantial agreement, then use the least squares results, because inferences based on least squares are at present better understood. However, if the results of the two analyses differ, then reasons for these differences should be identified. Observations that are downweighted in the robust fit should be carefully examined.

Some Mathematical Insights into Ridge Analysis

Consider the **B** matrix discussed in Chapter 6. Consider also the **P** matrix (orthogonal) which diagonalizes **B**. That is,

$$\mathbf{P'BP} = \boldsymbol{\Lambda}$$

$$= \begin{bmatrix} \lambda_1 & & & 0 \\ & \lambda_2 & & \\ & & \ddots & \\ 0 & & & \lambda_k \end{bmatrix}$$

where the λ_i are the eigenvalues of **B**. The solution **x** that produces locations where $\partial L / \partial \mathbf{x} = \mathbf{0}$ is given by

$$(\mathbf{B} - \mu\mathbf{I})\mathbf{x} = -\tfrac{1}{2}\mathbf{b}$$

If we premultiply $(\mathbf{B} - \mu\mathbf{I})$ by **P'** and post multiply by **P** we obtain

$$\mathbf{P'}(\mathbf{B} - \mu\mathbf{I})P = \boldsymbol{\Lambda} - \mu\mathbf{I}$$

because $\mathbf{P'P} = \mathbf{I}_k$. If $\mathbf{B} - \mu\mathbf{I}$ is negative definite, the resulting solution **x** is at least a local maximum on the radius $R = (\mathbf{x'x})^{1/2}$. On the other hand if $\mathbf{B} - \mu\mathbf{I}$ is positive definite, the result is a local minimum. Because

$$(\mathbf{B} - \mu\mathbf{I}) = \boldsymbol{\Lambda} - \mu\mathbf{I}$$

$$= \begin{bmatrix} \lambda_1 - \mu & & & 0 \\ & \lambda_2 - \mu & & \\ & & \ddots & \\ 0 & & & \lambda_k - \mu \end{bmatrix}$$

then if $\mu > \lambda_{max}$, $\mathbf{B} - \mu\mathbf{I}$ is negative definite and if $\mu < \lambda_{min}$, $\mathbf{B} - \mu\mathbf{I}$ is positive definite.

The above development does not prove that the solutions **x** are absolute maxima or minima. For further details see Draper (1963) or Myers (1976).

APPENDIX 5

Moment Matrix
of a Rotatable Design

In the development that follows, it is convenient in expressing the polynomial model to make use of *derived power vectors* and *Schlaifflian matrices*. If $\mathbf{z}' = [z_1, z_2, \ldots, z_k]$, then $\mathbf{z}'^{[p]}$, the derived power vector of degree p, is defined such that

$$\mathbf{z}'^{[p]}\mathbf{z}^{[p]} = (\mathbf{z}'\mathbf{z})^p$$

For example if $\mathbf{z}' = [z_1, z_2, z_3]$, then

$$\mathbf{z}'^{[2]} = \left[z_1^2, z_2^2, z_3^2, \sqrt{2}\,z_1 z_2, \sqrt{2}\,z_1\, z_3, \sqrt{2}\,z_2 z_3 \right]$$

and

$$\mathbf{z}'^{[1]} = \mathbf{z}'$$

If a vector $\mathbf{x}$ is formed from a vector $\mathbf{z}$ containing k elements through the transformation

$$\mathbf{x} = \mathbf{Hz} \tag{A.5.1}$$

then the Schlaifflian matrix $\mathbf{H}^{[p]}$ is defined such that

$$\mathbf{x}^{[p]} = \mathbf{H}^{[p]}\mathbf{z}^{[p]}$$

It is readily seen that if the transformation matrix $\mathbf{H}$ is orthogonal, then $\mathbf{H}^{[p]}$ is also orthogonal. One can write

$$\mathbf{x}'^{[p]}\mathbf{x}^{[p]} = \mathbf{z}'^{[p]}\mathbf{H}'^{[p]}\mathbf{H}^{[p]}\mathbf{z}^{[p]} \tag{A.5.2}$$

The left-hand side of Equation (A.5.2) is, by definition, $(\mathbf{x}'\mathbf{x})^p$. Because $\mathbf{H}$ is

675

orthogonal,

$$(\mathbf{x}'\mathbf{x})^P = (\mathbf{z}'\mathbf{z})^P = \mathbf{z}'^{[P]}\mathbf{z}^{[P]}$$

and thus the Schlaifflian matrix $\mathbf{H}^{[P]}$ is orthogonal. Another result which is quite useful in what follows is that, given two vectors x and z, each having k elements, then

$$(\mathbf{x}'\mathbf{z})^P = \mathbf{x}'^{[P]}\mathbf{z}^{[P]}$$

For a response function of order d, the estimated response $\hat{y}$ can be written in the form

$$\hat{y} = \mathbf{x}'^{[d]}\mathbf{b} \tag{A.5.3}$$

where for a point $(x_1, x_2, \ldots, x_k)$ we have

$$\mathbf{x}' = (1, x_1, x_2, \ldots, x_k)$$

and the vector $\mathbf{b}$ contains the least squares estimators $b_0, b_1, \ldots,$ and so on, with suitable multipliers. For example, for $k = 2$, $d = 2$, then $\mathbf{x}' = [1, x_1, x_2]$, and $\mathbf{b}'$ and $\mathbf{x}'^{[2]}$ are given by

$$\mathbf{b}' = \left[b_0, b_1/\sqrt{2}, b_2/\sqrt{2}, b_{11}, b_{22}, b_{12}/\sqrt{2} \right]$$

$$\mathbf{x}'^{[2]} = \left[1, \sqrt{2}x_1, \sqrt{2}x_2, x_1^2, x_2^2, \sqrt{2}x_1x_2 \right]$$

Thus, from Equation (A.5.3) we obtain

$$\text{Var}\left[\hat{y}(\mathbf{x})\right] = \mathbf{x}'^{[d]}[\text{Var }\mathbf{b}]\mathbf{x}^{[d]}$$

$$= \sigma^2 \mathbf{x}'^{[d]}[\mathbf{X}'\mathbf{X}]\mathbf{x}^{[d]} \tag{A.5.4}$$

where $\sigma^2(\mathbf{X}'\mathbf{X})^{-1}$ is the variance–covariance matrix of the vector $\mathbf{b}$.

Consider now a *second* point $(z_1, z_2, \ldots, z_k)$ which is the same distance from the origin as the point described by $(x_1, x_2, \ldots, x_k)$. Denote by $\mathbf{z}'$ the vector $(1, z_1, z_2, \ldots, z_k)$. There is, then, an orthogonal matrix $\mathbf{R}$ for which

$$\mathbf{z} = \mathbf{R}\mathbf{x} \tag{A.5.5}$$

where $\mathbf{R}$ is of the form

$$\mathbf{R} = \begin{bmatrix} 1 & \vdots & 0 & 0 & \cdots & 0 \\ 0 & \vdots & & & & \\ 0 & \vdots & & & & \\ \vdots & \vdots & & \mathbf{H}_{k \times k} & & \\ 0 & \vdots & & & & \end{bmatrix} \tag{A.5.6}$$

and **H** is an orthogonal matrix with the dimensions indicated in Equation (A.5.6). The variance of the estimated response at the second point is then

$$\text{Var}\left[\hat{y}(\mathbf{z})\right] = \sigma^2 \mathbf{z}'^{[d]}(\mathbf{X}'\mathbf{X})^{-1}\mathbf{z}^{[d]}$$

Let $\mathbf{R}^{[d]}$ be the Schlaifflian matrix of the transformation in Equation (A.5.5).

$$\text{Var}\left[\hat{y}(\mathbf{z})\right] = \sigma^2 \mathbf{x}'^{[d]}\mathbf{R}'^{[d]}(\mathbf{X}'\mathbf{X})^{-1}\mathbf{R}^{[d]}\mathbf{x}^{[d]}$$

$$= \sigma^2 \mathbf{x}'^{[d]}\left[\mathbf{R}'^{[d]}(\mathbf{X}'\mathbf{X})\mathbf{R}^{[d]}\right]^{-1}\mathbf{x}^{[d]} \qquad (A.5.7)$$

because $\mathbf{R}^{[d]}$ is orthogonal. For the design to be rotatable, var $\hat{y}$ is constant on spheres, which implies that for any orthogonal matrix **H** we have

$$\mathbf{X}'\mathbf{X} = \mathbf{R}'^{[d]}(\mathbf{X}'\mathbf{X})\mathbf{R}^{[d]} \qquad (A.5.8)$$

where **R** is of the form indicated in Equation (A.5.6). The requirement in Equation (A.5.8) essentially means that the moment matrix remains the same if the design is *rotated*—that is, if the rows of the *design* matrix, denoted by **D** in the equation

$$\begin{bmatrix} 1 & \vdots \\ 1 & \vdots \\ \vdots & \vdots & \mathbf{D} \\ 1 & \vdots \end{bmatrix} = \begin{bmatrix} 1 & x_{11} & x_{21} & \cdots & x_{k1} \\ 1 & x_{12} & x_{22} & \cdots & x_{k2} \\ \vdots & \vdots & \vdots & \vdots & \vdots \\ 1 & x_{1N} & x_{2N} & \cdots & x_{kN} \end{bmatrix} = \begin{bmatrix} \mathbf{x}'_1 \\ \mathbf{x}'_2 \\ \vdots \\ \mathbf{x}'_N \end{bmatrix} \qquad (A.5.9)$$

are rotated via the transformation

$$\mathbf{z}_i = \mathbf{R}'\mathbf{x}_i$$

It is easily seen that the rotated design will have moment matrix (apart from the constant N^{-1}) equal to the right-hand side of Equation (A.5.8).

Consider now a vector $\mathbf{t}' = [1, t_2, t_2, \ldots, t_k]$ of dummy variables. The utility of these variables is in the construction of a generating function for the design moments. Consider the quantity

$$\text{M.F.} = N^{-1}\mathbf{t}'^{[d]}\mathbf{X}'\mathbf{X}\mathbf{t}'^{[d]} \qquad (A.5.10)$$

The matrix $\mathbf{X}'\mathbf{X}$ is alternatively given by

$$\mathbf{X}'\mathbf{X} = \sum_{u=1}^{N} \mathbf{x}_u^{[d]}\mathbf{x}_u'^{[d]}$$

where the vector $x'_u = [1, x_{1u}, x_{2u}, \ldots, x_{ku}]$ refers to the uth row of the design matrix, augmented by 1—that is, the uth row of the matrix in Equation (A.5.9).

$$\text{M.F.} = N^{-1} \sum_u \left\{ t'^{[d]} x_u^{[d]} x_u'^{[d]} t^{[d]} \right\}$$

$$= N^{-1} \sum_u \left\{ t' x_u \right\}^{2d} \qquad (A.5.11)$$

From Equation (A.5.11), it is seen that upon expanding $t' x_u$ we have

$$\text{M.F.} = N^{-1} \sum_u \left\{ 1 + t_1 x_{1u} + t_2 x_{2u} + \cdots + t_k x_{ku} \right\}^{2d} \qquad (A.5.12)$$

When Equation (A.5.12) is expanded, the terms involve moments of the design through order $2d$. In fact, the coefficient of $t_1^{\delta_1} t_2^{\delta_2} \cdots t_k^{\delta_k}$ is

$$\frac{(2d)!}{\prod\limits_{i=1}^{k} (\delta_i)!(2d - \delta)!} \left[1^{\delta_1} 2^{\delta_2} \cdots k^{\delta_k} \right] \qquad (A.5.13)$$

where $\sum_{i=1}^{k} \delta_i = \delta \leq 2d$. For a rotatable design,

$$\text{M.F.} = N^{-1} t'^{[d]} (X'X) t^{[d]} = N^{-1} t'^{[d]} R^{[d]} X'X R^{[d]} t^{[d]}$$

$$= N^{-1} (t'R')^{[d]} X'X [Rt]^{[d]}$$

where R is a $(k + 1) \times (k + 1)$ orthogonal matrix introduced in Equation (A.5.6). This implies that for a rotatable design, an orthogonal transformation on t does not affect the M.F. Because M.F. is a polynomial in the t's (also involving the design moments), for a rotatable design, M.F. must be a function of $\sum_{i=1}^{k} t_i^2$. That is, it is of the form

$$\text{M.F.} = \sum_{j=0}^{d} a_{2j} \left[\sum_{i=1}^{k} t_i^2 \right]^j \qquad (A.5.14)$$

It is easily seen that the coefficient of $t_1^{\delta_1} t_2^{\delta_2} \cdots t_k^{\delta_k}$ in Equation (A.5.14) is zero *if any of the* δ_i *are odd*. For the case where all δ_i are even, the coefficient from the multinomial expansion of $[\sum_{i=1}^{k} t_i^2]^j$ is given by

$$\frac{a_\delta(\delta/2)!}{\prod\limits_{i=1}^{k} (\delta_i/2)!} \qquad (A.5.15)$$

The reader should now consider Equation (A.5.15) in conjunction with Equation (A.5.13), the former pertaining to the generating function for the moments *in general*, and the latter pertaining to the case of a rotatable design, with the value being zero for moments with any δ_i odd. Upon equating the two and solving for the moment, the result is as given

$$N^{-1} \sum_{i=1}^{N} x_{1u}^{\delta_1} x_{2u}^{\delta_2} \cdots x_{ku}^{\delta_k} = \frac{\lambda_\delta \prod_{i=1}^{k} (\delta_i)!}{2^{\delta/2} \prod_{i=1}^{k} (\delta_i/2)!} \tag{A.5.16}$$

for all δ_i even and

$$N^{-1} \sum_{u-1}^{N} x_{1u}^{\delta_1} x_{2u}^{\delta_2} \cdots x_{ku}^{\delta_k} = 0 \tag{A.5.17}$$

for any δ_i odd. Here, λ_δ is given by

$$\lambda_\delta = \frac{a_\delta 2^{\delta/2} (\delta/2)! (2d - \delta)!}{(2d)!} \tag{A.5.18}$$

If we consider Equations (A.5.16) and (A.5.18) for the second-order case we have $d = 2$ and thus $[i] = [ij] = [ijk] = [iij] = [iii] = 0$ for $i \neq j \neq k$ and $[ii] = \lambda_2$ (fixed by scaling) $[iiii]/[iijj] = 3$.

APPENDIX 6

Rotatability of a Second-Order Equiradial Design

Consider first the moment of x_1 of order δ ($\delta = 1, 2, 3, 4$). After multiplication by n_1, we obtain

$$\sum_{u=0}^{n_1-1} x_{1u}^\delta = \sum_{u=0}^{n_1-1} \left\{ \rho \cos\left(\theta + \frac{2\pi u}{n_1} \right) \right\}^\delta \tag{A.6.1}$$

Because $\cos \tau = [e^{i\tau} + e^{-i\tau}]/2$, Equation (A.6.1) can be written

$$\sum_{u=0}^{n_1-1} x_{1u}^\delta = \left(\frac{\rho}{2} \right)^\delta \sum_{u=0}^{n_1-1} \left[a\omega^u + a^{-1}\omega^{-u} \right]^\delta$$

where $\omega = e^{2i\pi/n_2}$ and $a = e^{i\theta}$. Using a binomial expansion,

$$\sum_{u=0}^{n_1-1} x_{iu}^\delta = \left(\frac{\rho}{2} \right)^\delta \sum_{t=0}^{\delta} \binom{\delta}{t} a^{\delta-2t} \sum_{u=0}^{n_1-1} \omega^{(\delta-2t)u} \tag{A.6.2}$$

An expression similar to Equation (A.6.2) for the moments of x_2 can be established with little difficulty. In considering the portion $\sum_{u=0}^{n_1-1} \omega^{(\delta-2t)u}$ of Equation (A.6.2) for $t = 0, 1, 2, \ldots, \delta$, it is first noted that

$$-\delta \le (\delta - 2t) \le \delta$$

The interest here is in moments of order $n_1 - 1$ and less because $n_1 \ge 5$. If $\delta - 2t = 0$, then

$$\sum_{u=0}^{n_1-1} \omega^{(\delta-2t)u} = n_1$$

On the other hand, it is not difficult to show that for $|\delta - 2t| \le n_1 - 1$ and nonzero, we obtain

$$\sum_{u=0}^{n_1-1} \omega^{(\delta-2t)u} = 0$$

Using Equation (A.6.2) for $\delta = 1$, $(\delta - 2t)$ takes on values -1 and $+1$ and thus $\sum_{u=0}^{n_1-1} x_{1u} = 0$. Likewise, for the case of $\delta = 3$, it is found that $\sum_{u=0}^{n_1-1} x_{1u}^3 = 0$. In a similar fashion, it can be shown that the odd moments of x_2 are zero. For $\delta = 2$, $\delta - 2t$ does that on a zero value when $t = 1$. Evaluating Equation (A.6.2)

$$n_1[11] = \frac{n_1 \binom{2}{1} \rho^2}{4}$$

and thus

$$[11] = \frac{\rho^2}{2}$$

For $\delta = 4$, the nonzero contribution in Equation (A.6.2) appears when $t = 2$. Thus the pure fourth moment is given by

$$[1111] = \frac{\binom{4}{2} \rho^4}{16}$$

$$= \frac{3\rho^4}{8}$$

Similar procedures can be used for the case of x_2 to show that

$$n_1[2^2] = \frac{\rho^2 n_1}{2}, \qquad n_1[2^4] = \frac{\rho^4 n_1}{8}$$

At this point, consider moments of the type $\sum_{u=0}^{n_1-1} x_{1u}^{\delta_1} x_{2u}^{\delta_2}$, where $\delta_1 + \delta_2 \le 4$. One can write

$$\sum_{u=0}^{n_1-1} x_{1u}^{\delta_1} x_{2u}^{\delta_2} = \sum_{u=0}^{n_1-1} \left\{ \rho \cos\left(\theta + \frac{2\pi u}{n_1}\right) \right\}^{\delta_1} \left\{ \rho \sin\left(\theta + \frac{2\pi u}{n_1}\right) \right\}^{\delta_2}$$

$$= \left(\frac{\rho}{2}\right)^{\delta_1+\delta_2} \left(\frac{1}{i}\right)^{\delta_2} \sum_{u=0}^{\delta_2 n_1 - 1} \{a\omega^u + a^{-1}\omega^{-u}\}^{\delta_1} \{a\omega^u - a^{-1}\omega^{-u}\}^{\delta_2}$$

$$(A.6.3)$$

where a and ω are as before and $i = \sqrt{-1}$. For $\delta_1 = 1$ and $\delta_2 = 1$, Equation (A.6.3) is easily seen to be zero after making use of the fact that $\sum_{u=0}^{n_1-1}\omega^{2u}$ and $\sum_{u=0}^{n_1-1}\omega^{-2u} = 0$. Likewise, $n_1[112]$ and $n_1[122]$ are found to be zero. For $n_1[1112]$, Equation (A.6.3) becomes

$$\left(\frac{\rho}{2}\right)^4\left(\frac{1}{i}\right)\sum_{u=0}^{n_1-1}\{a^3\omega^{3u} + a^{-3}\omega^{-3u} + 3a\omega^u + 3a^{-1}\omega^{-u}\}\{a\omega^u - a^{-1}\omega^{-u}\}$$

$$= \left(\frac{\rho}{2}\right)^4\left(\frac{1}{i}\right)\{-3n_1 + 3n_1\}$$

$$= 0$$

Similarly $n_1[1222]$ can be shown to be zero.

It remains now to develop the expression for $\sum_{u=0}^{n_1-1}x_{1u}^2 x_{2u}^2$. We can once again use Equation (A.6.3) to develop the following:

$$\sum_{u=0}^{n_1-1} x_{1u}^2 x_{2u}^2 = \left(\frac{\rho}{2}\right)^4(-1)\sum_{u=0}^{n_1-1}\{a^2\omega^{2u} + 2 + a^{-2}\omega^{-2u}\}\{a^2\omega^{2u} - 2 + a^{-2}\omega^{-2u}\}$$

$$= \frac{\rho^4}{16}(-1)\{n_1 - 4n_1 + n_1\}$$

$$= \frac{\rho^4 n_1}{8}$$

On the basis of [$iiii$] and [$iijj$] in the above developments, it is clear that the equiradial design is rotatable.

APPENDIX 7

Relationship Between D-Optimality and the Volume of a Joint Confidence Ellipsoid on $\boldsymbol{\beta}$

The response surface model represents a special case of the general linear model

$$\mathbf{y} = \mathbf{X}\boldsymbol{\beta} + \boldsymbol{\varepsilon}$$

and the $100(1 - \alpha)\%$ confidence ellipsoid on $\boldsymbol{\beta}$ under the assumptions $\boldsymbol{\varepsilon} \sim N(\mathbf{0}, \sigma^2\mathbf{I})$ is given by solutions to $\boldsymbol{\beta}$ in

$$\frac{(\mathbf{b} - \boldsymbol{\beta})'(\mathbf{X}'\mathbf{X})(\mathbf{b} - \boldsymbol{\beta})}{ps^2} \leq F_{\alpha, p, n-p}$$

or

$$(\mathbf{b} - \boldsymbol{\beta})'(\mathbf{X}'\mathbf{X})(\mathbf{b} - \boldsymbol{\beta}) \leq C \qquad (A.7.1)$$

where $C = ps^2 F_{\alpha, p, n-p}$. As a result, the volume of the confidence region is the volume of the ellipsoid

$$(\mathbf{b} - \boldsymbol{\beta})'(\mathbf{X}'\mathbf{X})(\mathbf{b} - \boldsymbol{\beta}) = C \qquad (A.7.2)$$

Suppose we consider the orthogonal matrix $\mathbf{Q}$ such that

$$\mathbf{Q}'(\mathbf{X}'\mathbf{X})\mathbf{Q} = \begin{bmatrix} \lambda_1^* & & & 0 \\ & \lambda_2^* & & \\ & & \ddots & \\ 0 & & & \lambda_p^* \end{bmatrix}$$

where the λ_i^* are the eigenvalues of $\mathbf{X}'\mathbf{X}$. We can write

$$\mathbf{X}'\mathbf{X} = \mathbf{Q} \begin{bmatrix} \lambda_i^* & & 0 \\ & \ddots & \\ 0 & & \lambda_p^* \end{bmatrix} \mathbf{Q}'$$

As a result, Equation (A.7.2) can be written in familiar form

$$\sum_{i=1}^{p} (b_i - \beta_i)^2 \lambda_i^* = C$$

Of course the volume of this ellipsoid is proportional to $[\prod_{i=1}^{P}(1/\lambda_i^*)]^{1/2}$. The quantity $[\prod_{i=1}^{P}(1/\lambda_i^*)]^{1/2} = 1/|\mathbf{X}'\mathbf{X}|^{1/2}$. As a result, the volume of the ellipsoid is inversely proportional to the square root of the determinant of $\mathbf{X}'\mathbf{X}$.

APPENDIX 8

Relationship Between Maximum Prediction Variance in a Region and the Number of Parameters

Consider the expression for hat diagonals discussed in Chapter 2. Given the linear model

$$y_i = \mathbf{x}_i'^{(m)}\mathbf{b} \qquad (i = 1, 2, \ldots, n)$$

the ith hat diagonal is given by

$$h_{ii} = \mathbf{x}_i'^{(m)}(\mathbf{X}'\mathbf{X})^{-1}\mathbf{x}_i^{(m)} \qquad (i = 1, 2, \ldots, n)$$

and, indeed,

$$\sum_{i=1}^{n} h_{ii} = \sum_{i=1}^{n} \mathbf{x}_i'^{(m)}(\mathbf{X}'\mathbf{X})^{-1}\mathbf{x}_i^{(m)} = p \text{ (number of parameters)}$$

See Myers (1990) or Montgomery and Peck (1992). Consider an experimental design and the corresponding scaled prediction variance

$$v(\mathbf{x}) = n\mathbf{x}'^{(m)}(\mathbf{X}'\mathbf{X})^{-1}\mathbf{x}^{(m)}$$

Suppose we consider the maximum value of $v(\mathbf{x})$ in some region R. The *average value* of $v(\mathbf{x})$ at the data points is p. Obviously then

$$\max_{\mathbf{x} \in R} v(\mathbf{x}) \geq p$$

In addition, if a design is G-optimal all hat diagonals are equal, thus implying that $v(\mathbf{x}) = p$ at all data points and

$$v(\mathbf{x}) \leq p$$

at all other locations in R.

APPENDIX 9

The Development of Equation (8.21)

$$V^r = \frac{N\psi}{\sigma^2} \int_{v_r} \text{Var } \hat{y}(\mathbf{x}) \, d\mathbf{x}$$

$$= \frac{N\psi}{\sigma^2} \int_{v_r} \mathbf{x}'^{(m)} (\mathbf{X}'\mathbf{X})^{-1} \mathbf{x}^{(m)} \, d\mathbf{x}$$

$$= \frac{N\psi}{\sigma^2} \int_{v_r} \text{tr} \Big[\mathbf{x}'^{(m)} (\mathbf{X}'\mathbf{X})^{-1} \mathbf{x}^{(m)} \Big] \, d\mathbf{x}$$

$$= \frac{N\psi}{\sigma^2} \int_{v_r} \text{tr} \Big[(\mathbf{X}'\mathbf{X})^{-1} \mathbf{x}^{(m)} \mathbf{x}'^{(m)} \Big] \, d\mathbf{x}$$

$$= \text{tr}(\mathbf{X}'\mathbf{X})^{-1} \left[\frac{N\psi}{\sigma^2} \int_{v_r} \mathbf{x}^{(m)} \mathbf{x}'^{(m)} \, d\mathbf{x} \right]$$

$$= \text{tr}(\mathbf{X}'\mathbf{X})^{-1} S$$

Determination of Data Augmentation Result (Choice of $\mathbf{x}_{r+1}$ for the Sequential Development of a D-Optimal Design)

Let us begin with a well-known result dealing with the determinant of a matrix. Given a square matrix

$$\mathbf{A} = \begin{bmatrix} \mathbf{A}_{11} & \mathbf{A}_{12} \\ \mathbf{A}_{21} & \mathbf{A}_{22} \end{bmatrix}$$

then if $\mathbf{A}_{22}$ is nonsingular

$$|\mathbf{A}| = |\mathbf{A}_{22}||\mathbf{A}_{11} - \mathbf{A}_{12}\mathbf{A}_{22}^{-1}\mathbf{A}_{21}| \tag{A.10.1}$$

or

$$|\mathbf{A}| = |\mathbf{A}_{11}||\mathbf{A}_{22} - \mathbf{A}_{21}\mathbf{A}_{11}^{-1}\mathbf{A}_{12}| \tag{A.10.2}$$

if $\mathbf{A}_{11}$ nonsingular.

Now let $\mathbf{X}_{r+1} = \begin{bmatrix} \mathbf{X} \\ \cdots \\ \mathbf{x}'_{r+1} \end{bmatrix}$ denote the $\mathbf{X}$ matrix after $\mathbf{x}_{r+1}$ has been selected. Obviously $\mathbf{X}$ is $\mathbf{X}_{r+1}$ apart from the selected point. It is easy to show that

$$(\mathbf{X}'\mathbf{X}) = \mathbf{X}'_{r+1}\mathbf{X}_{r+1} - \mathbf{x}_{r+1}\mathbf{x}'_{r+1}$$

687

and from Equation (A.10.1),

$$|\mathbf{X}'_{r+1}\mathbf{X}_{r+1} - \mathbf{x}_{r+1}\mathbf{x}'_{r+1}| = \begin{vmatrix} \mathbf{X}'_{r+1}\mathbf{X}_{r+1} & \mathbf{x}_{r+1} \\ \mathbf{x}'_{r+1} & 1 \end{vmatrix}$$

$$= |\mathbf{X}'_{r+1}\mathbf{X}_{r+1}| \, |1 - \mathbf{x}'_{r+1}(\mathbf{X}'_{r+1}\mathbf{X}_{r+1})^{-1}\mathbf{x}_{r+1}|$$

from Equation (A.10.2). Now the above implies that

$$|\mathbf{X}'\mathbf{X}| = |\mathbf{X}'_{r+1}\mathbf{X}_{r+1}| \, |1 - \mathbf{x}'_{r+1}(\mathbf{X}'_{r+1}\mathbf{X}_{r+1})^{-1}\mathbf{x}_{r+1}|$$

Thus, because $1 - \mathbf{x}'_{r+1}(\mathbf{X}'_{r+1}\mathbf{X}_{r+1})^{-1}\mathbf{x}_{r+1}$ is a scalar, we have

$$\frac{|\mathbf{X}'_{r+1}\mathbf{X}_{r+1}|}{|\mathbf{X}'\mathbf{X}|} = \frac{1}{1 - \mathbf{x}'_{r+1}(\mathbf{X}'_{r+1}\mathbf{X}_{r+1})^{-1}\mathbf{x}_{r+1}} \qquad (A.10.3)$$

A well-known result for regression analysis relates $\mathbf{x}'_{r+1}(\mathbf{X}'_{r+1}\mathbf{X}_{r+1})^{-1}\mathbf{x}_{r+1}$ to $\mathbf{x}'_{r+1}(\mathbf{X}'\mathbf{X})^{-1}\mathbf{x}_{r+1}$. Keep in mind that the former relates to variance of a predicted value at the location defined by $\mathbf{x}_{r+1}$ whereas $\mathbf{x}'_{r+1}(\mathbf{X}'\mathbf{X})\mathbf{x}_{r+1}$, relates to prediction variance at $\mathbf{x}_{r+1}$ for a fitted model *before the new point was placed in the data set.* From Myers (1990) or Montgomery and Peck (1992) we have

$$\mathbf{x}'_{r+1}(\mathbf{X}'\mathbf{X})^{-1}\mathbf{x}_{r+1} = \frac{\mathbf{x}'_{r+1}(\mathbf{X}'_{r+1}\mathbf{X}_{r+1})^{-1}\mathbf{x}_{r+1}}{1 - \mathbf{x}'_{r+1}(\mathbf{X}'_{r+1}\mathbf{X}_{r+1})^{-1}\mathbf{x}_{r+1}} \qquad (A.10.4)$$

Now consider Equations (A.10.3) and (A.10.4). One desires to select $\mathbf{x}_{r+1}$ in order to maximize $|\mathbf{X}'_{r+1}\mathbf{X}_{r+1}|$ for D-optimality. Because $\mathbf{x}'_{r+1}(\mathbf{X}'_{r+1}\mathbf{X}_{r+1})^{-1}\mathbf{x}_{r+1}$ is a hat diagonal, we obtain $\mathbf{x}'_{r+1}(\mathbf{X}'_{r+1}\mathbf{X}_{r+1})^{-1}\mathbf{x}_{r+1} \le 1$. One then must choose $\mathbf{x}_{r+1}$ so that $\mathbf{x}'_{r+1}(\mathbf{X}'_{r+1}\mathbf{X}_{r+1})^{-1}\mathbf{x}_{r+1}$ is maximized. From Equation (A.10.4), maximizing $\mathbf{x}'_{r+1}(\mathbf{X}'_{r+1}\mathbf{X}_{r+1})^{-1}\mathbf{x}_{r+1}$ is equivalent to maximizing $\mathbf{x}'_{r+1}(\mathbf{X}'\mathbf{X})^{-1}\mathbf{x}_{r+1}$. Thus one should choose the next point $\mathbf{x}_{r+1}$ at which the prediction variance [i.e., $\mathrm{Var}[\hat{y}(\mathbf{x}_{r+1})]/\sigma^2$] is maximized.

References

Allen, D. M. (1971), "Mean Square Error of Prediction as a Criterion for Selecting Variables," *Technometrics*, 13, 469–475.

Allen, D. M. (1974), "The Relationship Between Variable Selection and Data Augmentation and a Method for Prediction," *Technometrics*, 16, 125–127.

Andrews, D. F. (1974), "A Robust Method for Multiple Linear Regression," *Technometrics*, 16, 523–531.

Andrews, D. F., Bickel, P. J., Hampel, F. R., Huber, P. J. Rogers, W. II., and Tukey, J. W. (1972), *Robust Estimates of Location*, Princeton University Press, Princeton, NJ.

Bartlett, M. S., and Kendall, D. G. (1946), "The Statistical Analysis of Variance Heterogeneity and the Logarithmic Transformation," *Journal of the Royal Statistical Society, Series B*, 8, 128–150.

Beaton, A. E., and Tukey, J. W. (1974), "The Fitting of Power Series, Meaning Polynomials, on Band Spectroscopic Data," *Technometrics*, 16, 147–185.

Belsley, D. A., Kah, E., and Welsch, R. E. (1980), *Regression Diagnostics: Identifying Influential Data and Sources of Collinearity*, John Wiley & Sons, New York.

Biles, W. E. (1975), "A Response Surface Method for Experimental Optimization of Multiresponse Processes," *Industrial and Engineering Chemistry-Process Design and Development*, 14, 152–158.

Box, G. E. P. (1957), "Evolutionary Operation: A Method for Increasing Industrial Productivity," *Applied Statistics*, 6, 81–101.

Box, G. E. P. (1963), "The Effect of Errors in the Factor Levels and Experimental Design," *Technometrics*, 6, 247–262.

Box, G. E. P. (1988), "Signal-to-Noise Ratios, Performance Criteria, and Transformations," (with discussion), *Technometrics*, 30, 1–40.

Box, G. E. P., and Behnken, D. W. (1960), "Some New Three-Level Designs for the Study of Quantitative Variables," *Technometrics*, 2, 455–475.

Box, G. E. P., and Cox, D. R. (1964), "An Analysis of Transformations" (with discussion), *Journal of the Royal Statistical Society, Series B*, 26, 211–246.

Box, G. E. P., and Draper, N. R. (1959), "A Basis for the Selection of a Response Surface Design," *Journal of the American Statistical Association*, 54, 622–654.

689

Box, G. E. P., and Draper, N. R. (1963), "The Choice of a Second Order Rotatable Design," *Biometrika*, 50, 335–352.

Box, G. E. P., and Draper, N. R. (1969), *Evolutionary Operation*, John Wiley & Sons, New York.

Box, M. J., and Draper, N. R. (1971), "Factorial Designs, the $X'X$ Criterion, and Some Related Matters," *Technometrics*, 13, 731–742.

Box, G. E. P., and Draper, N. R. (1974), "Some Minimum Point Designs for Second Order Response Surfaces," *Technometrics*, 16, 613–616.

Box, G. E. P., and Draper, N. R. (1975), "Robust Designs," *Biometrika*, 62, 347–352.

Box, G. E. P., and Draper, N. R. (1987), *Empirical Model-Building and Response Surfaces*, John Wiley & Sons, New York.

Box, G. E. P., and Hunter, J. S. (1954), "A Confidence Region for the Solution of a Set of Simultaneous Equations with an Application to Experimental Design," *Biometrika*, 41, 190–199.

Box, G. E. P., and Hunter, J. S. (1957), "Multifactor Experimental Designs for Exploring Response Surfaces," *The Annals of Mathematical Statistics*, 28, 195–241.

Box, G. E. P., and Hunter, J. S. (1961a), "The 2^{k-p} Fractional Factorial Designs, Part I," *Technometrics*, 2, 311–352.

Box, G. E. P., and Hunter, J. S. (1961b), "The 2^{k-p} Fractional Factorial Designs, Part II," *Technometrics*, 3, 449–458.

Box, G. E. P., and Jones, S. (1989), "Designing Products that are Robust to the Environment," paper presented at ASA Conference, Washington, DC.

Box, G. E. P., and Jones, S. (1992), "Split-Plot Designs for Robust Product Experimentation," *Journal of Applied Statistics*, 19, 3–26.

Box, G. E. P., and Meyer, R. D. (1986), "Dispersion Effects from Fractional Designs," *Technometrics*, 28, 19–27.

Box, G. E. P., and Wetz, J. M. (1973), "Criterion for Judging the Adequacy of Estimation by an Approximation Response Polynomial," Technical Report No. 9, Department of Statistics, University of Wisconsin, Madison.

Box, G. E. P., and Wilson, K. B. (1951), "On the Experimental Attainment of Optimum Conditions," *Journal of the Royal Statistical Society, Series B*, 13, 1–45.

Byrne, D. M., and Taguchi, G. (1987), "The Taguchi Approach to Parameter Design," *Quality Progress*, December, 1987, 19–26.

Carroll, R. J., and Ruppert, D. (1988), *Transformation and Weighting in Regression*, Chapman & Hall, New York.

Cook, R. D. (1977), "Detection of Influential Observations in Linear Regression," *Technometrics*, 19, 15–17.

Cook, R. D. (1979), "Influential Observations in Linear Regression," *Journal of the American Statistical Association*, 74, 169–174.

Cornell, J. A. (1986), "A Comparison Between Two Ten-Point Designs for Studying Three-Component Mixture Systems," *Journal of Quality Technology*, 18, 1–15.

Cornell, J. A. (1990), *Experiments with Mixtures: Designs, Models, and the Analysis of Mixture Data*, 2nd edition, John Wiley & Sons, New York.

Cornell, J. A. (1995), "Fitting Models to Data from Mixture Experiments Containing Order Factors," *Journal of Quality Technology*, 27, 1, 13–33.

Covey-Crump, P. A. K., and Silvey, S. D. (1970), "Optimal Regression Designs with Previous Observations," *Biometrika*, 57, 551–566.

Cox, D. R. (1958), *Planning of Experiments*, John Wiley & Sons, New York.

Cox, D. R. (1971), "A Note on Polynomial Response Functions for Mixtures," *Biometrika*, 58, 155–159.

Crosier, R. B. (1984), "Mixture Experiments: Geometry and Pseudo-Components," *Technometrics*, 26, 209–216.

Daniel, C. (1959), "Use of Half-Normal Plots in Interpreting Factorial Two-Level Experiments," *Technometrics*, 1, 311–342.

Davison, J., Myers, R. H., and Lentner, M. (1995), "Split Plot and Response Surfaces," Technical Report (95-1), VPI & SU.

Del Castillo, E., and Montgomery, D. C. (1993), "A Nonlinear Programming Solution to the Dual Response Problem," *Journal of Quality Technology*, 25, 199–204.

Derringer, G., and Suich, R. (1980), "Simultaneous Optimization of Several Response Variables," *Journal of Quality Technology*, 12, 214–219.

Draper, N. R. (1963), "Ridge Analysis of Response Surfaces," *Technometrics*, 5, 469–479.

Draper, N. R. (1982), "Center Points in Second-Order Response Surface Designs," *Technometrics*, 24, 127–133.

Draper, N. R. (1985), "Small Composite Designs," *Technometrics*, 27, 173–180.

Draper, N. R., and John, J. A. (1988), Response Surface Designs for Quantitative and Qualitative Variables," *Technometrics*, 423–428.

Draper, N. R. and Lin, D. K. J. (1990), "Small Response Surface Designs," *Technometrics*, 32, 187–194.

Draper, N. R., and St. John, R. C. (1977), "Designs in Three and Four Components for Mixtures Models with Inverse Terms," *Technometrics*, 19, 117–130.

Dutter, R. (1977), "Numerical Solution of Robust Regression Problems, Computational Aspects, a Comparison," *Journal of Statistical Computation and Simulation*, 5, 207–238.

Dykstra, O., Jr. (1966), "The Orthogonalization of Undesigned Experiments," *Technometrics*, 8, 279–290.

Ellerton, R. R. W. (1978), "Is the Regression Equation Adequate—A Generalization," *Technometrics*, 20, 313–316.

Giovannitti-Jensen, A., and Myers, R. H. (1989), "Graphical Assessment of the Prediction Capability of Response Surface Designs," *Technometrics*, 31, 159–171.

Graybill, F. (1976), *Introduction to the Theory of Linear Statistical Models*, Duxbury Press, Boston.

Gunst, R. F., and Mason, R. L. (1979), "Some Considerations in the Evaluation of Alternative Prediction Equations," *Technometrics*, 21, 55–63.

Hackney, H., and Jones, P. R. (1969), "Response Surface for Dry Modulus of Rupture and Drying Shrinkage," *Am. Ceram, Soc. Bull.*, 46.

Hartley, H. O. (1959), "Smallest Composite Design for Quadratic Response Surfaces," *Biometrics*, 15, 611–624.

Hebble, T. L., and Mitchell, T. J. (1972), "Repairing Response Surface Designs," *Technometrics*, 14, 767–779.

Heinsman, J. A., and Montgomery, D. C. (1995), "Optimization of a Household Product Formulation Using a Mixture Experiment," *Quality Engineering*, 7, 583–600.

Hill, R. W. (1979), "On Estimating the Covariance Matrix of Robust Regression *M*-Estimates," *Communications in Statistics, Series A*, 8, 1183–1196.

Hill, R. C., Judge, G. G., and Fomby, T. B. (1978), "Testing the Adequacy of a Regression Model," *Technometrics*, 20, 491–494.

Himmelblau, D. M. (1971), *Process Analysis by Statistical Methods*, John Wiley & Sons, New York.

Hocking, R. R. (1978), "The Regression Dilemma: Variable Estimation, Coefficient Shrinkage, or Robust Estimation," paper presented at the ASQC Fall Technical Conference, Rochester, New York.

Hoerl, A. E. (1959), "Optimum Solution of Many Variables Equations," *Chemical Engineering Progress*, 55, 67–78.

Hoerl, A. E. (1964), "Ridge Analysis," *Chemical Engineering Symposium Series*, 60, 67–77.

Hoerl, R. W. (1985), "Ridge Analysis 25 Years Later," *The American Statistician*, 39, 186–193.

Hogg, R. V. (1979), "An Introduction to Robust Estimation," in *Robustness in Statistics*, edited by R. L. Launer and G. N. Wilkinson, Academic Press, New York.

Hoke, A. T. (1974), "Economical Second-Order Designs Based on Irregular Fractions of the 3^n Factorial," *Technometrics*, 17, 375–384.

Huber, P. J. (1964), "Robust Estimation of a Location Parameter," *Annals of Mathematical Statistics*, 35, 73–101.

Huber, P. J. (1973), "Robust Regression: Asymptotics, Conjectures, and Monte Carlo," *Annals of Statistics*, 1, 799–821.

Khuri, A. I., and Conlon, M. (1981), "Simultaneous Optimization of Multiple Responses Represented by Polynomial Regression Functions," *Technometrics*, 23, 363–375.

Khuri, A. I., and Cornell, J. A. (1987), *Response Surfaces*, Marcel Dekker, New York.

Kiefer, J. (1959), "Optimum Experimental Designs," *Journal of the Royal Statistical Society, Series B*, 21, 272–304.

Kiefer, J. (1961), "Optimum Designs in Regression Problems," *Annals of Mathematical Statistics*, 32, 298–325.

Kiefer, J., and Wolfowitz, J. (1959), "Optimum Designs in Regression Problems," *Annals of Mathematical Statistics*, 30, 271–294.

Koons, G. F., and Wilt, M. H. (1985), "Design and Analysis of an ABS Pipe Compound Experiment," *Experiments in Industry: Design, Analysis, and Interpretation of Results*, edited by R. D. Snee, L. B. Hare, and R. Trout, American Society for Quality Control, Milwaukee, 111–117.

Koshal, R. S. (1933), "Application of the Method of Maximum Likelihood to the Improvement of Curves Fitted by the Method of Moments," *Journal of the Royal Statistical Society, Series A*, 96, 303–313.

Kunugi, T., Tamura, T., and Naito, T. (1961), "New Acetylene Process Uses Hydrogen Dilution," *Chemical Engineering Progress*, 57, 43–49.

Lentner, M., and Bishop, T. (1993), *Experimental Design and Analysis*, Valley Book Company, Blacksburg, VA.

Lind, E. E., Goldin, J., and Hickman, J. B. (1960), "Fitting Yield and Cost Response Surfaces," *Chemical Engineering Progress*, 56, 62.

Lucas, J. M. (1974), "Optimum Composite Design," *Technometrics*, 16, 561–567.

Lucas, J. M. (1976), "Which Response Surface Design Is Best," *Technometrics*, 18, 411–417.

Lucas, J. M. (1989), "Achieving a Robust Process Using Response Surface Methodology," paper presented at ASA Conference, Washington, D.C.

Lucas, J. M., and Ju, H. L. (1992), "Split Plotting and Randomization in Industrial Experiments," *ASQC Quality Congress Transactions*, Nashville, TN.

Marquardt, D. M., and Snee, R. D. (1975), "Ridge Regression in Practice," *The American Statistician*, 29, 3–20.

McLean, R. A., and Anderson, V. L. (1966), "Extreme Vertices Design of Mixture Experiments," *Technometrics*, 8, 447–454.

Montgomery, D. C. (1991a), "Using Fractional Factorial Designs for Robust Process Development," *Quality Engineering*, 3, 193–205.

Montgomery, D. C. (1991b), *Design and Analysis of Experiments*, 3rd edition, John Wiley & Sons, New York.

Montgomery, D. C., and Peck, E. A. (1992), *Introduction to Linear Regression Analysis*, John Wiley & Sons, New York.

Montgomery, D. C., and Voth, S. R. (1994), "Multicollinearity and Leverage in Mixture Experiments," *Journal of Quality Technology*, 26, 96–108.

Myers, R. H. (1976), *Response Surface Methodology*, Blacksburg, VA, published by Author.

Myers, R. H. (1990), *Classical and Modern Regression with Applications*, 2nd edition, Duxbury Press, Boston.

Myers, R. H., and Carter, W. H., Jr. (1973), "Response Surface Techniques for Dual Response Systems," *Technometrics*, 15, 301–317.

Myers, R. H., and Milton, S. (1991), *A First Course in the Theory of Linear Statistical Models*, Duxbury Press, Boston.

Myers, R. H., Khuri, A. I., and Carter, W. H. (1989), "Response Surface Methodology: 1966–1988," *Technometrics*, 3, 137–157.

Myers, R. H., Khuri, A. I., and Vining, G. (1992), "Response Surface Alternatives to the Taguchi Robust Parameter Design Approach," *American Statistician*, 46, 2, 131–139.

Myers, R. H., Vining, G., Giovannitti-Jensen, A., and Myers, S. L. (1992), "Variance Dispersion Properties of Second-Order Response Surface Designs," *Journal of Quality Technology*, 24, 1–11.

Nair, V. N., and Pregibon, D. (1986), "A Data Analysis Strategy for Quality Engineering Experiments," *AT & T Technical Journal*, 74–84.

Nair, V. N., et al. (1992), "Taguchi's Parameter Design: A Panel Discussion," *Technometrics*, 34, 2, 127–161.

Nalimov, V. V., Golikova, T. I., and Mikeshina, N. G. (1970), "On Practical Use of the Concept of *D*-Optimality," *Technometrics*, 12, 799–812.

Nigam, A. K., Gupta, S. C., and Gupta, S. (1983), "A New Algorithm for Extreme Vertices Designs for Linear Mixture Models," *Technometrics*, 25, 367–371.

Notz, W. (1982), "Minimal Point Second Order Designs," *Journal of Statistical Planning and Inference*, 6, 47–58.

Piepel, G. F. (1982), "Measuring Component Effects in Constrained Mixture Experiments," *Technometrics*, 25, 97–101.

Piepel, G. F. (1988), "Programs for Generating Extreme Vertices and Centroids of Linearly Constrained Experimental Regions," *Journal of Quality Technology*, 20, 125–139.

Piepel, G. F., and Cornell, J. A. (1994), "Mixture Experiment Approaches: Examples, Discussion, and Recommendations," *Journal of Quality Technology*, 26, 177–196.

Pignatiello, J., and Ramberg, J. (1991), "Top Ten Triumphs and Tragedies of Genichi Taguchi," *Quality Engineering*, 4(2), 211–225.

Plackett, R. L., and Burman, J. P. (1946), "The Design of Optimum Multifactorial Experiments," *Biometrika*, 33, 305–325.

Pukelsheim, F. (1995), *Optimum Design of Experiments*, Chapman & Hall, New York.

Ramsay, J. O. (1977), "A Comparative Study of Several Robust Estimates of Slope, Intercept, and Scale in Linear Regression," *Journal of the American Statistical Association*, 72, 608–615.

Roquemore, K. G. (1976), "Hybrid Designs for Quadratic Response Surfaces," *Technometrics*, 18, 419–423.

Rousseeuw, P. J. (1984), "Least Median of Squares Regression," *Journal of the American Statistical Association*, 79, 871–880.

Rousseeuw, P. J., and Leroy, A. M. (1987), *Robust Regression and Outlier Detection*, John Wiley & Sons, New York.

Rozum, M. A. (1990), "Effective Design Augmentation for Prediction," Ph.D. Thesis, Virginia Tech.

Rozum, M., and Myers, R. H. (1991), "Variance Dispersion Graphs for Cuboidal Regions," paper presented at ASA Meetings, Atlanta, GA.

Scheffe, H. (1958), "Experiments with Mixtures," *Journal of the Royal Statistical Society, Series B*, 20, 344–366.

Scheffe, H. (1959), "Reply to Mr. Quenouille's Comments About my Paper on Mixtures," *Journal of the Royal Statistical Society, Series B*, 23, 171–172.

Scheffe, H. (1965), "The Simplex-Centroid Design for Experiments with Mixtures," *Journal of the Royal Statistical Society, B*, 25, 235–263.

Schmidt, S. R., and Launsby, R. G. (1990), *Understanding Industrial Designed Experiments*, Colorado Springs, CO.

Shoemaker, A. C., Tsui, K. L., and Wu, C. F. J. (1991), "Economical Experimentation Methods for Robust Design," *Technometrics*, 33, 415–427.

Snee, R. D., and Marquardt, D. W. (1974), "Extreme Vertices Designs for Linear Mixture Models," *Technometrics*, 16, 399–408.

Snee, R. D. (1985), "Computer-Aided Design of Experiments: Some Practical Experiences," *Journal of Quality Technology*, 17, 222–236.

Spendley, W., Hext, G. R., and Himsworth, F. R. (1962), "Sequential Application of Simplex Designs in Optimization and EVOP," *Technometrics*, 4, 441–461.

St. John, R. C., and Draper, N. R. (1975), "*D*-Optimality for Regression Designs: A Review," *Technometrics*, 17, 15–23.

St. John, R. C. (1984), "Experiments With Mixtures in Conditioning and Ridge Regression," *Journal of Quality Technology*, 16, 81–96.

Suich, R., and Derringer, G. C. (1977), "Is the Regression Equation Adequate—One Criterion," *Technometrics*, 19, 213–216.

Taguchi, G. (1986), *Introduction to Quality Engineering*, Asian Productivity Organization, UNIPUB, White Plains, NY.

Taguchi, G. (1987), *System of Experimental Design: Engineering Methods to Optimize Quality and Minimize Cost*, UNIPUB/Kraus International, White Plains, NY.

Taguchi, G., and Wu, Y. (1980), *Introduction to Off-Line Quality Control*, Central Japan Quality Control Association, Nagoya, Japan.

Vining, G. (1993), "A Computer Program for Generating Variance Dispersion Graphs," *Journal of Quality Technology*, 25, 45–58.

Vining, G. G., and Myers, R. H. (1990), "Combining Taguchi and Response Surface Philosophies: A Dual Response Approach," *Journal of Quality Technology*, 22, 38–45.

Welch, W. J., Yu, T. K., Kang, S. M., and Sacks, J. (1990), "Computer Experiments for Quality Control by Parameter Design," *Journal of Quality Technology*, 22, 15–22.

Welsch, R. E. (1975), "Confidence Regions for Robust Regression," *Proceedings of the Statistical Computing Section*, American Statistical Association, Washington, DC.

Yates, F. (1935), "Complex Experiments," *Journal of the Royal Statistical Society*, Supplement 2, 181–247.

Yates, F. (1937), *Design and Analysis of Factorial Experiments*, Technical Communication No. 35, Imperial Bureau of Soil Sciences, London.

Index

WILEY SERIES IN
PROBABILITY AND STATISTICS

ESTABLISHED BY WALTER A. SHEWHART AND SAMUEL S. WILKS
Editors
*Vic Barnett, Ralph A. Bradley, Nicholas I. Fisher, J. Stuart Hunter,
J. B. Kadane, David G. Kendall, David W. Scott, Adrian F. M. Smith,
Jozef L. Teugels, Geoffrey S. Watson*

*Now available in a lower priced paperback edition in the Wiley Classics Library.

*Now available in a lower priced paperback edition in the Wiley Classics Library.

*Now available in a lower priced paperback edition in the Wiley Classics Library.